OSEN BIBLIOGRAPHIC PROJECT
ON ANTISEMITISM

)tto Dov Kulka

: Sara Grosvald, Sylviane Stampfer

:
nder Avram, Yisrael (Elliot) Cohen,
gelberg, Mirjam Factor, Lily Fogel (z"l),
)aniel Romanovsky, Tamar Stern,
nate Wolfson, Manuel Zkorenblut

The Vidal Sassoon International Center
for the Study of Antisemitism
The Hebrew University of Jerusalem

ANTISEMITISM
An Annotated Bibliography

Volume 15
1999

Edited by
Susan Sarah Cohen

K · G · Saur München 2002

Die Deutsche Bibliothek – CIP-Einheitsaufnahme

Antisemitism : an annotated bibliography ; [the Felix Posen bibliographic project on antisemitism] / The Vidal Sassoon International Center for the Study of Antisemitism, The Hebrew University of Jerusalem. Ed. by Susan Sarah Cohen. – München : Saur.
Vol. 15. 1999–2002
ISBN 3-598-23714-6

Gedruckt auf säurefreiem Papier / Printed on acid free paper

Printed in Germany

Druck: Strauss Offsetdruck, Mörlenbach
Binden: Buchbinderei Schaumann, Darmstadt
ISBN 3-598-23714-6

CONTENTS

PREFACE

The Vidal Sassoon International Center for the Study of Antisemitism of the Hebrew University of Jerusalem, established by an international group of scholars in 1982, is an interdisciplinary research center which engages in research on antisemitism throughout the ages. The Felix Posen Bibliographic Project of the Center comprises a current on-line data base accessible through the Israel University Inter-Library Network, the Internet, and printed volumes.

The present edition of the bibliography includes works about antisemitism – books, dissertations, masters' theses, and articles from periodicals and collections – published in 1999, with supplementary material from 1996-1998 not published in the previous volumes. Material from 1997-1999 not entered now (ca. 1,700 items) will appear in the forthcoming volumes. The bibliography does not include newspaper articles, reviews, audio-visual material, and works of fiction, nor does it cover antisemitic publications (except for critical editions published for scholastic purposes, which include a scholarly introduction – e.g. the diaries of Joseph Goebbels, the speeches of Adolf Hitler).

For the purpose of this bibliography, antisemitism is defined as antagonism toward Jews and Judaism as expressed in writings (e.g. the New Testament, polemical literature, works of fiction), in the visual arts (e.g. art, caricatures, films), and in actions (e.g. massacres and pogroms, discriminatory legislation, the Holocaust). Works are included which deal with philosemitism and Jewish responses to antisemitism. In borderline cases (e.g. dealing with Christian-Jewish relations, anti-Israel and anti-Zionism, works of fiction and films), books and articles are listed when the author of the work relates to the subject of antisemitism.

The bibliography lists all the publications dealing with antisemitism in the Holocaust period (1933-1945) – i.e. works on the Jewish experience (memoirs, memorial books, etc.) as well as works on antisemitic ideology, policy, and attitudes. It also lists publications on postwar reactions to the Holocaust – e.g. Holocaust commemoration, memory and meaning, study and teaching; it does not deal with the psychological effects of the Holocaust on survivors and their descendants.

In regard to Soviet Jewry, works are included that deal with Soviet ideology and policy toward the Jews but not with the Jewish experience as such.

In regard to anti-Zionism, works are included that describe anti-Zionism as antisemitism, particularly those that discuss antisemitic accusations vaunted against Zionism and Israel, such as the "world Jewish conspiracy," Zionist collaboration with Nazis, the "Zionism equals racism" formula, and the comparison of Israelis with Nazis.

In regard to the Arab-Israeli conflict, works are included that deal with antisemitism in the Arab world but not with the political conflict between the State of Israel and its Arab neighbors.

The listings were compiled mainly from the holdings of the Jewish National and University Library in Jerusalem, which receives the majority of works on Jewish subjects published throughout the world. The works listed come from a diverse range of disciplines – history, psychology, sociology, anthropology, literature, art.

The bibliography is multilingual, containing works in the European languages, Hebrew, and Yiddish, and in other languages as they come to our attention. An English translation of the title is given in brackets for languages other than French, German, Italian, and Spanish.

The format of each entry follows the rules of the Chicago Manual of Style, with slight modification.

Each item is listed only once – a work dealing with two or more of the chronological or geographical subdivisions is listed in the appropriate "General" section. The detailed subject index at the end of the volume compensates for the "general" classifications.

The geographical chapter divisions are according to present-day boundaries, but exceptions have been made in the case of the former Czechoslovakia and USSR, which are still given as headings for items dealing with those countries in general.

In the subject index, names of rulers (kings, queens, emperors) are given in the original language (e.g. Enrique, Heinrich, Wilhelm). Names of places are divided – countries and capital cities are given in English, while villages, towns, and cities are given in the original language. Geographical locations are listed according to present-day boundaries (e.g. items on Breslau will be entered under Wrocław and in the chapter for Poland). Cross-references are given in such cases.

Authors of memoirs and autobiographies are listed in both the author and subject indexes.

Items which are a translation or a revised version of a work published previously are accompanied by a note stating where and when the original version appeared. If it was published in 1984 or after (when this bibliography began), the abstract for that item will be found in the appropriate volume of the bibliography. If it was not listed earlier, the abstract will accompany the later edition or translation.

The annotation expresses the view of the author(s) of the work described and not those of the annotator. The length of the annotation does not indicate the importance of a particular work.

The Bibliographic Project on Antisemitism is accessible via the Internet: HTTP://SICSA.HUJI.AC.IL

and via Telnet: HAR2.HUJI.AC.IL – username: SICSA; no password is necessary.

ACKNOWLEDGMENTS

The Vidal Sassoon International Center for the Study of Antisemitism wishes to thank the administration and the staff of the Jewish National and University Library for their support and cooperation, without which this volume could not have been produced.

Jerusalem S.S.C.
July 2002 — Av 5762

LIST OF PERIODICALS AND COLLECTIONS

Acta Musicologica. Basle: Baerenreiter
AJS Review. Waltham, MA: Association for Jewish Studies
Al-Masaq: Islam and the Medieval Mediterranean. Turnhout: Brepols
Alpayim: Ktav et rav-tehumi le-iyyun, hagut ve-sifrut. Tel-Aviv: Am Oved
Am Wedding haben sie gelebt: Lebenswege juedischer Buergerinnen und Buerger. Berlin: Metropol, 1998 (no. 566)
American Behavioral Scientist. Thousand Oaks, CA: Sage
American Jewish Archives Journal. Cincinnati, OH: American Jewish Archives
American Sociological Review. Washington, DC: American Sociological Association
The Americanization of the Holocaust, ed. Hilene Flanzbaum. Baltimore: Johns Hopkins University Press, 1999 (no. 1143)
The Americas: A Quarterly Review of Inter-American Cultural History. Washington, DC: Academy of American Franciscan History
Amicizia Ebraico-Cristiana: Bollettino. Firenze
Annali di Storia dell'Esegesi. Bologna: Dehoniane
Anthropological Quarterly. Washington, DC: Catholic University of America Press
Anti-Judaism and the Gospels, ed. William R. Farmer. Harrisburg, PA: Trinity Press International, 1999 (no. 61)
Antisemitism Research. The Hague: Center for Information and Documentation on Israel
Approaches to Ancient Judaism, New Series, Vol. 13, ed. Jacob Neusner. Atlanta: Scholars Press, 1998 (no. 69)
Archiv fuer Reformationsgeschichte. Guetersloh: Guetersloher Verlagshaus
Les archives de la Shoah, ed. Jacques Fredj. Paris: L'Harmattan, 1998 (no. 361)
Archives Juives: Revue d'histoire des Juifs de France. Paris: Liana Levi
Ariel Center for Policy Research: Policy Papers. Shaarei Tikva
Aschkenas: Zeitschrift fuer Geschichte und Kultur der Juden. Wien: Boehlau
Aspekte juedischen Lebens in Duesseldorf und am Niederrhein, eds. Angela Genger, Kerstin Griese. Duesseldorf: Mahn- und Gedenkstaette, 1997 (no. 192)
Association for Jewish Studies: Review. Waltham, MA see AJS Review
Association of Jewish Libraries: Convention (33rd, Philadelphia, 1998) see Proceedings of the 33rd Annual Convention of the Association of Jewish Libraries
Auf dem Wege zum christlich-juedischen Gespraech: 125 Jahre Evangelisch-lutherischer Zentralverein fuer Zeugnis und Dienst unter Juden und Christen, ed. Arnulf H. Baumann. Muenster: LIT, 1998 (no. 187)
Augustinianum. Roma: Institutum Patristicum Augustinianum

Aus der Geschichte lernen? Chancen der Aufklaerung nach Auschwitz, eds. Hanna Lehming, Siegfried von Kortzfleisch. Bad Segeberg: C.H. Waeser, 1997 (no. 40)
Australasian Journal of Philosophy. Oxford: Oxford University Press, for the Australasian Association of Philosophy
Australian Jewish Historical Society Journal. Sydney and Melbourne
Australian Journal of Jewish Studies. Sydney: Australian Association for Jewish Studies
Avotaynu: The International Review of Jewish Genealogy. Teaneck, NJ: Avotaynu
Barukh she-asani ishah? Ha-ishah be-yahadut – me-ha-tanach ve-ad yamenu, eds. David Yoel Ariel et al. Tel-Aviv: Yedioth Ahronoth, Sifrei Hemed, 1999 (no. 360)
Bayerische Akademie der Wissenschaften: Mathematisch-naturwissenschaftliche Klasse: Sitzungsberichte. Muenchen
Belaruski Histarychny Chasopis. Minsk: Belaruski Histarychny Chasopis
Il bene e il male dopo Auschwitz: Implicazioni etico-teologiche per l'oggi: Atti del Simposio Internazionale (Roma, 22-25 settembre 1997), eds. Emilio Baccarini, Lucy Thorson. Milano: Paoline, 1998 (no. 917)
Berastseiski Khranohraf: Sbornik navukovykh prats. Brest: Brestski Dzyarzhawny Universitet imya A.S. Pushkina
"Beseitigung des juedischen Einflusses...": Antisemitische Forschung, Eliten und Karrieren im Nationalsozialismus. Frankfurt a.M.: Campus, 1999 (no. 189)
Betrayal: German Churches and the Holocaust, eds. Robert P. Ericksen, Susannah Heschel. Minneapolis: Fortress Press, 1999 (no. 596)
Between History and Literature: Studies in Honor of Isaac Barzilay, ed. Stanley Nash. Tel-Aviv: Hakibbutz Hameuchad, 1997 (nos. 83, 103)
Bijdragen tot de Eigentijdse Herinnering. Brussel: Stichting voor de Eigentijdse Herinnering see Les Cahiers de la Mémoire Contemporaine
Bilder der Judenfeindschaft: Antisemitismus: Vorurteile und Mythen, eds. Julius H. Schoeps, Joachim Schloer. Muenchen: Bechtermuenz, 1999 (no. 29)
Bilder im Kopf: Auschwitz – einen Ort sehen = Obrazy w głowie: Oświęcim – ujęcia pewnego miejsca, eds. Knut Dethlefsen, Thomas B. Hebler. Berlin: Hentrich, 1997 (no. 1049)
Bishvil ha-Zikaron. Jerusalem: Yad Vashem
Biuletyn Żydowskiego Instytutu Historycznego. Warszawa: Żydowski Instytut Historyczny
Bogosloviye posle Osventsima i GULaga i otnosheniye k yevreyam i iudaizmu v Pravoslavnoi Tserkvi bolshevistskoi Rossii, ed. Natalia Pecherskaya. Sankt-Peterburg: Vysshaya Religiozno-Filosofskaya Shkola, 1997 (no. 1060)
Bogosloviye posle Osventsima i yevo svyaz s bogosloviyem posle Gulaga: Materialy vtoroi mezhdunarodnoi nauchnoi konferentsii, Sankt-Peterburg, 1998, ed. N. Pecherskaya. Sankt-Peterburg: Vysshaya Religiozno-Filosofskaya Shkola, 1999 (no. 1081)
Boletín de la Real Academia de la Historia. Madrid

Bulletin der Schweizerischen Gesellschaft fuer Judaistische Forschung: Beiheft zu "Judaica." Basel: Schweizerische Gesellschaft fuer Judaistische Forschung
Bulletin du Centre de Recherche Français de Jérusalem. Jérusalem: Centre de Recherche Français de Jérusalem
Cahier International sur le Témoignage Audiovisuel: Etudes sur le témoignage audiovisuel des victimes des crimes et génocides nazis. Bruxelles: Centre d'Etudes et de Documentation – Fondation Auschwitz
Les Cahiers de la Mémoire Contemporaine. Bruxelles: Fondation de la Mémoire Contemporaine
Cahiers de Littérature Orale. Paris: Publications Langues'O
Les Cahiers du Judaïsme. Paris: Alliance Israélite Universelle
Canadian Slavonic Papers/Revue canadienne des slavistes. Edmonton, Alta.: Canadian Association of Slavists
CEMOTI – Cahiers d'Etudes sur la Mediterranée Orientale et le Monde Turco-Iranien. Paris: AFEMOTI
El centinela de la fe: Estudios jurídicos sobre la Inquisición de Sevilla en el siglo XVIII, ed. Enrique Gacto Fernández. Sevilla: Universidad de Sevilla, 1997 (no. 127)
Central European History. Boston: Humanities Press
Christen und Juden: Perspektiven einer Annaeherung, ed. Wolfgang Kraus. Guetersloh: Chr. Kaiser, Guetersloher Verlagshaus, 1997 (no. 1080)
La Civiltà Cattolica. Roma: Compagnia di Gesù
Collectanea Theologica. Warszawa: Akademia Teologii Katolickiej
Comment devient-on dreyfusard?, eds. Janine Chêne et al. Paris: L'Harmattan, 1997 (no. 237)
Commentary. New York: American Jewish Committee
Comparative Literature. Eugene, OR: University of Oregon
Contemporary Jewry: A Journal of Social Scientific Inquiry. New York: Association for the Social Scientific Study of Jewry
La controversia judeocristiana en España (desde los orígenes hasta el siglo XIII): Homenaje a Domingo Muñoz León, ed. Carlos del Valle Rodríguez. Madrid: Consejo Superior de Investigaciones Científicas, 1998 (no. 53)
Creencias y culturas, eds. Carlos Carrete Parrondo, Alisa Meyuhas-Ginio. Salamanca: Universidad Pontificia; Tel-Aviv: Universidad de Tel-Aviv, 1998 (no. 81)
Crime & Delinquency. Thousand Oaks, CA: Sage
Crisis and Creativity in the Sephardic World, 1391-1648, ed. Benjamin R. Gampel. New York: Columbia University Press, 1997 (no. 19)
Cross Currents. New Rochelle, NY: Association for Religion and Intellectual Life
Dachauer Hefte: Studien und Dokumente zur Geschichte der national-sozialistischen Konzentrationslager. Dachau: Dachauer Hefte
Dapim le-Heker Tekufat ha-Shoah. Haifa: University of Haifa; Tel-Aviv: Hakibbutz Hameuchad
Le Débat. Paris: Gallimard

Demonizing the Other: Antisemitism, Racism, and Xenophobia, ed. Robert S. Wistrich. Amsterdam: Harwood Academic, for the Vidal Sassoon International Center for the Study of Antisemitism, Hebrew University, 1999 (no. 33)
Deutsche Zeitschrift fuer Philosophie. Berlin: Akademie Verlag
Dimensions: A Journal of Holocaust Studies. New York: International Center for Holocaust Studies, ADL
Dopet eller döden: Judiskt liv i det medeltida Spanien, eds. Hanne Trautner-Kromann, Salomon Schulman. Stockholm: Brutus Östlings Bokförlag Symposion, 1996 (no. 118)
Dorohoi – generaţii de iudaism şi sionism: Dorohoi, Săveni, Mihăileni, Darabani, Herţa, Rădăuţi-Prut (România), Vols. 3-4. ed. Shlomo David. Kiryath Bialik: Organizaţia Israelienilor Originari din Oraşul şi Judeţul Dorohoi, 1996-1998 (nos. 834, 835)
East European Jewish Affairs. London: Institute for Jewish Policy Research
Gli ebrei di Pisa (secoli IX-XX): Atti del Convegno internazionale Pisa, 3-4 ottobre 1994, ed. Michele Luzzati. Pisa: Pacini, 1998 (no. 46)
Eighteenth-Century Studies. Baltimore: Johns Hopkins University Press
ELH – English Literary History. Baltimore: Johns Hopkins University Press
The Emancipation of Catholics, Jews and Protestants: Minorities and the Nation State in Nineteenth-Century Europe, eds. Rainer Liedtke, Stephan Wendehorst. Manchester: Manchester University Press, 1999 (no. 230)
Ephemerides Theologicae Lovanienses. Leuven: Peeters
Esprit. Paris: Esprit
Estudios de Asia y Africa. México, D.F.: El Colegio de México
Estudios Migratorios Latinoamericanos. Buenos Aires: Centro de Estudios Migratorios Latinoamericanos
Ethical Theory and Moral Practice. Dordrecht: Kluwer Academic
Ethnicity and the Bible, ed. Mark G. Brett. Leiden: E.J. Brill, 1996 (no. 75)
Ethnology: An International Journal of Cultural and Social Anthropology. Pittsburgh, PA: University of Pittsburgh
European History Quarterly. London: Sage
European Judaism. Oxford: Berghahn Books; London: Leo Baeck College
The European Legacy. Basingstoke, Hants: Carfax
Evangelische Theologie. Guetersloh: Chr. Kaiser, Guetersloher Verlagshaus
Evolution of the Synagogue: Problems and Progress, eds. Howard Clark Kee, Lynn H. Cohick. Harrisburg, PA: Trinity Press International, 1999 (no. 59)
Exil: Forschung, Erkenntnisse, Ergebnisse 1933-1945. Maintal: E. Koch
Der Exodus aus Nazideutschland und die Folgen: Juedische Wissenschaftler im Exil, eds. Marianne Hassler, Juergen Wertheimer. Tuebingen: Attempto, 1997 (nos. 615, 1179)
The Expulsion of the Jews and Their Emigration to the Southern Low Countries (15th-16th C.), eds. Luc Dequeker, Werner Verbeke. Leuven: Leuven University Press, 1998 (no. 98)

A Feminist Companion to Reading the Bible: Approaches, Methods and Strategies, eds. Athalya Brenner, Carole Fontaine. Sheffield: Sheffield Academic Press, 1997 (no. 957)

Flucht nach Eretz Israel: Die Bricha und der juedische Exodus durch Oesterreich nach 1945, ed. Thomas Albrich. Innsbruck: Studien Verlag, 1998 (no. 967)

Frankfurter Judaistische Beitraege. Frankfurt a.M.: Gesellschaft zur Foerderung judaistischer Studien in Frankfurt am Main

Freiburger Rundbrief: Zeitschrift fuer christlich-juedische Begegnung. Freiburg: Freiburger Rundbrief

"Der Fremde im Dorf": Ueberlegungen zum Eigenen und zum Fremden in der Geschichte, eds. Hans-Juergen Boemelburg, Beate Eschment. Lueneburg: Institut Nordostdeutsches Kulturwerk, 1998 (no. 784)

Fritz Bauer Institut: Jahrbuch zur Geschichte und Wirkung des Holocaust, 1998/99 see "Beseitigung des juedischen Einflusses..."

Gebrochene Identitaeten: Zur Kontroverse um kollektive Identitaeten in Deutschland, Israel, Suedafrika, Europa und im Identitaetskampf der Kulturen, eds. Werner Gephart, Karl-Heinz Saurwein. Opladen: Leske + Budrich, 1999 (no. 1051)

Das Gedaechtnis der Dinge: KZ-Relikte und KZ-Denkmaeler, 1945-1995, ed. Detlef Hoffmann. Frankfurt a.M.: Campus, 1998 (no. 931)

Geduldet, geschmaeht und vertrieben: Salzburger Juden erzaehlen, eds. Daniela Ellmauer et al. Salzburg: Otto Mueller Verlag, 1998 (no. 175)

Die Gegenwart der Geschichte des Holocaust: Intergenerationelle Tradierung und Kommunikation der Nachkommen, eds. Christian Staffa, Katherine Klinger. Berlin: Institut fuer Vergleichende Geschichtswissenschaften, 1998 (no. 1065)

The German Army and Genocide: Crimes against War Prisoners, Jews, and Other Civilians in the East, 1939-1944, ed. Hannes Heer. New York: New Press, 1999 (no. 379)

German History: The Journal of the German History Society. London: Arnold

German Studies Review. Tempe, AZ: German Studies Association at Arizona State University

Germanisch-Romanische Monatsschrift. Heidelberg: C. Winter

Die Geschichte der Juden in Biebesheim. Riedstadt: Forum Verlag, 1997 (no. 690)

Geschichte der Juden in Heidelberg. Heidelberg: Verlag Brigitte Guderjahn, 1996 (no. 36)

Geschichte in Wissenschaft und Unterricht: Zeitschrift des Verbandes der Geschichtslehrer Deutschlands. Seelze: Erhard Friedrich Verlag

Ghetto Writing: Traditional and Eastern Jewry in German-Jewish Literature from Heine to Hilsenrath, eds. Anne Fuchs, Florian Krobb. Columbia, SC: Camden House, 1999 (no. 1162)

Ha-Reich ha-shlishi: Ma'azan histori, ed. Moshe Zimmermann. Jerusalem: Magnes Press, Hebrew University, 1999 (no. 485)

Ha-Tzionut: Me'assef le-toldot ha-tenu`ah ha-tzionit ve-ha-yishuv ha-yehudi be-Eretz-Yisrael. Tel-Aviv: Tel-Aviv University, Institute for Zionist Research; Hakibbutz Hameuchad
Hamizrah Hehadash. Jerusalem: Magnes Press, Hebrew University
Handbuch des deutschsprachigen Exiltheaters, 1933-1945. Vol. 1-2/2, eds. Frithjof Trapp et al. Muenchen: K.G. Saur, 1999 (no. 699)
Harmonie und Gewalt: Oesterreich, Europa und die Zukunft der Vergangenheit, eds. Wilfried Graf, Roswitha Riepl. Muenster: agenda, 1997 (nos. 433, 1056)
The Harvard Mosaic: A Review of Jewish Thought and Culture. Cambridge, MA: Harvard Mosaic
Harvard Theological Review. Cambridge, MA: Harvard University
Herkenning see Ter Herkenning: Tijdschrift voor christenen en joden
Hertz Jospa – Juif, résistant, communiste, ed. Jean-Philippe Schreiber. Bruxelles: EVO (Editions Vie Ouvrière), 1997 (no. 518)
Hesed ve-emet: Studies in Honor of Ernest S. Frerichs, eds. Jodi Magness, Seymour Gitin. Atlanta: Scholars Press, 1998 (no. 142)
Historia: Zeitschrift fuer Alte Geschichte. Stuttgart: Franz Steiner
The Historical Journal. Cambridge: Cambridge University Press
Historical Journal of Film, Radio and Television. Basingstoke, Hants: Carfax
Historical Reflections/Réflexions Historiques. Alfred, NY: Historical Reflections
Historical Social Research/Historische Sozialforschung. Koeln: Quantum
Historische Zeitschrift. Muenchen: R. Oldenbourg
History & Memory: Studies in Representation of the Past. Bloomington: Indiana University Press
History of European Ideas. Oxford: Pergamon
History Today. London: History Today
History Workshop Journal. Oxford: Oxford University Press
Holocaust and Genocide Studies. Cary, NC: Oxford University Press
Holocaust Füzetek. Budapest: Magyar Auschwitz Alapítvány
Das Holocaust-Mahnmal: Dokumentation einer Debatte, ed. Michael S. Cullen. Zuerich: Pendo, 1999 (no. 1093)
Hommes & Migrations. Paris: Hommes & Migrations
'Ilu: Revista de ciencias de las religiones. Madrid: Universidad Complutense
Im Herzen der Finsternis: Victor Klemperer als Chronist der NS-Zeit, ed. Hannes Heer. Berlin: Aufbau-Verlag, 1997 (no. 618)
Immigrants & Minorities. London: Frank Cass
Index on Censorship. London: Writers and Scholars International
Ingeborg Bachmann und Paul Celan: Poetische Korrespondenzen, eds. Bernhard Boeschenstein, Sigrid Weigel. Frankfurt a.M.: Suhrkamp, 1997 (no. 1155)
Insider/Outsider: American Jews and Multiculturalism, eds. David Biale et al. Berkeley: University of California Press, 1998 (no. 1037)
Integrazione e identità: L'esperienza ebraica in Germania e Italia dall'Illuminismo al fascismo, ed. Mario Toscano. Milano: FrancoAngeli, 1998 (no. 171)
Les intellectuels face à l'Affaire Dreyfus alors et aujourd'hui: Perception et impact de l'Affaire en France et à l'étranger, eds. Roselyne Koren, Dan Michman. Paris: L'Harmattan, 1998 (no. 226)

International Journal of Sociology. Armonk, NY: M.E. Sharpe
International Journal on Audio-Visual Testimony see Cahier International sur le Témoignage Audiovisuel
Internationales Josephus-Kolloquium, Muenster 1997: Vortraege aus dem Institutum Judaicum Delitzchianum, eds. Folker Siegert, Juergen U. Kalms. Muenster: LIT, 1998 (no. 1033)
Internationales Theodor Herzl Symposion (2nd, Vienna, 1997): Der Bericht. Wien: Ideenagentur Austria, 1997 (no. 168)
Israel Law Review. Jerusalem: Hebrew University
Israele – da Mosè agli accordi di Oslo. Bari: Dedalo, 1999 (no. 951)
Jahrbuch der Historischen Forschung in der Bundesrepublik Deutschland. Muenchen: R. Oldenbourg
Jahrbuch fuer Antisemitismusforschung. Frankfurt a.M.: Campus
The Jerusalem Review. Tel-Aviv: Ah'shav
Jewish Affairs. Houghton: South African Jewish Board of Deputies
Jewish History. Haifa: Haifa University Press
Jewish Journal of Sociology. London: World Jewish Congress
Jewish Perspectives on the Experience of Suffering, ed. Shalom Carmy. Northvale, NJ: Jason Aronson, 1999 (nos. 355, 437)
The Jewish Quarterly. London: Jewish Literary Trust
Jewish Social Studies. Bloomington: Indiana University Press
Jewish Studies. Jerusalem: World Union of Jewish Studies
Jewish Studies at the Turn of the Twentieth Century, Vol. I-II, eds. Judit Targarona Borrás and Ángel Sáenz-Badillos. Leiden: Brill, 1999 (no. 32)
Jewish Studies Quarterly. Tuebingen: Mohr Siebeck
Jewries at the Frontier: Accommodation, Identity, Conflict, eds. Sander L. Gilman, Milton Shain. Urbana: University of Illinois Press, 1999 (no. 156)
The Jews in Poland, Vol. II, ed. Sławomir Kapralski. Cracow: Judaica Foundation, Center for Jewish Culture, 1999 (no. 207)
The Jews in the Romanian History: Papers from the International Symposium, Bucharest, 30 September-4 October, 1996, ed. Ion Stanciu. Bucureşti: Silex, [1999] (no. 51)
Journal for the Psychoanalysis of Culture & Society see JPCS – Journal for the Psychoanalysis of Culture & Society
Journal of American Studies. Cambridge: Cambridge University Press
Journal of Church and State. Waco, TX: Baylor University, Dawson Institute of Church-State Studies
Journal of Contemporary History. London: Sage
Journal of Early Christian Studies. Baltimore: Johns Hopkins University Press
Journal of English and Germanic Philology. Champaign, IL: University of Illinois Press
Journal of Genocide Research. Basingstoke, Hants: Carfax
The Journal of Holocaust Education. London: Frank Cass
The Journal of Israeli History. London: Frank Cass
Journal of Jewish Thought and Philosophy. Amsterdam: Harwood Academic
Journal of Medieval History. Amsterdam: Elsevier

Journal of Northwest Semitic Languages. Stellenbosch: University of Stellenbosch, Department of Ancient Studies
Journal of the American Academy of Religion. Atlanta: Scholars Press
Journal of the Australian Jewish Historical Society see Australian Jewish Historical Society Journal
JPCS – Journal for the Psychoanalysis of Culture & Society. Gardiner, NY: Critical Press
Judaica Bohemiae. Praha: Státní Židovské Muzeum
Judaism. New York: American Jewish Congress
Judaism and Christianity in First-Century Rome, eds. Karl Paul Donfried, Peter Richardson. Grand Rapids, MI: William B. Eerdmans, 1998 (no. 73)
Juden im Ruhrgebiet: Vom Zeitalter der Aufklaerung bis in die Gegenwart, eds. Jan-Pieter Barbian et al. Essen: Klartext, 1999 (no. 570)
Juden in der Oberlausitz. Bautzen: Lusatia, 1998 (no. 37)
Juden und Christen zur Zeit der Kreuzzuege, ed. Alfred Haverkamp. Sigmaringen: Jan Thorbecke, 1999 (no. 90)
Juden – unsere aelteren Brueder im Glauben: Was haben die Juden mit der christlichen Oekumene zu tun?, ed. Klaus Gierse. Beckum: Kartellverband katholischer deutscher Studentenvereine, [1997] (no. 193)
Juedische Denker im 20. Jahrhundert, ed. Hanna Lehming. Hamburg: E.B.-Verlag, 1997 (no. 362)
Die juedische Gemeinde Hennef-Geistingen, ed. Gisela Rupprath. Hennef: Stadt Hennef, 1999 (no. 675)
Juedischer Nietzscheanismus, eds. Werner Stegmaier, Daniel Krochmalnik. Berlin: Walter de Gruyter, 1997 (no. 169)
Juedischer Republikanismus in Frankreich, ed. Christoph Miething. Tuebingen: Max Niemeyer, 1998 (no. 179)
Juedisches Leben in Moenchengladbach: Gestern und heute. Moenchengladbach: Stadtarchiv, 1998 (no. 982)
Jurbarkas: Istorijos puslapiai, ed. Arnoldas Piročkinas. Vilnius: Pradai, 1996 (no. 751)
Justice. Tel-Aviv: International Association of Jewish Lawyers and Jurists
Khronika-2000. Kyyiv: Fond Spryyannya Rozvytku Mystetstv Ukrayiny
Ki Baruch Hu: Ancient Near Eastern, Biblical, and Judaic Studies in Honor of Baruch A. Levine, eds. Robert Chazan et al. Winona Lake, IN: Eisenbrauns, 1999 (no. 1158)
Kibbutz galuyot: Aliyah le-Eretz Israel – mythos u-metzi'ut, ed. Dvora Hacohen. Jerusalem: Zalman Shazar Center for Jewish History, 1998 (nos. 293, 308)
Kirche und Israel: Neukirchener theologische Zeitschrift. Neukirchen-Vluyn: Neukirchener Verlag des Erziehungsvereins
Klio: Beitraege zur Alten Geschichte. Berlin: Akademie Verlag
Kompendium: Feministische Bibelauslegung, eds. Luise Schottroff, Marie-Theres Wacker. Guetersloh: Chr. Kaiser, 1998 (no. 57)
Die Konstruktion der Nation gegen die Juden, eds. Peter Alter et al. Muenchen: Wilhelm Fink Verlag, 1999 (no. 185)

Korni: Vestnik Narodnovo Universiteta Yevreiskoi Kultury v Tsentralnoi Rossii i Povolzhye. Saratov: Yevreiskii Mir

Korot: The Israel Journal of the History of Medicine and Science. Jerusalem: Magnes Press, Hebrew University

Kultur allein ist nicht genug: Das Werk von Elie Wiesel – Herausforderung fuer Religion und Gesellschaft, eds. Reinhold Boschki, Dagmar Mensink. Muenster: LIT, 1998 (nos. 811, 921)

Il Lager: Il ritorno della memoria: Atti del Convegno internazionale, 6-7 aprile 1995, Università degli Studi di Verona, eds. Gian Paolo Marchi, Giovanna Massariello Merzagora. Milano: ANED; Trieste: Edizioni Lint, 1997 (no. 933)

Land and Community: Geography in Jewish Studies, ed. Harold Brodsky. Bethesda, MD: University Press of Maryland, 1997 (no. 1052)

"L'chajim": Die Geschichte der Juden im Landkreis Darmstadt-Dieburg, ed. Thomas Lange. Darmstadt: Landkreis Darmstadt-Dieburg, 1997 (no. 643)

Leben in zwei Diktaturen: Victor Klemperers Leben in der NS-Zeit und in der DDR, eds. Christoph Wielepp, Hans-Peter Luehr. Dresden: Friedrich-Ebert-Stiftung, Buero Dresden, 1997 (no. 706)

Leben mit der Erinnerung: Juedische Geschichte in Prenzlauer Berg, eds. Gisela Reller et al. Berlin: Hentrich, 1997 (no. 667)

Legacy: The Yad Vashem Journal of Holocaust Education. Jerusalem: Yad Vashem

El legado material hispanojudío: VII Curso de Cultura Hispanojudía y Sefardí de la Universidad de Castilla-La Mancha, eds. A.M. López Álvarez, R. Izquierdo Benito. Cuenca: Ediciones de la Universidad de Castilla-La Mancha, 1998 (nos. 87, 115)

Leo Baeck Institute Year Book. London: Leo Baeck Institute

Leo Baeck – zwischen Geheimnis und Gebot: Auf dem Weg zu einem progressiven Judentum der Moderne, ed. Verena Muehlstein. Karlsruhe: Evangelische Presseverb. fuer Baden; Osnabrueck: secolo, 1997 (no. 420)

Lessons and Legacies II: Teaching the Holocaust in a Changing World, ed. Donald G. Schilling. Evanston, IL: Northwestern University Press, 1998 (no. 453)

Letters and Texts of Jewish History, ed. Norman Simms. Hamilton, N.Z.: Outrigger, for the Waikatό Jewish Studies Seminar, 1998 (no. 760)

Libraries & Culture: A Journal of Library History. Austin, TX: University of Texas Press

A Life in Jewish Education: Essays in Honor of Louis L. Kaplan, ed. Jack Fruchtman, Jr. Bethesda, MD: University Press of Maryland, 1997 (no. 184)

"Man haengt immer zwischen Himmel und Erde...": Juedische Emigrantinnen und Emigranten (1933-45) aus Bremen berichten, eds. Inge Marssolek, Wiebke Davids. Bremen: Selbstverlag des Staatsarchivs Bremen, 1997 (no. 648)

Massuah: Kovetz shnati le-toda`at ha-shoah ve-ha-gvurah. Tel-Aviv: Massuah

Materia Giudaica: Rivista dell'Associazione Italiana per lo Studio del Giudaismo. Firenze: Giuntina

Max Liebermann und die franzoesischen Impressionisten, eds. G. Tobias Natter, Julius H. Schoeps. Koeln: DuMont, 1997 (nos. 677, 678)
Media, Culture & Society. London: Sage
Medieval Encounters: Jewish, Christian and Muslim Culture in Confluence and Dialogue. Leiden: E.J. Brill
Medieval History Journal. New Delhi: Sage
Medizinische Wissenschaften und Judentum, eds. Nora Goldenbogen et al. Dresden: Verein fuer regionale Politik und Geschichte Dresden, 1996 (nos. 259, 698)
"Meinetwegen ist die Welt erschaffen": Das intellektuelle Vermaechtnis des deutschsprachigen Judentums: 58 Portraits, eds. Hans Erler et al. Frankfurt a.M.: Campus, 1997 (no. 597)
La mémoire contre la nuit. Bordeaux: Ed. du Passant, 1997 (nos. 337, 536)
Mennonite Quarterly Review. Goshen, IN: Mennonite Quarterly Review Publication Committee
Menora: Jahrbuch fuer deutsch-juedische Geschichte. Muenchen: Piper
Menora und Hakenkreuz: Zur Geschichte der Juden in und aus Schleswig-Holstein, Luebeck und Altona (1918-1998), eds. Gerhard Paul, Miriam Gillis-Carlebach. Neumuenster: Wachholtz Verlag, 1998 (no. 663)
Mexico le-ahar ha-mahapekhah: Hevrah u-politikah, 1910-1952, eds. Tzvi Medin, Raanan Rein. Tel Aviv: University Publishing Projects, 1999 (no. 904)
Meyerbeer – Wagner: Eine Begegnung, eds. Gunhild Oberzaucher-Schueller et al. Wien: Boehlau, 1998 (no. 252)
Midstream. New York: Theodor Herzl Foundation
Militaergeschichtliche Mitteilungen. Muenchen: Oldenbourg
Miscelánea de Estudios Árabes y Hebraicos. Granada: Universidad de Granada
Mit den Augen eines Kindes: Children in the Holocaust – Children in Exile – Children under Fascism, ed. Viktoria Hertling. Amsterdam: Rodopi, 1998 (no. 382)
Modern History Review. Deddington, Oxon: Philip Allan
Modern Judaism. Baltimore: Johns Hopkins University Press
Muensteraner Judaistische Studien: Wissenschaftliche Beitraege zur christlich-juedischen Begegnung, 2 see Internationales Josephus-Kolloquium
Múlt és Jövő. Budapest: Múlt és Jövő
Die Musik des osteuropaeischen Judentums – totalitaere Systeme – Nachklaenge: Kolloquium... 1994, ed. Frank Geissler. Leipzig: Dresdner Zentrum fuer Zeitgenoessische Musik; UniMedia, 1997 (no. 365)
Myosotis: Zeitschrift fuer Buchwesen. Mainz: Myosotis
"Nach der Kristallnacht": Juedisches Leben und antijuedische Politik in Frankfurt am Main, 1938-1945, ed. Monica Kingreen. Frankfurt a.M.: Campus, 1999 (no. 635)
Die Nacht hat zwoelf Stunden, dann kommt schon der Tag: Antifaschismus, Geschichte und Neubewertung, ed. Claudia Keller. Berlin: Aufbau Taschenbuch Verlag, 1996 (no. 984)
Nationalities Papers. Abingdon, Oxon: Carfax

Nationalsozialistische Vernichtungspolitik, 1939-1945: Neue Forschungen und Kontroversen, ed. Ulrich Herbert. Frankfurt a.M.: Fischer Taschenbuch Verlag, 1998 (no. 381)
New Literary History: A Journal of Theory and Interpretation. Baltimore: Johns Hopkins University Press
New Readings in John: Literary and Theological Perspectives: Essays from the Scandinavian Conference on the Fourth Gospel, Arhus 1997, eds. Johannes Nissen, Sigfred Pedersen. Sheffield: Sheffield Academic Press, 1999 (no. 70)
Nietzsche-Studien. Berlin: Walter de Gruyter
Nordisk Judaistik: Scandinavian Jewish Studies. Lund: Nordisk Judaistik
"Nun steht aber diese Sache im Evangelium...": Zur Frage nach den Anfaengen des christlichen Antijudaismus, ed. Rainer Kampling. Paderborn: Ferdinand Schoeningh, 1999 (no. 65)
Obrazy w głowie: Oświęcim – ujęcia pewnego miejsca see Bilder im Kopf: Auschwitz – einen Ort sehen
October. Cambridge, MA: MIT Press
El Olivo: Documentación y estudios para el diálogo entre judíos y cristianos. Madrid: Centro de Estudios Judeo-Cristianos
Orbis: Bulletin international de documentation linguistique. Louvain: Centre International de Dialectologia Générale
Orbis Litterarum: International Review of Literary Studies. Copenhagen: Munksgaard
The Orthodox Forum Series, 7 see Jewish Perspectives on the Experience of Suffering, ed. Shalom Carmy. Northvale, NJ: Jason Aronson, 1999
Oxford Art Journal. Oxford: Oxford University Press
Passover and Easter: Origin and History to Modern Times, eds. Paul F. Bradshaw, Lawrence A. Hoffman. Notre Dame, IN: University of Notre Dame Press, 1999 (no. 120)
Past & Present: A Journal of Historical Studies. Oxford: Oxford University Press, for the Past and Present Society
Patterns of Migration, 1850-1914, eds. Aubrey Newman, Stephen W. Massil. London: Jewish Historical Society of England, in association with the Institute of Jewish Studies, University of London, 1996 (nos. 228, 1176)
Patterns of Prejudice. London: Sage
Pe`amim: Pirkei iyyun be-moreshet Yisrael ba-mizrah. Jerusalem: Ben-Zvi Institute
Performance and Cultural Politics, ed. Elin Diamond. London: Routledge, 1996 (no. 1144)
Peter Weiss Jahrbuch. St. Ingbert: Roehrig Universitaetsverlag
Pidhotovchi materialy populyarnoyi entsyklopediyi "Ukrayinske yevreistvo," Pt. 3. Kyyiv: Instytut Yudayiky, 1998 (no. 215)
Plastika: Ktav ha-et shel ha-Seminar he-Hadash le-Limudei Tarbut Hazutit/Bikoret/Teoria. Tel-Aviv: Camera Obscura, Beit ha-Sefer le-Omanut; Tefen: Ha-Muzeon ha-Patuah
Polin: A Journal of Polish-Jewish Studies. Oxford: Blackwell
Political Crossroads. Albert Park, Australia: James Nicholas Publishers

Political Psychology. Malden, MA: Blackwell
Population and Development Review. New York: Population Council
Primo Levi: Il mestiere di raccontare, il dovere di ricordare: Atti del Convegno, Trento, 14 maggio 1997, ed. Ada Neiger. Fossombrone: Metauro, 1998 (no. 935)
Proceedings of the 33rd Annual Convention of the Association of Jewish Libraries, Philadelphia, PA, June 21-24, 1998, eds. Barbara Y. Leff, Laurel S. Wolfson. New York: Association of Jewish Libraries, 1999 (nos. 223, 745)
The Proceedings of the Tenth British Conference on Judeo-Spanish Studies, 29 June-1 July 1997, ed. Annette Benaim. London: Department of Hispanic Studies, Queen Mary and Westfield College, 1999 (no. 14)
Public Opinion Quarterly. Chicago: University of Chicago Press
Il racconto della catastrofe: Il cinema di fronte ad Auschwitz, eds. Francesco Monicelli, Carlo Saletti. Verona: Società Letteraria di Verona, 1998 (no. 1178)
Radical Philosophy: A Journal of Socialist and Feminist Philosophy. London: Radical Philosophy
Rambam: Tidsskrift for jødisk kultur og forskning. København: C.A. Reitzels Forlag
La Rassegna Mensile di Israel. Roma: Unione delle Comunità Israelitiche Italiane
Remembrance, Repentance, Reconciliation: The 25th Anniversary Volume of the Annual Scholars' Conference on the Holocaust and the Churches, ed. Douglas F. Tobler. Lanham, MD: University Press of America, 1998 (no. 307)
Res Publica: A Journal of Legal and Social Philosophy. Liverpool: Deborah Charles Publications
Research in Social Movements, Conflicts and Change. Greenwich, CT: JAI Press
Review of English Studies. Oxford: Oxford University Press
Revue de l'Histoire des Religions. Paris: Presses Universitaires de France
Revue des Etudes Juives. Paris: Peeters
Revue d'Histoire de la Shoah – Le Monde Juif. Paris: Centre de Documentation Juive Contemporaine
Revue d'Histoire Moderne et Contemporaine. Paris: Société d'Histoire Moderne et Contemporaine
Revue Internationale de Philosophie. Bruxelles: Revue Internationale de Philosophie
Rizhskii Almanakh. Riga: Daugava Rizhskii Almanakh
Romantisme: Revue du dix-neuvième siècle. Paris: SEDES
Roots and Branches: Explorations into the Jewish Context of the Christian Faith, eds. John Fieldsend et al. Bedford: PWM Trust; Centre for Biblical and Hebraic Studies, 1998 (no. 114)
Russian Nationalism, Past and Present, eds. Geoffrey Hosking, Robert Service. Basingstoke, Hants: Macmillan, 1998 (no. 1122)
Russkii Yevrei. Moskva: Obyedinyonnaya Redaktsiya "MEG"
Sachor: Zeitschrift fuer Antisemitismusforschung, juedische Geschichte und Gegenwart. Essen: Klartext Verlag
Scandinavian Journal of History. Oslo: Scandinavian University Press

Schweizerische Gesellschaft fuer Judaistische Forschung: Bulletin see Bulletin der Schweizerischen Gesellschaft fuer Judaistische Forschung
Seeds of Reconciliation: Essays on Jewish-Christian Understanding, ed. Katharine T. Hargrove. N. Richland Hills, TX: BIBAL Press, 1996 (no. 956)
Sens: Juifs et chrétiens dans le monde aujourd'hui. Paris: L'Amitié Judéo-Chrétienne de France
Shakespeare Survey. Cambridge: Cambridge University Press
Shalom Trieste: Gli itinerari dell'ebraismo, ed. Adriano Dugulin. Trieste: Comune di Trieste, 1998 (nos. 264, 285)
The Shekel. Tamarac, FL: American Israel Numismatic Association
La Shoah: Témoignage impossible?, eds. Alain Goldschlaeger, Jacques Lemaire. Bruxelles: Editions de l'Université de Bruxelles, 1998 (no. 929)
Shofar: An Interdisciplinary Journal of Jewish Studies. Lincoln: University of Nebraska Press
Shvut: Studies in Russian and East European Jewish History and Culture. Tel-Aviv: Tel-Aviv University, Diaspora Research Institute; Beer-Sheva: Ben-Gurion University of the Negev
SIDIC. Rome: Service International de Documentation Judéo-Chrétienne
Sino-Judaica: Jews and Chinese in Historical Dialogue: An International Colloquium, Nanjing, 11-19 October 1996, ed. Aharon Oppenheimer. Tel-Aviv: Tel-Aviv University, 1999 (no. 1021)
The Social Studies. Washington, DC: Heldref
Solidaritaet und Hilfe fuer Juden waehrend der NS-Zeit: Regionalstudien 1-3, eds. Wolfgang Benz, Juliane Wetzel. Berlin: Metropol, 1996-1999 (no. 323)
Storia d'Italia. Annali XI: Gli ebrei in Italia, [pt.] II, ed. Corrado Vivanti. Torino: Giulio Einaudi, 1997 (no. 47)
Strangers & Neighbors: Relations between Blacks & Jews in the United States, eds. Maurianne Adams, John H. Bracey. Amherst: University of Massachusetts Press, 1999 (no. 220)
Studia Judaica. Cluj-Napoca: "Sincron"
Studia nad Faszyzmem i Zbrodniami Hitlerowskimi. Wrocław: Wydawnictwo Uniwersytetu Wrocławskiego
Studia Rosenthaliana: Tijdschrift voor joodse wetenschap en geschiedenis in Nederland. Assen: Van Gorcum
Studies and Texts in Jewish History and Culture, IV see A Life in Jewish Education: Essays in Honor of Louis L. Kaplan
Studies in Contemporary Jewry. New York: Oxford University Press
Tarbiz: Riv`on le-mada`ei ha-yahadut. Jerusalem: Magnes Press, Hebrew University, Institute of Jewish Studies
Tel Aviver Jahrbuch fuer Deutsche Geschichte. Gerlingen: Bleicher, [for the] Universitaet Tel Aviv, Institut fuer Deutsche Geschichte
Les Temps Modernes. Paris: Les Temps Modernes
Ter Herkenning: Tijdschrift voor christenen en joden. 's-Gravenhage: Boekencentrum
Terezínské Listy: Sborník Památníku Terezín. Terezín: Památník Terezín
Terezínské Studie a Dokumenty. Praha: Nadace Terezínská iniciativa

Terrorism and Political Violence. London: Frank Cass
Text + Kritik: Zeitschrift fuer Literatur. Muenchen: Text + Kritik
Theater gegen das Vergessen: Buehnenarbeit und Drama bei George Tabori, eds. Hans-Peter Bayerdoerfer, Joerg Schoenert. Tuebingen: Max Niemeyer, 1997 (no. 1153)
Theodor Herzl, Visionary of the Jewish State, eds. Gideon Shimoni, Robert S. Wistrich. Jerusalem: Hebrew University, Magnes Press; New York: Herzl Press, 1999 (no. 231)
Theory, Culture & Society. London: Sage
Theresienstaedter Studien und Dokumente. Prag: Stiftung Theresienstaedter Initiative (Nadace Terezínská iniciativa) im Verlag Academia
Tikkun: A Bi-Monthly Jewish Critique of Politics, Culture and Society. Oakland, CA: Institute for Labor and Mental Health
Transmission et passages en monde juif, ed. Esther Benbassa. Paris: Publisud, 1997 (no. 15)
Der Traum von Israel: Die Urspruenge des modernen Zionismus, ed. Heiko Haumann. Weinheim: Beltz Athenaeum, 1998 (no. 224)
Tribuene: Zeitschrift zum Verstaendnis des Judentums. Frankfurt a.M.: Tribuene
Trumah: Zeitschrift der Hochschule fuer Juedische Studien, Heidelberg. Berlin: Metropol
Tsafon: Revue d'études juives du Nord. Lille: Tsafon
Two Nations: British and German Jews in Comparative Perspective, eds. Michael Brenner et al. Tuebingen: Mohr Siebeck, 1999 (no. 153)
Undzer Tsait. New York: Undzer Tsait
Varieties of Multiculturalism in Modern European History: The Case of the Jews, eds. Michael Heyd, Oded Heilbronner. Jerusalem: International Center for University Teaching of Jewish Civilization, Hebrew University, 1998 (no. 149)
Verlorene Nachbarschaft: Die Wiener Synagoge in der Neudeggergasse: Ein Mikrokosmos und seine Geschichte, eds. Kaethe Kratz et al. Wien: Mandelbaum, 1999 (no. 488)
Vestnik Yevreiskovo Universiteta. Moskva: Mosty Kultury; Ierusalim: Gesharim
Viator: Medieval and Renaissance Studies. Turnhout: Brepols
Le vie del mondo – Berlino, Budapest, Praga, Vienna e Trieste: Intellettuali ebrei e cultura europea dal 1880 al 1930, ed. Riccardo Calimani. Milano: Electa, 1998 (no. 233)
Vielseitigkeit des Alten Testaments: Festschrift fuer Georg Sauer zum 70. Geburtstag, eds. James Alfred Loader, Hans Volker Kieweler. Frankfurt a.M.: Peter Lang, 1999 (no. 145)
Vierteljahrshefte fuer Zeitgeschichte: Beilage. Muenchen: R. Oldenbourg, [for the] Institut fuer Zeitgeschichte
The Visigoths: Studies in Culture and Society, ed. Alberto Ferreiro. Leiden: Brill, 1999 (no. 86)
Vtoraya mirovaya voina: Diskussii, osnovnye tendentsii, rezultaty issledovanii, ed. Wolfgang Michalka. Moskva: Ves Mir, 1996 (no. 652)

Vtoroye dykhaniye: Literaturnyi almanakh. eds. Aleksandr Nakhshin, Mikhail Ronkin. Petach-Tikva: Krugozor, 1997 (no. 871)
War in History. London: Arnold
Washington Conference on Holocaust-Era Assets, November 30-December 3, 1998: Proceedings, ed. J.D. Bindenagel. Washington, DC: U.S. Government Printing Office, 1999 (no. 1042)
Weimarer Beitraege: Zeitschrift fuer Literaturwissenschaft, Aesthetik und Kulturwissenschaften. Wien: Passagen Verlag
Western Journal of Communication. Fresno, CA: Western States Communication Association
Western States Jewish History. Santa Monica, CA: Western States Jewish History Association
Women: A Cultural Review. Oxford: Oxford University Press
Women and the Holocaust: Narrative and Representation, ed. Esther Fuchs. Lanham, MD: University Press of America, 1999 (no. 292)
Women in the Inquisition: Spain and the New World, ed. Mary E. Giles. Baltimore: Johns Hopkins University Press, 1999 (no. 129)
Yad Vashem: Kovetz mehkarim. Jerusalem: Yad Vashem
Yad Vashem Studies. Jerusalem: Yad Vashem
Yahadut Bavel: Ktav-et le-heker toledot yehudei Bavel ve-tarbutam. Or Yehudah: Merkaz Moreshet Yahadut Bavel
Yalkut Moreshet. Tel-Aviv: Moreshet
Yehudei Brit ha-Mo`atzot be-Ma`avar. Jerusalem: Hebrew University
Yehupets: Khudozhnyo-publitsystychnyi almanakh Instytutu Yudayiky. Kyyiv: "Sfera"
Yevrei Belarusi: Istoriya i kultura: Sbornik statei. Minsk: Chetyre Chetverti
Yevrei i russkaya revolyutsiya: Materialy i issledovaniya, ed. Oleg V. Budnitskii. Moskva: Gesharim; Ierusalim: Hebrew University, Institute of Jewish Studies, 1999 (no. 276)
Yevrei v Rossii: Istoriya i kultura: Sbornik nauchnykh trudov, Vol. 5, ed. Dmitri Elyashevitch. Sankt-Peterburg: Peterburgskii Yevreiskii Universitet, 1998 (nos. 278, 532, 602, 996)
Yevreiskaya tsivilizatsiya: Problemy i issledovaniya: Materialy konferentsii, eds. Rashid M. Kaplanov et al. Moskva: Tsentr Nauchnykh Rabotnikov i Prepodavatelei Iudaiki v vuzakh "Sefer," 1998 (no. 158)
Yevreiskoye naseleniye Yuga Ukrainy: Yezhegodnik: Issledovaniya, vospominaniya, dokumenty, eds. F.G. Turchenko et al. Kharkov-Zaporozhye: Yevreiskii Mir, 1998 (no. 218)
Dos Yiddishe Vort. New York: Agudath Israel of America
Yod: Revue des études hébraïques et juives. Paris: Publications Langues' O
Zeitgeschichte. Innsbruck: Studien Verlag
Zeitschrift fuer Geschichtswissenschaft. Berlin: Metropol
Zeitschrift fuer Religions- und Geistesgeschichte. Leiden: E.J. Brill
Zmanim: Riv`on le-historia. Tel-Aviv: Tel-Aviv University
Znamya. Moskva: Znamya

Żydzi w Lublinie: Materialy do dziejów społeczności żydowskiej Lublina, Vol. II, ed. Tadeusz Radzik. Lublin: Wyd. Uniwersytetu Marii Curie-Skłodowskiej, 1998 (no. 777)
Żydzi w Polsce – swoi czy obcy?, eds. Waldemar Bukowski, Zdzisław Noga. Kraków: [s.n.], 1998 (no. 48)

BIBLIOGRAPHIES AND REFERENCE WORKS

0001 Belpoliti, Marco: *Primo Levi*. Milano: Bruno Mondadori, 1998. ix, 211 pp.

Presents a brief account of the life and work of Primo Levi, a lexicon, and a bibliography. The biography (pp. 1-17) contains terms (in bold letters) which are explained in the lexicon (pp. 21-192). The bibliography (pp. 193-206) is divided into three sections: a list of Levi's publications, works about him, and other works mentioned in the biography. *MZ*

0002 Benz, Wolfgang; Graml, Hermann; Weiss, Hermann, eds.: *Enzyklopaedie des Nationalsozialismus*. Stuttgart: Klett-Cotta, 1997. 900 pp.

Consists of three sections. "Handbuch" (pp. 9-341) contains 26 essays by various authors on different aspects of the National Socialist regime. "Lexikon" (pp. 343-815) contains brief encyclopedic entries, including "Antisemitismus," "Arisierung," "Juden," etc. "Register" (pp. 817-895) presents a biographical index of prominent German personalities. The following essays relate to the persecution of the Jews: Wippermann, Wolfgang: Ideologie (11-21); Kwiet, Konrad: Rassenpolitik und Voelkermord (50-65); Gruettner, Michael: Wissenschaft (135-153); Duemling, Albrecht: Musik (176-180); Eiber, Ludwig: Verfolgung [On the concentration camps and labor camps.] (275-295); Kreuter, Maria-Luise: Emigration (296-308). *SSC*

0003 Charny, Israel W., ed.: *Encyclopedia of Genocide. Vol. I-II.* Santa Barbara, CA: ABC-Clio, 1999. lxxxviii, 345; lv, 347-718 pp.

A reference work with major sections on the Armenian genocide and the Holocaust. Emphasizes the comparative study of the genocides of all peoples, the process of genocide, early warning systems for genocide, legal deterrents, and other means of prevention of genocide. See the entry "Antisemitism" (pp. 50-58) and the main entry for "The Holocaust" (pp. 296-338), as well as numerous other entries which contain information about the Holocaust – e.g. Art of the Holocaust, Concentration Camps and Death Camps, Denials of the Holocaust, Education about the Holocaust, Films of the Holocaust, Internet on the Holocaust and Genocide, Literature of the Holocaust, Medicine and the Holocaust, The Nazi Genocide State, Rescuers of Jews in Nazi Europe, Theology Responds to the Holocaust, Trials for Genocide and War Crimes, United States Holocaust Memorial Museum, Yad Vashem. *SSC*

0004

גרינברג, גרשון: הציבור הדתי ביישוב בארץ-ישראל בתקופת השואה: רשימת פרסומים (מאמרים וספרים) על השקפות בענייני האמונה היהודית והשואה, 1948-1938. רמת-גן: אוניברסיטת בר-אילן, המכון לחקר השואה ע"ש ארנולד וליאונה פינקלר, 1997. [4], 116 עמ'.

[Greenberg, Gershon: *Wartime Yishuv Religious Jewry Confronting the Holocaust: An Annotated Listing of Articles and Books Reflecting Jewish Religious Responses in Eretz-Israel during the Era of Catastrophe, 1938-1948*. Ramat-Gan: Bar-Ilan University, Arnold and Leona Finkler Institute of Holocaust Research, 1997. 116, iv pp.]

An annotated bibliography on Orthodox Jewish theology in Eretz-Israel during and immediately after the Holocaust. Lists 625 Hebrew and Yiddish articles from 25 periodicals, and some excerpts from books. *RE*

0005

גרינברג, גרשון: האמונה היהודית והשואה במחשבה היהודית הדתית בארה"ב בשנים 1948-1938: רשימת פרסומים (מאמרים וספרים) מוערת. רמת-גן: אוניברסיטת בר-אילן. המכון לחקר השואה ע"ש ארנולד וליאונה פינקלר, 1999. 104, IV עמ'.

[Greenberg, Gershon: *Faith and Holocaust as Issues of Jewish Religious Thought in the USA, 1938-1948: An Annotated Listing of Articles and Books*. Ramat-Gan: Bar-Ilan University, Arnold and Leona Finkler Institute of Holocaust Research, 1999. 104, iv pp.]

A bibliography of works by Jewish thinkers, published in the USA during the Holocaust period. The writers represent, for the most part, four Orthodox Jewish groups: Mizrahi, Agudath Israel, the Musar movement, and Habad. Lists seven books (p. 6) and ca. 400 articles (pp. 7-104) from 25 journals (in Hebrew, Yiddish, and English), as well as 12 research works on the topic. Each article cited includes a brief quote from the work. *RE*

0006 Jóźwik, Marek; Mahorowska, Teresa; Umińska, Apolonia, eds.: *Relacje z czasów Zagłady: Inwentarz: Archiwum ŻIH-INB, zespół 301. Tom I: Nr. 1-900 = Holocaust Survivor Testimonies: Catalogue: Jewish Historical Institute Archives, Record Group 301. Vol. I: No. 1-900*. Warszawa: Żydowski Instytut Historyczny, Instytut Naukowo-Badawczy, 1998. 320 pp. In Polish and English.

The first volume of the catalogue contains information about 900 archival units out of 7,200. The testimonies began to be collected in the autumn of 1944 by the Central Historical Commission which, in 1947, became the Jewish Historical Institute. Each entry contains a brief summary of the content of the file, in Polish and in English. *HV*

0007 Lande, Peter: What Happened to Shmuel and Rebeka during the Holocaust? *Avotaynu* 13, 4 (Win 1997) 22-26.

Contains guidelines and recommendations for persons intending to trace the fate of Holocaust victims or survivors. The International Tracing Center in Arolsen, Yad Vashem, and the United States Holocaust Memorial Museum are, although excellent, not the only collections of names of Holocaust victims. Describes, briefly, collections extant in France, Germany, Hungary, Poland, and other countries. *DR*

0008 Lerner, L. Scott, ed.: *The Dreyfus Affair in the Making of Modern France = L'affaire Dreyfus et l'émergence de la France moderne*. Woodbridge, CT: Primary Source Media, 1997. xxxiii, 80 pp. In English and French.

A guidebook to the 66 microfilm reels of the Dreyfus Collection at the Harvard University Library. The Collection contains rare books, pamphlets, periodicals, newspaper clippings, printed images, broadsides, and scrapbooks. It was assembled initially by Lee M. Friedman, given to the Library in 1957, and has been supplemented by other materials. The guidebook contains an annotated listing of selected titles (pp. xiv-xxviii) and an author/title main entry index (pp. 1-40) and reel index (pp. 41-80). *DR*

0009 Mueller, Rolf-Dieter; Ueberschaer, Gerd R.: *Hitler's War in the East, 1941-1945: A Critical Assessment*. Providence, RI: Berghahn Books, 1997. x, 405 pp.

A survey of literature on World War II on the Eastern front, in five parts, presenting bibliographical essays and a bibliographical list for each section. Part C (pp. 207-280), "The Ideologically Motivated War of Annihilation in the East," by Ueberschaer, deals with the historiography of the mass murder in the occupied Soviet territories and at the front line in 1941-45. Focuses mainly on the Nazi genocide of the Jews, but dwells also on the Nazi murder of Soviet POWs, Soviet murders of German POWs, and other military crimes. Pp. 253-280 contain a bibliography comprising ca. 500 items. *DR*

0010 Vatere, Eva: *Ebreji – mediķi Latvijā 1918-1996* [Jews – Medics in Latvia 1918-1996]. Rīga: Latvijas Medicīnas akadēmija, 1997. 258, [16] pp.

A biographical dictionary of more than 5,000 Jewish physicians, dentists, pharmacists, medical students, mid-level medical personnel, and other medicine-related professionals, who practiced in Latvia between 1918-96. Records the fate of many of them who perished in the Holocaust. *DR*

ANTISEMITISM THROUGHOUT THE AGES

GENERAL

0011 *Annali di Storia dell'Esegesi* 16, 2 (1999). With English summaries.

This issue, entitled "Giudaismo e antijudaismo," consists mainly of papers delivered at a seminar held in Bologna in February 1999. Contents: Tampellini, Stefano: Aspetti di polemica antigiudaica nell'opera di Esichio di Gerusalemme (con particolare attenzione al "Commentario al Levitico") (353-358); Facchini, Cristiana: Il "Vindex sanguinis" di Isaac Viva (Amsterdam 1681): Di una polemica sull'accusa di omicidio rituale [On the work by a Padua rabbi, Isaac Hayyim Cantarini (known as Isaacus Viva), aiming to reject the charge of ritual murder levelled against the Jews.] (359-378); Pesce, Mauro: Cassiodoro e gli ebrei: Provvedimenti politici e riflessione teologica: Dalle "Variae" al "Commento ai Salmi" (379-401); Perani, Mauro: Un'accusa di omicidio rituale contro gli ebrei di Ragusa-Dubrovnik nel 1622: Il "Ma`aseh Yeshurun" di Aharon ben David ha-Kohen, Venezia 1657 [On an account of the ritual murder accusation reported in the short story by Rabbi Aharon ben David from Ragusa, who was an eyewitness to the events; his story was published posthumously in "Zekan Aharon" (Venice, 1657).] (403-434); Sofia, Francesca: Cittadinanza e nazionalità all'alba della seconda emancipazione: Francesco Gambini e la questione ebraica [On the antisemitism of Gambini, an Italian lawyer who lived in Piedmont in the first half of the 19th century.] (435-457); Bonola, Gianfranco: Un no e un sì al "paragrafo ariano" nella chiesa (i "Gutachten" di Marburg ed Erlangen) [An Italian translation and explication of two German documents produced in 1933, following the Nazi seizure of power, by the Theological Faculties of the Universities of Marburg and Erlangen relating to whether the "Aryan Paragraph" should be extended to the Churches.] (459-483). *SSC*

0012 Aruffo, Alessandro: *La Chiesa e gli ebrei: Da Costantino alla Shoah.* Roma: Datanews, 1998. 151 pp.

Presents, in popular fashion, a history of the relationship between the Church and the Jews, from the 4th to the 20th centuries. Sustains the thesis that antisemitism would not have been possible had not secular anti-Jewish prejudices and myths taken root in the Christian community and in the ecclesiastical hierarchy. Pp. 129-151 contain the declaration of the German Bishops of 1995, the declaration of repentance of the French bishops of 1997, and the 1998 Vatican document "We Remember: A

Reflection on the Shoah," preceded by the introductory letter by Pope John Paul II. *MZ*

0013 Bayme, Steven: *Understanding Jewish History: Texts and Commentaries*. Hoboken, NJ: Ktav, in association with the American Jewish Committee, 1997. xvii, 443 pp.

Organized in 34 units, this volume is aimed at the general reader desiring a core course covering the main contours of Jewish history. Each unit is accompanied by textual readings, questions for discussion, and additional bibliography. The following chapters relate to antisemitism: The Church and the Jews [About the Early Church.] (115-125); Jewry and Islam (126-139); The Crusades and the Jews (164-175); Jews and Christian Spain (188-197); East European Jewry [Relates, also, to the Chmielnicki pogroms.] (223-232); Modern Anti-Semitism [In the 19th century.] (307-320); Reaction in Eastern Europe [On the extent and impact of the pogroms and the persecution of Jews in Russia between 1881-1914.] (321-331); The Holocaust [Discusses the Nazis, the victims, and the bystanders.] (383-403). *REK*

0014 Benaim, Annette, ed.: *The Proceedings of the Tenth British Conference on Judeo-Spanish Studies, 29 June-1 July 1997*. London: Department of Hispanic Studies, Queen Mary and Westfield College, 1999. xiv, 304 pp.

The following papers deal with antisemitism: Edwards, John H.: Ritual Murder in the Siglo de Oro: Lope de Vega's "El niño inocente de La Guardia" [Discusses, also, the Inquisition's interrogation and torture of Jews held responsible for the alleged ritual murder, and Spanish historians' accounts of the events.] (73-88); Lappin, Anthony: The Function and Meanings of the Jew's Goat as a Dramatic Symbol in Gil Vicente's "Auto da barca do inferno" (89-101); Kerem, Yitzchak: The Dual Portuguese Attitude to Sephardic Jewry throughout History: Disdain and Acceptance [Relates, also, to discrimination in the early modern period and in the Holocaust period.] (145-158); Rother, Bernd: Portuguese and Spanish Reactions to the Holocaust [On the governments' policies regarding refugees during World War II.] (159-167). *SSC*

0015 Benbassa, Esther, ed.: *Transmission et passages en monde juif*. Paris: Publisud, 1997. 609 pp.

A collection of papers prepared by a study group on the theme of transmission of Jewish tradition to the next generation. Partial contents: Poznanski, Renée: Des Juifs parlent aux Français: "Le document vert" de Henri Hertz (1942) [On the clandestine publication informing the French population of the persecution of the Jews.] (201-225); Melammed, Renée Levine: Les femmes crypto-juives face à l'Inquisition espagnole: Leur rôle dans la transmission du marranisme (229-245); Grynberg, Anne; Nicault, Catherine: Le Consistoire central en France sous l'Occupation: Une résistance spirituelle? (247-278); Wieviorka, Annette: Mémoriaux [On three Holocaust memorials erected in Paris between 1946-51.] (487-497). *TS*

0016 Eckert, Willehad Paul et al.: Antisemitismus, Schoa und Kirche: Studie eines theologischen Arbeitskreises. *Freiburger Rundbrief* 6, 4 (1999) 262-279.

Traces the problematic relationship of the Church and of Christians to the Jews from the New Testament through the 20th century, including the deicide accusation and the claim of supersession, pogroms and expulsions. Suggests that religious anti-Judaism made the public receptive to racist antisemitism. Thus the majority of churchmen and church members failed to protest against the Holocaust; some were among the perpetrators, although others (including Pope Pius XII) rescued Jews. Vatican II was a turning point in the relationship. Discusses the theological problem raised by the Holocaust. Points out that many Jewish victims held fast to their faith in God even while accusing him. Affirms that God's ways are inscrutable, but it is legitimate to ask where he was in Auschwitz, and Christians need not contradict Jews who say that the world is unredeemed. Outlines changes required in Christian teaching to bring about an attitude of respect for Judaism. *RW*

0017 Engel, David: The Concept of Antisemitism in the Historical Scholarship of Amos Funkenstein. *Jewish Social Studies* 6, 1 (Fall 1999) 111-129.

Discusses the impreciseness and, hence, the lack of usefulness of the word "antisemitism." Refers to it as an "inflated" term that has become a stereotype that posits a single phenomenon beyond time and space. The historian Amos Funkenstein distanced himself from such a global usage of the term in his rejection of a connection between medieval Christian anti-Judaism and post-emancipation political antisemitism. The latter, in its specific, German, historical context is considered to have been "atheist" and racist. For Funkenstein it would have been anachronistic to use the term in relation to the Middle Ages. Engel rejects Funkenstein's connection of 19th-century racist ideas with the mass killing of Jews by the Nazis. Concludes that the term "antisemitism" is not a useful analytic tool. Recommends replacing it by focused study in order to determine the ideas that the Nazis used to justify the genocide they perpetrated on the Jews and the sources of these ideas. *YC*

0018 Fugmann, Markus: *Moderner Antisemitismus*. Frankfurt a.M.: Verlag Moritz Diesterweg, 1998. 114 pp. (Brennpunkt Geschichte).

An overview of the history of antisemitism in Europe from the ancient period to the present, including its basis in beliefs, theories, and ideologies. Pp. 46-64 deal with Nazi ideology and the Holocaust. Pp. 89-104 contain extracts from sources, from Philo to Franz Stangl. *RW*

0019 Gampel, Benjamin Raphael, ed.: *Crisis and Creativity in the Sephardic World, 1391-1648*. New York: Columbia University Press, 1997. xvi, 413 pp.

A collection based on papers presented at an international symposium held in New York in 1992. Partial contents: Yerushalmi, Yosef Hayim: Exile and Expulsion in Jewish History [Appeared in German in "Ein Feld in Anatot" (1993).] (3-22); Beinart, Haim: Order of the Expulsion from Spain: Antecedents, Causes, and Textual Analysis (79-94); Tavares, Maria José Pimenta Ferro: Expulsion or Integration? The Portuguese Jewish Problem [Discusses the situation of the Jews of Portugal at the end of the 15th century – their choice of forced baptism or expulsion.] (95-103); Gampel, Benjamin Raphael: The Exiles of 1492 in the Kingdom of Navarre: A Biographical Perspective (104-117). *DR*

0020 Grinberg, Efraim: *Yevrei v stranakh mira* [Jews in the Countries of the World]. Ierusalim: Ierusalimskii Izdatelskii Tsentr, 1997. 249 pp.

Presents historical data on Jews in various countries, organized in chronological tables. Pt. 3 (pp. 61-153), "Yevreistvo Yevropy" ("European Jewry"), relates, also, to persecutions of Jews in the Middle Ages, the Holocaust, and anti-Jewish policies in some countries (Romania, USSR, etc.). *DR*

0021 Grunberger, Béla; Dessuant, Pierre: *Narcissisme, christianisme, antisémitisme: Etude psychanalytique*. Arles: Actes Sud, 1997. 492 pp.

Examines the sources of antisemitism in history, which are attributed to Judaism's rejection of Jesus as Messiah. Hypothesizes that the enigma of anti-Jewish feeling can be solved by comprehension of Christianity as a narcissistic religion. Concurs with theologians who incriminate Christianity for its maintenance of an endemic antisemitic culture, which varied in intensity in response to political, social, and religious upheavals. Argues that the Shoah was the culmination of 2,000 years of disdain for the Jew in religious instruction. Underscores the necessity to seek a fundamental reason for antisemitism, beyond a superficial measurement of its magnitude. Utilizes psychoanalytic theory to illustrate how persons like Hitler are prone to become antisemitic, based on a duality of the narcissistic vs. non-narcissistic personality. Presents examples of this duality, such as the material vs. spiritual, to explain the coexistence of Christ in human vs. divine form. This paradox led Christians to declare

Jews as symbols of evil, and unredeemable because of their rejection of baptism. Concludes that Christian narcissism is what led to an apocalypse, the Shoah. Pt. 6 (pp. 349-423) analyzes Nazi antisemitism. *TS*

0022 Guigue, Bruno: *Aux origines du conflit israélo-arabe: L'invisible remords de l'Occident*. Paris: L'Harmattan, 1998. 145 pp.

Analyzes the historical foundations on which the Arab-Israeli conflict is based. Traces the conflict to its theological roots, from the anti-Judaism of medieval Christianity in Europe (in contrast to the more beneficent attitude of Islam). Contends that Europeans were motivated to support the creation of a Jewish state primarily by guilt feelings about their persecution of the Jews throughout history, particularly during the Holocaust. Believes that the other reason for their support was a blind admiration of the socialist pioneering spirit they associated with Zionism. Argues that these two aspects led Western governments to turn a blind eye to Arab nationalist aspirations. *TS*

0023

איבלאגון, אברהם: יין הרק"ח [ריטמינוס, קנדייא, חנייא]: סיפורה של האנטישמיות ושל עלילות הדם במשך הדורות ויחס התורה לתופעה. בני ברק: יעקב טרנטו, 1998. 47 עמ'.

[Ivlagon, Avraham: *Wine of the Rokayah [Rethimnon, Candia, Khania]: The Story of Antisemitism and of the Blood Libels throughout the Generations and the Torah View of This Phenomenon*. Ed.: Ya`akov Trento. Bnei Brak: Ya`akov Trento, 1998. 47 pp.]

Edition of a manuscript written by Ivlagon (1846-1934), a rabbi in Crete, in response to blood libels which took place before and during his lifetime. Pp. 13-30 discuss the practice of human sacrifice in the ancient period and the biblical condemnation of it. Pp. 30-47 discuss specific blood libels and righteous persons who condemned them, mentioning in particular those which occurred in Izmir in 1913; in Corfu in 1890; in Tiszaeszlár in 1882; in Damascus in 1840; in Poland in 1749; in Blois in 1171; and in Russia in the early 20th century. The original manuscript is in the Cambridge University Library. *RE*

0024 Kats, Aleksandr S.: *Yevrei, Khristianstvo, Rossiya* [Jews, Christianity, Russia]. Sankt-Peterburg: Novyi Gelikon, 1997. 479 pp.

Pp. 7-108 discuss Jewish-Christian relations from early Christianity to the modern period. Contends that the Jews played a positive role in shaping this religion. It was the Christian "clerical aristocracy," from the 2nd century on, who suppressed the Jewish contribution to the advancement of Christianity and invented the myth of deicide, as well as other anti-Jewish myths. The rest of the book recounts the history of Russia from the

Khazars to the present, and reflects on the role of Jews in this history. Does not deny the disproportionately large role of Jews in the Revolution and later in the Soviet spheres of power. Contends that Jews greatly reinforced Soviet science, culture, the economy, and the military potential. Despite that, the postwar Stalinist leadership purged the party and apparatus of Jews. *DR*

0025 Keith, Graham: *Hated without a Cause? A Survey of Anti-Semitism.* Carlisle, Cumbria: Paternoster Press, 1997. xiii, 301 pp.

Refutes accusations made against Christianity and the Church that they have always been a source of antisemitism. States that hostility toward Jews existed in the ancient world long before Christianity, and its cause was Jewish religious and civic separatism. Contends that the New Testament contains no more recriminations against the Jews than the Bible; the former denies the chosenness of Jews, but it does not depict them as excessive sinners. The enmity toward Jews in the writings of early Christians (e.g. Chrysostom) is accounted for by their endeavors to separate Christianity from Judaism. In the medieval period, antisemitism was expressed by some clerics, but the Church never officially supported it. Luther may be blamed for antisemitism, but Protestantism did not advocate it. The Enlightenment was more hostile toward the Jews than was the Church. Antisemitism in 19th-20th-century Germany was mainly anti-Christian, and Hitler was as much anti-Christian as anti-Jewish. On the other hand, the Muslim roots of antisemitism (beside anti-Zionism) in the contemporary Arab world must not be ignored. Suggests that Christians must regard Jews as no different from other peoples, i.e. they are not immune to criticism. The Church has the right to try to evangelize the Jews, and these efforts must not be dismissed as antisemitism. *DR*

0026 Landauer, Gustav: *Dichter, Ketzer, Aussenseiter: Essays und Reden zu Literatur, Philosophie, Judentum.* Ed.: Hanna Delf. Berlin: Akademie Verlag, 1997. lviii, 290 pp.

A collection of essays by Landauer (1870-1919). The following deal with antisemitism: Kiew [Appeared in "Der Sozialist" 5 (1913). In reaction to the Beilis affair, deplores Christian ignorance and superstitiousness about Jews. Connects the blood libel to the belief that Jews desecrate Christ in the form of the host, piercing it in order to draw blood; thus they supposedly also draw the blood of Christian children, in whom Christ is present, in order to bake it into their matzot (which show suspicious perforations!).] (177-184); Ostjuden und Deutsches Reich [Appeared in "Der Jude" 1 (1916). Argues against the unwillingness of Germans, even German Jews, to allow mass immigration of empoverished, backward East European Jews.] (185-190); Der Kaufmann von Venedig [Appeared in "Der Jude" 2 (1917/18). Interprets the play as opposing Shylock, a "man who has no music in his soul," to Portia, who is all music, grace, and lightness; but Shakespeare also points to the source of Shylock's deformity (his brutishness and vengefulness) in the humiliations he suffers from Christians, notably Antonio. In Shylock, the reactions of the oppressed Jewish people take on heroic dimensions.] (194-221). *RW*

0027 Poliakov, Léon: *Istoria antisemitismului. Vol. 1-2* [The History of Antisemitism. Vol. 1-2]. Trans.: Lelia Băluş (vol. 1), Janina Ianoşi (vol. 2). Bucureşti: Hasefer, 1999. 296; 350 pp. Translation of the first two parts of "Histoire de l'antisémitisme" (Paris: Calmann-Lévy 1955-77). See the abstract for the Russian translation: "Istoriya antisemitizma: Epokha very" (Moskva: Lekhaim; Ierusalim: Gesharim, 1997).

Vol. 1 is sub-titled: De la Hristos la evreii de curte; vol. 2: De la Mahomed la marani.

0028 Popelyanskii, Yakov Yu.: *Razmyshleniya ob antisemitakh: S lyubov'yu...* [Reflection on Antisemites: With Love...] Kazan: Karpol, 1996. 136 pp.

Reflects on the nature of antisemitism. Its cause lies not in the Jews but in the antisemites: Jews may have their group traits, or ethnic character, but it is not Jewish traits which arouse prejudice. Distinguishes three kinds of antisemitic accusations against Jews: libel (accusing Jews of sins they did not commit), generalization of actual negative traits of some Jews to all of the Jewish people, and "Russophobia." Describes two types of antisemites. The first type, the most dangerous, consists of pathological persons for whom antisemitism is an obsession (for the most part they are either paranoiac or suffer from functional agnosia). The second type comprises those for whom antisemitism is marginal to their worldview (e.g. Voltaire, Mendeleyev, Marx, Dostoyevsky). Concludes that the way to overcome antisemitism is not by assimilation of the Jews, but by people gaining better knowledge of each other. *DR*

0029 Schoeps, Julius Hans; Schloer, Joachim, eds.: *Bilder der Judenfeindschaft: Antisemitismus: Vorurteile und Mythen*. Muenchen: Bechtermuenz, 1999. 310 pp. Originally published as "Antisemitismus: Vorurteile und Mythen" (Muenchen: Piper, 1995).

This edition contains the same articles, but their order of appearance in the book has been changed somewhat.

0030 Sibony, Daniel: *Le "racisme," ou, La haine identitaire*. [Paris]: Christian Bourgois, 1997. 411 pp.

A thoroughly revised version of his "Ecrits sur le racisme" (1988). Reflects on attitudes toward the "Other" and relates them to identity crisis.

Argues that racism is an effect of people's need for strong roots and identity, which can easily be achieved by rejection of the "Other." Analyzes Christian antisemitism as a collective hatred of Jews and Judaism; because Christianity is not an original religion and has its sources in the Hebrew Bible, Christian identity is blurred. Through the ages, Christianity did not admit that it originated in Judaism. Argues that the Jews were not hated because they were considered inferior, but because of their strong roots, rich tradition, laws, etc. Examines, as well, the psychological causes of Holocaust denial. Contends that Rassinier, himself a deportee, could not acknowledge the much stronger Jewish sufferings which blurred his own value as a victim. Holocaust denial is also a form of anguish. The culpability is so strong that it has to be denied. Concludes that racism cannot be extirpated because it is related to everyone's identity problems. People should just learn about it the better to resist it. *HV*

0031 Strecker, Christian: Paulus aus einer "neuen Perspektive": Der Paradigmenwechsel in der juengeren Paulusforschung. *Kirche und Israel* 11, 1 (1996) 3-18.

Surveys writings, mostly by American and British theologians since the 1960s, that seek to revise Luther's anti-Judaistic interpretation of Paul – an interpretation taken over by most later Lutheran theologians. Whereas Luther understood Paul as condemning "the works of the Torah," the "new" theologians argue that Paul's intention was solely to extend salvation also to the Gentiles who believed in Christ, without their having to obey the Torah. Paul's view of the role of the Torah in the salvation of the Jews is interpreted differently by the different theologians, but all agree that he did not condemn it, though he may have condemned certain anti-Gentile aspects of it. *RW*

0032 Targarona Borrás, Judit; Sáenz-Badillos, Ángel, eds.: *Jewish Studies at the Turn of the Twentieth Century. Vol. I-II.* Leiden: Brill, 1999. xiii, 635; xiv, 701 pp. On title page also: Proceedings of the 6th EAJS [European Association for Jewish Studies] Congress, Toledo, July 1998.

Partial contents: Vol. I: "Biblical, Rabbinical, and Medieval Studies": Alfonso, Esperanza: `Abd al-Karim al-Magili: Un paralelo magrebí a los acontecimientos de 1066 en Granada [On an antisemitic poem by Maghili (15th century), influenced by an antisemitic pamphlet by the Jewish convert to Islam Abd al-Haqq Islami, which incited the populace against the Jews in the Maghreb.] (370-378). Vol. II: "Judaism from the Renaissance to Modern Times": Ramras-Rauch, Gila: Ida Fink and Holocaust Literature (225-228); Meijers, Daniel: Europe's Last Pogrom? A Provisional Note on the Sociogenesis of Discrimination and Violence [On the pogrom in Kielce in 1946.] (330-336); Pilarczyk, Krzysztof: Zur

Zensurfrage der juedischen Buecher in Polen im 16. und 17. Jahrhundert (346-353); Kerem, Yitzchak: The Fate of Greek Sephardic Cultural Personalities in the Holocaust [Mainly on Jewish musicians from Salonika and their activities before the war and in Auschwitz.] (523-529). *SSC*

0033 Wistrich, Robert Solomon, ed.: *Demonizing the Other: Antisemitism, Racism, and Xenophobia*. Amsterdam: Harwood Academic, for the Vidal Sassoon International Center for the Study of Antisemitism, Hebrew University of Jerusalem, 1999. x, 373 pp.

A collection of papers delivered at the conference "The 'Other' as Threat: Demonization and Antisemitism," organized by the Vidal Sassoon International Center for the Study of Antisemitism, Hebrew University of Jerusalem, in June 1995. Partial contents: Wistrich, Robert Solomon: The Devil, the Jews, and Hatred of the "Other" (1-15); Schul, Yaacov; Zukier, Henri: Why Do Stereotypes Stick? [On the persistence of stereotypes, especially antisemitic ones, from a motivational-cognitive perspective.] (31-43); Amishai-Maisels, Ziva: The Demonization of the "Other" in the Visual Arts [From the ancient to the modern periods, with a focus on anti-Jewish images in Christian art.] (44-72 + 8 pp. of illustrations); Schwartz, Daniel R.: Antisemitism and Other -ism's in the Greco-Roman World [In the Roman Empire, with a focus on Alexandria in the first century.] (73-87); Lazarus-Yafeh, Hava: Jews and Christians in Medieval Muslim Thought (108-117); Zukier, Henri: The Transformation of Hatred: Antisemitism as a Struggle for Group Identity [In Western culture throughout the ages.] (118-130); Almog, Shmuel: The Borrowed Identity: Neo-Pagan Reactions to the Jewish Roots of Christianity [Surveys attitudes toward the Jewish component in Christianity throughout history, from the standpoint of Gentile identity.] (131-147); Cohen, Richard I. (Yerachmiel): Recurrent Images in French Antisemitism in the Third Republic (183-195); Kulka, Otto Dov: The Critique of Judaism in Modern European Thought: Genuine Factors and Demonic Perceptions [Discusses, for instance, the deists, Marx, Wagner, and Nazi ideology.] (196-209); Friedlaender, Saul: "Europe's Inner Demons": The "Other" as Threat in Early Twentieth-Century European Culture [On the view of the Jew as representing the threatening "Other" in radical right discourse (e.g. Nesta Webster, Georges Bernanos) and in Nazi ideology.] (210-222); Burrin, Philippe: Nazi Antisemitism: Animalization and Demonization (223-235); Epstein, Simon (Simcha): When the Demon Itself Complains of Being Demonized [On antisemites' responses to Jewish propaganda launched against them; discusses Nazi responses (Hitler, Goebbels, Gregor Strasser) between 1927-33.] (236-243); Felstiner, John: "All Poets Are Yids": The Voice of the "Other" in Paul Celan [An earlier German version appeared in "Dachauer Hefte" 11 (1995).] (244-256); Gutman, Yisrael: The Popular Image of the Jew in Modern Poland [19th-20th centuries.] (257-266); Shillony, Ben-Ami: The Flourishing Demon: Japan in the Role of the Jews? [On "Japan bashing and Jew baiting" in the West, deriving from demonization of the successful outsider.] (294-309); Yadlin, Rivka: Anti-Jewish Imagery in the Contemporary Arab-Muslim World [On antisemitic rhetoric, polemical writings, and reports in the press.] (310-322); Porat, Dina: The "Protocols of the Elders of Zion": New Uses of an Old Myth [Examines publication and distribution of the "Protocols" in different parts of the world in the 1990s.] (323-335); Benz, Wolfgang: The Motivations and Impact of Contemporary Holocaust Denial in Germany [From postwar West Germany to present-day Germany, with a focus on the case of Guenter Deckert.] (336-348); Wistrich, Robert Solomon: Xenophobia and Antisemitism in the New Europe: The Case of Germany (349-363). *SSC*

0034 Wojciechowski, Michał: More about the Anti-Jewish Bias in the New Testament Translations. *Collectanea Theologica* [Fasciculus specialis] (1998) 83-88.

A sequel to a previous article by the author, in Polish ["Collectanea Theologica" 63 (1993)], which examined selected texts from eleven modern Polish translations of the Bible. Shows here that biased translations have also appeared in English, French, and German. These translations weaken the theological appreciation of Israel or strengthen criticisms against her. *DR*

0035 Yerushalmi, Yosef Hayim: *Sefardica: Essais sur l'histoire des Juifs, des marranes & des nouveaux-chrétiens d'origine hispano-portugaise*. Trans.: Cyril Aslanoff et al. Paris: Chandeigne, 1998. 364 pp.

A collection of essays, all of them published previously. They discuss Jewish survival despite the expulsion from Spain in 1492 and the forced conversion in Portugal in 1497. Partial contents: Le judaïsme séfarade entre la Croix et le Croissant [Appeared in "Inquisition et pérennité" (1992).] (11-33); Le massacre de Lisbonne en 1506 et l'image du roi dans le "Shebet Yehudah" [Published in English as a separate booklet (Cincinnati: Hebrew Union College – Jewish Institute of Religion, 1976). Pp. 117-173 contain three appendices relating to primary sources, in German and Portuguese, on the events in Lisbon.] (35-173); Assimilation et antisémitisme racial: Le modèle ibérique et le modèle allemand [Published in English as a separate booklet (New York: Leo Baeck Institute, 1982). Compares the situation of the Conversos in Spain and Portugal in the 15th-18th centuries with that of the German Jews in the 19th-early 20th centuries.] (255-292). *TS*

GENERAL: BY COUNTRY

Germany

0036 *Geschichte der Juden in Heidelberg*. Heidelberg: Verlag Brigitte Guderjahn, 1996. 649 pp.

This volume is the result of a decision taken in November 1988 by the city council to produce a collection of essays on the history of the Jews of Heidelberg. The following deal also with antisemitism: Ziwes, Franz-Josef: Die Juden im mittelalterlichen Heidelberg (15-41); Krauss, Martin: Zwischen Emanzipation und Antisemitismus (1802 bis 1862) (154-216); Wennemuth, Udo: Zur Geschichte der Juden in Heidelberg in der Weimarer Republik [See pp. 429-439: "Antisemitismus und Abwehr."] (348-439); Moraw, Frank: Die nationalsozialistische Diktatur (1933-1945) (440-555); Preuss, Monika: "Heimweh ist nur Weh, nicht Heim": Private Erinnerungen an juedisches Leben vor der Vernichtung [Gathered from various testimonies and memoirs.] (581-602). *ID*

0037 *Juden in der Oberlausitz*. Bautzen: Lusatia, 1998. 192 pp.

Contents: Hartstock, Erhard: Geduldet, angesehen und verfolgt: Aus der Geschichte der Juden in der Oberlausitz [From the Middle Ages through the Holocaust.] (6-103); Otto, Roland: Im Zentrum der Oberlausitzer Juden – Goerlitz: Der Anteil der juedischen Buerger am wirtschaftlichen, politischen und kulturellen Leben in Goerlitz [Pp. 119-130 deal with the Holocaust period.] (104-133); Kaulfuerst, Hans Eberhard; Bockholt, André: Die letzten Zeugen Moses: Juedisches Leben in Bautzen, aufgezeichnet aus der Erinnerung von Zeitgenossen (134-149); Griebel, Katrin: Spuren juedischen Lebens in Zittau (150-189). *RW*

0038 Kieval, Hillel Joseph: Antisemitism and the City: A Beginner's Guide. *Studies in Contemporary Jewry* 15 (1999) 3-18.

Rejects the view which sees modern antisemitism as a criticism of Jewish responsibility for modernism and "urbanity"; challenges the idea that the Jews had a central role in the evolution of the modern city. Focusing on Germany, argues that since the late Middle Ages anti-Jewish systems of knowledge have been a continuous feature of city life, with the Jews being marginalized. This is linked to expulsions, ritual murder charges, and the view of the Jews as strangers in a Christian society that wanted to be cleansed of them. States that modern antisemitism is connected with the apparent success of emancipation, which was interpreted as an irreversible defeat for the urban modernity that excluded the Jews. The political aim of this antisemitism was to reclaim the city from the Jews.

German antisemitism in the 1870s-80s attempted to place limitations on the Jews, but did not demand a reversal of emancipation as did the Nazis later on. *YC*

0039 Lehmann-Brune, Marlies: *Der Koffer des Karl Zuntz: Fuenf Jahrhunderte einer juedischen Familie*. Duesseldorf: Droste, 1997. 279 pp.

A history of sixteen generations of the Zuntz family, which settled in Frankfurt am Main at the end of the 15th century. Ch. 3 (pp. 47-65), "Pfefferkorn und Fettmilch," relates to host desecration accusations and blood libels at the beginning of the 16th century, and the confiscation of Jewish books organized by Johannes Pfefferkorn. Antisemitic outrages by Vincenz Fettmilch against the Jews led to a pogrom and expulsion on 22 August 1614, after which the Jews were allowed to return in summer 1615. Ch. 10 (pp. 224-256), "Aus Hamburg vertrieben – Rueckkehr in das 'versprochene Land'," deals with the Holocaust period and its aftermath. Reports on the fate of members of the family – escapes, emigration, suicide, hiding, and deportation. Pp. 257-265 present genealogical information about the family. *ID*

0040 Lehming, Hanna; Kortzfleisch, Siegfried von, eds.: *Aus der Geschichte lernen? Chancen der Aufklaerung nach Auschwitz*. Bad Segeberg: C.H. Waeser, 1997. 126 pp. On title page also: Zeitkritische Beitraege der Evangelischen Akademie Nordelbien.

A collection of essays on the issue of learning lessons from history, especially when dealing with the Nazi past. Partial contents: Muenz, Christoph: Im Schatten der Erleuchtung: Die Juden im Zeitalter der Aufklaerung [Discusses, also, antisemitism during the period of Enlightenment.] (11-42); Ben-Chanan, Yaacov: Das Schweigen schuetzt – das Reden heilt [Reflections by a survivor on the Holocaust and its lessons.] (43-64); Keilson, Hans: Die Faszination des Hasses: Das Verhaeltnis von Juden und Christen in Deutschland: Ein Versuch [Appeared also in "Vergegenwaertigungen des zerstoerten juedischen Erbes" (1997).] (65-88). *ID*

0041 Margalit, Gilad: The Image of the Gypsy in German Christendom. *Patterns of Prejudice* 33, 2 (Apr 1999) 75-85.

Traces the evolution of the German Christian image of the Gypsy, from the 15th century (when the Gypsies appeared in Germany) to the postwar period. Although it became a commonplace to note similarities between Christian attitudes towards Jews and Gypsies, and to regard the Holocaust and the Nazi genocide of Gypsies as the same, close examination shows that both the Christian stances toward these groups and the characters of their genocide were different. The Gypsies were regarded as pagans (despite their claims of being Christians), while Judaism had a very important role in the Christian worldview. The Nazis' preoccupation with Gypsies is in no way comparable to the centrality of antisemitism to the Nazi agenda. Except for Georg Althaus, no Christian theologian tried to interpret the genocide of the Gypsies in religious terms. Pp. 84-85 contain a commentary on Margalit's article by Donald Kenrick. *DR*

0042 Meynert, Joachim: *Juedische Geschichte und regionale Identitaet: Beitraege zur Geschichte der juedischen Minderheit in Ostwestfalen.* Bielefeld: Verlag fuer Regionalgeschichte, 1998. 116 pp.

A collection of articles, most of them published previously. Partial contents: Judenfeindschaft und Antisemitismus: Ein historischer Ueberblick fuer die politische Bildung [Appeared in "Beitraege zur politischen, wirtschaftlichen und kulturellen Weiterbildung" I (Paderborn: Landesarbeitsgemeinschaft Arbeit und Leben Nordrhein-Westfalen; Kooperationsstelle Wissenschaft-Arbeit-Gesellschaft Universitaet-GH Paderborn, 1988).] (11-21); Juedische Minderheit und christliche Gesellschaft: Beruehrungspunkte und Konfliktzonen in der Bielefelder Geschichte [A lecture delivered in fall 1993 in Bielefeld.] (22-37); "Das hat mir sehr weh getan!" Juedische Jugend in Ostwestfalen-Lippe: Streiflichter 1933-1939 [Appeared in "Opfer und Taeter: Zum nationalsozialistischen und juedischen Alltag in Ostwestfalen-Lippe" (1990).] (51-68); In Volmerdingsen zu Hause: Das Schicksal der juedischen Familie Gruenenklee waehrend der Zeit des Nationalsozialismus [Appeared in "Herrschaft und Verfolgung zur Zeit des Nationalsozialismus in Bad Oeynhausen und Umgebung" (1989). The family was deported to the Riga ghetto; includes a description of the ghetto.] (69-84); Das Ende vor Augen: Die Deportationen der Juden aus Bielefeld [Appeared in "Verfolgung und Widerstand im Rheinland und in Westfalen" (1992). They were deported by the Gestapo; includes a description of one of their destinations, the Riga ghetto.] (85-98); "Heimkehr" auf Zeit: Ueberlegungen zur kulturellen und gesellschaftlichen "Normalisierung" des deutsch-juedischen Verhaeltnisses [Appeared in "Kultur und Gesellschaft der Bundesrepublik Deutschland" (Frankfurt a.M.: Peter Lang, 1996).] (99-106); "Gedaechtnisorte": Erinnerungskultur und Nationalsozialismus [A lecture delivered at the Universitaet Bielefeld in January 1998.] (107-116). *RW*

0043 Pracht-Joerns, Elfi: *Juedisches Kulturerbe in Nordrhein-Westfalen. Part I: Regierungsbezirk Koeln.* Koeln: J.P. Bachem, 1997. xv, 650 pp.

Surveys Jewish remains – synagogues and other communal buildings, cemeteries, private homes and businesses – in the Koeln district (which includes Koeln, Aachen, Bonn, and numerous smaller towns and villages where Jews once lived). For each locality gives a brief history of the Jewish community, including pogroms in the medieval and early modern periods, expulsions, and residence restrictions, up to the extinction of the community in the Nazi period. Also gives the history of each building up to its destruction in the "Kristallnacht" pogrom (or for economic reasons after the war). For cemeteries, mentions not only desecration by the Nazis but also more recently, by neo-Nazi vandals. *RW*

0044 Sippel, Heinrich: *Juedisches Leben in Schlitz*. Pulheim: [The Author], 1997. 99 pp.

Surveys the history of the Jews in Hesse, emphasizing antisemitic events, e.g. blood libels in Fulda in 1235 and accusations of well-poisoning in 1348-49, followed by pogroms. Relates more specifically to Jewish life in Schlitz between 1730-1939. Pp. 66-89 deal with the Nazi period. In 1933 there were 13 Jewish families in Schlitz (43 Jews). The persecutions led to the emigration of 13 Jews by the end of 1935, and another 23 by 1937, leaving only seven Jews in the town. There was no "Kristallnacht" pogrom there, partly because the synagogue was no longer in Jewish hands. The last Jew left on 1 June 1939. *ID*

Hungary

0045 Toronyi, Zsuzsanna: *Magyarországi zsidók: Vatikáni nyilatkozat a soáról és a MAZSIHISZ állásfoglalása* [Hungarian Jews: The Vatican Declaration on the Holocaust and the Statement of the Association of Hungarian Jewish Communities (MAZSIHISZ)]. Budapest: Körtánc Egyesület, 1998. 53 pp.

Pp. 3-44 present a summary of Jewish life in Hungary from the Middle Ages to the present, including references to antisemitism and the Holocaust, and explain various aspects of Jewish religious life and culture. Pp. 45-52 contain a Hungarian translation of the Vatican document "We Remember: A Reflection on the Shoah" (1998). Pp. 52-53 present the document issued by the Magyarországi Zsidó Hitközségek Szövetsége, which praised the Vatican document and called on the Hungarian Catholic Church to prepare a similar declaration relating to its specific role during the Shoah. *MA*

Italy

0046 Luzzati, Michele, ed.: *Gli ebrei di Pisa (secoli IX-XX): Atti del Convegno internazionale Pisa, 3-4 ottobre 1994*. Pisa: Pacini, 1998. viii, 433 pp.

Partial contents: Frattarelli Fischer, Lucia: Ebrei a Pisa fra Cinquecento a Settecento (89-115); Prosperi, Adriano: Ebrei a Pisa: Dalle carte dell'Inquisizione romana (117-157); Di Porto, Bruno: Gli ebrei a Pisa dal Risorgimento al Fascismo: Tra identità e integrazione [Pp. 332-340 deal with the fascist period.] (283-340); Savelli, Laura: Una città e i suoi ebrei: Pisa tra la due guerre mondiali (347-373); Tanti, Giovanna: L'applicazione delle leggi razziali a Pisa: Il caso dell'Università (381-390); Orsi, Pier Luigi: La demografia dell'ebraismo pisano (1938-1944) (391-399); Forti, Carla: Persecuzione e deportazione degli ebrei di Pisa (1943-1944) (401-425). *MZ*

0047 Vivanti, Corrado, ed.: *Storia d'Italia. Annali XI: Gli ebrei in Italia, [pt.] II.* Torino: Giulio Einaudi, 1997. xx, 893-1975 pp. Pt. I appeared in 1996.

Partial contents: Rosa, Mario: La Santa Sede e gli ebrei nel Settecento (1067-1087); Della Peruta, Franco: Gli ebrei nel Risorgimento fra interdizioni ed emancipazione (1133-1167); Della Seta, Simonetta; Carpi, Daniel: Il movimento sionistico [Also on the fascist period and World War II.] (1321-1368); Miccoli, Giovanni: Santa Sede, questione ebraica e antisemitismo fra Otto e Novecento (1369-1574); Sarfatti, Michele: Gli ebrei negli anni del fascismo: Vicende, identità, persecuzione ((1623-1764); Mazzamuto, Salvatore: Ebraismo e diritto dalla prima emancipazione all'età repubblicana (1765-1827); Luzzatto, Amos: Autocoscienza e identità ebraica [In the fascist period and after the war.] (1829-1900). *SSC*

Poland

0048 Bukowski, Waldemar; Noga, Zdzisław, eds.: *Żydzi w Polsce – swoi czy obcy?* [The Jews in Poland – Ours or Strangers?]. Kraków: [s.n.], 1998. 293 pp.

The catalogue of an exhibition organized in Poland, intended especially for high-school students, presenting the history of the Jews in Poland from the Middle Ages to the present. Pp. 4-96 contain essays summarizing the most important events in Polish-Jewish history, including aspects of intolerance and antisemitism throughout the ages. The exhibition was organized by the Stowarzyszenie Centrum Polsko-Niemieckie w Krakowie [Polish-German Center Society in Kraków]. Partial contents: Bukowski, Waldemar; Chrobaczyński, Jacek; Kuler, Andrzej; Rydel, Jan: Zagłada Żydów w Polsce (1939-1945) [The Destruction of the Jews in Poland (1939-1945)] (53-77); Trojański, Piotr: W okresie powojennym [In the Postwar Period. Deals, among other issues, with the pogrom in Kielce in 1946 and with the antisemitic campaign of 1968.] (78-97). Pp. 197-238 contain photographs relating to the Holocaust. *HV*

0049 Krakowski, Stefan: Poland's Destroyed Jewish Communities, Numismatically Remembered: Lublin. *The Shekel* 30, 3 (May-June 1997) 31-35.

Traces the history of the Jewish community in Lublin from the Middle Ages to the Holocaust, noting also antisemitic incidents throughout the centuries. Pp. 33-35 deal with the Holocaust period. Relates the establishment of the Judenrat in 1939 and the ghetto in 1941, and the deportations and killings between 1942-44. Several thousand Jews of Lublin who survived the war, mainly in the USSR, left Poland either in 1945-46 or in 1968. Includes two illustrations of emergency scrip used in the ghetto in place of official currency. *DR*

Romania

0050 Oişteanu, Andrei: *Mythos & logos: Studii şi eseuri de antropologie culturală*. Bucureşti: Nemira, 1997. 368 pp.

Pt. 2 (pp. 177-263), "'Evreul imaginar' versus 'evreul real'," contains four essays. "Imaginea evreului in cultura tradiţională românească" (pp. 177-228) appeared in abridged form in "Contribuţia evreilor din Romania la cultura şi civilizaţie" (1996) and in English in "Identitate/alteritate in spaţiul cultural românesc" (1996). "Stigmatul etnic" (pp. 229-245) deals with the history of the yellow badge and its use in Romania during the Holocaust. "'Infelix culpa': Naţionalism şi antisemitism în România interbelică" (pp. 246-256) contains comments on Leon Volovici's "Ideologia naţionalistă şi 'problema evreiască' in România anilor '30" (1995). Discusses the involvement of Romanian intellectuals in the antisemitic trends of the 1930s-40s, and criticizes the moderate intellectuals for not taking a stand against antisemitism. "Sindromul Anne Frank" (pp. 257-263) refers to the issue of Holocaust denial and the contribution of people such as Anne Frank and Steven Spielberg to Holocaust awareness. *AA*

0051 Stanciu, Ion, ed.: *The Jews in the Romanian History: Papers from the International Symposium, Bucharest, 30 September-4 October, 1996*. Trans.: Ioan Petru Vasca, Raluca Taraş. Bucureşti: Silex, [1999]. 248 pp. At head of title page: Institute of History "N. Iorga," Romanian Academy, Bucharest; Goldstein-Goren Center, Diaspora Research Institute, Tel Aviv University.

Partial contents: Barbu, Violeta: The Polemic Theology during the Second Half of the 18th Century: A Religious Dimension of the Alterity (19-26); Oişteanu, Andrei: The Image of the Jew in Romanian Traditional Culture: From Deicide to Infanticide (47-56); Gyémánt, Ladislau: The Limits of the Tolerance in Transylvania during the First Half of the 19th Century: The Jewish Case (70-78); Stan, Apostol: The Jews in Romania between

1866 and 1868: Immigration and Reaction of the Guarantor Powers (94-99); Taftă, Lucia: French Diplomatic Echoes to the Jewish Question in Romania (1871-1878) (124-129); Stanciu, Ion: The International Press on the Hay Note: Echoes and Significations (137-141); Oldson, William O.: The American Diplomatic Assessment of the Romanian "Jewish Problem" 1900-1941 (171-178); Demény, Lajos: C.G. Costa-Foru against the Anti-Semitism (179-188); Ioanid, Radu: The Fate of the Romanian Jews under the Antonescu Regime: Statistical Data, Preliminary Conclusions (201-204); Rotman, Liviu: The Politics of the Communist Regime concerning the Jews: Contradictions, Ambivalence and Misunderstanding (1945-1953) [Maintains that in January 1953 the Romanian communist leadership was preparing a political trial and anti-Jewish repressions modelled on the Slánský trial and the "Doctors' Plot."] (230-247). *DR*

Russia, USSR, CIS

0052 Klier, John Doyle: 'Traditional Russian Religious Antisemitism': A Useful Concept or a Barrier to Understanding? *The Jewish Quarterly* 174 (Sum 1999) 29-34.

Examining the centuries-long history of hostility toward Jews in Russia, from the medieval period on, contends that until the late 19th century Russian antisemitism was not religious but rather economic and xenophobic. The ideas that led to publications such as Brafman's "The Book of the Kahal" or "The Protocols of the Elders of Zion" were borrowed from the West. It was this secular antisemitism which influenced ecclesiastical antisemitism in modern Russia, and not vice versa. Concludes that antisemitic rhetoric in Russia today does not have a strong religious appeal, despite the occasional camouflage of religious-style rhetoric. Rather, contemporary antisemitism is a direct descendant of Soviet anti-Zionism. *DR*

Spain

0053 Valle Rodríguez, Carlos del, ed.: *La controversia judeocristiana en España (desde los orígenes hasta el siglo XIII): Homenaje a Domingo Muñoz León.* Madrid: Consejo Superior de Investigaciones Científicas, 1998. 357 pp.

Partial contents: Valle Rodríguez, Carlos del: El Concilio de Elvira (13-19); Jiménez Patón, Lorenzo: Prudencio y la tradición "adversus Iudaeos" (21-41); Barcala Muñoz, Andrés: La polémica antijudía en los "Tractatus Origenis" de Gregorio de Elvira (43-62); Valle Rodríguez, Carlos del: La Carta Encíclica del obispo Severo de Menorca (a. 418) (63-76); Díez

Merino, Luis: San Isidoro de Sevilla y la polémica judeocristiana (77-110); Valle Rodríguez, Carlos del: San Julián de Toledo (119-130); Lara Olmo, Juan Carlos: La polémica de Albaro de Córdoba con Bodón/Eleazar (131-159); Colomina Torner, Jaime: El antijudaísmo hispanogodo y sus posibles ecos en los textos litúrgicos e ildefonsinos (171-190); Valle Rodríguez, Carlos del: Pedro Alfonso y su "Diálogo" (201-222); Niclós, José Vicente: San Martín de León y la controversia con los judíos en el siglo XII (243-252); Valle Rodríguez, Carlos del: La disputa de Barcelona de 1263 (277-291); Valle Rodríguez, Carlos del: La disputa de Mallorca (a. 1286) (293-300); Ragacs, Ursula: Raimundo Martí, O.P. – biografía (301-308); Muñoz León, Domingo: San Pedro Pascual, obispo de Jaén (309-315); Hames, Harvey J.: Ramón Llull y su obra polémica contra los judíos (317-344). *MZ*

0054 Baltrusch, Ernst: Bewunderung, Duldung, Ablehnung: Das Urteil ueber die Juden in der griechisch-roemischen Literatur. *Klio* 80, 2 (1998) 403-421.

States that judgments pronounced on Jews and Judaism in Greek and Roman literature depend largely on the insight of the writer on the subject, ranging from admiration or tolerance to outright rejection. The more the knowledge of the writer about Judaism was limited, the more he gave way to negatively expressed fears and prejudices. In quantity, there is a prevalence of sources rejecting the Jews. However, contrary to modern interpretations, the Jewish religion itself was viewed positively, as well as the fact that the Jews had remained faithful to it. On the negative side was the view of their strange way of life, based on religious principles such as circumcision and the dietary laws. Notes analogies in form and content with modern antisemitism. *ID*

0055 Baltrusch, Ernst: Die Christianisierung des Roemischen Reiches: Eine Zaesur in der Geschichte des Judentums? *Historische Zeitschrift* 266, 1 (Feb 1998) 23-46.

Disputes the common conception that increasingly restrictive Roman legislation against the Jews, culminating in the codex and decrees of Justinian, was due to Christian influence. The Church combated the Jewish religion in order to convert Jews to Christianity and prevent the conversion of Christians to Judaism; it did not advocate the persecution of Jews. Restrictive legislation by the state began long before the Christianization of Rome, during the early years of the Empire. The goal of the emperors, pagan and Christian, was to unite all the peoples of the Empire under a centralized administration that treated all its subjects evenhandedly; the Jews' demand for special privileges necessitated by their religion was incompatible with this goal. The emperors' attempts to impose conformity resulted in three great Jewish uprisings in Judaea and in the Diaspora in the 1st-2nd centuries. *RW*

0056

בר-כוכבא, בצלאל: החיבור האנטי-יהודי של אפולוניוס מולון. תרביץ 69, 1 (אוקט'-דצ' 1999) 5-58.

[Bar-Kochva, Bezalel: The Anti-Jewish Treatise of Appolonius Molon. *Tarbiz* 69, 1 (Oct-Dec 1999) 5-58.]

Deals with a treatise against the Jews attributed to Appolonius Molon, a prominent orator who lived in Rhodes in the first half of the 1st century

BCE. He was identified in writings of the ancient period as hostile toward the Jews. Examines his attitude toward the Jews in the context of his views and the culture of his times. Regards the treatise as an ethnographic monograph in which Appolonius draws on pro-Jewish and anti-Jewish sources known in his times (e.g. Lysimachus, Timochares, Posidonius of Apamea) which describe the origins of the Jews, their customs and the actions of their leaders. Contends that Appolonius' adverse treatment of the Jews in his treatise does not necessarily reflect negative personal experiences with Jews. It may be regarded as a polemical response to contemporary works and to disciples of rival philosophical schools. *LFo*

0057 Butting, Klara: Das Buch Ester: Vom Widerstand gegen Antisemitismus und Sexismus. *Kompendium: Feministische Bibelauslegung,* eds. Luise Schottroff, Marie-Theres Wacker. Guetersloh: Chr. Kaiser, Guetersloher Verlagshaus, 1998. Pp. 169-179.

Examines the events in the Book of Esther from a feminist perspective. Claims that the refusal of Vashti and Mordechai to bow down before a sovereign authority enabled, first, the governmental organization of the submission of women, and second, the annihilation of the Jews. Those who are strong feel threatened by those who are different. Emphasizes the parallel between the submission of Persian women and the annihilation of Persian Jewry. In both cases there is a totalitarian pattern of thought and action; the hatred of the Jewish people adds a murderous dimension. *ID*

0058 Cohick, Lynn Harrison: Melito of Sardis's "Peri Pascha" and Its "Israel." *Harvard Theological Review* 91, 4 (Oct 1998) 351-372.

Most scholars have attributed the homily "Peri Pascha" to Melito, a 2nd-century Quartodeciman bishop of Sardis, mentioned in Eusebius, and regarded it as an anti-Jewish work. The discovery of the magnificent synagogue during the excavations in Sardis in 1958-75 seemingly substantiates the hypothesis that the large Jewish community in Sardis was the target of the homily. No historical evidence, however, and nothing in the homily itself confirms the assumptions either that Melito of Sardis was the author of "Peri Pascha," or that there was a big Jewish community in Sardis in his time, or that "Israel" referred to in the homily means Jews. More likely, the homily reflects an intra-Christian controversy (most probably part of the anti-Marcionite polemics), and "Israel," denounced by the unknown author, is an image invented by him for rhetorical purposes. *DR*

0059 Cohick, Lynn Harrison: Melito's "Peri Pascha": Its Relationship to Judaism and Sardis in Recent Scholarly Discussion. *Evolution of the Synagogue: Problems and Progress,* eds. Howard Clark Kee, Lynn H. Cohick. Harrisburg, PA: Trinity Press International, 1999. Pp. 123-140.

Evaluates four basic approaches to interpreting the homily "Peri Pascha," attributed to Melito of Sardis, in regard to what it might tell us about Melito's attitude toward the Jews of Sardis in the 2nd-3rd centuries. Considers evidence from archaeology (the synagogue discovered at Sardis) and from literature. Traces the development of scholarly interpretations and assumptions concerning the "Peri Pascha"'s anti-Judaism, from the late 1930s to the present. Cautions that one should not draw conclusions about Melito's relations to Jewish contemporaries, because Melito may not have been from Sardis, the synagogue evidence may not relate to Melito's period, the sermon may have been dealing more with biblical images than with contemporary Jews, and the theories about the socio-political status of the Sardis Jewish community may not be correct. *YC*

0060 Dunn, Geoffrey D.: Two Goats, Two Advents and Tertullian's "Adversus Iudaeos." *Augustinianum* 39, 2 (Dec 1999) 245-264.

Discusses Christian interpretation of Leviticus 16, about the sacrificial goat and the scapegoat. Both Tertullian and Justin Martyr disputed the Jewish view that Jesus could not be the Messiah because he did not come triumphantly; they claimed that the second advent will be triumphant, arguing that the scapegoat refers to Jesus' first coming and the sacrificial goat to his second advent. Tertullian derived his theory from the early Christian "Epistle of Barnabas." He combined the idea of Jesus' second coming with other arguments for Christianity's supersession of Judaism. Tertullian's interpretation not only consists of arguments directed against Judaism but also includes the antisemitic charge of Jewish responsibility for the death of Jesus. *YC*

0061 Farmer, William R., ed.: *Anti-Judaism and the Gospels*. Harrisburg, PA: Trinity Press International, 1999. viii, 311 pp.

A collection of papers, the result of a three-year research project on this topic organized at the University of Dallas. Contents: Levine, Amy-Jill: Anti-Judaism and the Gospel of Matthew (9-36). Pp. 37-62 contain responses to Levine by Philip L. Shuler and Warren Carter. Schmidt, Daryl D.: Anti-Judaism and the Gospel of Luke (63-96). Pp. 97-119 contain responses to Schmidt by David L. Balch and Allan J. McNicol. Rensberger, David: Anti-Judaism and the Gospel of John (120-157). Pp. 158-175 contain responses to Rensberger by Mark Goodwin and Thomas D. Lea. Wilken, Robert Louis: Something Greater Than the Temple [Contends that early Christianity was not uniformly anti-Jewish, and that it emphasized continuity with, as well as difference from, Judaism.] (176-202). Pp. 203-215 contain responses to Wilken by Everett Ferguson and D. Jeffrey Bingham. Tyson, Joseph B.: Anti-Judaism in the Critical Study of the Gospels (216-251). Pp. 252-264 contain responses to Tyson by

James O. Duke and Ellen T. Charry. Sanders, Ed Parish: Reflections on Anti-Judaism in the New Testament and in Christianity (265-286). *YC*

0062 Fleishman, Joseph: Why Did Ahasuerus Consent to Annihilate the Jews? *Journal of Northwest Semitic Languages* 25, 2 (1999) 41-58.

Based on the text of the Book of Esther, argues that King Ahasuerus agreed to the proposal of his chief officer Haman to annihilate the Jews of Persia because of Haman's "demonic ingenuity." The latter's persuasiveness was based on two major ideas: that the Jewish people were different (with their own laws) and that they posed a threat to the empire and to the king himself. Notes that the king accepted the charges against this people without investigation and without seeking advice. He was driven by his own prejudice and by the attitude of his people, who were ready to annihilate the Jews even after the instigator Haman had been killed. *YC*

0063 González Salinero, Raúl: La exclusión social de los judíos en el Imperio cristiano (ss. IV-V). *'Ilu* 4 (1999) 103-113.

In the 4th and 5th centuries, patristic literature and Church leaders condemned the Jews theologically and urged their isolation from society. The Church as an institution acted similarly through council decrees. The Roman state applied discriminatory legislation that made any form of assimilation impossible for the Jews. Asserts that this was the beginning of the progressive juridical degradation and isolation of Jews in Europe, which led to the rise of the medieval ghetto. *MZ*

0064 Grüll, Tibor: Zsidó-keresztény kapcsolatok a késő antikvitásban: Severus de Minorca levele a helyi zsidó közösség áttérítéséről [Jewish-Christian Relations in Late Antiquity: The Epistle of Severus of Minorca on the Conversion of the Local Jewish Community]. *Múlt és Jövő* 4 (1999) 76-91.

Analyzes the relationship between Jews and Christians in the early 5th century in the Roman Empire. Refers to the works of the Church Fathers and archaeological findings of Jewish inscriptions. Discusses, in detail, the epistle attributed to Bishop Severus of Minorca on the forced conversion of 540 Jews in Minorca in 418. Describes, also, the persecution of Jews and the razing of synagogues in 393 and 576. *MA*

0065 Kampling, Rainer, ed.: *"Nun steht aber diese Sache im Evangelium...": Zur Frage nach den Anfaengen des christlichen Antijudaismus.* Paderborn: Ferdinand Schoeningh, 1999. 364 pp.

Partial contents: Mussner, Franz: Was macht das Mysterium Israel aus? (15-30); Kampling, Rainer: "Und er ging seiner Gewohnheit am Sabbat in die Synagoge": Jesuanisches zur Frage nach dem Ursprung des christlichen Antijudaismus (53-72); Frankemoelle, Hubert: Antijudaismus im Matthaeusevangelium? Reflexionen zu einer angemessenen Auslegung (73-106); Blum, Matthias: Antijudaismus im lukanischen Doppelwerk? Zur These eines lukanischen Antijudaismus (107-149); Scholtisseck, Klaus: Antijudaismus im Johannesevangelium? Ein Gespraechsbeitrag (151-181); Dautzenberg, Gerhard: Alter und neuer Bund nach 2Kor 3 (229-249); Fiedler, Peter: Antijudaismus als Argumentationsfigur: Gegen die Verabsolutierung von Kampfesaeusserungen des Paulus im Galaterbrief (251-279); Oberlinner, Lorenz: Antijudaismus in den Pastoralbriefen? (281-299); Backhaus, Knut: Das wandernde Gottesvolk – am Scheideweg: Der Hebraeerbrief und Israel (301-320); Bee-Schroedter, Heike: Die Shoah als Herausforderung an das traditionelle Selbstverstaendnis historisch-kritischer Exegese: Religionspaedagogische Impulse fuer eine kontextuelle bibeltheologische Hermeneutik (321-363). *MZ*

0066 Koltun-Fromm, Naomi: Psalm 22's Christological Interpretive Tradition in Light of Christian Anti-Jewish Polemic. *Journal of Early Christian Studies* 6, 1 (Spr 1998) 37-57.

The early Christian writers Justin Martyr, Tertullian, and Aphraates composed similar arguments, claiming for Psalm 22 a christological interpretation. They held that the Jews misunderstood this Psalm as pertaining to King David, whereas in fact it pertained to Jesus and foreshadowed his suffering. This interpretation was a result of early Christian polemics and probably originated with Justin's anti-Jewish works, with Tertullian and Aphraates basing their works on Justin. Many Christian biblical interpretations were motivated by perceived threats of Jewish alternative meanings and were polemics against Jewish interpretations. Justin's exegesis became a model for later Christian polemicists. *REK*

0067 Lambrecht, Jan: Jewish Slander: A Note on Revelation 2,9-10. *Ephemerides Theologicae Lovanienses* 75, 4 (Dec 1999) 421-429.

Argues that, in the context of the passage discussed, the Greek word "blasphemia" means not blasphemy but denunciation to the pagan authorities. This hypothesis stems from the author's interpretation that Revelation 2,9-10 blames the Jews, whom it calls "a synagogue of Satan," for persecution of the Christians. The thrust of this passage and similar ones in other Christian theological writings is that the Christians are the true Jews, whom the so-called Jews will be forced to acknowledge. *YC*

0068 Lieu, Judith M.: *Image and Reality: The Jews in the World of the Christians in the Second Century*. Edinburgh: T & T Clark, 1996. xiv, 348 pp.

Examines the rhetorical function of Jews and Judaism in Christian texts of the 2nd century. The texts examined are from Asia Minor, an area which, at the time, encompassed both Jewish and diverse Christian communities. Discusses the "Martyrdom of Polycarp," Justin Martyr's "Dialogue with the Jew Trypho," the "Epistle of Barnabas," and Melito of Sardis' "Peri Pascha." Discerns, already in the 1st century, the tendency of Christian writers to portray the Jews as the archetypal enemies of Christians and their religion. *REK*

0069 Olson, Stanley: Does Anything Justify Anti-Judaism? The Evidence of Christian Judaism. *Approaches to Ancient Judaism. New Series, Vol. 13,* ed. Jacob Neusner. Atlanta: Scholars Press, 1998. Pp. 169-209.

Contends that the Church of the 1st century did not reject the Jewish Law and, in particular, did not regard Christian Judaism (i.e. belief in Jesus combined with following the Jewish Law) as an inferior form of Christianity. The decisive rejection of Christian Judaism, and therefore of the Jewish Law came later, with the Gentilization of the Church in the 2nd-4th centuries, and can be detected, for instance, in Augustine and Jerome. Contends that anti-Judaic Christian authors have misinterpreted Paul, especially his Epistles to the Romans. Paul neither preferred the Gentile Christians to the Jewish ones, nor despised the Jewish Law; unity, not uniformity of the Church, was his aim. Warns that despising the Jewish Law is tantamount to despising God and his will expressed at Sinai. The attempts to "Gentilize" Jesus are anti-historical. Christian anti-Judaism has led to antisemitism and shares responsibility for the genocide of Jews. *DR*

0070 Pedersen, Sigfred: Anti-Judaism in John's Gospel: John 8. *New Readings in John: Literary and Theological Perspectives: Essays from the Scandinavian Conference on the Fourth Gospel, Århus 1997,* eds. Johannes Nissen, Sigfred Pedersen. Sheffield: Sheffield Academic Press, 1999. Pp. 172-193.

Interprets anti-Jewish passages in the Gospel of John as not referring to all Jews but to those who adopted Christianity and then became "apostates." These Jews and the non-Jews who refused to accept Jesus as divine are those John referred to by the term "children of the Devil." *YC*

0071 Raspe, Lucia: Manetho on the Exodus: A Reappraisal. *Jewish Studies Quarterly* 5, 2 (1998) 124-155.

Based on a study of Manethonian fragments, as well as of writings attributed to Manetho in Josephus's "Contra Apionem," discusses the story of the Jews' exodus from Egypt as described in ancient Greek and Latin literature, according to which the Jews, afflicted with leprosy, were driven out of the country by the king of Egypt. States that the attribution of the origins of this story to the Egyptian priest Manetho (3rd century BCE) is debatable. Manetho's works, as quoted by Josephus, actually present two different versions of the origin of the Jews (one associates them with the Hyksos who invaded Egypt, and the other with lepers who lived in Egypt ca. 500 years later), neither of which is antisemitic. However, later historians (e.g. Lysimachus and Apion) wrote other versions of these stories and used them for anti-Jewish propaganda, in which the Jews became "the sworn enemy of mankind." Contends that although Manetho may have fathered antisemitic literature, this was not his intention. *RE*

0072 Rutgers, Leonard Victor: *The Hidden Heritage of Diaspora Judaism*. 2nd ed. Leuven: Peeters, 1998. 320 pp. First published as "The Hidden Heritage of Diaspora Judaism: Essays on Jewish Cultural Identity in the Roman World" (Kampen: Kok Pharos, 1997).

A collection of essays published previously. Ch. 8 (pp. 171-197), "Roman Policy towards the Jews: Expulsions from the City of Rome during the First Century C.E.," first appeared in "Classical Antiquity" 13 (1994). The present version contains an appendix: "Review of Botermann's Judenedikt der Kaisers Claudius (1996)" (pp. 191-197).

0073 Rutgers, Leonard Victor: Roman Policy towards the Jews: Expulsions from the City of Rome during the First Century C.E. *Judaism and Christianity in First-Century Rome,* eds. Karl Paul Donfried, Peter Richardson. Grand Rapids, MI: William B. Eerdmans, 1998. Pp. 93-116. First published in "Classical Antiquity" 13 (1994).

0074 Severus of Minorca: *Letter on the Conversion of the Jews*. Ed. & trans.: Scott Bradbury. Oxford: Clarendon Press, 1996. x, 144 pp.

Pp. 1-77 contain an introduction by Bradbury, relating that in 416, after the relics of St. Stephen arrived in Minorca, tensions arose between the Christian and Jewish communities of the island, whereas up to then relations between them had been cordial. The tensions grew, and riots erupted in 418, culminating with the conversion of the entire Jewish community of Minorca (540 persons) to Christianity. Severus, the bishop of Minorca, dispatched a letter to the bishops of the Roman Empire, describing the events in Minorca in 416-418 and closing with an exhortation to the bishops to carry out the conversion of the Jews in their own communities. The letter, known as "Epistula Severi" (pp. 79-125, in

Latin with an English translation), is a precious document, as it indicates the beginning of the Christian-Jewish conflict in late antiquity. Bradbury describes the historical background to the riots and the writing of the letter, and examines the questions of the authenticity and historical reliability of the letter. *DR*

0075 Stegemann, Wolfgang: Anti-Semitic and Racist Prejudices in Titus 1:10-16. *Ethnicity and the Bible*, ed. Mark G. Brett. Leiden: E.J. Brill, 1996. Pp. 271-294. Appeared in German in "Kirche und Israel" 11, 1 (1996).

0076 Stegemann, Wolfgang: Antisemitische und rassistische Vorurteile in Titus 1,10-16. *Kirche und Israel* 11, 1 (1996) 46-61.

Analyzes the passage, which warns against false teachings of which Jews or Jewish Christians and Cretans are guilty. Suggests that the author is using prejudices widely accepted in the ancient world to attack a deviant group within Christianity. *RW*

0077 Stern, Menahem, ed.: *Grecheskiye i rimskiye avtory o yevreyakh i iudaizme. Vol. 1: Ot Gerodota do Plutarkha* [Greek and Latin Authors on Jews and Judaism. Vol. 1: From Herodotus to Plutarch]. Ed. & trans.: N.V. Braginskaya. Moskva: Mosty Kultury; Jerusalem: Gesharim, 1997. xxi, 632 pp. Originally published as "Greek and Latin Authors on Jews and Judaism. Vol. 1: From Herodotus to Plutarch" (Jerusalem: Israel Academy of Sciences and Humanities, 1974).

An anthology of excerpts from ancient works on Jews and Judaism, in Greek and Latin, with a Russian translation, accompanied by comments by Stern. Inter alia, contains texts by Manetho, Apion, Seneca and others who express anti-Jewish views. *DR*

0078 Yavetz, Zvi: Latin Authors on Jews and Dacians. *Historia* 47, 1 (1998) 77-107.

Compares the treatment by Latin authors of two disliked Eastern "barbarians," the Dacians and the Jews, both of whom carried out failed rebellions against Rome. Some of the authors expressed an animus toward Jews that they did not have toward the Dacians, thus demonstrating that Romans did not consider the Jews to be merely another group of unruly barbarians. States that the views of Cicero and other writers were influenced by Greek Jew-haters, who concocted a Greco-Egyptian version of Jewish history in order to slander the Jews and convince the Romans to put an end to the Jewish state. Seneca's anti-Jewish writings may have stemmed from a visit to Egypt, after which he began to hate Oriental religions and was horrified by the growing number of Jewish proselytes in Rome. Tacitus, writing after the defeat of Judea, moderated the Greco-

Egyptian version of history, but he hated the Jews, probably due to resentment of their influence in Rome, and especially their proselytizing. His contemporary, the satirist Juvenal, slandered Jewish customs and religion for the same reason. Concludes that the treatment of the Dacians by these authors was more favorable because "they were not proselytizers, had no Diaspora, and never endangered Roman mores." *RE*

0079 Bardelle, Thomas: *Juden in einem Transit- und Brueckenland: Studien zur Geschichte der Juden in Savoyen-Piemont bis zum Ende der Herrschaft Amadeus VIII.* Hannover: Hahnsche Buchhandlung, 1998. xii, 396 pp. Based on the author's diss. – Universitaet Trier, 1994.

Describes the settlement of Jews in Savoy and Piedmont, and the conditions under which they lived, from the late 13th-early 15th centuries. Savoy was a land of refuge for Jews expelled from neighboring countries, especially France. Until the reign of Amadeus VIII (beginning in 1398) they lived relatively unmolested, except for a blood libel trial in Chambéry in 1329 (in which the Jews were acquitted) and persecution occasioned by the Black Death in 1349. At that time, masses of Jews imprisoned in Chambéry were lynched; Jews who survived to stand trial, in Chambéry and elsewhere, were sentenced to death by burning. Amadeus VIII issued a new legal code incorporating anti-Jewish canon law. In 1416-17, and again in 1426, he stood behind a hostile examination of Jewish sacred books carried out by Franciscans and the Inquisition with the aid of Jewish converts, which in 1417 resulted in a ceremonial burning of the Talmud. However, in his reign, as in previous ones, the penalty for alleged crimes such as deceit or usury by moneylenders, sexual misconduct with Christian women, possession of prohibited books, or blasphemy, was payment of a large sum into the Duke's coffers. *RW*

0080 Cano, María José: El antijudaísmo en la tradición historiográfica hispanohebrea. *Miscelánea de Estudios Árabes y Hebraicos* 47 [Sección hebreo] (1998) 37-56.

Examines Christian hostility toward Jews in medieval Spain as expressed in works by Jewish authors writing in the 16th century, such as "Kitzur zekher tzadik" by Joseph ben Zaddik, "Sefer ha-Kabbalah" by Abraham ben Solomon of Torrutiel, and "Sefer Shevet Yehudah" by Solomon ibn Verga. These works contain information and reflections about the persecution of the Jews and the expulsion from Spain in 1492. Asserts that according to these works antisemitism was deeply rooted in Spanish society and culture, and that the violence exercised against the Jews was directed from the Crown, the nobility and the Church. *MZ*

0081 Castaño González, Javier: Tensiones entre las comunidades judías y la monarquía en Castilla, c. 1447-1474: El nombramiento del Juez Mayor de las aljamas. *Creencias y culturas,* eds. Carlos Carrete Parrondo, Alisa Meyuhas Ginio. Salamanca: Universidad Pontificia; Tel-Aviv: Universidad de Tel-Aviv, 1998. Pp. 11-20.

Traditionally, the office of "Juez Mayor de las aljamas," the highest representative of the Jewish communities to the court of the king of Castile, was appointed by the king on the communities' recommendation. Taking advantage of Jewish weakness following the death of the incumbent, Rabbi Abraham Bienveniste (ca. 1435), as well as growing anti-Jewish hostility, the "condestable" (high constable) Álvaro de Luna obtained the appointment of Pedro de Luján, a non-Jew, as Juez Mayor, in order to have more control over the Jewish communities (ca. 1447). This measure triggered strong Jewish opposition, which resulted in the demotion of Luján and the appointment by the king of a council of five Jewish elders. *AA*

0082 Chazan, Robert: The First Crusade as Reflected in the Earliest Hebrew Narrative. *Viator* 29 (1998) 25-38.

Discusses the "Mainz Anonymous" narrative, which exists in only one manuscript; it was discovered in the late 19th century and dated to the 14th century. Focuses on its characteristic features, especially those which distinguish this text from Christian narratives on the same event. The anonymous Jewish author regards the massacre of Rhineland Jews which accompanied the Crusade as an unprecedented event. He admits to lofty ideals, albeit perverted ones, among the persecutors of the Jews; he regards the Crusade as a baronial-popular undertaking (the papal initiator of the campaign is nowhere in evidence); he emphasizes the wide range of Christians who partook in the enterprise; he states that the deicide accusation motivated the riots; and he highlights the Jewish readiness for martyrdom and unwillingness to convert. *DR*

0083 Chazan, Robert: The Trier Unit of the Lengthy Hebrew First-Crusade Narrative. *Between History and Literature: Studies in Honor of Isaac Barzilay,* ed. Stanley Nash. Tel-Aviv: Hakibbutz Hameuchad, 1997. Pp. 37-49.

Examines the account of events in Trier found in the longest version of the three extant Hebrew chronicles on the First Crusade, the one attributed to Solomon bar Simson. It reports that the Crusaders left Trier without harming the Jews, but then the Trier townsmen carried on an extended campaign of persecution and slaughter against the Jews. The Jews attempted to protect themselves through bribery and by appeal to the bishop. The bishop did try to protect them, but the Christian populace threatened to kill him as well; after some Jews were killed, the bishop successfully pressured most of those remaining to convert. Solomon bar Simson does not stress that the great majority of Trier's Jews were forcibly converted but, rather, that they were martyred. *REK*

0084

כהן, ג'רמי: ויכוח פריז השני והפולמוס היהודי-נוצרי של המאה השלוש עשרה. תרביץ 68, 4 (יולי-ספט' 1999) 557-579.

[Cohen, Jeremy: The Second Disputation of Paris and the Jewish-Christian Polemic in the Thirteenth Century. *Tarbiz* 68, 4 (July-Sept 1999) 557-579.]

Deals with Christian anti-Jewish activities in 13th-century France, as reflected in the stance of the Dominican friar Pablo Christiani from Spain, whose ideas influenced the anti-Jewish policy of King Louis IX. Focuses on the Hebrew manuscript recently found (published by Joseph Shatzmiller in 1994), relating a disputation between a friar Paul and Rabbi Abraham ben Samuel of Rouen which was conducted in Paris in the early 1270s, convened by Louis IX. Presents a summary of the arguments recorded in the Hebrew narrative, and assesses their significance. Offers additional support for Shatzmiller's identification of the friar with Pablo Christiani. Analysis of the friar's anti-Jewish arguments reveal his extensive debt to the works of Raymond Martini, who may also have been in Paris at the same time. Rabbi Abraham's response to these arguments, and parallels in contemporary Hebrew sources, situate the Parisian narrative within the evolution of Jewish anti-Christian polemic during the Middle Ages. *LFo*

0085 Edwards, John H.: The Debate about the Origins of the Spanish Inquisition. *Al-Masaq* 11 (1999) 1-13.

At the heart of the debate is the question whether the Conversos were genuine New Christians or crypto-Jews whose apostasy could have legitimately been viewed as heresy by the Church. Partisans of the latter view are historians Cecil Roth, Yitzhak Baer, and Haim Beinart. However, if the converts were genuine (as believed by Benzion Netanyahu, Norman Roth, and Henry Kamen), causes other than Judaizing have to be found for the establishment establishment of new tribunals in Aragón and Castile. Netanyahu and Roth found them in a "very modern looking antisemitism among the lower classes... together with 'realpolitik' on the part of the monarchs." Edwards suggests that the debate should avoid false dichotomies, since the Conversos were neither uniformly Jewish nor uniformly Christian. Understanding the origin of the Inquisition requires comparisons with earlier and contemporary experience elsewhere in Europe and the abandonment of a belief in abstract monoliths like Jew and Judaism, Christian and Christianity. In 15th-century Spain (as before and after), individuals really were confused about their faith and identity. *YC*

0086 González Salinero, Raúl: Catholic Anti-Judaism in Visigothic Spain. *The Visigoths: Studies in Culture and Society,* ed. Alberto Ferreiro. Leiden: Brill, 1999. Pp. 123-150.

Challenges the widely held view that Jews enjoyed tolerance during the reign of the Visigothic Arian monarchs. Shows how the situation of the Jews worsened starting with the conversion of the first Visigothic king to Catholicism in 589. The Church and Crown both desired religious unity; this led to social exclusion, the gradual reduction of Jewish civil rights, persecution, and bloodshed. Largely as a result of negative Christian attitudes, expressed in literature and in the various Councils of Toledo, penalties were introduced for the transmission of Jewish tradition or attempts to convert non-Jews. These included physical mutilation, expulsion, or death. Although measures against the Jews failed in the ultimate goal of converting them to Christianity, the anti-Jewish actions caused them much suffering. *YC*

0087 Gonzálvez Ruiz, Ramón: La constitución 10a del Concilio de Peñafiel (1302). *El legado material hispanojudío: VII Curso de Cultura Hispanojudía y Sefardí de la Universidad de Castilla-La Mancha,* eds. Ana María López Álvarez, Ricardo Izquierdo Benito. Cuenca: Ediciones de la Universidad de Castilla-La Mancha, 1998. Pp. 247-264.

Analyzes a conflict which emerged between canon law and Jewish communal legislation in medieval Spain as a result of the enactment of the 10th constitution of the Council of Peñafiel (1302). The constitution regulated that Jews and Moors who converted to Christianity would not lose their assets. This law stood in opposition to the autonomous legislation of the Jewish communities, which punished conversion to Christianity by loss of all property. States that there is no evidence that the 10th constitution was enforced during the 14th century (before the massacres of 1391) and suggests that the Jewish communities probably did not apply their internal laws after 1391, when they faced massive conversions of Jews to Christianity. Concludes that this was probably how the discord between the opposing laws was resolved. *MZ*

0088 Gow, Andrew Colin: "Ryzhiye yevrei": Apokaliptika i antisemitizm v Germanii v Sredniye veka i ranneye Novoye vremya ["The Red Jews": Apocalypticism and Antisemitism in Medieval and Early Modern Germany]. *Vestnik Yevreiskovo Universiteta* 20 (1999) 41-61.

The deterioration of the situation of the Jews in Europe in the 11th-15th centuries was determined not only by economic factors, but also by cultural ones, especially those pertaining to folk culture. The rise of apocalyptic expectations in the 11th century brought about the dissemination of popular chimeric fantasies connecting Jews with the end of days, Antichrist, Gog and Magog, etc. The legend of the "Red Jews" – a mighty and bellicose people, originating from the Ten Lost Tribes, having a kingdom somewhere in the East and hostile to Christianity – may be attributed to this period. The popular mind associated the "Red Jews" with the coming of Antichrist and identified them with Gog and Magog. The legend persisted until the 16th century. *DR*

0089 Haverkamp, Alfred: *Verfassung, Kultur, Lebensform: Beitraege zur italienischen, deutschen und juedischen Geschichte im europaeischen Mittelalter*. Eds.: Friedhelm Burgard et al. Mainz: Verlag Philipp von Zabern, 1997. xxiii, 552 pp. On title-page also: Dem Autor zur Vollendung des 60. Lebensjahres.

A collection of essays, all published previously. The following deal with antisemitism: Die Juden im mittelalterlichen Trier [Appeared in "Kurtrierisches Jahrbuch" 19 (1979).] (127-187); Die Judenverfolgungen zur Zeit des Schwarzen Todes im Gesellschaftsgefuege deutscher Staedte [Appeared in "Zur Geschichte der Juden in Deutschland des spaeten Mittelalters und der fruehen Neuzeit" (1981).] (223-297); Lebensbedingungen der Juden im spaetmittelalterlichen Deutschland [Appeared in "Zerbrochene Geschichte: Leben und Selbstverstaendnis der Juden in Deutschland" (1991).] (463-484). *ID*

0090 Haverkamp, Alfred, ed.: *Juden und Christen zur Zeit der Kreuzzuege*. Sigmaringen: Jan Thorbecke, 1999. xx, 371 pp. (Konstanzer Arbeitskreis fuer mittelalterliche Geschichte: Vortraege und Forschungen, 47).

Most of the articles are based on lectures held at a conference in Reichenau, March 1996. Partial contents: Cohen, Jeremy: The Hebrew Crusade Chronicles in Their Christian Cultural Context (17-34); Haverkamp, Eva: "Persecutio" und "Gezerah" in Trier waehrend des Ersten Kreuzzugs (35-71); Grossman, Avraham: The Cultural and Social Background of Jewish Martyrdom in Germany in 1096 (73-86); Yuval, Israel Jacob: Christliche Symbolik und juedische Martyrologie zur Zeit der Kreuzzuege (87-106); Lotter, Friedrich: "Tod oder Taufe": Das Problem der Zwangstaufen waehrend des Ersten Kreuzzugs (107-152); Hiestand, Rudolf: Juden und Christen in der Kreuzzugspropaganda und bei den Kreuzzugspredigern [Contends that the Church hierarchy never intended the Crusades to lead to persecution of Jews; pogroms were instigated by itinerant preachers, and were carried out in places where the pope's messages did not reach.] (153-208); Stacey, Robert C.: Crusades, Martyrdoms, and the Jews of Norman England, 1096-1190 (233-251); Mentgen, Gerd: Kreuzzugsmentalitaet bei antijuedischen Aktionen nach 1190 [On massacres, forced conversions, and expulsions.] (287-326); Patschovsky, Alexander: Feindbilder der Kirche: Juden und Ketzer im Vergleich (11.-13. Jahrhundert) (327-357). *RW*

0091 *Jewish History* 13, 2 (Fall 1999).

Five articles in this issue deal with the representation in Jewish historical texts and preservation in the Jewish collective memory of the anti-Jewish violence and massacres perpetrated by the Crusaders in the Rhineland during the First Crusade in 1096. Contents: Chazan, Robert: Christian and Jewish Perceptions of 1096: A Case Study of Trier [Compares the narratives on the same event – the assault and forced conversion of the bulk of the Jewish community in Trier in 1096 – described in the Christian chronicle "Gesta Treverorum" and in the Jewish "Solomon bar Simson Chronicle."] (9-22); Lasker, Daniel J.: The Impact of the Crusades on the Jewish-Christian Debate [Contends that the Jewish contemporaries of the 1096 massacres did not see them as a watershed event in Jewish-Christian relations.] (23-36); Marcus, Ivan G.: The Representation of Reality in the Narratives of 1096 [Examines the episode of Mistress Rachel from Mainz and her four martyred children as narrated in three Hebrew chronicles.] (37-48); Myers, David Nathan: "Mehabevin et ha-tsarot": Crusade Memories and Modern Jewish Martyrologies [On the reflection of the massacres of 1096 in Jewish historical writings of the 19th-20th centuries.] (49-64); Roemer, Nils: Turning Defeat into Victory: "Wissenschaft des Judentums" and the Martyrs of 1096 [Ennobling of Jewish suffering for the faith in 1096 in the writings of Graetz, Zunz, and other Jewish scholars of that period.] (65-80). *DR*

0092 Johnson, Willis Harrison: *Between Christians and Jews: The Formation of Anti-Jewish Stereotypes in Medieval England.* Diss. – University of California, Berkeley, 1997. 410 pp. Unseen.

Examines anti-Jewish stereotypes in medieval England in the context of medieval Christian culture. A study of ubiquitous exegetical images of heretics and Jews smitten by God in the anus finds some of the primary features of late medieval characterizations of Jews to have been determined by multiple causes. These images ultimately developed into the idea that Jewish men menstruate. 13th century allegations against Jews of coin clipping derived from Christians' discomfort with circumcision, but the accusation only became dangerous when it was reinforced by social and economic pressures. The collective punishment of London's Jews, and the decree of expulsion of the Jews in 1290 were understood by Jews as Divine confirmation of Jewish culpability in coin clipping. This ensured the last, ironic stage in the development of the stereotype: its acceptance among Jews. Examines, also, literary representations of Jews after the expulsion. Anti-Jewish tropes in late medieval English literature tended toward less negative, exegetically standard forms in the social absence of Jews. Representations of Jews by Chaucer, Langland, and the authors of "The Siege of Jerusalem" and the "Croxton Play of the Sacrament" lack the malice of anti-Jewish texts written 100 years earlier. An appendix contains a critical edition of MS Cambridge Add. 3037, Thomas of Monmouth's "The Life and Miracles of Saint William of Norwich."

0093 Johnson, Willis Harrison: The Myth of Jewish Male Menses. *Journal of Medieval History* 24, 3 (Sept 1998) 273-295.

Discusses the claims of some scholars (e.g. Sander Gilman) that medieval Christians believed that Jewish men menstruated and, therefore, Jews collectively were gendered feminine in Christian thought. States that this idea actually first appeared in the early modern period and that it was a reinterpretation of motifs dating back to late antiquity. References to a Jewish male bloody flux were derived from a traditional Christian belief that heretics and betrayers of Jesus had blood and/or guts coming out of their anuses. In the 12th century, the idea of anal bleeding was linked to Jewish deicidal blood guilt. The earliest mention of monthly bleeding by male Jews appeared in a 1503 account of ritual murder trials held in Tyrnau in 1494. *REK*

0094 Jordan, William Chester: Jews, Regalian Rights, and the Constitution in Medieval France. *AJS Review* 23, 1 (1998) 1-16.

Discusses the situation of the Jews in France in the 13th-early 14th centuries. The Jews, along with the serfs and the rights to coinage, were considered to be under royal authority. However, many local barons objected to these conditions. The decision made by Philip the Fair to expel the Jews in 1306 was a part of his efforts to retain regalian rights. These efforts also included coinage reform in the 1290s and in 1311, and an attack on the Knights Templar in 1307. States that although Philip shared with his barons the spoils taken from the Jews, the purpose of the expulsion was to demonstrate and enhance royal supremacy. *REK*

0095 Leroy, Béatrice: *Les Juifs du bassin de l'Ebre au Moyen Age: Témoins d'une histoire séculaire*. Biarritz: J & D Editions, 1997. 192 pp.

Chronicles the history of Spanish Jewry inhabiting the region of the Ebre basin, which comprised the provinces of Castile, Aragón, Navarra, and Catalonia. Jewish communities existed there from the first decades of the Christian era, but their status declined from the 7th century when the first community of Conversos was established due to oppression under Visigoth rule. Following the Muslim conquest, Jews were tolerated but their liberty was severely restricted when dhimmi status was imposed upon them. Ch. 6 (pp. 147-177), "Des violences à l'expulsion," traces the impact of the Inquisition on the Jews of the region and their expulsion in 1492 (from Navarra in 1498). *TS*

0096 Liberanome, Daniele: Gli Ebrei al tempo di Teodorico e il ruolo della Chiesa di Roma. *La Rassegna Mensile di Israel* 64, 3 [Sept-Dec 1998] (1999) 21-39.

Comments on the policy of Theodoric, King of the Ostrogoths, toward the Jews, when, during his rule in Italy (489-526), episodes of anti-Jewish hostility took place, based on the writings of Cassiodorus (ca. 490-580), a Roman monk who was secretary to the king. Between 507-511 and 523-526 the Church and the populace instigated acts of violence against Jews in several cities of northern Italy, violating and burning synagogues and accusing Jews of desecrating Christian symbols. Although at that time the papacy maintained a tolerant policy toward Jews, based on the teaching of St. Augustine, the local bishops did not. The Jews turned to Theodoric for protection. The king referred to traditional Roman law (not to canon law) in order to deal with their claims and preserve public order. *MZ*

0097 Lipton, Sara: *Images of Intolerance: The Representation of Jews and Judaism in the "Bible moralisée."* Berkeley: University of California Press, 1999. xvi, 241 pp.

Explores why Jews and Judaism are the fundamental theme in the two earliest surviving French illuminated Bible commentaries (13th century). While noting continuity with earlier anti-Jewish theological stereotypes, focuses on the forceful antisemitic visual images of this type of Bible. Stresses that the Jews are compared to heretics and misusers of the Christian Scriptures. Concludes that the Jews were understood as embodying sins or crimes, such as carnality, greed, pride, lust, and bloodthirstiness, considered threats to the body of Christendom. The qualities caricatured were not viewed as exclusively Jewish but rather as contagious qualities that might infect Christians. The negative qualities were associated with challenges that prosperity and intellectual speculation posed to the "successful" society of northern France. While not viewing the often demonic anti-Jewish thrust of the Bibles as intentionally antisemitic, relates to the exploitation of the available antisemitic imagery. The "Bible moralisée" is seen as a genre that challenges the distinctions between popular and elite, clerical and laic, in regard to anti-Judaism and other aspects of the Middle Ages. *YC*

0098 Maccoby, Hyam: The Tortosa Disputation, 1413-14, and Its Effects. *The Expulsion of the Jews and Their Emigration to the Southern Low Countries (15th-16th C.)*, eds. Luc Dequeker, Werner Verbeke. Leuven: Leuven University Press, 1998. Pp. 23-34.

From the mid-13th century there developed in Spain a conception that contemporary Jews were "talmudic Jews" and not the "biblical Jews" that Augustine had in mind. The 21-month long Tortosa Disputation (1413-14) was part of a campaign to convert Jews. Its chief promoter, Vicente Ferrer, carried on a campaign of terror which produced a considerable number of converts among the Jews of Aragón and Castile. Forced, insincere conversions (there had been many in 1391) produced a problem of insincere Christians, which the Inquisition was instituted to resolve. Concludes that the impetus to conversion came largely from the friars, especially the Dominicans. The massacres and forced conversions of 1391, the Tortosa Disputation of 1413-14, and the expulsion in 1492 all form part of a single pattern, which is the breakdown of the Augustinian scheme of toleration and the increasingly intense campaign to convert the Jews. *REK*

0099 Meyuhas Ginio, Alisa: *Fontes iudaeorum regni Castellae. Vol. 8: De bello iudaeorum: Fray Alonso de Espina y su "Fortalitium fidei."* Salamanca: Universidad Pontificia de Salamanca, 1998. 105 pp.

Presents an analytical summary of the third book of the "Fortalitium fidei" by the Franciscan friar Alonso de Espina (Castile, ca. 1458-1464). The entire work consists of five books dealing with the enemies of the Christian faith. Book three focuses on the Jews. The first six sections deal with the nature of the Jews and their arguments against the Christian religion. The next four sections describe the war carried out by Jews against Christianity. Section 11 deals with the obligations of the Jews in Christian society according to canon law, civil law, and royal legislation. Section 12 explains why the Jews will convert on the day of final judgment. Asserts that Espina, although he accepted the fact of Jews living in a Christian society, advocated complete separation from them. He considered the greatest danger to be the interaction between Christians and Jews as embodied in the Conversos. The expulsion and expropriation of the Jews was in his view the only true solution to preservation of the Christian world. *MZ*

0100 Meyuhas Ginio, Alisa: *La forteresse de la foi: La vision du monde d'Alonso de Espina, moine espagnol (?-1466).* Trans.: Zvi Rabi. Paris: Cerf, 1998. 249 pp. Translated from the Hebrew. Abridged from the author's diss. – Tel-Aviv University, 1988.

0101 Moses ben Nachman (Nachmanides): *La "Disputa di Barcellona," "Sefer vikkuach haRamban."* Ed. & trans.: Francesco Bianchi. [Roma]: Edizione Lamed, 1999. 93 pp.

See the editor's introduction (pp. 5-43) for the historical background to the disputation. See also the abstract for the French translation: "La dispute de Barcelone" (Lagrasse: Verdier, 1984). *SSC*

0102 Muñoz Solla, Ricardo: Estrategias de persuasión y oyentes judíos en dos sermones de san Vicente Ferrer. *El Olivo* 49 (Jan-June 1999) 25-43.

Analyzes the discourse of the Spanish preacher Vicente Ferrer (1350-1419), known for his successful campaigns to convert the Jews to Christianity in the late 14th-early 15th centuries, based on Pedro M. Cátedra's "Sermón, sociedad y literatura en la Edad Media: San Vicente Ferrer en Castilla (1411-1412)" (1994), containing some of Ferrer's sermons which were directed at the Jews. Considers the content of two anti-Jewish sermons as a textual expression of segregationism toward Jews and the form of his sermons as characterized by a theatrical and sensational style. His preaching increased the already hostile atmosphere surrounding the Jews. Concludes that the persecutions of 1391 and the decadence of the Jewish communities created the appropriate environment for the success of his sermons. *MZ*

0103 Netanyahu, Benzion: The Conversion of Don Samuel Abravanel. *Between History and Literature: Studies in Honor of Isaac Barzilay,* ed. Stanley Nash. Tel-Aviv: Hakibbutz Hameuchad, 1997. Pp. 81-110.

Discusses why Don Samuel Abravanel (grandfather of Isaac), a wealthy Jew in the court of Castile, who was a protector of his fellow Jews and supporter of their rabbis, converted to Christianity. Argues that, contrary to the commonly-held view, his conversion took place prior to the pogroms and forced conversions of 1391. Suggests that it was motivated by Abravanel's disgust and disillusionment with the secular and religious leadership of the Jewish community of Sevilla, as a result of infighting and mutual betrayal among Castilian court Jews, whose main function and source of wealth was tax farming. Describes pogroms in the half-century prior to 1391, and explains popular hatred of the Jews as largely the result of their role as tax agents of the king. *REK*

0104 Netanyahu, Benzion: *Los orígenes de la Inquisición en la España del siglo XV*. Trans.: Ángel Alcalá Galve, Ciriaco Morón Arroyo. Barcelona: Crítica, 1999. xix, 1269 pp. Originally published as "The Origins of the Inquisition in Fifteenth Century Spain" (New York: Random House, 1995).

0105 Ocker, Christopher: Ritual Murder and the Subjectivity of Christ: A Choice in Medieval Christianity. *Harvard Theological Review* 91, 2 (Apr 1998) 153-192.

Discusses blood libels and host desecration accusations in medieval Europe, stating that there were ca. 85 such cases in the 12th-16th centuries. Attributes the initial popularity of these accusations to teachings of theologians and interpretations of the New Testament which showed enmity between God and the Jews, as well as the blindness and sinfulness of the Jews and their contempt for Christianity. Popular legends describing alleged crimes by Jews, which focused on the need for blood in their rituals, also fanned religious hatred. The accusations were generally followed by massacres or pogroms, to avenge not only the supposed crime but also the death of Jesus. The desire to avenge Jesus' death was probably due to a growing popular affinity with him, and it continued in spite of pronouncements by some theologians that Jews did not bear collective guilt for his crucifixion and that they should not be killed but only restricted because of their contempt for Christianity. *RE*

0106 Picard, Marianne: *Juifs et judaïsme: Manuel d'histoire juive. Vol. 2: De 70 à 1492*. Paris: Biblieurope, 1998. 287 pp.

A textbook chronicling ancient and medieval Jewish history from the destruction of the Second Temple to the expulsion from Spain. Discusses Jewish life in the Roman Empire and under Islamic and Christian regimes. Relates, also, to persecution in medieval Europe. Contends that despite unceasing moral and physical oppression, study and adherence to tradition protected Jews throughout their wanderings. Includes excerpts from documents, and illustrations, chronological tables, and maps relevant to each era. *TS*

0107 Rabello, Alfredo Mordechai: The Legal Status of Spanish Jews during the Visigothic Catholic Era: From Reccared (586) to Reccesswinth (672). *Israel Law Review* 33, 4 (Fall 1999) 756-786.

The transition of the ruling Visigothic stratum from Arianism to Catholicism in 587 worsened the legal status of the Jews in Spain. Royal legislation, as well as canons of Church councils held in Spain between 586-672, aimed at the most rapid conversion of the Jews and tended to obstruct Jewish existence in Spain. Both the royal and ecclesiastical laws in fact introduced forced baptism, impeded possessing slaves and holding public positions by Jews, and impeded and then prohibited mixed marriages. The concerted royal and ecclesiastical policies led to the emergence of a class of converted Jews and their progeny, who also became victims of the official restrictions. Meanwhile, the inability of the kings to enforce the anti-Jewish laws and canons allowed some unbaptized Jews to continue their lives in Catholic Visigothic Spain. *DR*

0108 Ragacs, Ursula: Zur juedisch-christlichen Kontroverse im Mittelalter: Eine Spurensuche. *Frankfurter Judaistische Beitraege* 25 (Dec 1998) 105-119.

After Abraham ibn Daud, on the basis of the Talmud, posited a chronology for the life of Jesus that ruled out any connection between the

Crucifixion and the destruction of the Temple, medieval Christian scholars began to take a greater interest in the Talmud and to use talmudic passages in disputations against Jews. In the first disputation in Paris (1240), Nicholas Donin accused the Talmud of blasphemy against Christianity and against Jesus and Mary. The Jews then denied that the Talmud mentions Jesus at all. Later disputations centered on passages predicting the coming of the Messiah around the time of Jesus; the arguments then were mostly about chronology. *RW*

0109 Resnick, Irven Michael: *On Roots of the Myth of Jewish Male Menses in Jacques de Vitry's History of Jerusalem*. Ramat-Gan: Bar-Ilan University, Ingeborg Rennert Center for Jerusalem Studies, 1998. 27 pp. (International Rennert Guest Lecture Series, 3).

Traces the attempt by Christians to find a "motive" for alleged Jewish ritual murders of Christians from the 12th century, when there were more Christian polemical treatises against Jews than in all previous centuries. The search for an explanation of ritual murder led to a fusion of theological and natural scientific notions. Jewish males were believed to have a flux of blood via hemorrhoids, both due to theological considerations (see Mathew 27:25, "His blood is on us and on our children," and interpretations of Samuel 5:6-12 and Ps. 77:66) and contemporary medical ideas (that a defective diet and life-style, including melancholy, fear, and idleness, caused bleeding). These were combined in de Vitry's "History of Jerusalem" (ca. 1221), which was apparently the first literary source to state that, like women, Jewish men had monthly flows of blood. The conclusion drawn by Christians was that Jews believed that this ill and others could be cured by using Christian blood, rather than by conversion to Christianity where they might be saved by the blood of Christ. This "explanation" of the Jews' alleged need for Christian blood was actually cited at trials for ritual murder. *YC*

0110 Rodríguez Zahar, León: Tres aspectos del origen de la segregación de la gente del libro en el Islam: Hanifismo, "umma" y "dhimma." *Estudios de Asia y Africa* 34, 3 [110] (Sept-Dec 1999) 583-622.

With the establishment of Islam and during its early expansion in the 7th century, the relationship of the emerging Muslim states to religious minorities (including Jews) was regulated by the "dhimma," a contract imposed by the conquerors on the conquered, establishing rights, obligations, and restrictions. Asserts that the Jews were the first to be subjugated in the Arabian peninsula; they lost their land and independence, becoming tenants of the state. The situation worsened under the rule of the first caliphs. Explains that the "dhimma" evolved from the idea of "umma," the religious ethnic community which was rooted in pre-Islamic religions, known as "Hanifism." Muhammad united many Arab tribes under this idea, segregating "ummas" other than Muslims. *MZ*

0111 Rubin, Miri: *Gentile Tales: The Narrative Assault on Late Medieval Jews.* New Haven: Yale University Press, 1999. xiii, 266 pp.

A study of the host desecration accusation as a powerful narrative about Jews that developed in late medieval Europe, showing the use of a tale in the unfolding of violence against Jews. Presents examples of various cases which erupted in the 14th-15th centuries, most frequently in the German-speaking regions, since the expulsions from England and France left the Holy Roman Empire as the home of most European Jews outside of the Iberian Peninsula. The position of Jews provided an emblematic flash-point for the eruption of tensions between towns and lords. A strong administration could ensure that a trial rather than a massacre followed, but a trial could also lead to torture, punishment and banishment. The outcome depended on the persons involved, the specific political structure, and the position of the Jews in each locality. The host desecration accusation was perpetuated in folklore and in the public memory; artifacts were enshrined and became subjects of veneration. The tale also entered medieval literature and art. However, it did not survive the Reformation, which aroused skepticism regarding miracle stories. Includes illustrations. *SSC*

0112 Rumeu de Armas, Antonio: El Decreto de 1492 expulsando a los judíos de Castilla: Anomalías Cancillerescas. *Boletín de la Real Academia de la Historia* 195, 1 (Jan-Apr 1998) 1-17.

Explains the expulsion of the Jews from Spain in 1492 as a result of growing anti-Jewish hostility, as well as of religious and political considerations, namely the achievement of national unity. Discusses the extreme secrecy surrounding the expulsion decree, which was signed in Granada (Castile) on 31 March 1492 but made official by three Aragonese court officials and registered only by the Chancellery of Aragón (and not Castile). States that the decree, which was to be made public only on 29 April, stipulated the sealing of the Jewish houses and quarters, and the subsequent inventory of all Jewish property. In order to avoid any last-hour transfer of Jewish monies or valuables, and because he did not trust the Castilian bureaucracy, King Ferdinand (of Aragón) entrusted this "secret operation" to his closest assistants. Includes the text of the expulsion decree. *AA*

0113 Sicherman, Harvey; Gevaryahu, Gilad J.: Rashi and the First Crusade: Commentary, Liturgy, Legend. *Judaism* 48, 2 (Spr 1999) 181-197.

Demonstrates that Rashi (1040-1106), the great Jewish Bible and Talmud commentator, referred, albeit indirectly, to the contemporary Jewish tragedy, the massacres of the First Crusade. Rashi's selection of midrash for his commentary on Genesis 1 was probably intended to refute the papal claims to the Holy Land that launched the First Crusade. Rashi's choice of verses from Isaiah in his penitential poems also refer to Christian imagery and Crusader behavior. His use of metaphor was understood by Jews; it kept him safe from Christian reprisal. A legend that links Rashi with Godfrey of Bouillon, the Crusader conquerer of Jerusalem, may reflect historical events. An English translation of the legend as found in Guenzburg ms. 652 is appended (pp. 193-194). *YC*

0114 Smith, John C.P.: Settlement, Expulsion and Restoration: The Jews of England, 1066-1656. *Roots and Branches: Explorations into the Jewish Context of the Christian Faith,* eds. John Fieldsend et al. Bedford: PWM Trust; Centre for Biblical and Hebraic Studies, 1998. Pp. 181-204.

A historical essay embracing the life of Jews in England from 1066 to 1290, when they were expelled, and their readmission to England in 1656. Taken as a whole, Jewish life in medieval England was harsh; only the period of 1066-1130 was relatively peaceful for Jews. When the crusading spirit began to pervade England, the situation of the Jews worsened: they lived through arbitrary fines, a ritual murder accusation, and anti-Jewish violence in the mid-12th century. In 1189-90 the Jews of England were victims of acts of mass violence which reached a peak in the massacre at York. Another wave of violence and extortion broke out during the reign of Henry III, in the 13th century, when expulsions of Jews from various places in England began. The attempted reform of Jewish life initiated by his son Edward I ended with the expulsion of the Jews in 1290. *DR*

0115 Suárez Fernández, Luís Bilbao: Claves históricas del problema judío en España medieval. *El legado material hispanojudío; VII Curso de Cultura Hispanojudía y Sefardi de la Universidad de Castilla-La Mancha,* eds. Ana María López Álvarez, Ricardo Izquierdo Benito. Cuenca: Ed. de la Universidad de Castilla-La Mancha, 1998. Pp. 15-75.

Discusses the premises for Christian hostility toward Jews in medieval Spain, stating that it resulted from a blend of popular and religious prejudices and stereotypes. Asserts that the restricted social interaction between Jews and Christians left a wide space for the Christian populace to develop negative stereotypes about the Jews, such as cowardice, sorcery, and usury. The Church abandoned its tolerant stance, characteristic of the early Middle Ages (based on St. Augustine's teaching), and accused the Jews of deicide. The negative stereotype of the Jew was transferred to the Converso, based on the assertion that the Jew was intrinsically depraved, and that he could not overcome his depravity through conversion. The Black Death plague, and the later economic crisis, increased hostility against the Jews, leading the Church from segregation of the Jews to an attempt to find "a kind of final solution" in baptism, emigration, or death. The defensive attitude of the state contributed to the tragic outcome of 1391 and 1492. *MZ*

0116 Testuzza, Giorgio: Fuga di un "marrano"? Dalla Spagna a Roma: Juan Arias Davila vescovo di Segovia (1461, 1497). *La Rassegna Mensile di Israel* 64, 3 [Sept-Dec 1998] (1999) 41-52.

Relates to the fate of the Converso Juan Arias Davilá, bishop of Segovia between 1461-97. In 1486 he and his family were accused by the Inquisition of heresy and Judaizing. Although the judging of bishops was an exclusive prerogative of the Vatican, the Inquisition collected evidence against him and tried to gain acceptance from the Vatican to judge him in Spain. The Catholic Kings supported the claims of the Inquisition and urged Pope Innocent VIII to leave the case of Arias Davilá to the Spanish tribunals. The Pope refused, since it would have meant relinquishing his authority in favor of the emerging Spanish state. Arias Davilá left Spain in 1490 for Rome, where he died in 1497; between 1490-97 he remained the official bishop of Segovia. *MZ*

0117 Toch, Michael: *Die Juden im mittelalterlichen Reich*. Muenchen: R. Oldenbourg, 1998. x, 185 pp. (Enzyklopaedie deutscher Geschichte, 44).

Presents a synopsis of the history of German Jewry during the medieval period. Discusses aspects of life in the Jewish community, and Jewish-Christian relations. Persecution of the Jews was not prevalent until the 13th century; however, after 1250 it became a central part of Jewish life, culminating in the 14th century when most of the Jewish communities were devastated after the Jews were accused of causing the Black Death epidemic. Describes host desecration accusations, blood libels, expulsions, and anti-Jewish polemics and stereotypes. Surveys, also, general trends in the research on Jewish history in this period. *ID*

0118 Trautner-Kromann, Hanne; Schulman, Salomon, eds.: *Dopet eller döden: Judiskt liv i det medeltida Spanien* [Baptism or Death: Jewish Life in Medieval Spain]. Trans.: Öjevind Lång, Mose Apelblat. Stockholm: Brutus Östlings Bokförlag Symposion, 1996. 140 pp.

A collection of nine essays, based on papers presented at the conference "The Jews in Christian Spain: History and Culture before the Expulsion" held at Lund University, October 1992. Most of the papers were translated from English, Danish, or Hebrew. Partial contents: Assis, Yom Tov: De spanska judarna – från förföljelser till fördrivning (1391-1492) [The Spanish Jews – from Persecution to Expulsion (1391-1492)] (35-51);

Jensen, Kurt Villads: Kristen polemik mot judarna [Christian Polemic against the Jews] (75-83); Kaplan, Yosef: Marrarerna i 1400-talets Spanien [Marranos in 15th-Century Spain] (95-109). *SSC*

0119 Walz, Rainer: Die Verfolgungen von 1096 und die Ritualmordlegende: Die Debatte ueber die Thesen Israel J. Yuvals. *Aschkenas* 9, 1 (1999) 189-232.

After analyzing the reactions to the massacres of 1096 in contemporary chronicles and piyyutim, including the call for divine revenge, and the Jewish tradition of "Kiddush Hashem" and attitude to suicide, summarizes, for German readers, Yuval's thesis (published in "Zion" 58, 1993) on the causal connection between the killing of Jewish children by their parents in order to prevent their conversion and the Christian belief that Jews ritually murdered Christian children. Evaluates the subsequent arguments of Yuval's opponents, published in "Zion" and other journals. Concludes that the death of the Jewish children was only indirectly connected to the blood libel; as additional evidence of the Jews' stubborn rejection of Christianity, it reinforced, together with the perception of the Jews' longing for revenge, and other (true or fancied) attributions, the chimera of the Jews' hatred for Christians, of which the blood libel was an element. *RW*

0120 Yuval, Israel Jacob: Passover in the Middle Ages. *Passover and Easter: Origin and History to Modern Times,* eds. Paul F. Bradshaw, Lawrence A. Hoffman. Notre Dame, IN: University of Notre Dame Press, 1999. Pp. 127-160.

In the 13th century Christians began extending the blood libel to the charge of host desecration. In the latter, the charge against the Jews was based on a misinterpretation of Jewish rituals connected with the Sabbath or the Passover holiday – e.g. establishing the Sabbath boundary ("eruv"), which could involve the hanging of a matzah on a nail; eating of the "afikoman" (which could be confused with the eucharist) and of "haroset" (which required wine, whose symbolism invoked that of blood) at the Passover Seder. Like the contemporary "miracle" of the appearance of Jesus' blood on the host (caused, as we now know, by bacteria), the libels against the Jews may have taken place not "out of malice but out of Christian naiveté." *YC*

THE EARLY MODERN PERIOD (1493-1788)

0121 Alpert, Michael: The Secret Jews of 18th Century Madrid. *Revue des Etudes Juives* 156, 1-2 (Jan-June 1997) 135-171.

Discusses the revival of Inquisition activity in the first half of the 18th century; 144 autos-da-fé have been identified, of which 125 involved Judaizers. The most active courts were in the south of Spain. Ca. 1,500 crypto-Jews were punished by the Inquisition between 1700-1760. Examines the confessions of Madrid Jews tried in Toledo, which attest to a significant measure of Jewish culture and religious practice – belief in the Jewish God, observance of rites and festivals, endogamous marriage, etc. Concludes that it is difficult to believe that the accusations against the Jews were invented; rather, one may assume that there were several thousand secret Jews in Spain at the time who formed some sort of a community, with leaders, rules, and a sense of solidarity and identity. *SSC*

0122 Barata, José Oliveira: *História do teatro em Portugal (séc. XVIII): António José da Silva (o Judeu) no palco joanino* [History of the Theater in Portugal (18th Century): António José da Silva (the Jew) on King João's Stage]. Algés, Portugal: DIFEL, 1998. 379 pp.

Discusses the life and works of Silva (1701-1739, nicknamed "the Jew"), a prolific and popular Portuguese playwright during the reign of João V. Pp. 37-69 relate his brief and tragic life. He was born in Rio de Janeiro to a Converso family that already had problems with the Inquisition. In 1711 the family immigrated to Portugal. Silva studied law in Coimbra and was involved in a trial for Judaizing held in 1726. Released, but marked as "tainted," his legal career compromised, he began to write successful comedies for the theater. Silva was tried again in 1737, with all of his extended family, and sentenced to death; he died in an auto-da-fé in Lisbon in 1739. Interestingly enough, his theatrical career was not mentioned at all in the trials; the Inquisition apparently acted against him in order to break up two strong Converso clans (his and his wife's) united by marriage bonds. *AA*

0123 Bethencourt, Francisco: *La Inquisición en la época moderna: España, Portugal e Italia, siglos XV-XIX.* Trans.: Federico Palomo. Madrid: Akal, 1997. 566 pp. Originally published as "História das Inquisições: Portugal, Espanha e Italia" (Lisboa: Temas e Debates, 1996).

0124 Coelho, António Borges: *Cristãos-novos judeus e os novos argonautas* [Jewish New Christians and the New Argonauts]. Lisboa: Caminho, 1998. 247 pp. (Questionar a história,4).

Discusses the place of Portuguese Conversos in the Portuguese expansion into the New World and in Portuguese identity. Discusses, also, the treatment of Conversos in Portugal, especially the forced conversions and persecution by the Inquisition (e. g. the autos-da-fé in Faro in 1637, and in Lisbon in 1647). Relates to the Converso diaspora, and the pervasive presence of the "gente da nação" in colonial America and elsewhere. Deals, alşo, with the struggle of the Jesuit António Vieira against the Inquisition for better treatment of Conversos (ca. 1649). *AA*

0125 *Eighteenth-Century Studies* 32, 1 (Fall 1998).

The section "Forum: Jewish Questions" (pp. 83-110) contains three essays on the "Jewish question" in 18th-century France and Germany. Contents: Schechter, Ronald: The Jewish Question in Eighteenth-Century France [Contends that it was the symbolic Jew who preoccupied the French philosophers of the Enlightenment, because the Jew could conveniently become the symbolic figure of the anti-citizen in the polemics on state sovereignty. Examines the views of two antisemitic writers, François J.A. Hell and Philippe François de Latour-Foissac.] (84-91); Hess, Jonathan M.: Sugar Island Jews? Jewish Colonialism and the Rhetoric of "Civic Improvement" in Eighteenth-Century Germany [Deals with the Prussian project to achieve the moral regeneration of the Jews by resettling them in colonies, both in Prussia and overseas. Examines the anonymous "Letter of a German Jew to the President of the Congress of the United States of America," exposing this idea.] (92-100); Kassouf, Susan: The Shared Pain of the Golden Vein: The Discursive Proximity of Jewish and Scholarly Diseases in the Late Eighteenth Century [Examines the ways in which the studious lifestyles of Jews and scholars became associated with disease in late 18th-century Western and Central European discourse. The image of the sick traditional Jew was connected with the perception of femininity of the Jews and Jewish culture.] (101-110). *DR*

0126 Feldman, D.: Neizvestnoye sledstviye o "kazanskoi sekte zhidovstvuyushchikh" 1748-1749 gg. [The Unknown Investigation of the "Judaizing Sect in Kazan" 1748-1749]. *Vestnik Yevreiskovo Universiteta* 20 (1999) 296-323.

Presents a sample of documents pertaining to the state investigation of a Judaizing sect in Kazan, Tataria, most probably a fictitious sect. Pp. 296-306 contain an introduction, dealing with the struggle against the "Judaizing" movement in Russia from 1471, when such a movement was detected for the first time (in the city of Novgorod) to the early 19th century. Contends that Archbishop Gennadii of Novgorod invented a Judaizing sect in the city in order to ascribe the non-conformist tendencies of the local clergy to its influence. In the following epoch, the Russian authorities used the struggle against the Jews and "Judaizers" as a tool to preserve the unity of the Russian Church. Up to 1772, official Russian Judeophobia had a purely religious character. The investigation of 1748-49, which ended with the acquittal of the nine defendants, was conducted with unusual dispatch, which attests to the importance which the authorities attached to these cases. *DR*

0127 García Marín, José María: Judaísmo entre el poder y la envidia: El caso Avila ante la Inquisición. *El centinela de la fe: Estudios jurídicos sobre la Inquisición de Sevilla en el siglo XVIII,* ed. Enrique Gacto Fernández. Sevilla: Universidad de Sevilla, 1997. Pp. 233-307.

In 1710, the royal commissioner Diego de Avila and other members of his family were accused of Judaizing by the Inquisition of Sevilla. Discusses the Inquisitorial proceedings, focusing on how the tribunal handled evidence, testimonies, and torture. Compares canon law with royal law concerning the rights of the accused and the legal validity of confession under torture. Pp. 297-307 present the text of the posthumous sentence pronounced by the tribunal against Diego de Avila in 1723; he died in prison in 1715. *MZ*

0128 Gavriel, Mardell J.: *The Anti-Semitism of Martin Luther: A Psychohistorical Exploration.* Diss. – Chicago School of Professional Psychology, 1997. 126 pp. Unseen.

Examines Luther's antisemitism by alternately employing psychodynamic explanations of anti-Jewish hostility and a rationale of prejudice based on Albert Bandura's social learning theory. Psychodynamic theories of antisemitism identify personality factors, unconscious psychological conflict, and analytic interpretations of social history as influential factors of antisemitism. In the case of Luther, explores how superstitiousne ss, characterological anger, and anal personality traits may have predisposed Luther to the development of prejudice. Discusses the status of medieval Jews as "internal foreigners," their connection to a moral code, the practice of circumcision, their rejection of Christ, and the medieval perception of affiliation with the devil to show how these factors might have made them a particularly attractive projection screen for Luther's unconscious conflicts.

0129 Giles, Mary E., ed.: *Women in the Inquisition: Spain and the New World.* Baltimore: Johns Hopkins University Press, 1999. ix,402 pp. On back of title page: c1999. Published 1998.

A collection of essays dealing with the Inquisition and the status of women in early modern Spain and the New World. Three essays deal with Converso women who were persecuted by the Inquisition, based on their trial records. "Pt. I: The Inquisition and Jewish Converts": Starr-LeBeau,

Gretchen: Mari Sánchez and Inés González: Conflict and Cooperation among Crypto-Jews (19-41); Beinart, Haim: Inés of Herrera del Duque: The Prophetess of Extremadura (42-52); Melammed, Renée Levine: María López: A Convicted Judaizer from Castile (53-72). *SSC*

0130 Gomes, Francisco de Azevedo: *Falsificação da historia. Vol. II* [Falsification of History. Vol. II]. [Braga, Portugal]: The Author, 1998. 525 pp.

An alternative reading of the history of the Portuguese discoveries, colonization, slave trade, and Inquisition, stressing the usually forgotten aspects of persecution, cruelty, and violence. Ch. 4 (pp. 233-505), "Inquisição," deals with the forced conversion of the Jews in 1496-97 and the merciless persecution of Conversos by the Inquisition in Portugal and its territories (e.g. the trials of the bankers Duarte da Silva and Francisco Gomes Henrique, 1647). Pp. 412-418 discuss the segregationist statute of the Jews in Ceuta and Tanger, including stipulations for special Jewish clothing. Pp. 481-489 deal with Mário Saa's "A Invasão dos Judeus" (1924), which applauds the Inquisition's stand against the Conversos and is full of antisemitic stereotypes and accusations, such as the advent of the republic in Portugal in 1910 resulting in a "Jewish state." *AA*

0131 Haasis, Hellmut G.: *Joseph Suess Oppenheimer, genannt Jud Suess: Finanzier, Freidenker, Justizopfer.* Reinbek bei Hamburg: Rowohlt, 1998. 479 pp.

A biography of Oppenheimer (1698-1738). Describes his career and mode of life, and his interrogation and trial, dismissing much of the scandal spread about him by later writers as invention. Portrays Jewish life in his native town, Heidelberg, hedged in by many restrictions and subject to periodic violence and student pogroms. In his later life, Oppenheimer often met with resentment of his success because he was a Jew. In his trial there was no overt antisemitism, but as a Jew he was easier to victimize and received far harsher treatment than other officials who were probably more deeply involved than he was in financial corruption. Asserts that the trial was unlawful and the authorities had determined on the death sentence even before the interrogations began. *RW*

0132 Hagen, Kenneth: Luther's So-Called "Judenschriften": A Genre Approach. *Archiv fuer Reformationsgeschichte* 90 (1999) 130-158.

Contends that in order to interpret correctly Luther's writings about the Jews, described as "Jewish" or "anti-Jewish" in the 20th century, we must withdraw from the post-Holocaust context and place them in their historical, i.e. medieval context. Close examination of the four works (written by Luther between 1523-43) shows that they are neither addressed to the Jews nor directed against them. They are apologetic treatises addressed to Christians, defending Christian doctrine against "Jewish" denials and detractions, and aimed at strengthening Christian faith. In the context of these works, "Jew" means a non-believer or an adherent of a false belief. Luther adopted theological genres typical of medieval apologetic and explanatory literature. His "Jewish writings" should not be regarded as hate-instigating works, as they were perceived in the Nazi period. *DR*

0133 Ioly Zorattini, Pier Cesare, ed.: *Processi del S. Uffizio di Venezia contro ebrei e giudaizzanti. Vol. 14: Indici generali.* Firenze: Leo S. Olschki, 1999. 417 pp.

This volume contains an essay by Massimo Prada, "La lingua dei 'Processi' tra Cinque e Seicento" (pp. 15-86), a glossary, also by Prada (pp. 89-170), a bibliography (pp. 171-190), the indexes to volumes I-XIII – persons and places (pp. 191-389), and errata and corrigenda (pp. 391-416). *SSC*

0134 Katz, David S.: Shylock's Gender: Jewish Male Menstruation in Early Modern England. *Review of English Studies* 50 [200] (1999) 440-462.

After a comprehensive survey of the widely accepted belief in Jewish male menstruation, with its basis in religious and sexual stereotypes, argues that this idea and its link with ritual murder charges provide a new perspective from which to understand Shakespeare's Shylock. A parallel is drawn with a 17th-century Spanish work, where blood needed to heal Jewish male menstruation is the motive for ritual murder of Christians. Although the Jewish usurer in the Spanish story lives in Turkey, he is "clearly Shylock, driven in his lust for Christian blood by the divine affliction of male menstruation." The cultural stereotypes in this image, which involve blood, gender, and racism, anticipate modern racism. *YC*

0135 Knoch-Mund, Gaby: *Disputationsliteratur als Instrument antijuedischer Polemik: Leben und Werken des Marcus Lombardus, eines Grenzgaengers zwischen Judentum und Christentum im Zeitalter des deutschen Humanismus.* Tuebingen: Francke, 1997. x, 490 pp. Based on the author's diss. – Universitaet Bern, 1991.

Contains a critical edition of the "Disputation zwischen einem Juden und einem Christen" (composed between 1540-55; MS Muenchen, Bayerische Staatsbibliothek, cgm 972) by Lombardus, a converted German-Italian Jew, and discusses its context in the history of disputations and in the relationship between Jews, converts, and Christians in the 16th century. Lombardus's text incorporated previous ones, especially by Sebastian Muenster and Paul Fagius, and apparently was translated from the

Hebrew. It gives a fair hearing to the arguments of the Jew; in the end, however, the Christian condemns the Jews for their stubbornness, and their rejection and crucifixion of Christ. In 1573, Lombardus published a more aggressively anti-Jewish treatise on the customs of the Jews and "their curses on Jesus Christ and his Church and their inhuman usury." From surviving letters it appears that Christians continued to treat him as an outsider. The Church at this time strove for mass conversion of the Jews, but distrusted converts, in spite of the usefulness of their intimate knowledge of Judaism in anti-Jewish polemics. *RW*

0136 Koenig, Imke: *Judenverordnungen im Hochstift Wuerzburg (15.-18. Jh.).* Frankfurt a.M.: Vittorio Klostermann, 1999. xiii, 331 pp. Based on the author's diss. – Universitaet Frankfurt am Main, 1998.

Discusses legislation concerning Jews in the governmental policies of the Wuerzburg district from the end of the medieval period to the 18th century. Mentions letters of protection as bases for individual rights, taxation problems, residential conditions, clothing regulations, prohibition of usury, decisions concerning Jewish commerce, etc. Contends that most of the regulations were initiated by guilds and professional corporations whose members were interested in limiting the Jewish presence. Various pressure groups attempted to influence government decisions during the legislative procedures. Mentions, also, Jewish requests for regulations to protect them from aggression, especially since the 17th century. *ID*

0137 Kriegel, Maurice: Le parcours de Juan Luis Vives: Du milieu judaïsant à l'option érasmienne. *Revue de l'Histoire des Religions* 215, 2 (Apr-June 1998) 249-281.

Describes the condition of the Conversos in Spain in the context of the Inquisition's anti-Judaizers campaign in Valencia. The entire Vives family suffered heavily as a result of the trials. The philosopher Juan Luis Vives left Spain in 1509 and settled in Bruges. His writings echo his suffering because of his family's tribulations, as well as his feeling of permanent danger at a time when every Spaniard was suspected of heresy in Christian Europe. Using assumptions from Erasmian political thought, Vives viewed the Inquisition as an abusive institution generating social dissent and destruction, and sought to influence the Grand Inquisitor Alonso Manrique to reform and humanize it. *AA*

0138 Lipiner, Eliyahu (Elias): *Os baptizados em pé: Estudos acerca da origem e da luta dos cristãos-novos em Portugal* [Those Who Were Baptized Standing: Studies on the Origin and the Struggle of the Conversos in Portugal]. Lisboa: Vega, 1998. 492 pp.

Discusses the tragedy of forced conversion and life under the threat of the Inquisition for those who called themselves "baptizadas em pé" (i.e. Conversos). Relates the influx of Jews into Portugal after their expulsion from Spain in 1492, the high entry fees, the enslavement of illegal Jews, the abduction of some 2,000 Jewish children who were shipped to the island of São Tomé in 1493, the decree of expulsion in 1496 and the subsequent forced conversion in 1497 (preceded by a second abduction of Jewish children to be Christianized). Pp. 53-104 deal with the names of Jews prior to and after their conversion. Other chapters relate to the Sephardi diaspora, persecution by the Inquisition, and personalities such as Samuel Usque, António José da Silva, Antonio Serrão de Castro, Gonçalo Anes Bandarra. *AA*

0139 Luther, Martin: *Contro gli Ebrei: Versione latina di Justus Jonas (1544).* Ed.: Attilio Agnoletto. Trans.: Vittorio Dornetti. Milano: Terziaria, 1997. xxv, 234 pp.

Presents Luther's "Von den Juden und ihren Luegen" (1543), translated into Italian from the Latin version of Justus Jonas (1544). In his book, Luther refutes Jewish assertions which he considers to be boastful and false, e.g. that the Jews were the chosen people and that they had received from God the Law, the country of "Canaan," and Jerusalem; he accuses the Jews of being obstinate in not recognizing Jesus as the Messiah and blaspheming against him and his mother Mary. In the introduction (pp. ix-xxi), Agnoletto asserts that the Latin translation provided Luther with a broad public in Europe, consolidating the prejudices on which Christian antisemitic tradition was based. In the appendixes (pp. 195-226), Dornetti discusses Luther's homiletic technique, explains Luther's use of the word "carnal," and links Luther's main accusations to literary precedents from the Middle Ages. *MZ*

0140 Meyuhas Ginio, Alisa: The Inquisition and the New Christians: The Case of the Portuguese Inquisition of Goa. *Medieval History Journal* 2, 1 (Jan-June 1999) 1-18.

Demonstrates that in the Portuguese colony of Goa, the Inquisition, which lasted from 1560 to 1812, closely resembled the Inquisition in the mother country, including the attitudes toward and treatment of New Christians. The latter were persecuted for Judaizing, e.g. keeping the Jewish Sabbath and dietary laws. In Goa, New Christians constituted a disproportionate number of those executed after being condemned. As in Portugal, people were accused of Judaizing in order to settle personal scores or to confiscate their wealth. Old Christians continued to look down on New Christians even after many generations. Such discrimination led to social cohesion and endogamy among New Christians. *YC*

0141 Moulinas, René: Le chapeau jaune des Juifs d'Avignon et du Comtat. *Archives Juives* 31, 1 (1998) 9-18.

Surveys the history of legislation in the Papal States regarding distinctive clothing for Jews during the era preceding the French Revolution. Until then, Jews were required to wear a particular yellow hat which they regarded as degrading. Despite attempts by the Jewish community to revoke this law or modify the color of the hat, these requests were systematically refused. In 1768 France occupied the region and the pontifical governor, Durini, proposed initiatives to change the status of the Jews; for such innovations he was recalled to Rome in 1774. In 1790, France ratified the unification of Avignon with the new national government; thereafter, laws regarding special clothing for Jews were annulled. However, the Comtat region remained recalcitrant until 1791, when federalists used this issue as a symbol of loyalty to national unity. *TS*

0142 Novinsky, Anita Waingort: Political Zionism in the Portuguese Renaissance (Damião de Gois). *Hesed ve-emet: Studies in Honor of Ernest S. Frerichs*, eds. Jodi Magness, Seymour Gitin. Atlanta: Scholars Press, 1998. Pp. 419-429.

Contrasts two approaches to the forced conversion in Portugal in 1497, reflecting the views of two thinkers of the Portuguese Renaissance, the Jew Isaac Abravanel and the Christian humanist Damião de Gois. The former consoled his fellow Jews with the traditional religious view of redemption via the Messiah, who will be sent to redeem and gather the dispersed Jews in God's time. Gois, in his "Crónica do Felicissimo Rei D. Manuel" (1566), can be considered "the first political Zionist of modern times" due to his socio-political, secular point of view. He believed that King Manuel I did not offer the Jews the same alternatives to conversion that he offered the Arabs. The Jews were a people without lands and kingdoms. Thus, they became total victims because they were always foreigners everywhere and had no power. Gois himself fell victim to the Inquisition for his liberal views that often were close to those of the Conversos. *YC*

0143 Poole, Stafford: The Politics of "Limpieza de Sangre": Juan de Ovando and His Circle in the Reign of Philip II. *The Americas* 55, 3 (Jan 1999) 359-389.

The concept of "limpieza de sangre" arose in Spain as a reaction to the presence of a sizeable minority of Conversos in both Church and society. It was always a racial concept, although it was eventually applied to all those punished by the Inquisition and their descendants, as well as some other categories. In the 16th century, the strongest support for the racial purity statutes were found in the cathedral chapters and in the prestigious fraternities of the major universities. The monastic orders were inconsistent. By the end of the 16th century, when "purity of blood" became a common obsession in Spain, the broad class of educated bureaucracy ("letrados") made the "limpieza" an element in their struggle against the hereditary aristocracy. Using the example of Juan de Ovando, a cleric, an Inquisitor, and later the president of the Council of the Indies, shows the elasticity of the concept of "limpieza" in the late 16th century. Not obsessed with the "danger to Catholicism" emanating from those with Jewish ancestry, the "letrados" dishonestly and hypocritically manipulated the "limpieza de sangre" statutes for the sake of politics, career, and personal relationships. *DR*

0144 Putík, Alexandr: The Tumult of Mladá Boleslav (Jungbunzlau, Bumsla) in the Messianic Year 5426/1666. *Judaica Bohemiae* 34 (1999) 4-106.

A history of the pogrom that broke out on 23 April 1666 in Mladá Boleslav, Bohemia. Relations between Christians and Jews in the town deteriorated in the first half of the 17th century, due to the growth of the town's Jewish population and economic competition between the two communities, as well as the activities of the Counter-Reformation in the area. The pretext for the riot was the noise the Jews made on a Sabbath evening which coincided with the Easter service in a nearby Catholic church. No Jews were killed during the riots, but they suffered great material losses. Using documents of the inquiries conducted by the Bohemian regency and local and ecclesiastic authorities, describes the development of the pogrom and the course of the inquiry in the subsequent two years. As a result of the riots, the town's Jewish community was fined. Draws attention to the messianic expectations of Jews in 1666 – the year of Shabbetai Zvi – as a factor in the tumult. Pp. 69-106 contain documents from the inquiries, in Latin, Czech, and German. *DR*

0145 Raddatz, Alfred: Johannes Eck 1541 und Luther 1543: Ueber die Behandlung der Juden. *Vielseitigkeit des Alten Testaments: Festschrift fuer Georg Sauer zum 70. Geburtstag,* eds. James Alfred Loader, Hans Volker Kieweler. Frankfurt a.M.: Peter Lang, 1999. Pp. 369-373.

Points out the striking similarity in content and wording between Eck's refutation of Osiander and Luther's "Von den Juden und ihren Luegen." Both accuse the Jews of indolence and parasitism and propose making them do hard and demeaning labor; both allege that the Jews, if they could, would willingly kill all the Christians. Suggests that Eck's influence may account for Luther's reversal of his previous pro-Jewish attitude. *RW*

0146 Schechter, Ronald: Rationalizing the Enlightenment: Postmodernism and Theories of Anti-Semitism. *Historical Reflections/Réflexions Historiques* 25, 2 (Sum 1999) 279-306.

Criticizes the postmodernist conception according to which the Enlightenment, with its extreme rationalism and abhorrence of the "Other," invented the discourse on the Jews which was adopted by contemporary and later antisemitic writers, and ultimately led to Auschwitz. This viewpoint was expressed earlier by Adorno and Horkheimer in their "Dialectic of Enlightenment" (1944) and by Arthur Hertzberg in his "French Enlightenment and the Jews" (1968). In Hertzberg's view the main culprit was Voltaire, who denounced the Jews as ignorant, irrational, persistently primitive, and troublingly close to nature. Contends that Voltaire's writings concerning Jewish issues were directed primarily against the Bible and the Church. Close examination of texts by his contemporaries – antisemites like François Hell or Philippe François de Latour-Foissac, and antisemitic passages in Abbé Grégoire's "Essai" – show stereotyping of Jews as rational, calculating, and detached from nature; they do not reveal any influence by Voltaire. Calls for a re-evaluation of the relationship between the Enlightenment and the Jews. *DR*

0147 Shoulson, Jeffrey S.: "Proprietie in This Hebrew Poesy": George Wither, Judaism, and the Formation of English National Identity. *Journal of English and Germanic Philology* 98, 3 (July 1999) 353-372.

Discusses the role of Hebrew, Judaism, and Jews in 17th-century England, focusing on how Hebrew studies shifted from a "politically, religiously, and intellectually liberating" force to a means to justify new forms of prejudice and repression. Wither (1588-1667) was famous among his contemporaries for his metrical translation of the Psalms and for his prose writings about them. Like other English Protestants of his time, he had a very ambiguous attitude towards Judaism and the Jews. He recognized the continuity between biblical Judaism and Christianity, and the importance of the former for English identity; however, he felt the need, as did the early Christians, to distance his society from Judaism and Jews. The latter factor engendered antisemitic attitudes. For Wither, the Psalms embodied this continuity and distance, which are indicated by the use of the term Old Testament for the Hebrew Bible. *YC*

0148 Sotomayor y Valenzuela, Luis Joseph de: *Brève relation de l'expulsion des Juifs d'Oran en 1669.* Trans. & ed.: Jean-Frédéric Schaub. Saint-Denis: Bouchène, 1998. 77 pp. Original title: "Breve relación y compendioso epitome de la general expulsión de los Hebreos...."

Pp. 7-26 contain the editor's introduction to this text which was written in 1670. Describes the figure of its author (an army captain), the situation in Oran before the expulsion, and its causes as presented in the text. In 1669 Oran (Algeria) was a Spanish colony. The town's population numbered 6,000, mostly Christians. There were about 450 Jews, but they possessed more than one thousand Muslim slaves, so Jewish influence was significant. The "Brève relation" was written on the order of Marquis de Los Vélez, the governor of Oran. Sotomayor gives mainly religious reasons for the expulsion, writing that it was "a just action," carried out for the "dignity of the Catholic region" and the "purity of the Catholic faith." Schaub explains that there were also economic reasons (e.g. envy of Jewish goods), as well as political ones. The Spanish authorities feared that the Jews could make an alliance against them with various tribes or even with the Turkish authorities of Alger. States that all extant sources on this expulsion are of Christian origin and favor the expulsion. The expelled Jews settled in Livorno and Villefranche (Nice). *HV*

0149 Wyczanski, Andrzej: Some Reflections on Accusations against the Jews in Poland in the Early Modern Period. *Varieties of Multiculturalism in Modern European History: The Case of the Jews,* eds. Michael Heyd, Oded Heilbronner. Jerusalem: International Center for University Teaching of Jewish Civilization, Hebrew University, 1998. Pp. 63-72.

There was a small but steady growth in ritual murder accusations, and to a lesser extent in host profanation accusations, against Jews in Poland in the 16th-18th centuries. This growth, and the outbursts of popular Judeophobia which accompanied the ensuing trials, can be explained neither by the economic situation in the country nor by the legal situation of the Jews nor by identifying Jews with heretics or Protestants. It is the accusations of witchcraft (and the witch-hunts) which can be compared to the accusations against and persecution of the Jews. In the eyes of both common people and the lower clergy, witches and Jews were agents of the Devil. Popular fear of Jews resembled fear of witches. The fear was channelled as aggression against Jews. In any event, the prejudices and persecution against Jews in Poland were similar to those in other Catholic countries. *DR*

The Modern Period (1789-1999)

General

0150 Adorno, Theodor W.: *"Ob nach Auschwitz noch sich leben lasse": Ein philosophisches Lesebuch.* Ed.: Rolf Tiedemann. Frankfurt a.M.: Suhrkamp, 1997. 569 pp.

A collection of essays and extracts, all published previously, on a wide range of topics. The editor's introduction, "'Nicht die Erste Philosophie sondern eine letzte': Anmerkungen zum Denken Adornos" (pp. 7-27), summarizes the essence of Adorno's philosophy and notes the impact of the Holocaust on his thought. Parts of the following deal with antisemitism and the Holocaust: Was bedeutet: Aufarbeitung der Vergangenheit [Appeared in the protocols of a conference with the same title (Frankfurt a.M.: Deutscher Koordinierungsrat der Gesellschaften fuer Christlich-Juedische Zusammenarbeit, 1960).] (31-47); Erziehung nach Auschwitz [Appeared in "Zum Bildungsbegriff der Gegenwart" (Frankfurt a.M.: M. Diesterweg, 1967).] (48-63); Das beschaedigte Leben [Selections from "Minima Moralia" (Berlin: Suhrkamp, 1951).] (65-129); Die Wunde Heine [Appeared in "Texte und Zeichen" 2 (1956).] (251-255); Elemente des Antisemitismus: Grenzen der Aufklaerung [Written with Max Horkheimer. Appeared in "Dialektik der Aufklaerung" (Amsterdam: Querido, 1947).] (433-471). *RW*

0151 Arvidsson, Stefan: Aryan Mythology as Science and Ideology. *Journal of the American Academy of Religion* 67, 2 (June 1999) 327-354.

Discusses attempts at creating a religion for the "Indo-European race," or at "reconstructing" an Aryan mythology undertaken in the 19th-20th centuries. These attempts were accompanied by reflections on the inherent distinctiveness between the Aryan and Semitic religions and mythologies, and even on the inability of the Semites to create a myth (Max Mueller, Ernest Renan, etc.). The call to create a national German religion came from Paul de Lagarde; he was the first to oppose the Germans not to "Rome" but to the Jews. The task of creating an Aryan religion, if it did not go along with paganism, meant an attempt to rid Christianity of its Jewish-Semitic roots, which was ventured by Renan and de Lagarde. The "voelkisch," and later Nazi, version of the opposition between "Aryan" and "Semitic" religions was different from that of Mueller-Renan – the former saw Semitic religion as life-rejecting and unheroic. The downfall of the Third Reich did not put an end to the attempts to "recreate" an Aryan mythology. *DR*

0152 Boym, Svetlana: Conspiracy Theories and Literary Ethics: Umberto Eco, Danilo Kiš and "The Protocols of Zion." *Comparative Literature* 51, 2 (Spr 1999) 97-122.

Discusses "The Protocols of the Elders of Zion" in terms of a conspiratorial worldview and a persistence of paranoia, as demonstrated by the continued popularity of the "Protocols" after they were conclusively revealed to be a forgery. Discusses two modernist works, Eco's novel "Foucault's Pendulum" (1988) and Kiš's short story "The Book of Kings and Fools" (1983), that explore the delights and dangers of paranoiac thinking in literature and life. The "Protocols" is an example of this phenomenon in literature that has had and continues to have a lethal influence on life. *YC*

0153 Brenner, Michael; Liedtke, Rainer; Rechter, David, eds.: *Two Nations: British and German Jews in Comparative Perspective.* Tuebingen: Mohr Siebeck, 1999. ix, 504 pp. (Schriftenreihe wissenschaftlicher Abhandlungen des Leo-Baeck-Instituts, 60).

Papers (and responses) delivered at a conference held at Clare College, Cambridge, in September 1997. They compare the historical experiences of British and German Jews from the 18th century to the 1930s, with an emphasis on recent historiography on these two countries. The following deal with antisemitism: Ruerup, Reinhard: Jewish Emancipation in Britain and Germany (49-61); Cesarani, David: Jewish Emancipation: From Teleology to a Comparative Perspective: A Comment on Reinhard Ruerup (63-66); Kushner, Tony (Antony): Comparing Antisemitisms: A Useful Exercise? (91-109); Rahden, Till van: In Defence of Differences: A Comment on Tony Kushner (111-115); Gartner, Lloyd P.: East European Jewish Migration: Germany and Britain [Deals, also, with immigration restrictions and xenophobia.] (117-133); Maurer, Trude: Changing Conditions, Changing Responses: A Long-Term Perspective on Immigrant and Native Jews: A Comment on Lloyd P. Gartner (135-139); Feldman, David: Jews and the State in Britain [Pp. 154-160 discuss discriminatory measures against Jews as "aliens" in the early 20th century.] (141-161); Clark, Christopher M.: The Jews and the German State in the Wilhelmine Era [Discusses, also, discriminatory practices.] (163-184); Hoffmann, Christhard: Boundaries of Citizenship: A Comment on David Feldman and Christopher Clark (185-192); Endelman, Todd M.: Jewish Self-Hatred in Britain and Germany [Notes that expressions of self-hatred were rare in Britain, but depressingly common in Germany.] (331-363); Mendes-Flohr, Paul Robert: Jewish Self-Hatred and the Dialectics of Assimilation: A Comment on Todd M. Endelman (365-369); Robertson, Ritchie: The Representation of Jews in British and German Literature: A Comparison (411-441); Timms, Edward: The "Enemy

Within" and the "All-Purpose Foreigner": A Comment on Ritchie Robertson (443-449). *SSC*

0154 Ferrand, Pierre: The Continued Impact of "The Protocols." *Midstream* 44, 2 (Feb-Mar 1998) 17-20.

"The Protocols of the Elders of Zion" became popular in the aftermath of World War I; it was used by generals fighting the Bolshevik regime in Russia, and by German right-wingers (including the Kaiser in exile in Holland) to explain their defeat in the war. Although the work was internationally popular, it was eventually refuted and marginalized in England, the USA, and even in Germany during the Weimar regime. It became part of official gospel in Germany after Hitler's rise to power. In recent years, in the West, accusations of Jewish conspiracies have not spawned pogroms because antisemitic thugs have not received support from governments or "respectable" public opinion. The situation is different in the Muslim world, where dissemination of the "Protocols" is sponsored by governments and supported by spiritual and intellectual leaders. *REK*

0155 Finzi, Roberto: *Anti-Semitism: A Century in Focus.* Trans.: Maud Jackson. Moreton-in-Marsh, Gloucestershire: Windrush Press, 1999. 127 pp. Originally published as "L'antisemitismo: Dal pregiudizio contro gli ebrei ai campi di sterminio" (Firenze: Giunti – Casterman, 1997).

0156 Gilman, Sander L.; Shain, Milton, eds.: *Jewries at the Frontier: Accommodation, Identity, Conflict.* Urbana: University of Illinois Press, 1999. 401 pp.

Partial contents: Leveson, Marcia: The Enemy Within: Some South African Jewish Writers [Deals with South African antisemitism and Jewish self-hatred as evinced in the works of Sarah Gertrude Millin, Nadine Gordimer, and Jillian Becker.] (242-258); Stratton, Jon: The Color of Jews: Jews, Race, and the White Australia Policy (309-334); Shain, Milton: Ethnonationalism, Anti-Semitism, and Identity Politics: The North American and South African Experiences (335-350). *DR*

0157 *Justice* 18 (Fall 1998).

This issue includes three lectures delivered at a conference on "Anti-Semitism and Holocaust Denial towards the 21st Century," convened by the International Association of Jewish Lawyers and Jurists in Salonika, June 1998. The conference commemorated the Jewish community of Salonika, which was almost wiped out by the Holocaust. Contents: Nener, Itzhak: [Opening remarks – on the deportation of Salonika's Jews to Auschwitz and on the dangers of antisemitism and Holocaust denial.] (26-

27, 44); Ahlmark, Per: Sweden: From Indifference to Protecting Memory [On actions taken in Sweden in the 1990s to commemorate the Holocaust and combat Holocaust denial.] (28-31); Margaritis, George: Anti-Semitism in Greece [In the 19th-20th centuries.] (32-34). *DR*

0158 Kaplanov, Rashid M.; Mochalova, V.V.; Chulkova, L.A., eds.: *Yevreiskaya tsivilizatsiya: Problemy i issledovaniya: Materialy konferentsii* [Jewish Civilization: Problems and Research: Proceedings of a Conference]. Moskva: Tsentr Nauchnykh Rabotnikov i Prepodavatelei Iudaiki v vuzakh "Sefer," 1998. 319 pp.

Papers presented at the Fifth International Conference on Jewish Studies held in Moscow in 1997. Partial contents: Klier, John Doyle: The Balta Pogrom and Its Aftermath [The pogrom broke out in 1882 in Balta, Ukraine.] (116-125); Milyakova, Lidiya V.: Novaya vlast i yevreiskii vopros v Polshe v 1945-1947 gg. [The New Regime and the Jewish Question in Poland in 1945-1947] (134-138); Elyashevich, Dmitrii A.: Yevreiskii teatr v Rossii i tsenzura [The Jewish Theater in Russia and Censorship. On the decree of 1883 prohibiting stage performances in Yiddish.] (256-266); Murchie, Alison: The Imperial War Museum's Holocaust Exhibition: A Major Exhibition for the New Millennium (295-299). *DR*

0159 Katz, Jacob: World War I – a Crossroads in the History of European Jewry. *Yad Vashem Studies* 27 (1999) 11-21. Appeared in Hebrew in "Yad Vashem: Kovetz mehkarim" 27 (1999).

Translation of a keynote lecture delivered in Hebrew in March 1989 at the conference "European Jewry in View of Fascism and National Socialism" (the proceedings were not published) held at Yad Vashem. Reflects on the impact of the World War I and interwar periods on Jewish history. It was that war which destroyed the faltering equilibrium between Jews and non-Jews in Europe and eventually brought about the Holocaust. *DR*

0160

כ"ץ, יעקב: מלחמת העולם הראשונה – פרשת דרכים בתולדות יהודי אירופה. יד ושם 27 (1999) 9-16.

[Katz, Jacob: World War I – a Crossroads in the History of European Jewry. *Yad Vashem* 27 (1999) 9-16.] Appeared in English in "Yad Vashem Studies" 27 (1999).

0161 Kempter, Klaus: *Die Jellineks 1820-1955: Eine familienbiographische Studie zum deutschjuedischen Bildungsbuergertum.* Duesseldorf: Droste, 1998. viii, 631 pp. Based on the author's diss. – Universitaet Heidelberg, 1996.

Traces the history of all of the family, but focuses on Adolf Jellinek, preacher and leader of the Jewish community in Vienna, and his son Georg and Georg's son Walter, prominent jurists at the University of Heidelberg. Adolf Jellinek belonged to a group of Christians and Jews in Leipzig who worked for equal rights. He believed that better knowledge of the Other would remove prejudice, and adhered to this belief even when antisemitism in Austria burgeoned from the 1880s onward. At that time he published articles and pamphlets against antisemitism, but left the leadership of the fight to the more militant Joseph Samuel Bloch. When the University of Vienna blocked his advancement, Georg moved to the University of Heidelberg, where his son Walter taught after him. Walter considered himself a Protestant German. When the Nazis came to power, he at first justified their actions, even their antisemitism. He attempted to prove that the Jellineks were descended from Bohemian peasants who had converted to Judaism. While his case was being processed (until 1941), he was partially exempted from anti-Jewish measures. But in 1935 he had been dismissed from the university and his works were banned. His brother died after imprisonment and by the Gestapo; his sister was deported to Theresienstadt, but survivėd. He himself was protected by his marriage to an "Aryan." *RW*

0162 Romano, Sergio: *Lettera a un amico ebreo.* Milano: Longanesi, 1997. 152 pp.

A collection of brief essays on various subjects. The three essays on pp. 55-74 discuss the case of Arnaldo Momigliano, a Jewish Italian scientist and a fascist, who fell victim to the Racial Laws of 1938. On pp. 93-97, "Gli equivoci dell'antisemitismo," explains the differences between the terms antisemitism, judeophobia, and banal hostility. The three essays on pp. 119-133 discuss the attitude of the Catholic Church during the Holocaust, the views of John Paul II, the Carmelite convent at Auschwitz, and the canonization of Edith Stein. Pp. 135-139, "La conquista del passato," warn against disconnecting the Holocaust from its historical context and against imposing collective guilt on the Germans. *MZ*

0163 Rubinstein, William David; Rubinstein, Hilary L.: *Philosemitism: Admiration and Support in the English-Speaking World for Jews, 1840-1939.* Basingstoke, Hants: Macmillan Press, 1999. xiii, 276 pp.

Pt. I (pp. 1-108), "Philosemitic Reactions to Jewish Crises," deals with reactions against antisemitism of religious and political leaders, as well as others, mainly in England, but also in the U.S., Australia, and Canada, in 1840-1945. Focuses on public meetings, articles in and letters to the press, and fundraising, not only in capitals but also in the periphery. Attempts were made to oppose the antisemitism in Damascus (1840), Morocco (1859), the Vatican (the Mortara Affair, 1858-59), Romania (the 1870s),

Russia (1881-1906 and the Beilis case in 1911-13), France (the Dreyfus Affair), and the U.S. (the Leo Frank case). Describes how the British public, during World War II, showed strong and widespread opposition to Hitler's antisemitism and how England proved itself willing to accept a large number of refugees from Germany and Austria. Expresses the opinion that, later, only winning the war against Hitler could have saved more Jews. Pt. II (pp. 109-185), "A Typology of Philosemitism," indicates how broad a range of Anglo-Saxon ideologies were compatible with opposition to antisemitism. The epilogue (pp. 187-203) expresses the view that the establishment of the State of Israel has reduced antisemitism and that there is reason to believe that this phenomenon will disappear. *YC*

0164 Rubinstein, William David; Rubinstein, Hilary L.: Philosemitism in Britain and in the English-Speaking World, 1840-1939: Patterns and Typology. *Jewish Journal of Sociology* 40, 1-2 [1998] (1999) 5-47.

First, presents reactions in Britain and other English-speaking countries to the Damascus, Mortara, and Dreyfus Affairs, to Russian pogroms in 1881-82 and 1905, and to antisemitism after World War I and in the Nazi era. The roster of non-Jews, including political and religious leaders and intellectuals, who took part in rallies, petitions, etc., contains hundreds of names. Then, presents a typology of philosemitism consisting of four categories: Liberal (which considers religion a private matter); Protestant (which was affected by the Bible and identification either with the ancient Israelites or Jewish suffering); Gentile Zionist; and Elitist/Conservative (which stresses Jewish contributions to society). Concludes that, despite the reality and force of antisemitism, there is another, much less well-known, side of the coin in the modern English-speaking world, one "perhaps fully as large and as important as hostility toward the Jews." *YC*

0165 Sammons, Jeffrey L., ed.: *Die Protokolle der Weisen von Zion: Die Grundlage des modernen Antisemitismus – eine Faelschung: Text und Kommentar*. Goettingen: Wallstein, 1998. 128 pp.

The editor's introduction (pp. 7-26) points out that the "Protocols" is an obvious fiction. Traces the text back through Hermann Goedsche's novel "Biarritz" (1868) to Maurice Joly's "Dialogue aux Enfers" and possibly an unknown previous French text. The "Protocols" appeared first in Russia in various versions, the most influential being that by Sergei Nilus. He associated the Jewish conspiracy with the Antichrist, with absolute evil in cosmic combat against the divine order. Traces the many versions and editions of the "Protocols" published in the 1920s-30s, noting that it was especially influential in Germany; Hitler saw in it a proof of the Jewish peril and a tool for Nazi propaganda. Also surveys refutation of the authenticity of the "Protocols," including the proceedings of the Swiss trial in 1934-35. Asserts that after 1945 the "Protocols" almost disappeared in the West, but it was revived in Russia and in the Arab states and is now disseminated on the Internet. The editor's postscript (pp. 113-118) places the "Protocols" in the context of anti-revolutionary propaganda and beliefs in a Jewish conspiracy that arise in crisis periods and are again thriving worldwide. Pp. 119-127 contain an extract from "Biarritz." *RW*

0166 Schindel, Robert: Schweigend ins Gespraech vertieft: Anmerkungen zu Geschichte und Gegenwart des juedisch-nichtjuedischen Verhaeltnisses in den Taeterlaendern. *Text + Kritik* 144 (Oct 1999) 3-8.

Recalls the total identification of German and Austrian Jews with the culture of their countries, and their blindness to antisemitism and the approaching Holocaust. Describes these countries' rejection of the survivors, and the beginning of dialogue in the second generation, soon broken off because of sympathy for the Palestinians. Asserts that there never was, and cannot now be, German-Jewish or Austrian-Jewish symbiosis, and that Jews must maintain the memory of the real Holocaust against all the empty chatter that obscures it. *RW*

0167 Stegemann, Ekkehard W.: "The Sudden Appearance of the Jews in 'Secular' History": Zionism and Christianity. *Jewish Studies* 38 (1998) 57-72. Appeared in German in "Kirche und Israel" 12 (1997).

0168 Stegemann, Ekkehard W.: Zionismus und Christentum. *Kirche und Israel* 12, 2 (1997) 106-112. Appeared also in the "Internationales Theodor Herzl Symposion" (2nd, Vienna, 1997).

Christianity's attitude toward Zionism is seen to be ambivalent in two respects. Christianity failed to recognize the role its own anti-Judaism played in encouraging modern political antisemitism, and it rejected the establishment of the Jewish state because the Church continued to view itself as the true Israel. Christian theologians, both in Herzl's time and after 1948, expected the ingathering of the Jews to lead them to Jesus. Protestants as well as Catholics found it hard to accept a Jewish state. Christians admitted guilt not for antisemitism but for ineffective missionizing among the Jews. In the 1960s Christians began acknowledging their responsibility for antisemitism and realizing the need for new attitudes toward Zionism and Israel. Still, traditional ambivalences have resulted in anti-Jewish stereotypes being transferred to the Jewish state. Expresses hope that Zionism will be accepted as the liberation of the Jews rather than as the continuation of Christian messianic history. *YC*

0169 Stegmaier, Werner; Krochmalnik, Daniel, eds.: *Juedischer Nietzscheanismus.* Berlin: Walter de Gruyter, 1997. xxxi, 476 pp.

Based on papers delivered at a conference in Greifswald, September 1995. Partial contents: Hartwich, Wolf-Daniel: Judentum als "Rasse" bei Nietzsche und Freud: Zur Rezeption einer anthropologischen Kontroverse des 19. Jahrhunderts (288-302); Aschheim, Steven Edward: Nietzsche, Antisemitismus und Massenmord [Appeared in English in his "Culture and Catastrophe" (1996).] (384-404); Zimmermann, Moshe: Nietzsche in Israel [Primarily on the controversy over whether Nietzsche provided ammunition for antisemitism and Nazism.] (405-423). *RW*

0170 Stockhammer, Robert: Van Helsing, die E.T.s, die Juden und die Nazis: Marginalien zum Verhaeltnis von Okkultismus und Antisemitismus. *Jahrbuch fuer Antisemitismusforschung* 6 (1997) 285-291.

Discusses the book "Geheimgesellschaften und ihre Macht im 20. Jahrhundert oder Wie man die Welt nicht regiert: Ein Wegweiser durch die Verstrickungen von Logentum mit Hochfinanz und Politik," by Jan van Helsing (pseud.), published in Playa del Ingles, Canary Islands, in 1995. States that, historically, occultism was often associated with antisemitism; occultists had a share in the genesis of "The Protocols of the Wise Men of Zion." As the epithet "wise" suggests, their antisemitism was tinged with respect for Jewish secret wisdom, but also with hostility toward a dangerous rival. Van Helsing sees almost every international organization or movement as part of a conspiracy to dominate the world, and finds Jews active in them all. However, he leaves the reader free to reject this last detail; the important point for him is the conspiracy, with or without Jews. He believes that Hitler and the Thule Gesellschaft were occultists who saw through the machinations of the Jews, but fought them with the wrong means: in their hatred, they used the same weapons they condemned in the satanic Jewish God. *RW*

0171 Toscano, Mario, ed.: *Integrazione e identità: L'esperienza ebraica in Germania e Italia dall'Illuminismo al fascismo.* Trans.: Marina Astrologo. Milano: FrancoAngeli, 1998. 267 pp.

Papers delivered at a conference in Rome, November 1993. Some of the articles were translated into Italian for this volume. Partial contents: Moro, Renato: Le Chiese, gli ebrei e la società moderna: L'Italia [Part of this article appeared in "Storia e Problemi Contemporanei" 7 (1994).] (167-182); Jochmann, Werner: Chiese, ebrei e società moderna (183-193); Berding, Helmut: L'antisemitismo nella società moderna: Continuità e discontinuità [Appeared in German in "Nation und Gesellschaft in Deutschland" (1996).] (194-209); Toscano, Mario: L'uguaglianza senza diversità: Stato, società e questione ebraica nell'Italia liberale (210-235); Voigt, Klaus: I profughi ebrei in Italia (1933-1945) (244-267). *MZ*

0172 Tyson, Joseph B.: *Luke, Judaism, and the Scholars: Critical Approaches to Luke-Acts.* Columbia, SC: University of South Carolina Press, 1999. xi, 196 pp.

Examines the history of critical scholarship on two New Testament texts, the Gospel of Luke and the Acts of the Apostles, with particular attention to the ways in which scholars have interpreted the ancient authors' views on Jews and Judaism. There was a strong tradition in New Testament scholarship of the 19th century to portray 1st-century Judaism as moribund and casuistic, emphasize its negative qualities, depreciate the importance of its positive aspects, and deny a continuity between Judaism and Christianity. This tradition influenced such early German scholars as Ferdinand Christian Baur (1792-1860), Adolf von Harnack (1851-1930), and Adolf Schlatter (1852-1938), who read Luke-Acts as an anti-Judaic document. Both Ernst Haenchen (1894-1975) and Hans Conzelmann (1915-1989) questioned this tradition on many points; however, they emphasized negative aspects of Judaism in their treatment of Luke and Acts. The Holocaust was a turning point in critical scholarship of the New Testament, and since 1945 the Norwegian Jacob Jervell and a number of American scholars have assessed Luke's attitude toward Judaism more positively. *DR*

0173 Wagner, Gottfried: *Twilight of the Wagners: The Unveiling of a Family's Legacy.* Trans.: Della Couling. New York: Picador, 1999. viii, 310, [16] pp. Originally published as "Wer nicht mit dem Wolf heult: Autobiographische Aufzeichnungen eines Wagner-Urenkels" (Koeln: Kiepenheuer & Witsch, 1997).

0174 Wyrwa, Ulrich: "Holocaust": Notizen zur Begriffsgeschichte. *Jahrbuch fuer Antisemitismusforschung* 8 (1999) 300-311.

Traces the meaning of the word "holocaust" from its use for "burnt offering" in the Greek and Latin translations of the Bible, to its first use for mass slaughter in English, Spanish, and Italian sources in the 19th century. In 1882 an Italian historian used it to describe an anti-Jewish pogrom. It was applied to the Nazi genocide of the Jews as soon as it was known: in 1942, for instance, by the London "News Chronicle"; in 1943 by Sir Herbert Louis Samuel in a speech in the House of Lords; in 1945 in an American publication on restitution. The first scholarly publication to use it was the "Yad Vashem Bulletin" in 1957; English-language encyclopedias began to use the heading in the 1970s. In Germany the term took hold much more slowly, beginning with the TV series of that name.

Notes that many, both Jews and non-Jews, object to the term because it seems to identify mass murder with a sacred rite. *RW*

1789-1999: By Country

Austria

0175 Ellmauer, Daniela; Embacher, Helga; Lichtblau, Albert, eds.: *Geduldet, geschmaeht und vertrieben: Salzburger Juden erzaehlen.* Salzburg: Otto Mueller Verlag, 1998. 288 pp.

Sixteen testimonies of Jews who spent part of their lives, usually their childhood, in Salzburg. They describe the town in prewar days, their families, and the Jewish community, as well as their lives after emigration and the fate of their relatives. Many mention antisemitism at school, social isolation even before the Anschluss, and, after it, internment of fathers in Dachau, Aryanization of businesses, expulsion to Vienna to make Salzburg "judenrein," and emigration. Hans Jacoby (pp. 89-110) also relates how he survived in hiding in the Netherlands with the help of the resistance. The introduction (pp. 7-22), "Vom Neubeginn zur Vertreibung der Salzburger Juden (1867-1938)," sketches the history of the Jews in Salzburg in modern times. After the expulsion of 1498, it was only in 1867 that the new constitution forced the town to accept its first Jewish resident. Liberalism was short-lived; from the 1880s, and increasingly after World War I, Jews were excluded from society. Summarizes elements that appear frequently in the reminiscences. *RW*

0176 Gombrich, Ernst Hans: *Juedische Identitaet und juedisches Schicksal: Eine Diskussionsbemerkung.* Eds.: Emil Brix, Frederick Baker. Trans.: Dorothea McEwan. Wien: Passagen, 1997. 78 pp. Originally published as the first volume of "Occasions" (London: Oesterreichisches Kulturinstitut, 1997).

The introduction, by Brix (pp. 11-29), discusses Gombrich's views on Jewish identity and Jewish fate in Austria as presented at a seminar organized by the Oesterreichische Kulturinstitut in November 1996 in London on the theme "Fin-de-siècle Vienna and Its Jewish Cultural Influences." Pp. 33-64 contain the text of Gombrich's two lectures: "Zum Wiener Kunstleben um 1900" (pp. 33-54) and "Betrachtungen zur Tragoedie des Judentums" (pp. 55-64). In the latter part, he contends that the concept of Jewish culture was invented by Hitler and his followers. This provocative view is based on the ambiguity he attributes to the term "Jew." For him, the concept of Judaism means religion, exclusively. The life style of East European Jews had nothing in common with that of assimilated German and Austrian Jews. He claims that only the Nuremberg Laws offered a definition which included all shades of German and Austrian Jewry, based on the doubtful notion of race, while in the self-comprehension of assimilated Jews there was no such idea of a common Jewish culture. Pp. 65-72 contain a summary, by Baker, of the discussion which followed Gombrich's lectures. *ID*

0177 Lettl, Joseph: *Kirche und Judentum in Wien: Der Wandel des Verhaeltnisses seit 1782.* Klosterneuburg: Mayer, 1997. 240 pp.

Discusses changes in the attitudes of the Catholic Church in Austria toward Jews after 1782. Contends that the Church's loss of power, after the period of Enlightenment and the resulting Jewish emancipation, was followed by embittered spiritual arguments which impregnated, also, the Viennese press, especially during the second half of the 19th century. Discusses the increase in traditional Christian antisemitism at the end of the 19th-beginning of the 20th century. Traces, also, Hitler's ambivalent approach to Catholicism and his increasing hatred of Jews. Mentions the inner conflict of Pope Pius XII, who deplored the plight of the Jews during the Holocaust, but did not make any public protest. Pp. 143-175 deal with the issue of the Church and the Jews during the Nazi period. Traces the stages of persecution of Jews in Vienna. Relates actions of rescue untertaken by the Church, headed by Cardinal Innitzer, for converts to Catholicism, who numbered, in 1941, 7,917 persons. Pp. 176-208 trace the Church's attitude toward the Jews in the postwar period. Notes the harmful potential of persisting antisemitism, in historiography, stereotypes, and myths. *ID*

0178 Sottopietra, Doris: *Variationen eines Vorurteils: Eine Entwicklungsgeschichte des Antisemitismus in Oesterreich.* Wien: Passagen, 1997. 264 pp.

Traces the rise of racist antisemitism in 19th-20th century Europe. Analyzes sociological factors in its diffusion and instrumentalization and the transition from ideology to action in Nazism. In Habsburg Austria and in the First Republic, all the political parties except the Social Democrats were antisemitic. Explains the Holocaust as a product of the existing antisemitic ideology, which became a substitute religion, and an efficient bureaucracy and technology. Describes the postwar repression of guilt among all Austrians as shown in opinion polls of the 1970s-80s. These also show strong antisemitism as gradually decreasing, but moderate antisemitism and traditional stereotypes still widespread. Austria is the only Western country in which antisemitism was exploited by a major political party (in Waldheim's election campaign). This aroused a discussion on Austria's share in the Holocaust. Though subsequent leaders have expressed contrition, a large percentage of the population still clings to the old stereotypes. Discusses, also, the influence of antisemitism and the Holocaust on Zionism, the new Israeli identity, and Israel's fear of its Arab neighbors. *RW*

France

0179 Miething, Christoph, ed.: *Juedischer Republikanismus in Frankreich.* Tuebingen: Max Niemeyer Verlag, 1998. xii, 184 pp.

Based on papers delivered at a section of the 24th Deutscher Romanistentag, Muenchen 1995. Partial contents: Loewe, Siegfried: Bernard Lazare: Von der Assimilation zum Zionismus [In his early writings, Lazare reviled non-assimilated Jews, using antisemitic stereotypes; but, after awakening to the antisemitism in French society, he became a militant advocate of a Jewish nation combatting antisemitism by fighting for freedom and social justice.] (1-17); Gipper, Andreas: Julien Benda und der Patriotismus des Universalen: Bemerkungen zu einer juedisch-republikanischen Synthese [Criticizes Benda for rejecting Jewish particularism in favor of an abstract universalism, which he saw as rooted in Jewish tradition, and for ignoring antisemitism or interpreting it (even during the Dreyfus Affair and the Holocaust) as only an expression of anti-universalism.] (61-77); Spire, Marie-Brunette: Das politische Denken André Spires [Influenced by his experience of antisemitism and his conviction that the Jews must defend their honor.] (107-122); Uran, Steven: Antisemitismus und Republikanismus im Algerien der Kolonialzeit [Among the French settlers.] (123-137); Coenen-Mennemeier, Brigitta: "Nous sommes tous des Juifs allemands": Mai 68 und die literarische Behandlung der "question juive" [On the treatment of antisemitism in Patrick Modiano's "La Place de l'Etoile" and on the work of Marguerite Duras, particularly "Détruire dit-elle."] (138-153); Arnold, Rafael: Un Juif français – un Français juif? Juedische Identitaet und Selbstdarstellung in Raymond Arons "Mémoires" [Aron was forced to face his Jewish identity by a series of antisemitic events in France and by Sartre's book on antisemitism.] (154-163). *RW*

0180 Pierrard, Pierre: *Juifs et catholiques français: D'Edouard Drumont à Jacob Kaplan (1886-1994).* Paris: Cerf, 1997. 456 pp. An earlier edition appeared as "Juifs et catholiques français: De Drumont à Jules Isaac (1886-1945)" (Paris: Fayard, 1970).

Catholic antisemitism spread in France after the publication of Drumont's "La France juive" (1886) and of the antisemitic journal "La Libre Parole" from 1892, which he edited. Towards the end of the 19th century, numerous nationalist, anti-Masonic and antisemitic leagues were established. Surveys the publications, organizations, and leaders, dealing with their anti-Dreyfusard views. The most important Catholic antisemitic and anti-Dreyfusard movement was Charles Maurras' Action Française, which continued its activities into the interwar period, promoting political antisemitism. Relates, also, to the group of Christian Dreyfusards led by Charles Péguy. During the period of World War II, many French Catholics were fascinated by the personality of Pétain. The "Statute of the Jews" of 1941 did not provoke any official reaction from Catholic authorities. Catholic resistance during the war was centered in Lyon, where "L'Amitié chrétienne," an organization which helped persecuted Jews, was founded. Relating to the postwar period, discusses the beginnings of Christian-Jewish dialogue, the activities and writings of Jules Isaac, the founding in 1948 of "Amitié judéo-chrétienne," and other initiatives which led eventually to the "Nostra aetate" declaration. The last chapter presents people and organizations involved in Christian-Jewish dialogue in the post-Vatican II period. *HV*

0181 Pinard, Joseph: *Antisémitisme en Franche-Comté: De l'affaire Dreyfus à nos jours.* Besançon: Cêtre, 1997. 310 pp.

Examines the response of the regional press in Franche-Comté to events involving Jews, particularly the Dreyfus Affair, the Beilis case, the rise of French socialism in the 1930s, and the Holocaust. Traces the history of the Ligue Antisémitique de Vesoul and the publication of its monthly journal "Vesoul Antijuif" from 1899-1900 in protest against Zola's defense of Dreyfus. Discusses the influence of Abbé Brugerette, who defied clerical pressure in his support of Dreyfus. Documents the growing hostility in the late 19th century between Jews and the military in the town of Belfort, which often culminated in duels. Contrasts antisemitic references to Jewish politicians in "L'Eclair Comtois" with its publication of editorials and letters protesting persecution of Jews during the Nazi occupation. Concludes that tensions between Jews and officials of the Franche-Comté may be exacerbated by the current controversy over the growing influence of a former Nazi, Roland Gaucher, in the Front National. His vociferous protest against the "Jewish international conspiracy" was emphasized during his candidacy in the regional elections of 1992, and highlighted in his book, "La montée du Front National" (1997). *TS*

0182 Vérard, René: *Jean Pierre-Bloch, un français du monde entier.* Orléans: Corsaire, 1997. 251 pp.

A biography of the French Jew, journalist, and politician (1905-1995), based on his writings. A member of the Socialist Party, Pierre-Bloch fought all his life against racism and xenophobia. In 1929 he became a member of the LICA (Ligue Internationale Contre Antisémitisme). In 1930 he began to write for the socialist daily "Populaire." Describes political life at the end of the 1930s, and struggles in the Socialist Party in 1938-39 between the pro-Munich and anti-Munich wings. In 1941 Pierre-Bloch fled to England and joined the French government-in-exile. Deals, as well, with his postwar anti-racist activities. In 1968 Pierre-Bloch became the president of the LICA and director of the journal "Le Droit de Vivre." Pp. 193-231 relate to the situation in Algeria from 1870 to the 1960s, and the involvement of Pierre-Bloch in the struggle for equal rights for both Muslims and Jews. *HV*

0183 Wistrich, Robert Solomon: The Dreyfus Centenary. *The Jerusalem Review* 3 (1998-1999) 190-198.

Discusses the political significance and consequences of the Dreyfus Affair, stating that it presaged some of the horrors of the 20th century. It gave birth to the radical right and to an explosion of antisemitism in France, and "ultimately created the intellectual seeds of fascism in Europe." Antisemitism became an integrative ideology, uniting Frenchmen into a national front that transcended class divisions. The Affair strengthened the myths of the traitorous Jew and of the international Jewish syndicate, which supposedly attempted to exonerate Dreyfus. Anti-Jewish racist rhetoric of the time led to riots in many French cities and to a pogrom in French Algeria. States that Vichy's collaboration with the Nazis was the culmination of the anti-Dreyfusard tradition, an "indigenous French form of fascism which is far from exhausted." *REK*

Germany

0184 Aberbach, Moshe: The Failure of Jewish Emancipation in Europe. *A Life in Jewish Education: Essays in Honor of Louis L. Kaplan,* ed. Jack Fruchtman, Jr. Bethesda, MD: University Press of Maryland, 1997. Pp. 99-137.

Argues that the failure of Jewish emancipation in Europe, especially in Germany, was tragic: Jews did not understand that European society was intolerant and basically wanted the Jews to assimilate. Antisemitism was universal, characterizing both Left and Right, and the Protestant, Catholic, and Eastern Orthodox Churches. Many German intellectuals shared antisemitic presuppositions. German professors, even "liberal" ones like Mommsen, refused to recognize Jews as a national-religious entity. Political antisemites, like Karl Lueger, were influenced by racist antisemitism that advocated eliminating the Jews. Still, assimilated Jews refused to abandon their belief in emancipation. Predictions of disaster were ignored. Incompetent Jewish leaders, both lay and religious, either advised against emigration to Palestine (both Orthodox and Reform leaders opposed Zionism) or feared that supporting emigration activity would be seen as unpatriotic. The majority of European Jews were no wiser than their leaders. *YC*

0185 Alter, Peter; Baersch, Claus-Ekkehard; Berghoff, Peter, eds.: *Die Konstruktion der Nation gegen die Juden.* Muenchen: Wilhelm Fink Verlag, 1999. 290 pp.

A collection based on papers delivered at a conference of the Salomon Ludwig Steinheim-Institut fuer deutsch-juedische Geschichte and the Lehrstuhl fuer Neuere und Neueste Geschichte of Duisburg University, Duisburg, November 1996. All of the articles deal wholly or primarily with Germany. Partial contents: Mosse, George Lachmann: Die Juden im Zeitalter des modernen Nationalismus [Traces the evolution in Germany from a liberal, universalist patriotism that could include the Jews to a racist nationalism that excluded them.] (15-25); Rohrbacher, Stefan: Deutsche Revolution und antijuedische Gewalt (1815-1848/49) [Including pogroms.] (29-47); Dunk, Hermann Walter von der: Antisemitismus zur Zeit der Reichsgruendung: Unterschiede und Gemeinsamkeiten: Ein Inventar [Compares Germany with France and the Netherlands.] (65-91); Heid, Ludger: Sozialistischer Internationalismus, sozialistischer Zionismus und sozialistischer Antisemitismus (93-117); Hammerstein, Notker: Professoren in Kaiserreich und Weimarer Republik und der Antisemitismus (119-136); Bergmann, Werner: Nationalismus und Antisemitismus im vereinigten Deutschland [Based on polls conducted in the 1990s.] (137-155); Berghoff, Peter: "Der Jude" als Todesmetapher des "politischen Koerpers" und der Kampf gegen die Zersetzung des nationalen "Ueber-Lebens" [In 19th-century Germany.] (159-172); Jeismann, Michael: Der letzte Feind: Die Nation, die Juden und der negative Universalismus [In 19th-early 20th-century Germany.] (173-190); Baersch, Claus-Ekkehard: Die Konstruktion der kollektiven Identitaet der Deutschen gegen die Juden in der politischen Religion des Nationalsozialismus (191-223); Volkov, Shulamit: Nationalismus, Antisemitismus und die deutsche Geschichtsschreibung [Appeared in "Nation und Gesellschaft in Deutschland" (1996).] (261-271); Pulzer, Peter G.J.: Warum scheiterte die Emanzipation der Juden? (273-284). *RW*

0186 Backhaus, Fritz: "Hab'n Sie nicht den kleinen Cohn geseh'n?" Die Bilderwelt antisemitischer Postkarten vom Kaiserreich bis in die NS-Zeit: Ein Ausstellungsprojekt. *Jahrbuch fuer Antisemitismusforschung* 6 (1997) 313-323.

Discusses the function of picture postcards in spreading antisemitic stereotypes. The postcards made fun of Jewish names, physiognomy, body language, speech, and habits, making them so familiar that the Nazis could later summon them up by a mere hint (e.g. calling an enemy

"Isidore," a "Jewish" name). A popular series was based on a comic song, "Der kleine Cohn," on a Jew caught by his wife in a sexual escapade. *RW*

0187 Baumann, Arnulf H., ed.: *Auf dem Wege zum christlich-juedischen Gespraech: 125 Jahre Evangelisch-lutherischer Zentralverein fuer Zeugnis und Dienst unter Juden und Christen.* Muenster: LIT, 1998. 231 pp. (Institutum Judaicum Delitzschianum: Muensteraner Judaistische Studien, 1).

The Zentralverein was founded in 1871 as the Centralverein fuer Mission unter Israel. This collection examines its activities and evolutions from the 19th century to the present. Partial contents: Haufler-Musiol, Karin: 125 Jahre Zentralverein: Ein historischer Ueberblick (11-46); Lichtenberger, Hermann: Das christlich-juedische Gespraech: Wandlungen und Perspektiven (132-152); Roloff, Juergen: Folgerungen aus dem christlich-juedischen Gespraech fuer die Ekklesiologie (184-205). *RW*

0188 Beckmann, Volker: *Juedische Buerger im Amt Versmold: Deutsch-juedische Geschichte im westlichen Ravensberger Land.* Ed.: Rolf Westheider. Bielefeld: Verlag fuer Regionalgeschichte, 1998. 116 pp.

Traces the history of the Jewish community of Versmold in Kreis Halle, Westphalia, with 65 Jews in 1858 and 30 in 1930. In 1899 it was rumored that the new synagogue could not be built without Christian blood and that a Jew was waiting in the cellar of the Jewish department store for a Christian to fall through a trap door to be killed. Although the authorities tried to stop the rumor, people became afraid to buy in Jewish shops and Jews were attacked on the street. Surveys antisemitic organizations and parties in Germany from the late 19th century to 1933, and the organizations that combated antisemitism, in which Versmold Jews may have participated. Nazi and SS groups were organized in 1930, and in September of that year the Nazi party received 40.3% of the votes. After 1933 there was vandalization of Jewish homes, the spread of antisemitic propaganda, and an enforced boycott of Jewish shops. Traces the fate of families and individuals. *RW*

0189 *"Beseitigung des juedischen Einflusses...": Antisemitische Forschung, Eliten und Karrieren im Nationalsozialismus.* Frankfurt a.M.: Campus, 1999. 309 pp. (Fritz Bauer Institut: Jahrbuch zur Geschichte und Wirkung des Holocaust, 1998/99).

Partial contents: Papen, Patricia von: Schuetzenhilfe nationalsozialistischer Judenpolitik: Die 'Judenforschung' des "Reichsinstituts fuer Geschichte des neuen Deutschland" 1935-1945 [Under the direction of Walter Frank and Wilhelm Grau.] (17-42); Schiefelbein, Dieter: Das "Institut zur Erforschung der Judenfrage Frankfurt am Main":

Antisemitismus als Karrieresprungbrett im NS-Staat [The institute was sponsored by Rosenberg and directed during its first year by Wilhelm Grau.] (43-71); Wojak, Irmtrud: Das "irrende Gewissen" der NS-Verbrecher und die deutsche Rechtsprechung: Die "juedische Skelettsammlung" am Anatomischen Institut der "Reichsuniversitaet Strassburg" [On the trial in 1970 of Bruno Karl Beger and Wolf-Dietrich Wolff for selecting Auschwitz prisoners to be killed for August Hirt's collection of skeletons.] (101-130); Heimann-Jelinek, Felicitas: Zur Geschichte einer Ausstellung: "Masken. Versuch ueber die Schoa" [On an exhibition at the Jewish Museum of Vienna in 1997 of death masks of Jews which had been prepared, together with their skulls, for Josef Wastl's collection at the Vienna Natural History Museum in 1942. The viewers were filmed and at the end of the exhibition made to see themselves reacting to the Holocaust.] (131-145); Heschel, Susannah: Deutsche Theologen fuer Hitler: Walter Grundmann und das Eisenacher "Institut zur Erforschung und Beseitigung des juedischen Einflusses auf das deutsche kirchliche Leben" (147-167); Gregor, Birgit: Zum protestantischen Antisemitismus: Evangelische Kirchen und Theologen in der Zeit des Nationalsozialismus (171-200); Mommsen, Hans: Die Stellung der Militaeropposition im Rahmen der deutschen Widerstandsbewegung im "Dritten Reich" [Including their attitude toward the massacre of Jews.] (201-217); Kater, Michael Hans: Das Boese in der Medizin: Nazi-Aerzte als Handlanger des Holocaust [Including the development of racial "science," experiments on concentration camp prisoners, and victimization of Jewish physicians.] (219-239); Perels, Joachim: Antisemitismus in der Justiz nach 1945? [On the acquittal in two postwar trials of Nazi judges who had sentenced Jews to death for alleged sexual relations with "Aryan" women.] (241-252); Hampicke, Evelyn; Loewy, Hanno: "Juden ohne Maske": Vorlaeufige Bemerkungen zur Geschichte eines Kompilationsfilmes [On a Nazi film compiled from motion pictures featuring Jewish actors playing criminals or perverts, produced to accompany the exhibition "Der ewige Jude" (1937, Munich) and later released as a propaganda film.] (255-274); Gallwitz, Tim: "Unterhaltung – Erziehung – Mahnung": Die Darstellung von Antisemitismus und Judenverfolgung im deutschen Nachkriegsfilm 1946 bis 1949 (275-304). *RW*

0190 Boelke-Fabian, Andrea: "In der Falle der Ambivalenz": Ueberlegungen zu Identitaet und Projektion in Theodor Lessings Essay "Der Juedische Selbsthass" am Beispiel der Fallstudie ueber Maximilian Harden. *Zeitschrift fuer Religions- und Geistesgeschichte* 50, 3 (1998) 219-241.

Discusses Lessing's ambivalent attitude toward his Jewish origins on the basis of his book "Der Juedische Selbsthass" (1930), which presents the biographies of six Jewish personalities, revealing the tragedy of the self-hating Jew. In particular in his study of Maximilian Harden (1861-1927),

Lessing projects his own remaining unconscious identity conflicts into Harden's biography. Criticizing Harden's Jewish self-hatred permits Lessing to adopt a defensive attitude, not perceiving his own Jewish self-hatred. Besides, he does not mention Harden's or his own conversion to Christianity. *ID*

0191 Feld, Willi: *Judentum und Stadtentwicklung im Vergleich: Burgsteinfurt in Westfalen – Koeslin in Pommern.* Burgsteinfurt: Vereinigte Landsmannschaften Burgsteinfurt, [1997]. 80 pp.

Surveys and compares the history of Jews in Burgsteinfurt (Westphalia) and in Koeslin (Pomerania), from the 18th century until World War II. Discusses the aspirations of the Jews to emancipation, integration, and assimilation, their successes and their failures. Pp. 55-66 deal with the development of modern antisemitism from 1870, perceiving the Jew as a scapegoat for various problems in society. The Jew was victimized by stereotyped prejudices, fears, and aversions. The antisemitism reached a crisis point during the Weimar Republic, and led finally to the annihilation of European Jewry in the Holocaust. *ID*

0192 Genger, Angela; Griese, Kerstin, eds.: *Aspekte juedischen Lebens in Duesseldorf und am Niederrhein.* Duesseldorf: Mahn- und Gedenkstaette, 1997. 277 pp.

A collection of essays. The following include material on pre-Nazi antisemitism, and on Nazi persecution and Jewish reactions: Genger, Angela: Juden in Duesseldorf: Eine Einfuehrung (6-23); Prieur, Jutta: Juden am Unteren Niederrhein im 19. und 20. Jahrhundert (24-47); Sparing, Frank: "Im Mittelpunkt des Unterrichts steht das Juedische": Juedische Schulen in Duesseldorf (106-115); Klerks, Annette: "Wir sind voll Unruhe. Und das ist unser Glueck": Juedische Jugendgruppen in Duesseldorf 1902 bis 1938 (116-129); Zielke, Heiko: "Unsere Kraft unserem Volk": Makkabi und der juedische Sport in Duesseldorf 1924 bis 1938 (130-141); Griese, Kerstin: Die zionistische Bewegung in Duesseldorf: Von einer aktiven Minderheit zur wichtigen Auswanderungsinstanz (142-155); Sparing, Frank: Ostjuden in Duesseldorf (156-165); Id.: Juden in der Duesseldorfer Wirtschaft [Includes boycotts and Aryanization.] (166-175); Woelk, Wolfgang: Ausgeschaltet, vertrieben, ermordet: Juedische Aerzte in Duesseldorf (176-185); Kleinbongartz, Sigrid: Juedische Kuenstler in der Kuenstlergruppe Das Junge Rheinland in Duesseldorf [Includes their fate in the Nazi period.] (186-197); Jakobs, Hildegard: Juedische Musik? [On the exclusion of Jewish performers from concerts for the general public, and their activity in the Juedische Kulturbund and in the community.] (198-207); Piel, Ingo: Ein kritischer Kommentator: Siegfried Thalheimer (1899-1981), Journalist und Publizist [Thalheimer and Heymann published an anti-Nazi weekly, "Westland," in the Saar; both also wrote and lectured in defense of the Jews.] (212-219); Id.: "Dem juedischen Schicksal Widerstand leisten": Fritz Heymann (1897-1944), Journalist und Schriftsteller (220-225); Lekebusch, Sigrid: Evangelische Kirche und die Juden im Rheinland (234-241); Brzosa, Ulrich: Die Last der Geschichte tragen: Der Duesseldorfer Katholizismus zwischen traditionellem Antijudaismus und modernem Antisemitismus (242-255); Ehlert, Holger: Die Heine-Denkmaeler in Duesseldorf: Der beschwerliche Weg von Cararas Marmorbruechen nach Duesseldorf [On objection to a monument to Heine, partly due to antisemitism.] (256-269). *RW*

0193 Gierse, Klaus, ed.: *Juden – unsere aelteren Brueder im Glauben: Was haben die Juden mit der christlichen Oekumene zu tun? 27. Wuerzburger KV-Tage, 10.-12. Januar 1997.* Beckum: Kartellverband katholischer deutscher Studentenvereine, [1997]. 40 pp. (KV-Sonderdruck, N.F. 15).

The papers discuss the attitude of the Church towards Judaism throughout the ages and reflect on the common root of the two religions, which obliges Christianity to develop a new perspective on the Jews. Contents: Markus, Itzhak: Die Verheissung an das Volk Israel: Geschichte und Realisation heute (5-12); Mussner, Franz: Was haben die Juden mit der christlichen Oekumene zu tun? (13-25); Schwierz, Israel: Judentum in Deutschland 1945-1997 (26-32); Loehr, Wolfgang: Der KV und der Antisemitismus [From the late 19th century to post-World War II.] (33-40). *ID*

0194 Grupp, Peter: Juden, Antisemitismus und juedische Fragen im Auswaertigen Amt in der Zeit des Kaiserreichs und der Weimarer Republik: Eine erste Annaeherung. *Zeitschrift fuer Geschichtswissenschaft* 46, 3 (1998) 237-248.

States that the number of Jews and persons of Jewish descent in the Foreign Service of Wilhelmine and Weimar Germany is difficult to ascertain, as is the personal attitude of the officials toward Jews; probably they shared the usual rather confused antisemitism of the German upper classes, but they did not allow it to influence their work. Intervention on behalf of Jews – in Russia, Romania, Poland, Palestine – was determined by German national interests, not by sympathy for Jews. This pragmatic approach enabled the officials to adapt smoothly to the Nazi regime. *RW*

0195 Herf, Jeffrey: Not So Boring after All: Recent Trends in Political History of Twentieth Century Germany. *Tel Aviver Jahrbuch fuer Deutsche Geschichte* 28 (1999) 13-31.

Notes the revival of political history in a broadened form that recognizes the impact of politics on human life (e.g. the life of Jews in Nazi

Germany) and the importance in politics of culture, ideas, and ideology (e.g. antisemitism). Discusses recent works by such authors as Mosse, Friedlaender, Breitmann (on Himmler), Gerhard Weinberg (on the history of the Second World War) and others who exemplify this trend, particularly in attempting to explain the genesis of the Final Solution and place it in context as a central Nazi war aim. Also discusses works analyzing the interplay of politics and ideology in dealing with the Nazi past, again particularly the Holocaust, and in postwar manifestations of antisemitism in both West and East Germany. *RW*

0196 Kilcher, Andreas B.: Was ist "deutsch-juedische Literatur"? Eine historische Diskursanalyse. *Weimarer Beitraege* 45,4 (1999) 485-517.

Traces three approaches to the title question: those of the "voelkisch" antisemites and the Zionists, who perceived literature written by Jews as Jewish, and that of the assimilationists who considered it part of German literature. The leading antisemitic historian of literature was Adolf Bartels; but Eugen Duehring, Richard Wagner, and many others held similar views. Bartels saw it as his duty to identify authors of even partial Jewish descent in order to root out their pernicious influence on German literature. He contended that his task was made difficult by the Jews' skill at mimicry and a Jewish conspiracy to infiltrate German culture, but their Jewishness gave itself away in their un-German language. The Jewish assimilationist approach was to a large extent a response to antisemitism. The Holocaust seemed to put an end even to the concept of German-Jewish literature. But with a new generation of Jewish writers in German, the concept has been revived – a welcome development, provided that it stays clear of the danger of renewed ideologization. *RW*

0197 Markus, Stefanie; Scholz, Christian: Invasion der "hosenverkaufenden Juenglinge"? Die Diskriminierung der Ostjuden in Monarchie und Republik. *Sachor* 7 (1997) 40-55.

Discusses the history of the term "Ostjuden," pointing out that it subsumed Jews who differed widely from one another under one, negative, stereotype. The number of Jews who fled to Germany from the hardships and persecutions of Eastern Europe is hard to determine; many were only in transit to other countries. In any case, their appearance and manners were considered repulsive; they aroused fear of economic competition and of a Germany ruined by an influx of aliens. There were demands to deport them, and the government repeatedly obliged. During World War I, Germany recruited Poles, among them many Jews, for labor in Germany; at the end of the war these were joined by large numbers of refugees. Antisemites accused them of criminality, sexual depravation, and the spread of disease. There were pogroms, the largest one in Berlin in November 1923. Prussia and Bavaria tried to stop Jewish immigration from the East; Prussia placed unwanted immigrants in internment camps under inhuman conditions. *RW*

0198 Plenkov, O.Yu.: *Mify natsii protiv mifov demokratii: Nemetskaya politicheskaya traditsiya i natsizm* [Nationalist Myths vs. Democratic Myths: The German Political Tradition and Nazism]. Sankt-Peterburg: Russkii Khristianskii Gumanitarnyi Institut, 1997. 570 pp.

Discusses the intellectual evolution of German nationalist and conservative thought from the early 19th century, focusing on the genesis of Nazi ideology. Pp. 262-299, "Nemetskii antisemitizm," trace the development of German antisemitism. Contends that both in Prussia and Imperial Germany, the ruling elites dissociated themselves from antisemitism, which was propagated by fringe political groups and some writers. The antisemitism in Imperial Germany was nationalist rather than racial. Nazi antisemitism cannot be regarded as an outgrowth of pre-World War I German antisemitism: neither Hitler nor the rank-and-file Nazi supporters read Treitschke or Marr. It was the wartime and post-World War I developments that brought about the specific Nazi antisemitism. Among the factors were the influx of culturally alien "Ostjuden" into the country, and the political radicalization of many Jews of Central Europe in the wake of the Russian Revolution. Contends that Nazi antisemitism was more instrumental than sincere. Pp. 299-326 discuss the evolution of German racism and its influence on antisemitism. *DR*

0199 Schoell-Glass, Charlotte: *Aby Warburg und der Antisemitismus: Kulturwissenschaft als Geistespolitik.* Frankfurt a.M.: Fischer Taschenbuch Verlag, 1998. 317 pp.

Examines Warburg's scholarly work in relation to his lifelong concern with antisemitism. As a student in Germany in the 1880s he was aware of the antisemitism around him; as a Jew he could not hope to become a reserve officer (the key to advancement) or a university professor. He considered making it his life mission to "solve the Jewish question." The problem worsened during and after World War I. Aby supported the actions of his banker brother Max, who in 1916 submitted a memorandum against antisemitism, and in 1923-27 fought a libel action against the antisemite Theodor Fritsch. Aby kept a file of newspaper clippings and other material on antisemitism: pogroms in Eastern Europe, blood libels, the rise of the Nazis. But in his lectures and publications and the activities of his library, antisemitism hardly ever appears explicitly. Argues that his intention was rather to combat antisemitism indirectly by fighting irrationalism and promoting reason and enlightenment. The appendix (pp. 251-279) contains the texts of Warburg's letter to his mother on antisemitism (1889); Max Warburg's memorandum "Die Judenfrage"

(1916); and a "friendly" letter to Max by Boerries von Muenchhausen on the "otherness" of Jews and their danger to German culture, with Aby's comments (1917). *RW*

0200 Schoeps, Hans-Joachim: De vlucht in de haat: Van anti-judaïsme tot antisemitisme [The Flight into Hatred: From Anti-Judaism to Antisemitism]. *Ter Herkenning* 25,4 (1997) 254-266. Appeared in German in "Kurt Tucholsky und das Judentum" (1996).

Discusses the evolution of traditional anti-Judaism into modern antisemitism in 19th-20th-century Germany. Emancipation proved to be most dangerous for the Jew. The more he dressed like his neighbor, the more rights he won, the more he spoke its language, the more the surrounding society hated him. The Jew became the arch-enemy, who threatened economic and political life. Christian Jew-hatred was based on slandering the Jew, but not on annihilating him. Jews could be redeemed if they embraced Christianity. Traces the development of German antisemitism, from Wagner to Hitler. The Germans accepted Nazism as a kind of religion, to be followed strictly. After 1945, everyone believed that there was no more antisemitism in Germany. However, this is not true. 15% of the German population is openly antisemitic, and another 30% admit to some form of antisemitism. The way to fight this prejudice is by conveying correct information in the framework of the family, the church, the school, and the workplace. *MF*

0201 Spoettel, Michael: *Max Weber und die juedische Ethik: Die Beziehung zwischen politischer Philosophie und Interpretation der juedischen Kultur.* Frankfurt a.M.: Peter Lang, 1997. 151 pp.

Weber contrasted ancient Israel (before the Babylonian exile) with later Judaism; whereas the former was masculine and militaristic, the latter is feminine with a pariah mentality. The Jewish religion is centered exclusively on law and intellectual casuistry. Jews arouse hostility by their exclusiveness and the dual morality that allows them to exploit non-Jews. Resentment and a wish for revenge lead them to work for the overthrow of the existing social order in the naive hope of establishing Utopia on earth. But Jewish ethics, although not favoring asceticism, sanctifies suffering, non-resistance, and pacificism. All this was opposed to Weber's social-Darwinist worldview that without conflict there is no progress and no freedom. Weber was, besides, a militant nationalist, convinced of the justice of the German cause in the First World War. He perceived as dangerous the role of Jews in the communist regimes in Munich and elsewhere (pacifists staging a bloody revolution!) and in the German parliamentary committee investigating German war guilt. Yet he opposed the radical right and its antisemitism. *RW*

0202 Toury, Jacob: *Deutschlands Stiefkinder: Ausgewaehlte Aufsaetze zur deutschen und deutsch-juedischen Geschichte.* Gerlingen: Bleicher, 1997. 232 pp. (Schriftenreihe des Instituts fuer Deutsche Geschichte, Universitaet Tel Aviv, 18).

A collection of essays, all published previously. Partial contents: Juedische Aspekte der Reichsbannergruendung [Appeared in "Juden und deutsche Arbeiterbewegung bis 1933" (1992); in English: "Leo Baeck Institute Year Book" 37 (1992).] (93-113); Der Prozess der Lokal-Emanzipation: Die Herausbildung juedischer Buergerrechte in deutschen Ortschaften [Appeared in the "Leo Baeck Institute Year Book" 22 (1977). Demonstrates that opposition to full citizenship rights for Jews in 19th-century Germany was often stronger in local communities than in state governments.] (127-158); Gab es ein Krisenbewusstsein unter den Juden waehrend der "Guten Jahre" der Weimarer Republik, 1924-1929? [Appeared in "Tel Aviver Jahrbuch fuer deutsche Geschichte" 17 (1988).] (191-214). *RW*

Hungary

0203 Fejtö, François: *Hongrois et Juifs: Histoire millénaire d'un couple singulier (1000-1997): Contribution à l'étude de l'intégration et du rejet.* Paris: Balland, 1997. 411 pp. On title page also: En collaboration avec Gyula Zeke.

A history of the Jews in Hungary, focusing on the modern period. Describes the process of emancipation; the Jews were granted equal rights in 1867. Analyzes antisemitism which arose in Hungary at the end of the 19th century, and similarities and differences with German and Austrian antisemitism. Deals with the integration of Hungarian Jewry at the beginning of the 20th century, which was extremely successful. After World War I, antisemitism again increased. In 1920 the Hungarian parliament enacted a numerus clausus law; many Jewish students emigrated. Examines the deteriorating situation of Hungarian Jews during Horthy's regime, and the anti-Jewish laws of 1938-39 and 1941. In March 1944, with the Nazi occupation of Hungary, Eichmann organized the deportations to Auschwitz. Ca. 100,000 Jews survived this action because the war ended. Surveys, briefly, the situation of the Jews after the war. Concludes that the integration of Jews in Hungary led to a very strong symbiosis which can be compared to the history of the Jews in France and not to those in Eastern Europe. Due to this symbiosis, the great majority of Holocaust survivors opted to remain in Hungary after the war. Today, antisemitism is a marginal phenomenon in Hungary. *HV*

0204 Komlós, Aladár: *Magyar-zsidó szellemtörténet a reformkortól a holocaustig. Vol. 1-2* [Hungarian-Jewish Humanities from the Era of Reform to the Holocaust. Vol. 1-2]. Budapest: Múlt és Jövő, 1997. xx, 323; 392 pp.

Vol. 1, pp. 147-167, "Az antiszemitizmus kitörése: Zsidóábrázolások az irodalomban" ("The Outbreak of Antisemitism: The Representation of Jews in Literature"), relates that antisemitism broke out in Hungary in the 1880s. Member of parliament Győző Istóczy published antisemitic pamphlets, delivered antisemitic parliamentary speeches, and took other antisemitic initiatives. However, the major writers of the period were philosemitic – e.g. Sándor Petőfi, János Arany, Mór Jókai, Kálmán Mikszáth. Vol. 2 contains a selection of works written by Jewish Hungarian writers in the 1920s-40s. Pp. 312-318, "In memoriam..." [appeared in "Ichud Füzetek" (1947)], commemorates Jewish writers who perished in the Holocaust. The appendix (pp. 323-381) contains speeches and essays on the Jewish question, by literary personalities, published in the 1920s-30s. *MA*

0205 Ranki, Vera: *The Politics of Inclusion and Exclusion: Jews and Nationalism in Hungary.* New York: Holmes & Meier, 1999. xviii, 269 pp. Based on the author's diss. – University of New South Wales, 1996.

The case of Hungarian Jews in the 19th-20th centuries, first included in the Hungarian nation and then, after 1919, excluded from it, is a poignant illustration of the changing politics of nationalism and the process of exclusion to the point of genocide. Hungarian nationalism, which emerged in the 19th century, construed the nation as a cultural entity and tended to include Hungarian-speaking, assimilated Jews in the dominant Magyar minority of the Kingdom. The treaty of Trianon changed the formula of Hungarian nationalism and made the Jews not only outsiders, but also a social hindrance. Antisemitism, which was a rather marginal phenomenon in the 19th century, became a cornerstone of the official ideology under Horthy. Horthy's Hungary cooperated willingly with the Nazis in the "solution of the Jewish question." The anti-Jewish laws of 1938-41 were enacted without Nazi pressure and were supported by popular opinion. The regime was also instrumental in the deportation of Jews in 1944. Thus, the Holocaust in Hungary was a direct outcome of the country's interwar nationalism. Tragically, the Jews of Hungary relied on the "assimilationist social contract" up to the end of the war; they believed in dialogue with the authorities, and did not protest. *DR*

Lithuania

0206 *Jahrbuch fuer Antisemitismusforschung* 6 (1997).

The section entitled "Litauen," is based on lectures delivered at a conference organized by the Zentrum fuer Antisemitismusforschung in Berlin, 1996. Contents: Kuchenbecker, Antje: Juden und Judenfeindschaft in Litauen [Summarizes the lectures delivered at the conference.] (13-20); Donskis, Leonidas: Antisemitismus in Litauen: Tradition und heutige Erfahrung (21-31); Zingeris, Markas: Juedische Identitaet in einer nichtjuedischen Umwelt nach dem Holocaust (32-43). *RW*

Poland

0207 Kapralski, Sławomir, ed.: *The Jews in Poland. Vol. II.* Trans. and ed. of the English text: Michael Jacobs. Cracow: Judaica Foundation, Center for Jewish Culture, 1999. 415 pp.

A collection of articles, dedicated to Rafael Scharf on his 85th birthday. Partial contents: Jedlicki, Jerzy: The End of the Dialogue: Warsaw 1907-1912 [On debates conducted in Jewish and in non-Jewish circles regarding the place of the Jews in Polish society and the possibility of their assimilation.] (111-123); Tomaszewski, Jerzy: Anti-Semitism of Men and Anti-Semitism of Things: Inside Vladimir Jabotinsky's World of Ideas [Discusses Jabotinsky's views on antisemitism, and analyzes the causes of impoverishment of the Jews in interwar Poland.] (137-146); Rudnicki, Szymon: The Jews' Battle in the Sejm for Equal Rights [In 1919-31.] (147-163); Stola, Dariusz: Ignacy Schwarzbart's Lost Battle with Emigrationism [As the Polish government's solution to the "Jewish problem."] (183-198); Leichter, Sinai: Gebirtig's Ghetto Songs [On poems and songs written by the Yiddish poet Mordechai Gebirtig after the Nazi occupation and in the Kraków ghetto, before his deportation to Bełżec and death in June 1942.] (285-289); Turski, Marian: Individual Experiences in Diaries from the Łódź Ghetto (293-299); Engel, David: Who Is a Collaborator? The Trials of Michał Weichert [On the wartime activities of the Jew who was appointed by the Nazis as head of the Juedische Unterstuetzungsstelle fuer das Generalgouvernement (Jewish Aid Office), operating out of Kraków, and his postwar trials. Weichert (1890-1967) was a Yiddish theatrical director and critic; he was tried by a special criminal court in January 1946 on charges of serving the Germans, and exonerated; in 1949 he voluntarily submitted to a trial by a Jewish civic tribunal which found him guilty of collaboration. Subsequently, he was ostracized by the Jewish community. In 1958 he emigrated to Israel; he died in 1967 in Tel-Aviv.] (339-370); Gross, Jan Tomasz: The Wages of Fear: Confronting Stereotypes about Polish-Jewish Relations after the

War [Relating to widespread anti-Jewish sentiments in postwar Poland, discusses the Special Commission established by the Central Committee of Polish Jews for the purpose of self-defense after the Kielce pogrom; it functioned between July 1946-March 1947, and collected documentation on attacks against Jews; quotes cases from reports to the Commission found in the archives of ŻIH.] (371-384); Musiał, Stanisław: Black Is Black: The Sin of Anti-Semitism Cannot Be Trivialized [Appeared in Polish in "Tygodnik Powszechny" (16 November 1997). Discusses an antisemitic statement made by Henryk Jankowski, a parish priest in Gdańsk, in 1997 ("A Jewish minority in the Polish government cannot be tolerated, because the nation is afraid of it"), and the dismissive reactions of the Polish Church and politicians. States that the attempt to prove that black is white will convince no one; calls on the Church to forcefully condemn antisemitism as a sin.] (385-392); Hoffman, Eva: In the Era of Memory [The epilogue to her book "Shtetl" (1997) on the Jews of Brańsk; relates, also, to the horrors of the Holocaust years and to antisemitism in postwar Poland.] (393-403). *SSC*

0208 Loew, Andrea: "Zwischen den Fronten": Das polnische Judentum und der Antisemitismus im 19. und 20. Jahrhundert. *Sachor* 7 (1997) 22-39.

Traces political antisemitism in Poland, comparing it to Germany. In Poland, the Jews were seen as unassimilable and an obstacle to a unified Polish ethnic state. (Most of them were indeed unwilling to assimilate.) During the period before World War I, the Jews were accused of siding with Poland's occupying powers. Nationalistic antisemitic propaganda was propagated, especially through Jan Jeleński's weekly "Rola" and the political party Endecja, comparable to Treitschke's "Preussische Jahrbuecher" and the Christian Social Party in Germany. During the war, the Jews were accused of collaborating with the Russians or the Germans. After the war, Jews were killed in pogroms. Poles resented international pressure to sign a convention on minority rights; they wanted a purely Polish state. The Endecja radicalized its propaganda; the government, in competition with the Endecja, adopted increasingly antisemitic policies. Although during the Holocaust many Poles rescued Jews, most were indifferent or collaborated with the Nazis. After the war, returning Jews met with hostility. Jews were now seen as leading supporters of the hated communist regime. Even in the 1990s antisemitism has not disappeared; among the leading agitators is the priest Henryk Jankowski. *RW*

0209 Tazbir, Janusz: Conspiracy Theories and the Reception of "The Protocols of the Elders of Zion" in Poland. *Polin* 11 (1998) 171-182.

States that conspiracies have always existed throughout history, but a "conspiracy vision of history" attributes to conspiracy a decisive significance, devaluing or denying other causative factors. Surveys various conspiracy beliefs in Europe since the Middle Ages, the most important being the belief that Jesuits desired to dominate the world. The pamphlet "The Secret Councils of the Society of Jesus," by the Polish ex-Jesuit Hieronim Zahorowski (1582-1634), was very popular and became a model for later pamphlets written about different groups. Surveys anti-masonic pamphlets which appeared during the French Revolution, which always contained antisemitic remarks. Discusses the success of "The Protocols of the Elders of Zion" in 20th-century Europe. Focuses on its influence on Endecja historians in Poland, who explained various events in Polish history through a vision of conspiracy – e.g. Poland's partitions were provoked by Freemasons led by Jews. Some literary works also exploited conspiracy themes, as did the communist regimes in Poland and in the USSR. Concludes that the popularity of conspiracy theories lies in simple explanations for complex historical processes. *HV*

Russia, USSR, CIS

(see also Ukraine)

0210 Aren, Aleks: *Krov i pepel* [Blood and Ashes]. Ierusalim: Lira, 1997. 152 pp.

A collection of essays by Aren written in the 1990s, some of them published previously in Israel. Partial contents: Velizhskaya tragediya [The Tragedy in Velizh] (24-30); Neizvestnaya stranitsa sudebnoi dramy (posle protsessa Beilisa) [An Unknown Page of the Judicial Drama (after the Beilis Trial). On the trial held in June 1914, in St. Petersburg, of 25 lawyers who issued a declaration against the frame-up of the Beilis Affair; also on the trial of O.Yu. Vipper, the prosecuting attorney at the Beilis trial, held in September 1919 in Moscow.] (31-41); Yevreiskiye pogromy i tsarskii sud [Anti-Jewish Pogroms and the Tsarist Law Court] (31-41); Davno eto bylo [It Was Long Ago. An earlier version appeared in "Alef" (1996). Based on conversations with Leib Trakhtenberg, relates his experiences in the Kopaigorod ghetto, Transnistria.] (53-64); On lyubil svoi narod (ob I. Erenburge) [He Loved His People (about I. Ehrenburg). On Ehrenburg as a fighter against antisemitism.] (97-102). *DR*

0211 Geizer, Matvei M.: *Sem svechei* [Seven Candles]. Moskva: Gesharim, 1999. 335, [24] pp.

Contains short historical essays dealing with antisemitism in Russia and the USSR and with the Holocaust, interviews conducted by Geizer with prominent people, and his recollections of celebrities whom he knew. Amongst the historical essays, there is the story of his and some members of his family's survival in the ghetto of Bershad in Transnistria (pp. 13-26). In many of his interviews, Geizer raised the issue of antisemitism.

Among the topics which he discussed with his interviewees are the following: whether Khrushchev was antisemitic, the murder of Mikhoels, and the struggle conducted by the lawyer Genri Reznik against antisemitism in present-day Russia. *DR*

0212 Heifets, Mikhail: Tsareubiistvo 1918 goda: Ierusalimskaya versiya prestupleniya i falsifitsirovannovo sledstviya (fragmenty knigi) [The Murder of the Tsar in 1918: A Jerusalem Version of the Crime and the Falsified Investigation (Excerpts from the Book)]. *Russkii Yevrei* 4 (Sum 1997) 33-36.

Excerpts from the author's book (1991), presenting a brief summary of its contents. *DR*

0213 Roberts, Priscilla: Jewish Bankers, Russia, and the Soviet Union, 1900-1940: The Case of Kuhn, Loeb and Company. *American Jewish Archives Journal* 49, 1-2 [1997] (1999) 9-37.

The U.S. banking house Kuhn, Loeb and Company and its partners were frequently accused of pro-Bolshevik sympathies. Antisemitic propaganda alleged that this bank, and especially its senior partner Jacob Schiff, had secretly financed the Bolshevik Revolution. Schiff was sensitive to Jewish issues; in the period before the Russian Revolution he was engaged in alleviating Jewish suffering under the tsarist and other antisemitic regimes. Before and during World War I he staunchly opposed giving loans to Russia because the Jews did not have equal rights there. After the Revolution, Schiff supported the short-lived liberal government and granted a loan to it. However, he and his partners (Warburg, Kahn, et al.) were anti-communist as well as pro-Jewish. Therefore, the Civil War in Russia placed Schiff in a quandary: he was forced to support the White Russian movement despite its antisemitism. The bitter reality is that Kuhn, Loeb and Company were unable to exert anything but marginal leverage on both the tsarist and Soviet regimes. *DR*

0214 Stetsovskii, Yurii I.: *Istoriya sovetskikh repressii. Vol. 1* [A History of Soviet Repressive Measures]. [Moskva?]: Obshchestvennyi fond "Glasnost," 1997. 600 pp.

Pp. 479-552 discuss repressive measures motivated by antisemitism, from the Civil War (1918-20) until the post-Soviet period. The Red Army, along with other armies in the Civil War, perpetrated pogroms against Jews. Covert antisemitism was a factor in Soviet policy toward the Jews in the 1920s-30s. It became a motive for repressive measures during and after World War II, when it became clear that it was easier to mobilize the people against a national enemy than against a class enemy. Dwells on the affairs of the Jewish Anti-Fascist Committee and the "Doctors' Plot."

From the late 1980s, antisemitic repressive measures have been enforced through Russian extreme nationalist organizations. Discusses violent acts perpetrated by nationalists against Jews and other "national enemies." Some of the perpetrators were brought to trial; almost all of these ended with acquittal of the defendants or symbolic punishments. The authorities secretly favor the racist propaganda and extreme nationalism, and the fascists enjoy impunity. *DR*

Ukraine

0215 *Pidhotovchi materialy populyarnoyi entsyklopediyi "Ukrayinske yevreistvo." Pt. 3* [Preliminary Materials for the Popular Encyclopedia "Ukrainian Jewry." Pt. 3]. Kyyiv: Instytut Yudayiky, 1998. 236 pp.

Articles written by various authors, in popular form, dealing with various aspects of Ukrainian Jewish history. The following deal with the Holocaust and postwar antisemitism in Soviet Ukraine: Podolskii, Anatolii Ye.: Yevreyi Ukrayiny v roky natsystskoyi okupatsiyi: Istoriohrafiya problemy [The Jews of Ukraine in the Years of the Nazi Occupation: The Historiography of the Problem] (155-162); Kovba, Zhanna: Hreko-Katolytska Tserkva Skhidnoyi Halychyny v period Katastrofy yevropeiskoho yevreistva [The Greek-Catholic Church of Eastern Galicia in the Period of the Holocaust of European Jewry] (163-172); Mitsel, Mikhail: Antyyevreiska polityka Kompartiyi Ukrayiny v period vid XX z'yizdu [The Anti-Jewish Policy of the Communist Party of Ukraine in the Period after the 20th Party Congress] (173-187). *DR*

0216 Rabinovych, Yakov Y.: *Na zlami vikiv: Do 1000-richchya prozhyvannya yevreyiv v Ukrayini* [On the Edge of Centuries: To the 1000-Year Anniversary of Jewish Presence in Ukraine]. Kyyiv: [The Author], 1998. 360 pp.

A history of Jews in Ukraine, concentrating on the period after the partitions of Poland. Dwells on the anti-Jewish policy of the tsars, including in Russian Ukraine, the activities of the Black Hundred in Ukraine, and the Beilis case. Chs. 10 and 11 (pp. 114-194) deal with the Holocaust in Ukraine (focusing mainly on Galicia), Jewish resistance, and the participation of Ukrainian Jews in the Soviet Army. The rest of the book describes the Stalinist repression against the Jews in the postwar period, liquidation of Jewish cultural institutions in Ukraine, the deterioration of the Ukrainian Jewish community and its revival after 1992, when Ukraine gained independence. *DR*

0217 Rabinovych, Yakov Y.: *Yevrei mizh stolittyamy nadiy (politolohichne doslidzhennya): Do 1000-richchya prozhyvannya yevreyiv v Ukrayini*

[The Jews in Centuries of Hope (Politological Research): To the 1000-Year Anniversary of Jewish Presence in Ukraine]. Ivano-Frankyvsk: Ivano-Frankyvskyi derzhavnyi tekhnichnyi universytet nafty i hazu, 1997. 275 pp.

A history of the Jews in Ukraine, focusing on the 19th-20th centuries. Emphasizes that antisemitism was alien to the Ukrainian people: its violent outbursts in the past took place in the wake of great social disasters or were incited from abroad (mainly from Poland and Russia). Dwells on the antisemitic propaganda and the pogroms of the early 20th century; stresses that their main instigator was the Black Hundred, which had chapters in the large Ukrainian cities. Dwells on the tragic fate of Jewish agricultural settlements in Ukraine. First, they were decimated by the famine of 1932-33; then they were annihilated by the Nazis in 1941-42. Dwells on the Jewish resistance during World War II. Describes the history of the "Doctors' Plot" affair (without focusing specifically on Ukraine). *DR*

0218 Turchenko, F.G. et al., eds.: *Yevreiskoye naseleniye Yuga Ukrainy: Yezhegodnik: Issledovaniya, vospominaniya, dokumenty* [The Jewish Population in Southern Ukraine: A Yearbook: Research, Memoirs, Documents]. Kharkov-Zaporozhye: Yevreiskii Mir, 1998. 320 pp.

Papers presented at two colloquia: "The Jewish Population of the Southern Ukraine" and "First Zaporozhye Colloquium on the Problems of the Jewish Population in the Southern Ukraine," both held in Zaporozhye in 1996. Partial contents: Yekelchik, Sergei: Nekotorye teoreticheskiye problemy sovremennoi istoriografii Katastrofy [Some Theoretical Problems of Present-Day Holocaust Historiography] (22-31); Kozyrev, A.: Antiyevreiskie narodnicheskie proklamatsii nachala 80-ykh godov XIX veka v Ukraine [Anti-Jewish Narodnik Proclamations of the Early 1880s in Ukraine] (133-137); Shchupak, I.: Yevrei Zaporozhya: Gody okkupatsii goroda 1941-1943 [The Jews of Zaporizhzha: The Years of the City's Occupation 1941-1943. Mainly about Righteous Gentiles.] (262-266). *DR*

0219 Tverdokhleb, Nikolai F.: *Yevrei Kremenchuga: Istoricheskii ocherk* [The Jews of Kremenchug: Historical Essay]. Kremenchug: Piramida, 1999. 146 pp.

A historical essay on Jews in Kremenchug from 1776, when the first Jews settled in the town, until the present. Outlining tsarist anti-Jewish policies, stresses that the Ukrainian intelligentsia always opposed antisemitism. In October 1905 and during thr Civil War in Ukraine the Jews of Kremenchug experienced pogroms. In summer 1942, the Nazis murdered

8,000 Jews of the city. There were non-Jews in Kremenchug (including its mayor) who risked their lives to rescue Jews during the Holocaust. *DR*

USA

0220 Adams, Maurianne; Bracey, John H., eds.: *Strangers & Neighbors: Relations between Blacks & Jews in the United States.* Amherst: University of Massachusetts Press, 1999. xv, 862 pp.

A collection of analyses and documents relating to the history of Black-Jewish relations in the U.S. for 300 years. Partial contents: Hacker, Andrew: Jewish Racism, Black Anti-Semitism [Appeared in "Reconstruction" 1, 3 (1992).] (18-23); Williams, Oscar R., Jr.: Historical Impressions of Black-Jewish Relations prior to World War II [Appeared in "Negro History Bulletin" 40 (1977).] (34-42); Schappes, Morris Urman; Williams, Oscar R., Jr.: Historical Impressions of Black-Jewish Relations prior to World War II: Another Comment and Response [Appeared in "Negro History Bulletin" 41 (1978).] (43-49); Drescher, Seymour: The Role of Jews in the Transatlantic Slave Trade [Appeared in "Immigrants & Minorities" 12 (1993).] (105-115); Austen, Ralph A.: The Uncomfortable Relationship: African Enslavement in the Common History of Blacks and Jews [Appeared in "Tikkun" 9 (1994).] (131-136); Foner, Philip Sheldon: Black-Jewish Relations in the Opening Years of the Twentieth Century [Appeared in "Phylon" 36 (1975).] (237-244); Hertzberg, Steven: Jews and Blacks [Appeared in his "Strangers within the Gate City: The Jews of Atlanta, 1845-1915" (1978).] (245-260); Levy, Eugene: "Is the Jew a White Man?" Press Reaction to the Leo Frank Case, 1913-1915 [Appeared in "Phylon" 35 (1974).] (261-270); Dinnerstein, Leonard: Leo M. Frank and the American Jewish Community [Appeared in "American Jewish Archives" 20 (1968).] (271-282); Aptheker, Herbert: From "The Souls of Black Folks": A Comparison of the 1903 and 1952 Editions [Appeared in "Negro History Bulletin" 34 (1971). Shows how W. E. B. Du Bois attempted to eliminate possible antisemitic connotations in his book.] (283-286); Hellwig, David J.: Black Images of Jews: From Reconstruction to Depression [Appeared in "Societas" 8 (1978).] (300-315); Syrkin, Marie: Anti-Semitic Drive in Harlem [Appeared in "Congress Weekly" 8 [35] (1941). Discusses fascist antisemitic propaganda among Blacks in New York City.] (380-383); Capeci, Dominic J., Jr.: Black-Jewish Relations in Wartime Detroit: The Marsh, Loving, Wolf Surveys and the Race Riot of 1943 [Appeared in "Jewish Social Studies" 47 (1985).] (384-406); Weisbord, Robert G.; Stein, Arthur: Negro Perceptions of Jews between the World Wars [Appeared in "Judaism" 18 (1969).] (409-426); Wedlock, Lunabelle: Comparisons by Negro Publications of the Plight of the Jews in Germany with That of the Negro in America [Appeared in "Howard University Studies in the Social

Sciences" 3 (1942).] (427-443); Reddick, Lawrence Dunbar: Anti-Semitism among Negroes [Appeared in "Negro Quarterly" 1 (1942).] (449-456); Cripps, Thomas: African Americans and Jews in Hollywood: Antagonistic Allies [Appeared in "Struggles in the Promised Land: Toward a History of Black-Jewish Relations in the United States" (1997).] (457-471); Greenberg, Cheryl: Negotiating Coalition: Black and Jewish Civil Rights Agencies in the Twentieth Century [Appeared in "Struggles in the Promised Land" (1997).] (476-494); Sacks, Karen Brodkin: How Did Jews Become White Folks? [Appeared in "Race" (New Brunswick, NJ: Rutgers University Press, 1994).] (500-519); Carson, Clayborne, Jr.: Blacks and Jews in the Civil Rights Movement [Appeared in "Jews in Black Perspectives" (1984).] (574-589); Hill, Herbert: Black-Jewish Conflict in the Labor Context: Race, Jobs, and Institutional Power [A shorter version appeared in "Race Traitor" (New York: Routledge, 1996).] (596-619); Ferretti, Fred: New York's Black Anti-Semitism Scare [Appeared in "Columbia Journalism Review" 8 (1969).] (645-659); Karp, Walter; Shapiro, H.R.: Exploding the Myth of Black Anti-Semitism [Appeared in "The Public Life" (21 February 1969) and "Black Anti-Semitism and Jewish Racism" (1969).] (660-668); Hentoff, Nat: Blacks and Jews: An Interview with Julius Lester [Appeared in "Evergreen Review" 13 [65] (1969). Discusses the Ocean Hill-Brownsville Black-Jewish conflict, and antisemitic comments by Blacks about Jews, including on Lester's WBAI-FM radio program.] (669-680); Cohen, Kitty O.: Black-Jewish Relations in 1984: A Survey of Black U.S. Congressmen [Appeared in "Patterns of Prejudice" 19 (1985).] (688-701); Dreyfuss, Joel: Such Good Friends: Blacks and Jews in Conflict [Appeared in the "Village Voice" (27 August 1979).] (712-719); Willis, Ellen: The Myth of the Powerful Jew: The Black-Jewish Conflict, Part 2 [Appeared in the "Village Voice" (3 September 1979).] (720-728); Reed, Adolph L., Jr.: Blacks and Jews in the Democratic Coalition [Appeared in his "The Jesse Jackson Phenomenon" (1986).] (729-740); Branch, Taylor: Blacks and Jews in the Chicago Heartland [Appeared in "Esquire" (May 1989).] (741-762); Smith, Barbara: Between a Rock and a Hard Place: Relationships between Black and Jewish Women [Appeared in "Yours in Struggle: Three Feminist Perspectives on Anti-Semitism and Racism" (Ithaca, NY: Firebrand Books, 1984).] (765-780); Bell, Derrick: Great Expectations: Defining the Divide between Blacks and Jews [A lecture delivered at Baruch College, CUNY, Spring 1994). Discusses recent antisemitic statements by Black leaders, and the need to speak out against them.] (802-812); Lester, Julius: Blacks and Jews: Where Are We? Where Are We Going? [An earlier version appeared in "Blacks and Jews: Alliances and Arguments" (1994).] (813-822). Schoem, David; Stevenson, Marshall: Teaching Ethnic Identity and Intergroup Relations: The Case of Black-Jewish Dialogue [Appeared in "Teachers College Record" 91 (1990). On a class given at the University of Michigan on Black-Jewish relations.] (823-836). *YC*

1789-1918: General

0221 Abitbol, Michel: L'Affaire Dreyfus et la montée de l'antisémitisme colonial. *Archives Juives* 31, 2 (1998) 75-87.

Contends that internal politics within Algeria caused the violent acts of antisemitism during the years of the Dreyfus Affair. Disagrees with those historians who believe that it was essentially events in France which fomented anti-Jewish sentiment in Algeria. Motivated by the desire to distance the Jews from French political life, the annulment of the Crémieux Decree was the rallying point of the European press in the colony. Especially "La Libre Parole" incited the masses to attack the Jews as pariahs – both unable to become French and unwilling to integrate into Arab culture. Contends that the culmination of antisemitic agitation came in 1898 with the victory of Max Régis for the mayoralty of Algiers. Assesses the impact of the economic situation on hostility towards the Jews, who were seen as competitors of their Algerian counterparts. Concludes that Algerian colonial antisemitism persisted until after the end of World War II, when nationalism posed an even greater threat to French hegemony. *TS*

0222 Engel, Valerii V.: *"Yevreiskii vopros" v russko-amerikanskikh otnosheniyakh: Na primere "pasportnovo" voprosa 1864-1913* [The "Jewish Question" in Russian-American Relations: A Case Study of the "Passport" Question 1864-1913]. Moskva: Nauka, 1998. 133, [11] pp.

The "passport question" served as the cause for the American denunciation in 1912 of the Russian-American Treaty on Trade and Navigation of 1832. The conflict began in 1864, when a Jewish American (born in Russian Poland) was arrested in Russia as a defector from the army, although he had immigrated to the USA years before. In the following years Russia tried systematically to apply her restrictive legislation to Jewish Americans visiting Russia. The USA regarded this situation as incompatible with American principles and tried to influence Russia to change her behavior. However, due to pragmatic considerations, the USA abated its pressure in the 1900s. The initiative in the struggle against discrimination in Russia passed to the American Jewish Committee. A campaign to denounce the Russian-American Treaty was begun by Louis Marshall, and succeeded in December 1911. Dwells on the forces in the USA interested in breaking the Treaty and on the reasons for Russia's refusal to revise her policy. The causes of the "passport conflict" lay in the political and ethno-cultural differences between Russia and the USA. The denunciation of the Treaty damaged the Russian economy and caused an outburst of anti-Jewish sentiment in Russia. *DR*

0223 Grand, Roslyn: U.S.A. Support of Its Jewish Community as Reflected in American Judaica: Selected from the Rare Book Collection of Roslyn Grand. *Proceedings of the 33rd Annual Convention of the Association of Jewish Libraries, Philadelphia, PA, June 21-24, 1998*, eds. Barbara Y. Leff, Laurel S. Wolfson. New York: Association of Jewish Libraries, 1999. Pp. 166-171.

Focuses on efforts of the Jewish community to realize its civil rights in the post-Revolution USA, citing two original sources in Grand's collection. "Speeches on the Jew Bill 1829" reflects a struggle in the Maryland legislature from 1816-26 that finally established Jews' civil rights under the Constitution and put an end to legal restrictions. One of the highlights of this "emancipation" struggle was the speech of H.M. Brackenridge in 1818 stating that "no person ought by any law to be molested... on account of his religious persuasion." The second document, "Discriminations in Switzerland against Citizens of the U.S. of the Hebrew Persuasion," relates to practices enforced by Swiss cantons preventing Jews from being citizens or owning land, and restricting their trade, marriages, and residence. Efforts by American Jews and by U.S. consul in Switzerland Theodore Fay eventually led to a reaffirmation of U.S. refusal to allow discrimination against its citizens on the basis of religion. In 1866 the Swiss federal constitution banned discrimination by Swiss cantons against Jews. The Bavarian government apparently followed this example. Includes a selected bibliography of material in the Grand collection. *YC*

0224 Haumann, Heiko, ed.: *Der Traum von Israel: Die Urspruenge des modernen Zionismus.* Weinheim: Beltz Athenaeum, 1998. 329 pp.

A collection of articles compiled on the occasion of the 100th anniversary of the First Zionist Congress, held in Basel in 1897. The following deal with antisemitism: Kury, Patrick: Die ersten Zionistenkongresse aus der Sicht der damaligen Baseler Publizistik [See pp. 236-242, "Zionismus als Vorwand antijuedischer Agitation."] (232-249); Hagemeister, Michael: Die "Protokolle der Weisen von Zion" und der Basler Zionistenkongress von 1897 [A shorter version appeared in "Der erste Zionistenkongress von 1897" (1997).] (250-273); Ruethers, Monica: Von der Ausgrenzung zum Nationalstolz: "Weibische" Juden und "Muskeljuden" (319-329). *RW*

0225 Hirsch, Waltraud: *Eine unbescheidene Charakterologie: Geistige Differenz von Judentum und Christentum: Otto Weiningers Lehre vom bestimmten Charakter.* Frankfurt a. M.: Peter Lang, 1997. 414 pp. Based on the author's diss. – Universitaet Tuebingen, 1995.

Analyzes Weininger's conception of character, especially the dichotomy between the Jew and the Aryan-Christian in every individual. Weininger ascribed to the Jewish element in character everything he considered negative, and derived it from the Jew's lack of inner unity and spirituality. He believed that the individual must be introspective in order to recognize the Jew within himself; he must hate and subdue him. This self-hatred is superior to self-love. As the Jew is not introspective, he cannot be self-hating. Antisemitism, for Weininger, is the projection of Christian self-hatred onto the other. *RW*

0226 Koren, Roselyne; Michman, Dan, eds.: *Les intellectuels face à l'Affaire Dreyfus alors et aujourd'hui: Perception et impact de l'Affaire en France et à l'étranger: Actes du colloque de l'Université Bar-Ilan, Israël, 13-15 décembre 1994.* Paris: L'Harmattan, 1998. 351 pp.

Partial contents: Wistrich, Robert Solomon: Three Dreyfusard Heroes: Lazare, Zola, Clémenceau (13-41); Yagil, Limor: Bernard Lazare et l'Affaire Dreyfus: La réaction originale d'un intellectuel anarchiste et d'un Juif dans la France de la fin du XIXe siècle (43-61); Pagès, Alain: Emile Zola contre l'antisémitisme: Réflexions sur la question juive... (63-74); Weinberg, Henry Z.: L'affaire et l'antisémitisme: Le cas d'Urbain Gohier (75-89); Delmaire, Danielle: La perception de l'Affaire Dreyfus dans le milieu des intellectuels catholiques français: Vers une controverse? (91-104); Kornberg, Jacques: Herzl, the Zionist Movement, and the Dreyfus Affair (107-119); Schwarzfuchs, Simon: Abraham Ludvipol, intellectuel Juif d'Europe Orientale, et l'affaire Dreyfus (121-132); Bacharach, Zvi (Walter): The Dreyfus Affair and German Antisemitism at the Close of the Nineteenth Century (135-144); Preisman, Leonid: L'affaire Dreyfus et l'opinion publique russe (145-158); Zucker, Bat-Ami: Response of American Black Intellectuals to the Dreyfus Affair (159-169); Sehayik, Shaul: Coverage of the Dreyfus Affair in the Arab Press [In Egypt and Lebanon. Appeared in Hebrew in "Michael" 14 (1997).] (171-185); Griffiths, Richard: Maurice Barrès: Intelligence and "the Intellectuals" (209-221). *HV*

1789-1918: Africa

South Africa

0227 Shain, Milton: Imperialist Judaism in Africa: The Anglo-Boer War and the Image of the Jew in South Africa. *Jewish Affairs* 54, 3 (Spr 1999) 43-47.

In the 1890s, a sinister image of the South African Jew as connected to a network of international finance emerged both among the British left and in South Africa. The abortive Jameson Raid in Transvaal in 1895 served for many as a demonstration of the connection between Jewish finance and British imperialism. The Anglo-Boer war enhanced the image of the "cosmopolitan" (i.e. Jewish) capital behind the British conflict with the Boers. John A. Hobson enunciated clearly the view of the conspiratorial nature of international Jewish capital. The idea of warmongering Jewish international finance became popular in South Africa. It was at this time that the popular cartoon character Hoggenheimer, the quintessential Jewish parvenu, emerged in the South African press. *DR*

0228 Shain, Milton: Representations of the Jew in Cape Colony c. 1850-1885: Confirmation of the Interactionist Model of Antisemitism? *Patterns of Migration, 1850-1914,* eds. Aubrey Newman, Stephen W. Massil. London: Jewish Historical Society of England, 1996. Pp. 207-220.

A paper delivered at a conference in London, 1993. Tests the "interactionist" model of antisemitism (which holds that the actions of Jews contribute to antisemitism) in regard to the Cape Colony in South Africa in 1850-85, and finds it only partially valid. There were very few expressions of antisemitism in the press before the 1880s, and Jews were generally perceived by journalists as good citizens who practiced their religion privately. This view changed in the 1880s, when anti-Jewish stereotypes became more common. Although this was partially due to an influx of Jewish immigrants from Russia and other countries at that time, it was mainly a result of gradually worsening economic conditions: in the 1860s all merchants were attacked in the press, in the 1870s Jewish merchants were singled out for criticism, and the economic crisis of the 1880s was accompanied by anti-Jewish stereotypes. States that only lower-class Jews were looked upon unfavorably by the public, and that the negative stereotypes of Jews were a reflection of class prejudice. *RE*

1789-1918: Europe

General

0229 Clemesha, Arlene: *Marxismo e judaísmo: História de uma relação difícil* [Marxism and Judaism: The History of a Difficult Relationship]. São Paulo: Boitempo, 1998. 223 pp.

States that antisemitism is not a monopoly of right-wing parties, and that Marxism has always had antisemitic aspects. Discusses Marxist and socialist discourse on "the Jewish question" from the early writings of Marx through the Second International (1889-1914), and the socialist movement in tsarist Russia (including the Bund) up to World War I. Analyzes the writings of personalities such as Karl Marx, Karl Kautsky,

Otto Bauer, Abram Léon, and Lenin on Jews, the Jewish people, and their role and future in society, as well as the phenomenon of antisemitism. *AA*

0230 Liedtke, Rainer; Wendehorst, Stephan, eds.: *The Emancipation of Catholics, Jews and Protestants: Minorities and the Nation State in Nineteenth-Century Europe.* Manchester: Manchester University Press, 1999. x, 223 pp.

A collection of essays based on a lecture series on Jewish emancipation held by the Oxford Centre for Hebrew and Jewish Studies in the autumn of 1994. Partial contents: Cesarani, David: British Jews [England manifested an "antisemitism of toleration," related to the question of whether Jews could be Englishmen.] (33-55); Malino, Frances: French Jews [The emancipation of French Jews by the Revolution suffered a setback in 1808, then progressed until the Dreyfus Affair.] (83-99); Clark, Christopher M. : German Jews [German Jewish emancipation proceeded unevenly, since Germany was not unified; the failure of acculturation to broaden the Jewish occupational structure was targeted by German antisemitism.] (122-147); Luzzatto Voghera, Gadi: Italian Jews [Acculturation, encouraged by an Italian Jewish elite, was paralleled by a protracted political and legal emancipation that was crowned by the abolition of the Roman ghetto only in 1870.] (169-187). *YC*

0231 Shimoni, Gideon; Wistrich, Robert Solomon, eds.: *Theodor Herzl, Visionary of the Jewish State.* Jerusalem: Hebrew University, Magnes Press; New York: Herzl Press, 1999. xxvi, 351 pp.

A collection based on papers presented at an international conference marking the 100th anniversary of Herzl's book "Der Judenstaat," held in Jerusalem and Haifa in March 1996. Antisemitism of the late 19th century, and scandalous affairs with strong antisemitic implications (like the Panama Scandal and the Dreyfus Affair), greatly influenced Herzl and made him a Zionist. Partial contents: Kornberg, Jacques: The Construction of an Identity [Dwells on the role of "Jewish self-hatred" in Herzl's formative years.] (15-26); Gutwein, Daniel: The Development of Herzl's Theory of Antisemitism [Based on his experiences in Austria and in France.] (80-98); Berkowitz, Michael: Herzl and the Stock Exchange [Focuses on the Panama Scandal.] (99-111); Schoeps, Julius Hans: The Role of the Dreyfus Affair (112-125). *DR*

Austria

0232 Geller, Jay: The Godfather of Psychoanalysis: Circumcision, Antisemitism, Homosexuality, and Freud's "Fighting Jew." *Journal of the American Academy of Religion* 67, 2 (June 1999) 355-385.

Connects the key Freudian concepts of childhood sexuality and fear of circumcision with Freud's identity as a Jew and his personal and professional experience of antisemitism. Focuses on the famous Case of Little Hans (1906), who was analyzed in terms of a phobia regarding fear of castration, a fear which Freud associated with antisemitism. Freud's analysis concluded that the boy was normal, implying, that if Hans was allowed to grow up as a Jew he would have the resources to struggle as a Jew, presumably against the antisemitism in his environment. *YC*

0233 Pollak, Michael: L'antisemitismo di Otto Weininger: Una autovalutazione morale dello scenario intellettuale. *Le vie del mondo – Berlino, Budapest, Praga, Vienna e Trieste: Intellettuali ebrei e cultura europea dal 1880 al 1930,* ed. Riccardo Calimani. Milano: Electa, 1998. Pp. 107-120. Appeared in German in "Otto Weininger – Werk und Wirkung" (1984).

Belarus

0234 Funk, Yulia V.: *Yevrei Belarusi v kontse XIX-nachale XX v.* [Jews of Belarus in the Late 19th-Early 20th Centuries]. Minsk: Belorusskii Nauchno-Issledovatelskii Institut Dokumentovedeniya i Arkhivnovo Dela, 1998. 105 pp.

Ch. 1-3 (pp. 7-45) present a concise history of the Jews in Russia, focusing on restrictive legislation against the Jews, and especially its impact on the Jews of Belarus. Ch. 5 (pp. 60-94) deals, inter alia, with pogroms in Belarus and with the expulsion of Jews in 1914-15, during the German advance in this area. *DR*

0235 Mekhedko, Vladimir: Gomelskii pogrom 1903 goda: Etnicheskii konflikt ili grazhdanskaya voina? [The Homel Pogrom of 1903: An Ethnic Conflict or a Civil War?]. *Yevrei Belarusi* 3-4 (1998) 35-48.

Examines the pogrom which erupted in Homel on 1-2 September 1903. After the Kishinev pogrom in March of that year, the atmosphere in the city was charged. A Jewish self-defense group was formed by members of the Bund and Poalei-Zion. This self-defense was perceived by the city elites as part of the revolutionary movement, identified by many officials with the Jews. The September pogrom was not a spontaneous occurrence: it was prepared by the city leadership, with the aim of suppressing revolutionary activities; it was officials who incited the railway workers to anti-Jewish riots. The rioters had no religious motives. The pogrom may not be categorized as an ethnic clash or a "civil war," but as something in-between. *DR*

Denmark

0236 Lausten, Martin Schwarz: Tolerance and Enlightenment in Denmark: The Theologian Christian Bastholm (1740-1819) and His Attitude toward Judaism. *Nordisk Judaistik* 19, 1-2 (1998) 123-139.

A paper presented at a conference on "Toleration and the Enlightenment" in Cambridge, 1997. Surveys the social position of the Jews in Denmark in the 17th-18th centuries, characterized by both tolerance and restrictions. Bastholm was a leading theologian in Enlightenment Denmark. He wrote a three-volume history of the Jews (1777-82). He called for tolerance of the Jews, demanded that they receive access to higher education and public office, and distanced himself from Christian mission to the Jews. However, it is in the supplements to the third volume of his "History" where he relates to the relationship between Judaism and Christianity; although he attempted to be impartial, he condemned the Jewish religion and advocated conversion of the Jews. In 1813 a violent literary conflict broke out in Denmark when a German antisemitic tract was published in Danish. It was followed by other pamphlets and short tracts, some of them defending the Jews. A typesetter composed a compendium of Bastholm's "History" in which he emphasized the antisemitic features. Bastholm did not protest against this publication. On the contrary, in June 1813 he himself published an article in the newspaper "Dagen" attacking the Jews as harmful to the state. Concludes that Bastholm's attitude was characterized by both tolerance and prejudice. *DR*

France

0237 Chêne, Janine; Aberdam, Edith; Aberdam, Daniel, eds.: *Comment devient-on dreyfusard?* Paris: L'Harmattan, 1997. 131 pp.

A collection of papers delivered at a conference held at the Université Pierre Mendès France in Grenoble, 1994. Partial contents: Weil, Nicolas: L'historiographie de l'affaire Dreyfus (21-30); Bredin, Jean-Denis: Emile Zola, la vérité en marche (31-44); Lindenberg, Daniel: L'improbable ralliement de Jaurès (45-59); Prochasson, Christophe: Sur l'antisémitisme fin de siècle: Georges Sorel et l'affaire Dreyfus (61-76); Durand, Jean-Dominique: Léon XIII, un pape dreyfusard? (77-95); Delmaire, Danielle: Les Juifs de France à l'épreuve de l'affaire Dreyfus (97-113); Sahel, Claude: Comment Bernard Lazare est-il devenu dreyfusard? Genèse d'une "decision" (115-126). *HV*

0238 Compagnon, Antoine: *Connaissez-vous Brunetière? Enquête sur un antidreyfusard et ses amis.* Paris: Seuil, 1997. 285 pp.

A biography of Ferdinand Brunetière (1849-1906), the French literary critic, publicist, and director of the journal "Revue des Deux Mondes." Discusses his opinions on state, religion, the Dreyfus Affair, and Jews. Brunetière acted against revision of the Dreyfus trial and was one of the founders of the Ligue de la Patrie Française (1899). However, after a month he left the Ligue, considering it antisemitic. He strongly criticized Drumont's "La France juive" and was opposed to racial antisemitism but, at the same time, considered that the Jews, Freemasons and Protestants had too much influence in France. Discusses his ambiguous personality and the apparent paradoxes in his views. States that in his lifetime he was not viewed as an antisemite, either by Jews or by antisemites. His case shows that it was not necessary to equate anti-Dreyfusard attitudes with antisemitism. Brunetière, as a partisan of a strong state and a convinced Catholic, was opposed to anti-militarism and the campaign of intellectuals against the army and, as a consequence, he also opposed the revision of the Dreyfus trial. *HV*

0239 Dainville-Barbiche, Ségolène de: *Les scellés de l'affaire Dreyfus: Inventaire des papiers saisis en 1894 au domicile de Dreyfus.* Paris: Centre Historique des Archives Nationales, 1997. 53 pp.

Presents the catalog containing an inventory of documents pertaining to the Dreyfus Affair recently discovered in the Musée de l'Histoire de la France. Includes notes from Dreyfus' student years at the Ecole de Guerre (1890-92) and his travel notes. Examines the reasons that certain confidential documents, such as these, were requisitioned from the Dreyfus home by the French Court of Appeals in 1894 but never returned. The introduction (pp. 7-16) analyzes the Dreyfus Affair in the context of the tension which characterized French-German relations during that period. *TS*

0240 Forth, Christopher E.: Bodies of Christ: Gender, Jewishness and Religious Imagery in the Dreyfus Affair. *History Workshop Journal* 48 (Fall 1999) 17-39.

Deals with ambivalences of Dreyfusard intellectuals regarding Jews and their own religious identity. The Dreyfusards rejected the prevalent image of the feminine (i.e. weak) body of the Jew in favor of a heroically masculine image. They projected this positive male image onto an image of Jesus on the cross as a symbol of the suffering Dreyfus. Thus, while defending the political martyr, they minimized his Jewishness. Christian antisemitism was not eliminated but transformed, by depicting the antisemitic crowd as the ones who wanted Dreyfus' (Jesus') death. This crowd viewed the Dreyfusard intellectuals as Jews, whereas in the eyes of the liberal intellectuals, Jews such as Dreyfus were viewed as men. *YC*

0241 Gelfand, Toby: Dr. Germain Sée (1818-1896): "Israélite de Coeur." *Korot* 12 [1996-1997] (1998) 9-34.

Germain Sée, a highly regarded Parisian physician of Alsatian origin, was the first Jew to become professor of medicine or chief physician at a major Paris hospital. Though not formally observant, Sée identified himself as a Jew and was identified by others as such. In accord with conventional stereotypes, Sée was regarded by police spies as money-driven and conniving. With the rise of antisemitism in France from 1880, he was a subject of vicious caricature in Léon Daudet's satirical novel "Les Morticoles" (1894), which depicted him as a monstrous physician obsessed with money and power. Sée vigorously attacked ideas about "race" prevalent among medical men and others in the late 19th century (e.g. that there are characteristically "racial" or Jewish diseases). *REK*

0242 Hoedl, Klaus: Der "Juif Errant" in der franzoesischen medizinischen Literatur. *Aschkenas* 9, 1 (1999) 109-131.

"Pathological vagabondage" was a much-feared social-psychiatric phenomenon in late 19th-century France. It was thought to be caused by a tendency to epilepsy, hysteria, or neurasthenia, and to be a symptom of hereditary degeneration. Some psychiatrists considered Jews especially susceptible to this "disease," notably Henry Meige (expressed in his dissertation of 1893 based on a study of five Jewish patients at the Salpêtrière). Most psychiatrists, however, freely used the figure of speech "Wandering Jew" for vagabonds, without any intention of identifying them as actual Jews. Still, Jews were seen as similar to vagabonds in their alleged tendency to hysteria and neurasthenia, physical weakness, unfitness for army service, lack of willpower, and pathological religious enthusiasm. These characteristics, adding up to effeminacy, were also associated with other supposedly inferior groups, such as Blacks. *RW*

0243 Johnson, Martin Phillip: *The Dreyfus Affair: Honour and Politics in the "Belle Epoque."* Basingstoke, Hants: Macmillan, 1999. xi, 171 pp.

A history of the Dreyfus Affair, focusing mainly on the period from the case of the "bordereau" and the trial in 1894 to the Rennes trial in 1899. The Affair, which shook the whole of France and tore apart its society, began with a mistake, but continued as a chain of illegalities and falsifications. Dreyfus' Jewishness was not the only reason why he was chosen to be the traitor, but it also played a role in a country where antisemitism and race consciousness reached alarming proportions. Stresses the complexity of the Affair; not all the anti-Dreyfusards were antisemitic extremists, and many details of the case remain mysterious up to the present. For all sides, the Affair became a litmus test of patriotism or republicanism, of rationality or morality. Discusses the crucial role of

Mathieu Dreyfus and Zola, and also of Colonel Henry's forgery, in the rehabilitation of Alfred Dreyfus. The Affair marginalized, but also racialized, antisemitism in France; on the other hand, it definitively established the Republic. *DR*

0244 Kaplan, Robert Elliot: Making Sense of the Rennes Verdict: The Military Dimension of the Dreyfus Affair. *Journal of Contemporary History* 34,4 (Oct 1999) 499-515. Adapted from ch. 8 of his book "Forgotten Crisis: The Fin-de-Siècle Crisis of Democracy in France" (Oxford: Berg, 1995).

0245 Kleeblatt, Norman L.: Autour du corps d'Alfred Dreyfus. *Les Cahiers du Judaïsme* 2 (Sum 1998) 37-42.

Examines the image of the public degradation of Dreyfus in 1895 as reflected in the French antisemitic press of that era. This humiliation was designed to warn Jews of the futility of any attempt to rise to the upper echelons of the French military. Contends that the science of phrenology was used to prove Dreyfus' guilt. Assesses the portrayal of the Jew as traitor in various caricatures; these served to proclaim the racist precept of the Jew's physical incapacity to serve his country, and his deficient virility due to circumcision. Believes that Dreyfus was used as a scapegoat to conceal the illicit relationship between two European military attachés. Concludes that European Jewry was unduly influenced by these negative stereotypes and that their reaction could be found in the new ideal of the physically robust Zionist pioneer. *TS*

0246 Quinlan, Sean: The Racial Imagery of Degeneration and Depopulation: Georges Vacher de Lapouge and 'Anthroposociology' in Fin-de-Siècle France. *History of European Ideas* 24, 6 (1998) 393-413.

As an anthroposociologist and a critic of contemporary French society, Vacher de Lapouge was concerned with two dominant social obsessions of the Third Republic: fear of population decline and of racial degeneration. In particular, one of the undesirable results of the elimination of the mythical "Homo Europaeus" from the French population was, in his view, the rise of the Jews. The disruption of traditional racial bonds facilitated the rise of a Jewish "oligarchy" or "plutocracy." Vacher de Lapouge believed in the inherent parasitic nature of the Jews, regarded the Jews as the most dangerous competitors of the Aryans, and predicted future wars between these "races." His views undoubtedly influenced German racists of the 1890s-1900s, including Hitler. *DR*

0247 Tillier, Bertrand: La caricature antisémite pendant l'Affaire Dreyfus. *Hommes & Migrations* 1216 (Nov-Dec 1998) 93-103.

Surveys the history of antisemitic stereotypes in the art and literature of France, such as the figures of Judas and the Wandering Jew. Discusses caricatures of the Jew as peddler, usurer, and traitor – incapable of internalizing French cultural values. Interprets the cartoons by Caran d'Ache and Jean-Louis Forain, founders of the newspaper "Psst...!," dedicated to the portrayal of the Jew during the early years of the Dreyfus Affair. Considers these images as examples of mythical figures used for xenophobic purposes within a cultural, religious, and political framework with which the French public could identify. *TS*

Germany

0248 Bereswill, Mechthild; Wagner, Leonie: Public or Private? Antisemitism and Politics in the Federation of German Women's Associations. *Journal of Genocide Research* 1, 2 (June 1999) 157-168.

Examines the conflict that erupted in 1915-16 within the Federation of German Women's Organizations (Bund deutscher Frauenvereine) between the General German Women's Association (ADF) and the Jewish Women's League (JFB). In her speech on the occasion of the 50th anniversary of the ADF, Helen Lange did not mention the JFB as a member of the Federation. A prolonged correspondence between the ADF and JFB followed. Argues that this conflict was more than a personal clash between Lange and Bertha Pappenheim of the JFB. The Federation pursued a policy of homogeneity in the women's movement in Germany; this homogeneity was presented as confessionally neutral, but was in fact interpreted as based on Christianity. The JFB was seen by the Federation leadership as violating the movement's cultural unity. Concludes that underlying this conflict was the BDF's resentment of the visibility of Jewish women in the movement. Compares the conflict within the women's movement in 1915-16 to a similar conflict in 1904-5. *DR*

0249 Bernhardt, Hans-Michael: *Bewegung und Beharrung: Studien zur Emanzipationsgeschichte der Juden im Grossherzogtum Mecklenburg-Schwerin 1813-1869.* Hannover: Hahnsche Buchhandlung, 1998. xii, 373 pp. Based on the author's diss. – Technische Universitaet, Berlin, 1996.

In Mecklenburg-Schwerin there were never more than a few thousand Jews, because of its largely rural character and residence restrictions. Traces the ups and downs of their emancipation, connecting it with the balance of power between the Grand Duke, on the one hand, and a legislative body consisting mainly of the feudal landowners on the other. Grand Duke Friedrich Franz I (1785-1837) was a product of the Enlightenment and supported rights for Jews; Friedrich Franz II (1842-1883) opposed them. The powerful landowners consistently blocked rights for Jews. They argued that Jews were harmful, avaricious, and dishonest, and that they had no place in a Christian state. The towns at first refused residence and citizenship to Jews out of fear of competition; but with the rise of a bourgeoisie with more enlightened views, and Jewish integration in this bourgeoisie, many towns recognized the value of Jews in promoting industry and trade and gave them equal rights. The Jews themselves, encouraged by brief periods of emancipation – in 1813-17 by the Grand Duke, and in 1849-50 by the short-lived democratic parliament – fought for their rights, supported by liberals who argued for the equality of man and separation of Church and State. But emancipation finally came only when imposed from the outside, by the law of the North German Federation in 1869. *RW*

0250 Fleischer, Michael: *"Kommen Sie, Cohn": Fontane und die "Judenfrage."* Berlin: [The Author], 1998. 393 pp.

The attitude of Theodor Fontane (1819-1898) toward Jews was ambivalent. He had many Jewish friends who throughout his life did him favors which he never requited, and was dependent on Jewish publishers, reviewers, and readers. He perceived Berlin Jews as the mainstay of German culture. However, he was always conscious of the taint of their race. He believed in their domination of culture and the economy, and feared their domination of the state. He sided with Treitschke in the "Antisemitenstreit" and became more antisemitic as he aged. Whereas formerly he advocated assimilation, he came to believe that Jews were unassimilable and their emancipation had been a fatal mistake. In his letters he liked to describe the foibles of others; particularly venomous barbs were aimed at Jews, including his best friends. In his novels, he avoided antisemitism out of consideration for his Jewish public, but they contained many antisemitic allusions and dubious Jewish characters. The editors of his letters in 1904 censored the antisemitic passages; in the 1930s they were reinstated. *RW*

0251 Gilman, Sander L.: La "science de la race" à l'origine de la chirurgie esthétique. *Les Cahiers du Judaïsme* 2 (Sum 1998) 43-48.

Traces the origins of plastic surgery to the search for a solution to feelings of physical inferiority of Jews in Germany at the close of the 19th century. Describes the career of Jacques Joseph (1865-1934), a German Jew and orthopedic surgeon, who joined a dueling society and regarded his numerous scars as symbols of true integration into German society. Contends that Joseph's mission was to alter the noses or ears of Jewish patients for whom these organs had become sources of antisemitic ridicule. Surgery to reverse circumcision was the result of fear of femininity as much as a choice of social integration. Concludes that plastic surgery, as opposed to reconstructive surgery, was associated with the Jew's wish to become totally assimilated into German society. *TS*

0252 Jahrmaerker, Manuela: Wagners Aufsatz "Das Judenthum in der Musik" im Spiegel zeitgenoessischer Reaktionen. *Meyerbeer – Wagner: Eine Begegnung,* eds. Gunhild Oberzaucher-Schueller, Marion Linhardt, Thomas Steiert. Wien: Boehlau, 1998. Pp. 120-141.

Summarizes Wagner's essay, interpreting the last two sentences, which call on the Jews to participate in "this self-destructive bloody combat" ending in salvation through dissolution ("Untergang"), as a demand that both sides, Jews and Germans, change their identities. Surveys reactions to the essay in 1850: most seem to betray some anti-Jewish prejudice, but revolt at what they see as an antisemitism far beyond the norm of the period. One critic even suspects that the essay was meant as a satire on antisemitism. The others expose the illogic of Wagner's arguments, and hint at his personal motivation – his envy of Meyerbeer. The appendix to this collection (pp. 142-296) contains background documents (extracts from letters, essays, and diaries by and about Wagner) concerning his relationship with Meyerbeer and his attitude toward Jews. The text of "Das Judenthum in der Musik" appears on pp. 239-259. *RW*

0253 Kowalska, Elżbieta M.: "Grosse Schwester" oder "Judenpack"? Ueber das Verhaeltnis Wilhelm von Humboldts zur Frage der Emanzipation der Juden. *Orbis* 40 (1998) 35-50.

Humboldt has always been considered as having been pro-Jewish because of his close friendships with Jews (particularly Henriette Herz and Rahel Levin Varnhagen) as well as his advocacy of emancipation. Yet Hannah Arendt pointed to his use of crude antisemitic expressions in some of his letters referring to Rahel. To explain this contradiction, suggests that Humboldt related to Jews and non-Jews alike primarily as individuals, and only secondarily as representatives of a nationality. Thus he continued, much of his life, to love and admire Herz, "the big sister," and admit her moral influence on his development; but Varnhagen became distasteful to him, and then he saw her negative qualities as Jewish. While he attributed his intervention against a discriminatory law in 1814 to "a last spark of my piety toward Herz," he also remarked that he only loved the Jews "en masse," not as individuals. His advocacy of emancipation was part of his liberal political philosophy. *RW*

0254 Leopold, David: The Hegelian Antisemitism of Bruno Bauer. *History of European Ideas* 25,4 (1999) 179-206.

Discusses the views expressed by Bauer (1809-1882), who opposed the removal of legal discrimination against Jews by the state. Focuses on the period when he was a left Hegelian (ca. 1840-45). Bauer published two influential works in 1843, "Die Judenfrage" and "Die Faehigkeit der heutigen Juden und Christen, frei zu werden." His contempt for Jews and Judaism should be seen as undermining his reputation as a liberal. Despite his atheism, he did not criticize religions objectively but favored Christianity for its more "progressive" role in history. He believed that Judaism had no raison d'être, criticizing it, inter alia, for being exclusivistic and hypocritical. Bauer denied any Jewish contribution to Western civilization since the Talmud. He objected to Jewish emancipation, considering Jews "unsuited for citizenship." He viewed the Jews as responsible for their own estrangement in the modern world and, thus, deserving neither pity nor consideration. Later, in "Das Judenthum in der Fremde" (1863), he adopted a biological, racial definition of the Jew. *YC*

0255 Lerousseau, Andrée: Philosophie allemande et destin (1770-1850). *Tsafon* 32 (Win 1997-1998) 69-91.

Analyzes writings by German philosophers relating to the "Jewish question" in the period of the Jewish Enlightenment (Haskalah) in Germany. Quotes from works by Kant, Hegel, Lessing, Fichte, Feuerbach, and others, noting their total misunderstanding of Judaism. Kant described Judaism as a simple superstition, and Feuerbach asserted that Judaism is egoism in the form of religion. States that in the Haskalah period there was no dialogue between Jewish and German intellectuals, and that not one of the Germans showed an interest in the attempts to reform Judaism. Concludes that German philosophers rejected the idea of the integration of the Jews in modern society, and that German philosophy on the Jews formed a bridge between traditional Christian anti-Judaism and modern antisemitism. *HV*

0256 Loeffler, Hans F.: *Walther Rathenau – ein Europaeer im Kaiserreich.* Berlin: Berlin Verlag Arno Spitz, 1997. 202 pp.

Ch. 5 (pp. 143-162), "Antisemitismus und Zionismus bei Walther Rathenau," refutes the view, propagated by critics and biographers, of Rathenau as an example of Jewish self-hatred. Describes the rise of antisemitism in Wilhelmine Germany and analyzes its causes. This antisemitism caught the assimilated Jewish intellectuals unprepared, and caused many of them to seek their Jewish roots while not sacrificing their identification with Germany. Rathenau's "Hoere, Israel!" (1897) and "Staat und Judentum" (1911) must be read in this context. While he urged Jews to educate themselves to become better Germans, he accused society of rejecting the Jews, and the state of requiring baptism before admitting Jews to the higher civil service or the officer corps. He considered this material incentive for conversion repugnant and claimed that if the Jews were baptized, prejudice and hatred would simply be redirected toward the converts. He argued that the state could not afford to dispense with the services of any of its citizens, including the Jews. *RW*

0257 Niemeyer, Christian: Nietzsches rhetorischer Antisemitismus. *Nietzsche-Studien* 26 [1997] (1998) 139-162.

Discusses Nietzsche's early antisemitism, influenced by Wagner and perhaps adopted as a sign of loyalty to him, and his later development into an anti-antisemite. This occurred as he distanced himself from Wagner and encountered Jews that he admired. At first he tried to make light of what he saw as his own error by exonerating Wagner, explaining his antisemitism as actually Jewish self-hatred (because of the rumors about his parentage) or as due to the influence of the "Wagnerites." Shows that Wagner was a rabid antisemite long before any such influence. Discusses Wagner's antisemitic pronouncements and his unsuccessful attempts to convert Ludwig II to his view. Nietzsche gave vehement expression to his loathing for German nationalism and antisemitism, including Wagner's. Suggests (following Richard Lonbach) that Nietzsche saw Wagner's antisemitism as part of his romantic attachment to the past, whereas Nietzsche envisioned a future whose culture would embrace all peoples, and which could not dispense with the Jewish genius. *RW*

0258 Schmidt, Michael: "Wie ein roter Faden": Fontanes Antisemitismus und die Literaturwissenschaft. *Jahrbuch fuer Antisemitismusforschung* 8 (1999) 350-369.

Discusses Theodor Fontane's antisemitism, which Michael Fleischer, in his study "Kommen Sie, Cohn" (1998) traced "like a red thread" in the author's personal correspondence. One of Fontane's antisemitic letters was even reprinted in the SS journal "Das Schwarze Korps" in 1935. But in general, literary historians, including the editors of Fontane's correspondence with his wife (1998) minimize this trait in the writer. In his published works, Fontane was much more cautious in expressions of antisemitism; on occasion he even demolished antisemitic stereotypes, as in the person of the generous moneylender Baruch Hirschfeld in "Der Stechlin." Discusses Fontane's antisemitism in the context of his character and his social-political milieu. *RW*

0259 Schneck, Peter: Juedische Aerzte in Preussen und die Berliner Medizinische Fakultaet im 19. Jahrhundert. *Medizinische Wissenschaften und Judentum,* eds. Nora Goldenbogen et al. Dresden: Verein fuer regionale Politik und Geschichte Dresden, 1996. Pp. 7-13.

Traces the slow progress of emancipation in Prussia and the laws and regulations restricting Jewish rights, and points to segments of the population that opposed or supported emancipation. Although many Jews studied and practiced medicine, almost the only liberal profession open to them, they were not appointed to the teaching staff of the School of Medicine at the University of Berlin. The first Jewish Privatdozent was appointed in 1810, the second only in 1847. The first and only Jew to become a professor in the 19th century was appointed in 1872. *RW*

0260 Schumann, Peter: Theodor Fontane und die Juden. *Geschichte in Wissenschaft und Unterricht* 49, 9 (Sept 1998) 530-543.

Refutes allegations that Fontane was antisemitic. His writings, especially his letters, do contain uncomplimentary descriptions of Jews; but he was equally critical of other groups. Also, he never took a single-minded stand on any subject. He opposed antisemitism, had many Jewish friends, and expressed his appreciation of Jewish culture and the aristocratic character of many Jews whom he considered superior to the average German aristocrat. *RW*

0261 Smith, Helmut Walser: The Discourse of Usury: Relations between Christians and Jews in the German Countryside, 1880-1914. *Central European History* 32, 3 (1999) 255-276.

Explores how the simile of the usurious Jew, and its central assumption that Jews and Christians possessed radically different and religiously specific conceptions of work and trade, affected rural relations between Christians and Jews. The trading Jew (mainly cattle trade) was considered non-productive in a society based on hard work. Antisemitic propaganda contrasted the cunning, deceitful Jew with the simple and honest German peasant. Rural Jews were regarded as nothing but usurers. This prejudice was shared by many state officials, who schemed to oust Jews from the countryside and to replace their credit with that of the banks. Examination of the reality in rural Germany shows that these images were either wrong or exaggerated. Most of the usurers were Christian and most of the Jews were traders. The Jews' life style, moral standards, and mentality were not much different from those of the peasants; and the Jews' clients were not unsophisticated simpletons. Rural Christian-Jewish relations in 1880-1914, although far from idyllic, were nevertheless better than has previously been assumed in various publications. *DR*

0262 Zimmerman, Andrew: Anti-Semitism as Skill: Rudolf Virchow's "Schulstatistik" and the Racial Composition of Germany. *Central European History* 32,4 (1999) 409-429.

Virchow was a well-known opponent of antisemitism. Contends that, paradoxically, the anthropological study designed by Virchow and conducted under his guidance in the 1870s in schools throughout

Germany provided an important practical basis for German racial antisemitism. Virchow's survey led him to the conclusion that there was a separate and single "Jewish race" in Germany, and it made antisemitic racism a popular skill even before it became a generally accepted doctrine. The survey taught Germans not so much to think as to experience themselves in terms of "whiteness" and "brownness," and to recognize racial distinctions between Jews and non-Jews in Germany. *DR*

Great Britain

0263 Tombs, Robert: "Lesser Breeds Without the Law": The British Establishment and the Dreyfus Affair, 1894-1899. *Historical Journal* 41, 2 (1998) 495-510.

From 1897 on, the British embassy in Paris, Queen Victoria and her court, the prime minister, and, with some reservations, the press (led by "The Times") became impassioned sympathizers with Alfred Dreyfus and bitter critics of his persecutors. Traces the development of their reaction to the Affair and the main motifs of the pro-Dreyfus discourse in Britain. Defense of Dreyfus was accompanied by criticism of the French judicial system and the entire French civilization as corrupt; contrast between the moral virtues of the Dreyfusard "minority" and the depraved anti-Dreyfusard "majority"; enmity of anti-Dreyfusards against Britain; and a stress on the role of Catholicism as a pillar of anti-Dreyfusard fanaticism. The Dreyfusard position was congenial to the British, because it demonstrated the superiority of British or Anglo-Saxon political principles over the "continental." A few dissenting voices stemmed from Catholics and Irishmen on the one hand, and from antisemitic socialists on the other. *DR*

Italy

0264 Di Fant, Annalisa: La ricezione dell'"Affaire Dreyfus" sulla stampa cattolica italiana di Trieste. *Shalom Trieste: Gli itinerari dell'ebraismo,* ed. Adriano Dugulin. Trieste: Comune di Trieste, 1998. Pp. 297-306.

Asserts that in Trieste, as well as in the rest of Italy, the Catholic press ("L'Amico," "La Ricreazione," "L'Avvenire") opposed the revision in 1898 of Dreyfus' verdict by the French court, considering Dreyfus' alleged treason as part of a plot of Jewish money and power, in association with the liberal press. The outcome of the Affair, in the view of the Catholic press, would reveal and prove the excessive and dangerous influence of the Jews in society. *MZ*

0265 Fabbrici, Gabriele: Tra antiche e nuove paure: Gli ebrei reggiani nel 1797: Il caso di Benedetto Finzi di Correggio. *Materia Giudaica* 4 (1998) 14-20.

Recounts episodes of hostility against Jews in the Reggio Emilia region in 1797, under Napoleonic rule. The French invasion of Italy (1796) resulted in the enactment of a liberal constitution known as the "Constituzione Cispadana," which promulgated the emancipation of the Jews. The Christian world and the aristocracy reacted against these reforms. In Correggio, a Jew, Benedetto Finzi, was accused of inducing a Christian child to steal wheat from a mill. The authorities rejected the accusation, arousing a suspicion amongst the populace of their being bribed by the Jews. Afterwards, other Jews were accused of offending Christians during prayers and at processions. The tensions led to violence, until the intervention of the French army restored public order. *MZ*

Poland

0266 Lewalski, Krzysztof: "Zwiastun Ewangeliczny" – pismo Polaków-ewangelików w Królestwie Polskim wobec kwestii żydowskiej na przelomie XIX i XX wieku [The Attitude of "Zwiastun Ewangeliczny," a Periodical of Polish Protestants in the Kingdom of Poland, towards the Jewish Question at the Turn of the 20th Century]. *Biuletyn Żydowskiego Instytutu Historycznego* 192 (Dec 1999) 33-40.

Discusses the Lutheran-Evangelical community in Poland, which was composed mainly of German immigrants (from 1815 on). Its most important publication was the journal "Zwiastun Ewangeliczny" ("Evangelical Harbinger"), published in Warsaw. Surveys Jewish themes found in the journal. The main theme was the necessity of missionizing to the Jews. The writers of the articles believed that only through conversion would the Jews cease to be enemies of Christianity. Discusses, as well, views on Zionism, which was sometimes met with approval and sometimes seen as an obstacle to missionary work. Another important issue was boycott of the Jews, promoted by Endecja circles at the beginning of the 20th century. Protestant circles criticized the idea of boycott for fear that they too would be excluded by Endecja from the "native Polish Christian group." However, before World War I, Polish-Jewish relations deteriorated so much that boycott became a criterion of Polishness; the journal then joined the anti-Jewish campaign. It ceased publication in 1914. *HV*

0267 Nicholls, Stephen C.J.: *The Burning of the Synagogue in Neustettin: Ideological Arson in the 1880s*. Brighton: University of Sussex, 1999. 19 pp. (Centre for German-Jewish Studies: Research Paper, 2).

On 18 February 1881 the synagogue in Neustettin (now Szczecinek), Pomerania, was set on fire. The destruction of the synagogue was received with either indifference or cynical approval by the Christian majority and the conservative press. The authorities arrested five Jews who were falsely accused of committing the arson in order to obtain the insurance money; after two trials, four of the Jews were imprisoned. The arson was followed by riots in the town in summer 1881 and other disturbances in the vicinity. The arson and the ensuing riots were undoubtedly provoked by Ernst Henrici, who visited Neustettin five days earlier with a series of incendiary antisemitic speeches. The economic depression of the 1870s in rural Pomerania made its inhabitants vulnerable to the influence of political agitators. *DR*

0268 Penkalla, Adam: Opinie o wystąpieniach antyżydowskich w Kielcach w dniach 11 i 12 listopada 1918 roku [Opinions on the Antisemitic Disturbances in Kielce on 11-12 November 1918]. *Biuletyn Żydowskiego Instytutu Historycznego* 191 (Sept 1999) 55-69.

States that the anti-Jewish disturbances in Kielce took place in reaction to a meeting of the Jews in Kielce related to the independence of Poland. Four Jews were killed; ca. 250 people, mainly Jews, were injured. On 4 December, Jewish members of the City Council submitted a protest to the Council in which the authorities were criticized for lack of adequate police action during the events. This protest was discussed at a meeting of the Council on 11 December. Both the protest and minutes of the meeting are presented (pp. 56-68). The minutes show a diversity of opinions expressed by Polish and Jewish members of the Council. Concludes that the Polish members of the Council mainly represented the approach of Endecja, reflecting the prevailing attitude among the elites in the cities of Poland. A statement issued by the Council condemned the pogrom, but also contended that the Jews "provoked" the Poles by demanding autonomy in the new state of Poland which would then become "Judeo-Polonia." *HV*

0269 Weeks, Theodore R.: Fanning the Flames: Jews in the Warsaw Press, 1905-1912. *East European Jewish Affairs* 28, 2 (1998-1999) 63-81.

The revolutionary years 1905-6 marked a major rupture in relations between Poles and Jews in Russian Poland. These relations had deteriorated after the Polish insurrection of 1863, when the Poles were subjected to various administrative restrictions and humiliating measures, but the Jews were not. The influx of Jews from the Pale following the May Laws of 1882 was perceived by the Poles as an attempt to Russify the country with "these Lithuanians." Jewish economic success and the rise of Jewish nationalism exacerbated relations even more. Jews were seen by Poles, even liberals, as a foreign group. After 1906, the mainstream Warsaw press targeted all the Jews, not specific Jewish parties, and regarded plans to assimilate the Jews as unfeasible. The conflict with the Jews reached its peak in 1912, when it became clear that the result of the election to the Duma in Warsaw would depend on the Jewish vote. In that year, the tone of the liberal "Prawda" was no different from that of the oldest antisemitic newspaper "Rola." *DR*

0270 Weeks, Theodore R.: Poles, Jews, and Russians, 1863-1914: The Death of the Ideal of Assimilation in the Kingdom of Poland. *Polin* 12 (1999) 242-256.

Polish-Jewish relations, which were good in the period of the 1863 uprising against the Russian authorities, worsened greatly by 1914. Notes that these relations were not completely harmonious in 1863 nor did they degenerate irreparably in 1914. Several factors contributed to the change in relations: a deterioration of the legal and national situation of the Poles in the Kingdom of Poland, accompanied by an improvement in the economic and legal situation of the Jews in the early 1860s; an influx of Jewish immigrants from Russia, who were regarded as agents of Russification; the emergence of a new kind of antisemitism in Poland, based on national and economic factors; and the rise of Jewish national identity, which disheartened those Poles who expected full assimilation of the Jews. The abortive revolution of 1905 was a major watershed in Polish-Jewish relations; inter alia, Poles vented their frustration on the Jews. *DR*

0271 Weeks, Theodore R.: Polish-Jewish Relations 1903-1914: The View from the Chancellery. *Canadian Slavonic Papers* 40, 3-4 [Sept-Dec 1998] (1999) 233-249.

On the basis of documents now located in the Russian State Historical Archive (RGIA) in St. Petersburg, concludes that the tsarist government did not foster hostility between Poles and Jews in Russia's Polish provinces. However, Russian officials also did not sympathize with either national group. Their reports testify to the deterioration of relations between Poles and Jews, and their determination to maintain order and prevent violence. Russia lacked a consistent policy of Russification. In 1912, when Polish antisemitism was stirred up due to a conflict concerning Polish representation in the fourth Duma, Russian authorities banned the resultant anti-Jewish economic boycott in order to dampen nationalist tensions. Russia's ultimate lack of success in defusing such friction may have been a factor in the fall of the Empire. Russia was as much concerned about the Polish threat as about the Jewish threat; however, it feared disorder more than either. Finally, the Russian Empire clearly was unable to put an end to Polish antisemitism before 1914. *YC*

0272 Weeks, Theodore R.: *Zwischen zwei Feinden: Polnisch-juedische Beziehungen und die russische Behoerden zwischen 1863 und 1914.* Leipzig: Simon-Dubnow-Institut fuer Juedische Geschichte und Kultur, 1998. 27 pp.

Discusses relations between Poles and Jews in the part of Poland which was under Russian administration before World War I, and the attitudes of both towards the Russian authorities. States that although a major part of the Jewish population lived in these ten Polish provinces, this area was not dealt with much in studies on Jews in the Russian Empire. During the Polish revolt (1863-64) Jews and Poles united against the Russian occupier, but at the turn of the century Polish-Jewish relations were tense; Russian authorities did not sympathize with either group, and only attempted to keep the peace. Relates to the increasing tension in Polish-Jewish relations, and discusses the pogrom in Kishinev in 1903, the revolution in 1905 and its suppression, the boycott against the Jews in 1912, and the establishment of rural cooperatives which were meant to "liberate the population from Jewish exploitation." Claims that the failure of the Russian Empire to resolve ethnic conflicts accelerated the decline of the Empire. *ID*

0273 Wiech, Stanisław: Les répercussions de l'Affaire Dreyfus dans la presse régionale polonaise: L'exemple de la "Gazette de Kielce." *Revue des Etudes Juives* 158, 1-2 (Jan-June 1999) 213-226. Appeared in Polish in the "Biuletyn Żydowskiego Instytutu Historycznego" 167-168 (1993).

Russia

(see also Belarus)

0274 Boysen, Iris: Die revisionistische Historiographie zu den russischen Judenpogromen von 1881 bis 1906. *Jahrbuch fuer Antisemitismus-forschung* 8 (1999) 13-42.

Compares the traditional interpretation of the pogroms by Russian-Jewish historians, influenced by Dubnow and a nationalistic ideology, to the more differentiated one of the current generation. Traditional historiography attributes the pogroms to manipulation by government officials and government-backed secret societies; contemporary historiography casts doubt on such influence by the weak central government. The reactions of the civil authorities, the police, and the military to the pogroms differed from place to place – a circumstance inconsistent with orders from above – and were characterized mainly by helplessness. Besides, government officials (e.g. Plehve) subordinated their personal antisemitism to their concern for law and order. Historians now find the causes for the pogroms mainly in socioeconomic conditions. *RW*

0275 Budnitskii, Oleg V.: V.A. Maklakov i yevreiskii vopros [V.A. Maklakov and the Jewish Question]. *Vestnik Yevreiskovo Universiteta* 1 [19] (1999) 42-94.

Vasilii Maklakov, a leader of the right wing of the Russian pre-revolutionary Kadet party, harbored some anti-Jewish views, despite the fact that he fought against crude forms of antisemitism throughout his life. He believed that the individual could be antisemitic (e.g. boycott Jewish trade), but not the state. Supporting the idea of equality of rights for the Jews, he maintained that Russian society was not mature enough for this and toppled the bill for Jewish emancipation in the Duma in 1916. As a defender at the Beilis trial, Maklakov insisted that the goal of the court was to vindicate Beilis, but not all of Jewry, of the ritual murder accusation. During the civil war, while the entire Kadet party succumbed to the common antisemitic mood and condoned the pogroms of Denikin's army, Maklakov tried to convince Denikin to condemn the pogroms, at least symbolically. Dwells, also, on the correspondence between Maklakov and Shulgin in the 1920s. Shulgin regarded the Russian Revolution as a Jewish act intended to destroy the Russian state; he insisted on collective responsibility of the Jews for that, but spoke against pogroms. Maklakov saw the revolution as Russian, and opposed the principle of collective responsibility. *DR*

0276 Budnitskii, Oleg V., ed.: *Yevrei i russkaya revolyutsiya: Materialy i issledovaniya* [Jews and the Russian Revolution: Materials and Research Papers]. Moskva: Gesharim; Ierusalim: Hebrew University, Institute of Jewish Studies, 1999. 479 pp.

A collection of articles (some published previously) and historical documents pertaining to Jewish participation in the Russian revolutionary movement and to antisemitism. Partial contents: Budnitskii, Oleg V.: V chuzhom piru pokhmelye: Yevrei i russkaya revolyutsiya [A Hangover at Someone Else's Feast: Jews and the Russian Revolution. Appeared in "Vestnik Yevreiskovo Universiteta v Moskve" 13 (1996).] (3-21); Budnitskii, Oleg V.: S.G. Svatikov i yevo borba na "yevreiskom fronte": Vvedeniye k publikatsiyam [S.G. Svatikov and His Fight on the "Jewish Front": An Introduction to the [Following] Publications.] (22-28); Svatikov, Sergei G.: Sozdaniye "Sionskikh protokolov" po dannym ofitsialnovo sledstviya [The Making of the "Zion Protocols" According to the Official Inquiry. This manuscript, written in the 1930s, is published here for the first time.] (163-222); Budnitskii, Oleg V., ed.: B.I. Nicolaevsky – Vere Kon (Broido) [A letter written by Nicolaevsky to Kon in August 1964 on the origins of the "Protocols of the Elders of Zion," found in the Stanford University archives.] (223-230); Budnitskii, Oleg V.: Belye i yevrei (po materialam rossiiskovo posolstva v Parizhe i lichnovo archiva V.A. Maklakova) [The Whites and the Jews (According to the Materials of the Russian Embassy in Paris and Maklakov's Personal Archive). An earlier English version appeared in "Jews in Eastern Europe" 28 (1995).] (269-290); Krichevskii, Lev Yu.: Yevrei v apparate VChK-OGPU v 20-ye gody [Jews in the VChK-OGPU Apparatus in the Twenties. Appeared in "Vestnik Yevreiskovo Universiteta v Moskve" 8 (1995).] (320-350); Budnitskii, Oleg V.: "Yevreiskii vopros" v emigrantskoi publitsistike 1920-1930-kh [The "Jewish Question" in the Émigré Journalism of the 1920s-30s. An earlier version appeared in "Vestnik Yevreiskovo Universiteta v Moskve" 7 (1994).] (353-373); Budnitskii, Oleg V.: "Ostavim svyatochnye temy i pereidyom k yevreiskomu voprosu" (iz perepiski V.A. Maklakova i V.V. Shulgina) ["Let Us Drop Christmas Topics and Pass on to the Jewish Question" (from the Correspondence between V.A. Maklakov and V.V. Shulgin). Pp. 383-398 contain Shulgin's critical response to an article by Maklakov, complaining that he downplayed the Jewish role in the Russian Revolution; in his letters to Maklakov he maintained that the Jews must repent for all the evils they inflicted on Russia.] (374-442). *DR*

0277 Elyashevich, Dmitrii A.: *Pravitelstvennaya politika i yevreiskaya pechat v Rossii 1797-1917: Ocherki istorii tsenzury* [Government Policy and Jewish Printing in Russia 1797-1917: Essays on the History of Censorship]. Sankt-Peterburg: Mosty Kultury; Ierusalim: Gesharim, 1999. 792 pp.

Russian censorship of Jewish printing was contingent on tsarist policy toward the Jews. From 1797-1917 this policy went through several phases: acquaintance with Judaism, attempts to "improve" the Jews, and, from 1881, official war against the Jews, which in 1914-16 grew to outright persecution. Censorship followed the same stages. Before 1881, the Russian censorship regarded the Jews as a dark superstitious mass, which needed to be reeducated. Thus, its main target was rabbinic literature (e.g. the Talmud). Many censors were "enlightened" Jews, who were eager to "improve" their co-religionists; however, their reports provided Russian officials with anti-Jewish arguments. After 1881, the censorship's main target became secular Jewish literature – in Russian, Yiddish, or Hebrew – mainly due to the anti-government sentiments in the Jewish press, which focused on antisemitism, pogroms, etc. Special targets were the Yiddish language and "leftist" (including Zionist) publications. 1916 saw the outright suppression of Jewish printing. *DR*

0278 Elyashevich, Dmitrii A., ed.: *Yevrei v Rossii: Istoriya i kultura: Sbornik nauchnykh trudov, Vol. 5* [Jews in Russia: History and Culture: Collected Articles. Vol.5]. Sankt-Peterburg: Peterburgskii Yevreiskii Universitet, 1998. 393 pp.

Partial contents: Elyashevich, Dmitrii A.: Yevreiskaya pechat, politika i tsenzura v Rossii, 1797-1917: K postanovke voprosa [The Jewish Press, Policy, and Censorship in Russia, 1797-1917: On the Formulation of the Problem] (38-100); Nathans, B.: Jews, Law, and the Legal Profession in Late Imperial Russia (101-127); Klier, John Doyle: I.S. Aksakov and the Jewish Question, 1862-1886 [A Russian version appeared in "Vestnik Yevreiskovo Universiteta v Moskve" 17 (1998).] (155-169); Jignea, Clara L.: Protsessy po delam o Kishinyovskom pogrome 1903 goda: Vidnye rossiiskiye advokaty protiv sudebnoi vlasti [Trials of Cases Relating to the Kishinev Pogrom: Prominent Russian Advocates against the Power of the Judiciary] (175-210). *DR*

0279 Ganelin, Rafail Sh.; Kelner, Viktor Y.; Lukoyanov, I.V.: *Delo Mendelya Beilisa: Materialy Chrezvychainoi sledstvennoi komissii Vremennovo pravitelstva o sudebnom protsesse 1913 g. po obvineniyu v ritualnom ubiistve* [The Mendel Beilis Case: Materials of the Extraordinary Inquiry Commission of the Provisional Government on the Court Trial of 1913 on the Charge of Ritual Murder]. S.-Peterburg: Dmitrii Bulanin, 1999. 393 pp. At head of title page: Rossiiskii Yevreiskii Kongress, Gosudarstvennyi Arkhiv Rossiiskoi Federatsii.

After the March revolution of 1917, the Extraordinary Commission of Inquiry for the investigation of the miscarriage of justice committed by the top officials of the tsarist regime was established. This Commission opened an investigation of the Beilis trial which was held in 1911-13, aiming to ascertain to what degree the tsarist government had been involved with this trial and the unlawful acts committed by its officials in the course of the affair. The Commission of Inquiry failed to finish its work because of the October (Bolshevik) revolution. Pp. 3-25 contain the article by G. M. Reznik, "Sud nad M. Beilisom i mif o ritualnom ubiistve" ["The Court Trial of M. Beilis and the Myth of Ritual Murder"], describing the history of the Beilis case. Pp. 26-49 contain an introduction by the editors dealing with the activities of the Extraordinary Commission of Inquiry in 1917. Pp. 53-370 contain documents, mainly protocols of interrogations of participants of the trial, produced by the Commission. *DR*

0280 Gurovich, Mikhail Z.: Istoriya krovavovo naveta v Rossii (XVIII-XIX v.v.) [A History of the Blood Libel in Russia (18th-19th Centuries)]. *Korni* 7-8 (1997) 78-97.

A lecture delivered at the Jewish lecture hall in Nizhnii Novgorod, 1996. Argues that the ritual murder accusation originated not in the 12th century but in the Hellenistic period and continued into the Christian era. Throughout the centuries, decrees by Church authorities and monarchs condemned this libel, in Europe as well as in the Middle East, but could not stop it. In Russia, the first rumors of a blood libel emerged in the Chernihiv area in 1704. The first trial in Russia took place in Senno (Belarus) in 1799. A wave of such trials broke out in Russia's western areas in the wake of Napoleon's Russian campaign in 1812. Contends that the libel was of Polish-Ukrainian origin, and the first initiators of such trials (e.g. in Grodno and Velizh) were Polish Catholics. However, in 1853 a case occurred in Saratov, southeastern Russia. In the 1880s, the blood libel re-emerged in Russia, Europe, and Turkey; many pogroms of that period were accompanied by rumors of ritual murder. There were accusations even in the Soviet and post-Soviet periods. *DR*

0281 Kozhinov, Vadim V.: *"Chernosotentsy" i Revoliutsiya (zagadochnye stranitsy istorii)* ["Black Hundreders" and Revolution (Enigmatic Pages of History)]. 2nd, rev. ed. Moskva: [The Author], 1998. 228 pp. An earlier version appeared as vol. I of "Zagadochnye stranitsy istorii XX veka" (Moskva: Prima V, 1995).

A collection of essays, mainly aiming to justify the Black Hundred organizations in tsarist Russia. Ch. 4 (pp. 83-110), "Pravda o pogromakh" ("The Truth on the Pogroms"), criticizes generally accepted views on the pogroms in Russia. Contends that the pogroms were not a Russian invention: they took place also in Western Europe in the Middle Ages. The causes of the Russian pogroms were neither racial nor religious, but were purely economic. Neither the Black Hundred nor the authorities were involved in the bloodiest pogroms, in 1903-6. Only half of the victims in the Russian pogroms were Jews; the rest were non-Jewish leftists or were killed by the Jewish self-defense. Holds that the Black Hundred should not be regarded as a predecessor of the Nazis. Ch. 5 (pp. 111-134) continues to whitewash the Black Hundred from the accusation of participation in the pogroms. Dismisses Western criticism of some books published in Russia which are perceived worldwide as antisemitic. *DR*

0282 Kurganov, Yefim: Vasilii Rozanov i bibleiskii mir [Vasilii Rozanov and the World of the Bible]. *Russkii Yevrei* 4 (Sum 1997) 28-31.

Contends that Rozanov was neither an antisemite nor a philosemite. His main idea was that Christianity had led world civilization to a blind alley, and he called for a re-judaization of Christianity. He admitted the great contribution of the Jews to world culture; he admired the Bible, regarding it as a brilliant product of the Semitic spirit. Concludes that in assessing all of Rozanov's writings about the Jews, one must take into account that he was not a scholar, but rather a poet and a mystic. *DR*

0283 Lyubchenko, V.: "Pogrom visit v vozdukhe": Obshchestvennye nastroyeniya v Kiyeve posle pokusheniya na P.A. Stolypina (po materialam perlyustratsii) ["A Pogrom Is Hanging in the Air": Popular Moods in Kiev after the Attempt on P.A. Stolypin's Life (Based on the Police Examination of Private Letters)]. *Vestnik Yevreiskovo Universiteta* 19 (1999) 272-286.

On 1 September 1911, in the Kiev City Theater, the Russian-Jewish anarchist D. Bogrov fatally wounded the Russian Prime Minister, Pyotr Stolypin, with two shots. On 5 September Stolypin died. During nine days, 1-9 September, a pogrom seemed to be imminent. It did not occur, ostensibly because of the decisive preventive measures taken by the authorities. Pp. 277-284 present a sample of excerpts from private letters examined by the police, which show the level of pogromist sentiments in the city and its surroundings. Pp. 272-276 contain an introduction, describing the events. Suggests that, according to the letters, the measures taken by the authorities were not able to prevent the pogrom; most probably, it did not take place due to chance. Some minor anti-Jewish acts of violence did take place at the time. *DR*

0284 Rozental, Isaak S.: Moskva nachala XX veka: Yevrei, vlast, obshchestvo [Moscow of the Early 20th Century: Jews, Authority, Society]. *Vestnik Yevreiskovo Universiteta* 1 [19] (1999) 95-119.

The policy of the Moscow city and regional authorities toward the Jews was determined by their eagerness to preserve the purely Russian ethnic character of "Russia's second capital city." Acting against the Jews, the Moscow administration tried to present itself as an exponent of popular sentiments. Examines the attitudes of various groups in Moscow toward the Jews and the official politics that concerned them. The Right demanded more radical measures against the Jews, while the liberal bourgeoisie and intelligentsia deplored the antisemitic policy. The common people were greatly influenced by the "patriotic" Right. At the same time, the ideas for "solving the Jewish question" were different, even amongst the Right – from the gradual assimilation of all Jews to their elimination from Russia. The war that began in 1914 brought on an upsurge of patriotism and an influx of Jewish refugees into Moscow, which intensified antisemitic sentiments in Moscow. *DR*

Slovenia

0285 Verginella, Marta: Storie d'altri: Cenni sull'antisemitismo sloveno. *Shalom Trieste: Gli itinerari dell'ebraismo,* ed. Adriano Dugulin. Trieste: Comune di Trieste, 1998. Pp. 307-314.

States that hostility against Jews in Slovenia reached its highest point in the last twenty years of the 19th century, after civil and political emancipation in the Austro-Hungarian Empire was completed (1867) and Slovenian nationalism was on the rise. Antisemitism was displayed by the Church and the press, and in campaigns initiated by intellectuals and politicians. It was based on traditional stereotypes and on religious anti-Judaism. The Jews were identified with German (Austrian) culture (cosmopolitan and capitalist); consequently, they were seen as a danger to Slovenian national aspirations and to the economy. Slovenian antisemitism also echoed accusations against Jews in Austria after the economic crisis of 1873 and the antisemitic propaganda in Poland and Russia, which depicted the Jews as exploiters of the Slavic people. *MZ*

1919-1945: General

0286 Carpozi, George, Jr.: *Nazi Gold: The Real Story of How the World Plundered Jewish Treasures.* Far Hills, NJ: New Horizon Press, 1999. xx, 460 pp.

Discusses Nazi expropriation of Jewish property – gold, artworks, insurance policy payouts, etc. Focuses on the profit made by such neutral countries as Switzerland (particularly its banks), Sweden, Spain, Portugal, Turkey, and Argentina, which acquired Nazi gold in return for crucial supplies that enabled Germany to continue fighting the war. The Vatican received gold looted by the Ustasha fascists in Croatia. France, Great Britian, and the USA are faulted for having failed to return gold and other Jewish property that came into their hands. The Allies were also derelict in regard to pressuring other countries to restore Jewish property. Clarifies how insurance companies colluded with the Nazis and, after the war, refused to honor Jewish claims. Ch. 20 (pp. 295-317), "Pandora's Box" deals with art that the Nazis looted from Jews; notes recent efforts to identify such works and return them to their rightful owners. Concludes with discussion of steps taken in the USA in the 1990s to pressure Swiss banks and relevant insurance companies to honor claims by survivors or their heirs. *YC*

0287 Coffart, Delphine: "Le Voyage des Damnés" ou l'Odyssée des Juifs allemands dans l'Atlantique, 1939. *Tsafon* 32 (Win 1997-1998) 5-13.

Describes the affair of the ship "St. Louis," carrying 907 German Jewish refugees, which left the port of Hamburg on 13 May 1939 destined for Havana. The Cuban authorities refused to allow the refugees to disembark. Eventually, they were forced to return to Europe and were dispersed in Holland, Belgium, France, and Great Britain. France accepted the refugees only in transit. Surveys French press accounts from June 1939, both Jewish and non-Jewish. The French press was very neutral in describing the events, which can be explained by the delicate relations between France and Germany in 1939. The Jewish press showed gratitude toward and faith in the French authorities, which from the perspective of later events can be seen as derisory. *HV*

0288 Edelheit, Abraham Jacob: *The Yishuv in the Shadow of the Holocaust: Zionist Politics and Rescue Aliya, 1933-1939.* Boulder, CO: Westview Press, 1996. xv, 347 pp. Based on the author's diss. – City University of New York, 1992.

Discusses the efforts of the Yishuv (the Jewish community in Palestine) to rescue Jews from Nazi-occupied Europe, mainly Germany, between 1933-

39. Focuses on the question of whether internal Jewish politics (e.g. differences between the Mapai party and the Revisionists) hampered the rescue, opining that this was not the case. Although the Revisionists understood the threat to European Jewry earlier than Mapai, and demanded immediate mass rescue, Mapai is shown to have changed its policy, over time increasing its support for mass immigration of endangered Jews to Palestine. Concludes that since six million Jewish lives were lost, the rescue effort was a failure but does not blame the Zionists. Instead, cites three factors for the failure: lack of financial resources to absorb immigrants, the apparent improvement in conditions in Germany in 1936-37 that deterred Jews from leaving when it was possible, and British restriction of Jewish immigration to Eretz-Israel. The Yishuv is credited with the rescue of half of the 500,000 Jews saved, or twice as many as were allowed to immigrate to the U.S. Ultimately, the failure of rescue was due to Jewish powerlessness. *YC*

0289 Eisenstein, Paul Steven: Holocaust Memory and Hegel. *History & Memory* 11, 2 (Fall-Win 1999) 5-36.

There is an opinion widely accepted by scholars that Hegel cleared a conceptual space amenable to the perpetration of the Holocaust. According to Hegel's philosophy, everything particular (like Jews) must disappear and give way to the general in the historical process. Similarly, historians pursuing Absolute Knowledge must ignore a particularity such as the Holocaust. Scholars rejecting this philosophy have moved away from attempts to construct historical master-narratives of the Holocaust to the production of a plurality of narratives. Memorializing of the Holocaust has superseded its history. However, Hegel also stated that the particular rather than the universal lies in the historical domain, and that the particular in history is already universal. Illustrates this with David Grossman's novel "See under: Love," in which the only way to reach Absolute Knowledge of the Holocaust proved to be "The Complete Encyclopedia of Kazik's Life," narrating the fate of a boy – an individual victim of the genocide. *DR*

0290 Eshkoli (Wagman), Hava (Eva): Yishuv Zionism: Its Attitude to Nazism and the Third Reich Reconsidered. *Modern Judaism* 19, 1 (Feb 1999) 21-40.

Based on a lecture delivered at a conference held at Yad Vashem, Jerusalem, November 1997. Post-Zionist historians accuse the Zionist leadership in Eretz Israel in the 1930s of ideological identification with Nazism, active contacts with the Nazis, and avoidance of a militant stance against Nazism. Shows that the argumentation used by the post-Zionists is superficial and based on marginal examples. The people who felt ideological affinity with Nazism never took a central place in the

Revisionist movement or in the Yishuv leadership at all. At an early stage, the Zionists indeed attempted to "extract good from evil" and to exploit the Nazi will to oust Jews from Germany for their own goals, e.g. the Zionist Executive contracted the Transfer Agreement with the Germans. Meanwhile, the Zionist reaction to this agreement, as well as other contacts with Germany, was mixed. Examines the press reaction to it in the Land of Israel. The attitude of the Zionist leaders toward the contacts with the Nazis changed after the enactment of the Nuremberg Laws, and especially after the "Kristallnacht" pogrom. *DR*

0291 Friling, Tuvia: Nazi-Jewish Negotiations in Istanbul in Mid-1944. *Holocaust and Genocide Studies* 13, 3 (Win 1999) 405-436. Adapted from ch. 8 of his book (in Hebrew) "Hetz ba-arafel" (1998). An earlier Hebrew version of this article appeared in "Iyyunim Bitkumat Israel" 4 (1994).

In May 1944 Joel Brand brought a Nazi proposal from Budapest to Istanbul, aiming to ransom Hungarian Jews for 10,000 trucks and other goods. The proposal included an invitation to Menahem Bader, a Yishuv emissary in Istanbul, to go to Berlin to negotiate on the basis of Brand's proposal. Traces activities of the Jewish Agency Executive concerning the proposal. Focuses on reactions of the JAE, the Jewish leadership in Palestine, and the Yishuv emissaries in Istanbul. Reports on contacts between the Jewish representatives and British, Hungarian, and American intelligence agents in Istanbul and elsewhere. Close examination of the events in Istanbul in May-June 1944 does not support the opinion that Ben-Gurion had no interest in the rescue of European Jews and, therefore, in Brand's proposal. Although he did not believe in dealing with the Nazis, Ben-Gurion was prepared to negotiate with them behind the backs of the Allies if the latter would reject the "blood for trucks" deal. *DR*

0292 Fuchs, Esther, ed.: *Women and the Holocaust: Narrative and Representation.* Lanham, MD: University Press of America, 1999. xv, 143 pp. (Studies in the Shoah, 22).

Partial contents: Garber, Zev: The Problem of Edith Stein: Jewess and Catholic Saint [A longer version appeared in his "Shoah, the Paradigmatic Genocide" (1994).] (1-7); Kellenbach, Katharina von: Reproduction and Resistance during the Holocaust [Discusses decisions of Jewish women to bear a child or not have an abortion as resistance to genocide.] (19-32); Nowak, Susan E.: In a World Shorn of Color: Toward a Feminist Theology of Holocaust Testimonies [Focuses on the Holocaust memoir "Rena's Promise" (1995) by Rena Kornreich Gelissen.] (33-46); Lentin, Ronit: Re-occupying the Territories of Silence: Israeli Daughters of Shoah Survivors between Language and Silence [Discusses Israeli society's image of itself as "masculine" vs. the alleged passive "femininity" of

Holocaust victims.] (47-61); Wickersham, Erlis Glass: Women as Agents of Suffering and Redemption in the Poetry of Nelly Sachs (63-87); Baron, Lawrence: Women as Resistance Fighters in Recent Popular Films: The Case of Hanna Senesh and Helene Moszkiewiez [On "Hanna's War" (1988) and "A Woman at War" (1991).] (89-96); Fuchs, Esther: The Construction of Heroines in Holocaust Films: The Jewess as Beautiful Soul (97-112); Kafka, Phillipa: Selection from Margaret Engel (Kohn) Keat's Memoir: 1939-1941 [On Prague under Nazi occupation, and her flight to Shanghai in 1940 and to the U.S. in 1941. From a manuscript.] (113-126); Fuks, Zila: Life in Hiding: Razek's Barn: An Excerpt from "Silent Screams" [Translation of an excerpt from her Hebrew memoir, "Za`akot lelo kol" (1988), about hiding in the Polish village of Sterdyń.] (127-133). *YC*

0293

גלבר, יואב: מ"אל תעלו!" עד חוק השבות: לבטים ותמורות בגישה הציונית לעלייה. קיבוץ גלויות: עלייה לארץ ישראל – מיתוס ומציאות, עורכת דבורה הכהן. ירושלים: מרכז זלמן שזר לתולדות ישראל, 1998. עמ' 249-282.

[Gelber, Yoav: From "Do Not Go Up!" to the Law of Return: Inner Struggles and Changes in the Zionist Attitude toward Aliyah. *Kibbutz galuyot: Aliyah le-Eretz Israel – mythos u-metzi'ut* [Ingathering of Exiles: Aliyah to the Land of Israel – Myth and Reality], ed. Dvora Hacohen. Jerusalem: Zalman Shazar Center for Jewish History, 1998. Pp. 249-282.]

Based on a paper delivered at a conference held at Bar-Ilan University, Ramat-Gan, 1994. Discusses various opinions regarding aliyah in the 20th century, beginning with that of the Va`ad ha-Po`el ha-Zioni in 1919, which discouraged aliyah on the grounds that Eretz-Israel was economically unable to absorb immigrants. In the 1920s the Yishuv leadership encouraged aliyah only of people of means or those willing to do difficult work. This attitude gradually changed during the 1930s due to Nazi antisemitism. In 1935, with the enactment of the Nuremberg Laws, the Jewish Agency was pressured to accept more German immigrants, but in 1936 Great Britain further limited the number of Jewish immigrants due to Arab pressure. While illegal immigration was initially seen as a way to strengthen the Yishuv, after the "Kristallnacht" pogrom in 1938 it was seen more as a way to rescue Jews. In 1942 David Ben-Gurion urged the Jewish Agency to save as many Jews as possible, without any conditions, but by then it was difficult to save any. Only ca. 50,000 Jews managed to reach Eretz-Israel between 1939-44, ca. 12,000 of them illegally. *RE*

0294 Greenberg, Gershon: A Musar Response to the Holocaust: Yehezkel Sarna's "Le'teshuva ule'tekuma" of 4 December 1944. *Journal of Jewish Thought and Philosophy* 7, 1 (1997) 101-138.

Discusses a 1944 speech given in Jerusalem by Rabbi Sarna (1889-1969), a leading representative of the Musar movement in Palestine. He stressed ethics as a response of the Orthodox Jews to the murder of their fellow Jews in Europe and the destruction of Torah centers. Sarna basically sublimated empirical history into metahistory by using traditional religious terms: "hurban" (destruction) connected the Holocaust with the destruction of the Temples, "din" (strict justice) related to the tremendous suffering of the Jews, "rahamim" (mercy) was linked with "teshuvah" (returning to God) that would result in final "geulah" (redemption). Sarna stressed the continued objective presence of God despite His "hester panim" (hiding) and His continuing love that made "teshuvah" possible and "geulah" inevitable. He also emphasized the unbroken bond between God and the Jewish people. His ideas were echoed by other Musar leaders of the Slobodka school and of the Novogrudok school, who denied the relevance of empirical history to the relationship between God and the Jewish people, and advocated the ideal of "bitahon" (trust in God) despite the horrors of history. *YC*

0295 Hochstadt, Steve: The Social History of Jews in the Holocaust: The Necessity of Interviewing Survivors. *Historical Social Research* 22, 3-4 [83] (1997) 254-274.

Stresses the importance of using the oral testimonies of survivors. This type of evidence expands our perspective on the Holocaust, "democratizing" it by including simple people, as well as leaders, and by providing the view of victims, whereas much of Holocaust history thus far has been based on perpetrator testimony or documents. Further, oral testimony of survivors provides emotional cues to Jewish reactions to various kinds of suffering. Sensitivity is required on the part of the interviewer, who is a partner in the interview process. Discusses his interviews with Jews who found refuge from Nazi persecution in Shanghai, who relate how they were treated by the Japanese occupiers. *YC*

0296 Langer, Howard J.: *The History of the Holocaust: A Chronology of Quotations.* Northvale, NJ: Jason Aronson, 1997. xv, 306 pp.

A collection of quotations from books, articles, historical documents, public pronouncements, eyewitness accounts, etc., covering the history of antisemitism, the Holocaust, and the aftermath of the Holocaust. Pp. 1-35, "Prelude," cover the history of antisemitism from early Christianity until 1932. Pp. 37-213, "Holocaust," cover the history of the Nazi persecution and genocide of Jews, as well as the world's reaction to it in 1933-45. Pp. 217-263, "Aftermath," deal with postwar events pertaining to the Holocaust and reflections on it. *DR*

0297 Lustiger, Jean-Marie: Singularité de la Shoah. *Sens* 50, 2 (1998) 67-72.

A speech delivered at the University of Witten, Germany, 8 July 1997, on the occasion of conferring an honorary doctorate on Prof. Saul Friedlaender (first published in "Etudes," January 1998). Praises and comments on Friedlaender's research on the Holocaust. Adds to Friedlaender's analysis of human beings' destructive instinct [in his "Reflets du nazisme" (1982)] his own reflections based on study of the Bible. Underlines the uniqueness of the Holocaust and gives a theological interpretation, that it stemmed from a need to kill the people (the Jews) who gave moral law to the world (the Ten Commandments). The revolt against the Commandments brought on the murder of God's people. *HV*

0298 Morris, Benny: Response of the Jewish Daily Press in Palestine to the Accession of Hitler, 1933. *Yad Vashem Studies* 27 (1999) 363-407. Appeared in Hebrew in "Yad Vashem: Kovetz mehkarim" 27 (1999).

The Jewish press in Palestine reacted to the rise of political antisemitism in Germany even before 1933. It followed the Nazi accession to power in 1933 with anxiety and succeeded in conveying this sense to the reader. However, most papers ("The Palestine Post" was the only exception) focused, in their coverage and analysis of events in Germany, on their impact on Zionism and Palestine and viewed the developments through the prism of Zionist or pan-Jewish interests. The anti-Nazi boycott and the Transfer Agreement with Germany were the two burning issues in the Jewish world in 1933. While Revisionist Zionists supported the boycott, the General Zionists and the Left opposed it as harmful for the interests of the Yishuv and European Jewry. Similarly, the press considered rescuing German Jews mainly in terms of their usefulness for the Yishuv. The publications also reflect the inability of the Yishuv to offer substantive aid to German Jews. *DR*

0299

מוריס, בני: העיתונות היהודית היומית בארץ-ישראל נוכח עליית היטלר לשלטון, 1933.
יד ושם 27 (1999) 287-320.

[Morris, Benny: Response of the Jewish Daily Press in Palestine to the Accession of Hitler, 1933. *Yad Vashem* 27 (1999) 287-320.] Appeared in English in "Yad Vashem Studies" 27 (1999).

0300 Moyn, Samuel: Judaism against Paganism: Emmanuel Levinas's Response to Heidegger and Nazism in the 1930s. *History & Memory* 10, 1 (Spr 1998) 25-58.

Argues that Levinas underwent a personal and philosophical crisis when Heidegger, his mentor, came out in support of Nazism in 1933. Levinas's "guilty conscience" may have catalyzed his development from an orthodox Heideggerian to a thinker who saw a link between Heidegger's philosophy of "Being" and the imminentist, body-oriented philosophy of Nazism. Levinas discussed Nazism as a pagan assault on the freedom of Judeo-Christianity. Despite its attraction for him, the brutal political implications of Heidegger's thought made it unacceptable. For Levinas, Judaism came to represent the antipode of Nazism-Paganism-Heideggerism. "Being" was replaced by "Being Jewish." This Judaism was suprahistorical and universalistic, with the Jew symbolizing the rejection of all "paganism" in favor of a transcendentally-based ethics. *YC*

0301 Pizzuti, Giuseppe Mario: *L'eredità teo-logica del pensiero occidentale: Auschwitz.* Soveria Mannelli: Rubbettino, 1997. 237 pp.

Asserts that modern European thought is based on the contamination of the idea of God by the philosophy of being. This resulted in a mode of thought typical of the Occident, insensible to comprehension of the Other, the different. Ch. 3 (pp. 101-142), "Auschwitz," analyzes the Shoah as the revelatory outcome of that pattern of thought. *MZ*

0302 Raszelenberg, Patrick: The Khmers Rouges and the Final Solution. *History & Memory* 11, 2 (Fall-Win 1999) 62-93.

Demonstrates that the mass murders of the civilian population in Cambodia during the Khmer Rouge regime (1975-79) can not be equated with the Nazi genocide of Jews and did not constitute a genocide at all. Tuol Sleng, the central Khmer Rouge interrogation center, which after 1979 was converted into a memorial following a pattern of camp-site memorials in post-World War II Poland, was in no way a "Cambodian Auschwitz." *DR*

0303 Rice, Rondall R.: Bombing Auschwitz: US 15th Air Force and the Military Aspects of a Possible Attack. *War in History* 6, 2 (Apr 1999) 205-229.

Analyzes military aspects of the proposed bombing of Auschwitz in 1944. Had such an attack been ordered, the task would most likely have fallen to the US 15th Air Force, as a unit prepared for bombing oil facilities in Silesia. Disagreeing with Richard Foregger and James Kitchens, shows that such an operation, although difficult, was feasible and might have been relatively effective. By August 1944, enough information was, or could have been, available to plan and carry out the attack. In the case of bombing Birkenau, the aircraft would have had greater distance from the German gunners' positions in Auschwitz than there was during the bombing of the oil facilities. The bombing could have been accurate enough. The reason that the bombing did not take place was the lack of political will. Had the political will existed, and had the U.S. War Department seriously studied this possibility and exerted pressure on the military leaders, who were reluctant to take risks, it would have overcome all objections, as the Warsaw air-drop operations in August-September 1944 demonstrated. *DR*

0304 Rosenfeld, Gavriel D.: The Politics of Uniqueness: Reflections on the Recent Polemical Turn in Holocaust and Genocide Scholarship. *Holocaust and Genocide Studies* 13, 1 (Spr 1999) 28-61.

The question of the uniqueness of the Holocaust emerged in the USA in the late 1970s-80s in reaction to the double process of historicization and politicization of the Holocaust. The main exponents of the uniqueness conception were Yehuda Bauer, Deborah Lipstadt, Daniel Goldhagen, and especially Steven Katz, who claimed that the Holocaust was the only genuine case of genocide. In 1996, several scholars challenged this conception, among them two historians of the genocide of American Indians – David Stannard and Ward Churchill. The former stated that the most massive genocide was that of native Americans, and even conflated the supporters of the Holocaust uniqueness conception with Holocaust deniers. Both claimed that the uniqueness conception serves current Israeli and Zionist political goals. Without dismissing Stannard's and Churchill's stances outright, criticizes their numerous exaggerations. Stresses that the uniqueness conception emerged as a defense against attempts to "normalize" the Holocaust, and that it was study of the Holocaust that gave the incentive for the study of other forms of genocide. *DR*

0305 Surette, Leon: *Pound in Purgatory: From Economic Radicalism to Anti-Semitism*. Urbana: University of Illinois Press, 1999. xii, 314 pp.

Based on an extensive discussion of economic theory, including Social Credit and other "crankish" ideas that appealed to Ezra Pound, argues that this poet, who became a mouth-piece for fascist Italy, was not always an antisemite. He became one around 1936, influenced by the Depression and antisemites such as Father Coughlin and the maverick economic thinkers Arthur Kitson and Hugo Fack. Ch. 11 (pp. 239-260), "The Jewish Conspiracy," follows Pound's "descent into the paranoia of conspiracy theory and antisemitism." Views Pound's antisemitism as stemming from his muddled economic ideas and, hence, as an intellectual failure rather than a mental or emotional problem. Pound identified Jews with international finance and believed that they wanted to dominate the world. He associated them, on the one hand, with usury and, on the other, with the Bolsheviks. He also blamed World War II on the Jews and the bankers. Ch. 12 (pp. 261-289), "From Rome to Washington," makes clear that Pound remained virulently antisemitic through his years of incarceration for treason and beyond. Even in 1953 he still viewed Hitler and Mussolini as victims of the Jews. *YC*

0306

טבת, שבתי: השנים הנעלמות והחור השחור. תל-אביב: דביר, 1999. 374 עמ'.

[Teveth, Shabtai: *The Vanished Years and the Black Hole*. Tel-Aviv: Dvir, 1999. 374 pp.]

A biography of the life of David Ben-Gurion. Pp. 127-242 deal with the "black hole," the Holocaust period. Disputes accusations that he was indifferent to the plight of European Jewry and even passively cooperated with the Nazis. Contends that his concern for European Jewry was the basis for all his actions in the 1930s; he discussed and planned their mass emigration, which was for the most part thwarted by the Arabs and British immigration restrictions. It is true that Ben-Gurion used the Nazi threat to encourage aliyah, but he also used the Arab riots of 1936 to promote Zionism, believing that problems could strengthen the nation. Ben-Gurion and the Jewish Agency attempted to negotiate the rescue of Jews through 1942; however, by 1943 he realized that rescue attempts were futile and he concentrated on efforts to protect the Yishuv and prepare for Jewish immigration after the war. Pp. 225-242 discuss the activities of Nathan Schwalb and Rabbi Michael Dov Weissmandel. *RE*

0307 Tobler, Douglas F., ed.: *Remembrance, Repentance, Reconciliation: The 25th Anniversary Volume of the Annual Scholars' Conference on the Holocaust and the Churches*. Lanham, MD: University Press of America, 1998. xx, 215 pp. (Studies in the Shoah, 21).

Papers delivered at the conference held in 1995 at Brigham Young University in Salt Lake City, Utah. Partial contents: Culbertson, Elaine F.; Libowitz, Richard Lawrence: Teach the Teachers: Asking the Questions [Discusses a workshop developed for training teachers to teach about the Holocaust.] (19-29); Epstein, Joel J.: "Nazism and the Holocaust": Evolution of a Course, 1979-1994 [At Olivet College.] (31-46); Grunfeld, Uriel Jeremiah: Holocaust, Movies and Remembrance – the Pedagogical Challenge [Focuses on pedagogical use of the film "Schindler's List."] (47-56); Haas, Gerda: How Himmler Tried to Save His Life and Saved Mine Instead [Haas recounts her release from Theresienstadt to Switzerland in early 1945, due to a deal between Himmler and Musy, a Swiss ex-president.] (57-64); Halperin, Joan Ungersma: Ressentiment and Reconciliation between the Hidden Child and the Children of Rescuers [Relates the case of a child of a Dutch family of rescuers who resents the Jewish family being rescued.] (65-75); Herrmannová, Margit: On an Alien Planet [Recounts one year, 1943-44, spent by the author in Family Camp B II b in Auschwitz-Birkenau.] (77-88); Hornshøj-Møller, Stig:

"Der ewige Jude": An X-Ray of the Decision That Led to the Holocaust [Argues that the film "Der ewige Jude" led Hitler to his decision to perpetrate the Final Solution.] (89-105); Kitterman, David H.: Refusing to Kill in the Midst of the Holocaust: The Case of Klaus Hornig [Hornig, a member of a German police battalion stationed in Poland, refused orders to kill Russian POWs; he was sent to jail and then to Buchenwald, which he survived.] (107-126); Klein, Ilona: Primo Levi and Bruno Piazza: Auschwitz in Italian Literature [On their testimonies of their experiences in Auschwitz.] (127-135); Morahg, Gilead: Liberating the Camps: The Concentrationary Experience in Recent Israeli Literature [Focuses on David Grossman's novel "See under: Love."] (137-147); Oben, Freda Mary: Edith Stein and Etty Hillesum: Together at Westerbork [Discusses spiritual parallels between the two women, who were both at Westerbork in August 1942.] (149-163); Saidel, Rochelle Genia: Ravensbrueck Women's Concentration Camp: Before and after Liberation (165-175); Siegel, Ben: Why No Holocaust Novel, Sholem Asch? [Discusses readers' disappointment that Asch did not write a major novel about the Holocaust, although he did write stories on the topic.] (177-186); Teichman, Milton; Leder, Sharon: Remembering for the Future: Truth-Telling and Lamentation [Based on the introduction to the authors' anthology, "Truth and Lamentation: Stories and Poems on the Holocaust" (1994).] (187-194); Littell, Marcia Sachs: Breaking the Silence: A History of Holocaust Education in America (195-212). *YC*

0308

ויץ, יחיעם: הידיעות על ההשמדה באירופה ושאלת ממדי העלייה. קיבוץ גלויות: עלייה לארץ ישראל — מיתוס ומציאות, עורכת דבורה הכהן. ירושלים: מרכז זלמן שזר לתולדות ישראל, 1998. עמ' 233-248.

[Weitz, Yechiam: Information on the Destruction in Europe and the Question of the Dimensions of Aliyah. *Kibbutz galuyot: Aliyah le-Eretz Israel – mythos u-metzi'ut* [Ingathering of Exiles: Aliyah to the Land of Israel – Myth and Reality], ed. Dvora Hacohen. Jerusalem: Zalman Shazar Center for Jewish History, 1998. Pp. 233-248.]

Based on a paper delivered at a conference held at Bar-Ilan University, Ramat-Gan, 1994. Discusses changes in the attitudes of Jews in Eretz-Israel, and especially of David Ben-Gurion, toward aliyah during World War II. States that although Ben-Gurion's plan for European aliyah in the 1930s was mainly political – to ensure the creation of a Jewish state by having a Jewish majority – in the late 1930s he increasingly expressed concern for the fate of European Jews. In 1941 he presented a plan to the Jewish Agency for a massive immigration after the war, regardless of financial or any other considerations, and in 1944 he was even more determined to bring as many Jews as possible to Eretz-Israel. There were still political reasons for his determination, but there were ideological and humanitarian reasons as well, both for him and for the Jews in the Yishuv: the Holocaust proved that the existence of a Jewish state was necessary, and the Jews in the Yishuv felt that they were obligated to help their brethren in Europe. *RE*

0309 Zweig, Ronald W.: Feeding the Camps: Allied Blockade Policy and the Relief of Concentration Camps in Germany, 1944-1945. *The Historical Journal* 41, 3 (Sept 1998) 825-851.

Contends that sending relief parcels (food, medicine, clothing) to Jewish and other civil prisoners in the Nazi camps was a lost opportunity which could have saved thousands of lives. Various projects to send parcels were proposed in 1944 and early 1945; the main channel of relief might have been the International Red Cross. However, the Allied blockade was a stumbling block to all the relief projects. In the USA, public opinion was critical of the blockade; the War Refugee Board was created, and Washington was prepared to allow the ICRC to expand its relief efforts from POWs to civilians. In mid-1944, Britain was also willing to agree to a limited food relief program for camp internees. However, only small amounts of food were dispatched to civilian prisoners because the relief agencies failed to convince the Supreme Headquarters of the Allied Expeditionary Force (which was in charge of the situation) that improving the rations of the camp inmates would not strengthen the German capability to resist. *DR*

1919-1945: Europe

General

0310 Abram, Alvin: *The Light after the Dark*. Scarborough, Ont.: Abbeyfield Publishers, 1997. xix, 278 pp.

Contains biographies of six Holocaust survivors who emigrated to Canada. Ch. 1 (pp. 1-38), "An Eye for an Eye," recounts the story of Zalman (Eugene) Katz, most of whose family was killed in a mass murder in the Dzisna ghetto in Belarus. He fought with partisans and then with the Soviet Army. Ch. 2 (pp. 41-72), "The Promises," reports how 11-year-old Michael Kutz escaped a mass murder of 12,000 Jews in Nesvizh, Belarus, on 30 October 1941; he also fought with the partisans. Ch. 3 (pp. 75-126), "We Will Meet Again," tells how Moishe Perlmutter escaped from the Nazi-occupied Ukrainian village of Mulinsk and hid in the forest for three years. After the war he and other Jewish children fought off a pogrom in Zakopane. Ch. 4 (pp. 130-165), "I Have a Mission," recounts the survival of Dubi Arie, born in Warsaw in August 1939. In September his mother fled with him and his brother to Russia; in 1948 they emigrated to Israel. In 1974, Arie and his family moved to Toronto, where he discovered his mission as a Jewish artist. Ch. 5 (pp. 167-209), "I Knew My Enemy," is a memoir by Marjan Rosenberg, born in 1926 in Szydlowiec, who survived a mass murder in his hometown and then labor camps. After escaping, he passed as a Christian. Ch. 6 (pp. 211-246), "Don't Worry – It Will Be Good," recounts the story of Batia Schmidt and her young daughter Feiga (now Batia Malamud and Faigie Libman), who survived the Kovno ghetto, the Stutthof concentration camp, and a death march. *YC*

0311 Abravanel, Freddy H.: Bergen-Belsen: Ma déportation (août 1943) jusqu'à mon retour en Grèce (août 1945). *Revue d'Histoire de la Shoah* 165 (Jan-Apr 1999) 208-221.

Memoirs by a Greek Jew, born in Salonika in 1924. Describes the occupation of the city by the Nazis and the deportation of the Jews. He and his family, possessors of Spanish passports, were deported to Bergen-Belsen along with ca. 360 other Jews from Salonika. Describes their transport to the camp and conditions of life in the camp. After the liberation of Bergen-Belsen, Abravanel and his family returned to Greece. *HV*

0312 Aly, Goetz: *"Final Solution": Nazi Population Policy and the Murder of the European Jews.* Trans.: Belinda Cooper, Allison Brown. London: Arnold, 1999. viii, 305 pp. Originally published as "'Endloesung': Voelkerverschiebung und der Mord an den europaeischen Juden" (Frankfurt a.M.: S. Fischer, 1995).

0313

אנוליק, בנימין: הצלה מתוך המערכת הנאצית. דפים לחקר תקופת השואה 14 (1997) 327-337.

[Anolik, Binyamin: Rescue from within the Nazi System. *Dapim le-Heker Tekufat ha-Shoah* 14 (1997) 327-337.]

Describes rescue acts of members of the SS, the Gestapo, the militia and police forces, the Wehrmacht, the economic and industrial establishment, and the occupation authorities. The information was compiled from testimonies and diaries, and from the files of Yad Vashem's public committee for recognition of the "Righteous among the Nations." Some of these rescuers also perpetrated crimes against Jews. Not one of them has been recognized as "Righteous among the Nations." *LFo*

0314 Auschwitz: *La Solution finale*. Paris: L'Histoire, 1998. 114 pp. (Les Collections de l'Histoire, hors série 3).

Partial contents: Favez, Jean-Claude: D'où vient l'antisémitisme allemand (8-13); Friedlaender, Saul: Hitler et les Juifs [Appeared in "L'Histoire" 11 (1979).] (14-18); Friedlaender, Saul; Sales, Véronique: La Nuit de cristal: Récit d'un pogrom [An interview with Friedlaender; appeared in "L'Histoire" 218 (1998).] (20-21); Burrin, Philippe: Vers la Solution finale (22-25); Azéma, Jean-Pierre: Auschwitz: Comment fonctionne un camp de la mort (28-33); Pressac, Jean-Claude: Enquête sur les chambres à gaz (34-41); Burrin, Philippe: Les allemands, un peuple de bourreaux? (44-48); Levillain, Philippe; Bermond, Daniel: Le silence du Vatican: Entretien avec Philippe Levillain (49-51); Matard-Bonucci, Marie-Anne: Italie: Fascisme et antisémitisme d'état (52-55); Kaspi, André: Vichy a-t-il sauve les Juifs? (56-59); Marrus, Michael Robert: Pierre Laval et les enfants juifs (60-63); Kaspi, André: Les Alliés savaient-ils? (64-67); Wieviorka, Annette: L'insurrection du ghetto de Varsovie (68-69); Wieviorka, Annette: La gloire des justes (70-75); Bédarida, François: Comment écrire l'histoire du génocide (78-81); Wieviorka, Annette: Le jour où les Alliés découvrent les camps (82-89); Wieviorka, Annette: Nuremberg: Les Nazis devant leurs juges (90-95); Wieviorka, Annette: Adolf Eichmann ou la banalité du mal (96-99); Rousso, Henry: Portrait d'un faussaire: Paul Rassinier (100-101); Rebérioux, Madeleine: Faut-il des lois contre les négationnistes? (102-103). *TS*

0315 Bak, Sofie Lene: Altruisme og Holocaust: Jødeforfølgelserne i Danmark og Italien: En sammenligning [Altruism and the Holocaust: A Comparative Study of the Persecution of Jews in Denmark and Italy]. *Rambam* 8 (1999) 74-87. With an English summary.

The similarity in the number of Jews and the degree of assimilation in Denmark and Italy form the basis for this study. In both countries most of the Jews survived, aided by their fellow countrymen. But the Danish rescue operation was over in a few weeks and carried relatively limited risks, whereas the Italian escape routes were few and the Jews had to be kept hidden for longer periods of time. Moreover, the help that Italian Jews received was opposed to the policies of the regime. In contrast, both the reaction to the persecution of Jews and the rescue operation in Denmark was of a collective nature. Socio-psychological studies of altruism during the Holocaust have mainly focused on the background of individuals in an often hostile environment and may provide an explanation for the Italian situation, whereas the Danish collective action still lacks a plausible explanatory model. *SSC*

A collection of memoirs and documents on the history of the Jews in the city of Câmpulung Moldovenesc, Bukovina, mainly after 1900. Pp. 192-348 deal with the Holocaust period. They describe the anti-Jewish measures and the deportation in October 1941 of ca. 1,800 Jews from Câmpulung and its environs to Transnistria, and the suffering in the ghettos and camps there (e.g. Shargorod, Tulcin, Moghilev). Vol. 2 (pp. 353-750) presents memoirs of survivors about their families, arranged alphabetically by family name, including their fate in the Holocaust. *AA*

0317 Bartov, Omer: Victimes et absences. *Les Cahiers du Judaïsme* 4 (Spr 1999) 103-114.

Reflects on memories of a traumatic childhood of Jews and non-Jews, analyzing similarities and differences in their psychological states. Gives examples of memoirs or autobiographical stories by Albert Camus, Heinrich Boell, Otl Aicher, Georges Perec, Saul Friedlaender, and Shlomo Breznitz. Deals with descriptions of the war period in memoirs by German writers who were anti-Nazi. Mentions that both Jews and those Germans who did not support Nazism felt victimized. Finds similarities in the works of Boell, Aicher, Friedlaender, and Breznitz, in the way they reconstruct the past. All of them led a "double life." Friedlaender and Breznitz survived in Catholic monasteries. In contrast, Boell returned to Germany after the war, to his country and his mother tongue. The Jewish writers did not return to their hometowns; they lost their parents and their ties with their childhood and its memory, which caused a loss of part of their identity. *HV*

0318 Baumel, Judith Tydor: Gender and Family Studies of the Holocaust: A Historiographical Overview. *Women* 7, 2 (Fall 1996) 114-124.

The first books on women, children, and family life under Nazism appeared immediately after World War II. They were mostly memoirs and diaries of women survivors; in addition, two studies on children in the Holocaust were conducted by Ernst Papanek and Mark Dvorzetsky. In the 1950s-60s little systematic research was done on women and children in the Holocaust, although several collections of testimonies were published. The earlier memoirs of women do not deal with gender-related problems. Only since 1970 have there been women's memoirs strongly emphasizing gender-related experiences, not only in numerous collections of testimonies but also in academic studies. There is a group of historians involved in gender and family Holocaust research, conceptualizing these topics. Contends that gender and family studies of the Holocaust are still in their infancy. *DR*

0319 Baumgartner, Andreas: *Die vergessenen Frauen von Mauthausen: Die weiblichen Haeftlinge des Konzentrationslagers Mauthausen und ihre Geschichte.* Wien: Verlag Oesterreich, 1997. 249 pp.

Begins with a discussion of the role of women in the Third Reich according to Nazi ideology, and surveys the situation of women who were persecuted and sent to concentration camps. Relates briefly to the internment of women in Ravensbrueck, Auschwitz, and Bergen-Belsen. Focuses on women interned in Mauthausen from 1942, and on the establishment of a special women's section in the camp in September 1944, where 4,065 women were registered, among them 885 Jews. In the satellite camps Lenzing and Amstetten, 597 Jewish women were interned, while in St. Lambrecht, Mittersill, and Hirtenberg the number of Jews is not indicated. Mauthausen was classified as a third-grade camp, equivalent to an extermination camp. The Jews were the most ill-treated, and were more likely to be sent to labor commandos which meant almost certain death. Describes the death marches of thousands of Hungarian Jews, most of them women and children, to Mauthausen in March 1945; the number of those who died is not known. Notes that many women died before and during the liberation of the camp as a result of overcrowding, lack of sanitary conditions, and epidemics. *ID*

0320 Beck-Klein, Grete: *Was sonst vergessen wird: Von Wien nach Schanghai, England und Minsk: Juedische Schicksale 1918-1996.* Konstanz: Hartung-Gorre, 1997. 113 pp.

Memoirs of a Jewish woman, born in 1918 in Vienna. Relates the fate of her family (Beck) in the Holocaust. During the "Kristallnacht" pogrom, Grete's brother was arrested and imprisoned; he was released on condition that he leave Austria in two weeks. With much effort, her mother obtained a visa for him to Shanghai; her father also left for Shanghai in April 1939, while Grete and her sister were sent to England. Her mother stayed behind to take care of her own parents; she was deported to Minsk in June 1942 and murdered there. Grete's father died in Shanghai in February 1943. Grete married in England and emigrated to Israel in 1950. *ID*

0321 Bendremer, Jutta T.: *Women Surviving the Holocaust, in Spite of the Horror.* Lewiston, NY: Edwin Mellen Press, 1997. 118 pp.

Relates the stories of ten female Holocaust survivors now living in the Akron-Canton area of Ohio, based on interviews conducted with them by the author. Four of them were from Poland, two from Hungary, two from Czechoslovakia, the others from Germany, Romania, and Austria. The interviewees related their wartime and postwar experiences; some of them were in ghettos, some in Nazi camps. They were later able to lead productive, satisfying lives despite the horror of their wartime experiences. Pp. 109-114 contain the form of the questionnaire given to each woman, the list of questions used for the interviews, and data resulting from the above. *DR*

0322 Benz, Wolfgang: *The Holocaust: A German Historian Examines the Genocide*. Trans.: Jane Sydenham-Kwiet. New York: Columbia University Press, 1999. xi, 186 pp. Originally published as "Der Holocaust" (Muenchen: C.H. Beck, 1995).

0323 Benz, Wolfgang; Wetzel, Juliane, eds.: *Solidaritaet und Hilfe fuer Juden waehrend der NS-Zeit: Regionalstudien 1-3*. Berlin: Metropol, 1996-1999. 283; 331; 323 pp. Vols. 1-2 appeared in 1996; vol. 3 in 1999.

Contents: Vol. 1: "Polen, Rumaenien, Griechenland, Luxemburg, Norwegen, Schweiz": Benz, Wolfgang; Wetzel, Juliane: Moeglichkeiten und Formen der Hilfe fuer verfolgte Juden: Einleitende Bemerkungen (7-17); Kosmala, Beate: Ungleiche Opfer in extremer Situation: Die Schwierigkeiten der Solidaritaet im okkupierten Polen (19-97); Hausleitner, Mariana: Rumaenische Sonderwege – Pogrome und Hilfsaktionen: Ueberlebungschancen unter dem antisemitischen Regime (99-134); Spengler-Axiopoulos, Barbara: "Wenn ihr den Juden helft, kaempft ihr gegen die Besatzer": Der Untergang der griechischen Juden (135-185); Hoffmann, Serge: Luxemburg – Asyl und Gastfreundschaft in einem kleinen Land (187-204); Hoffmann, Christhard: Fluchthilfe als Widerstand: Verfolgung und Rettung der Juden in Norwegen (205-232); Picard, Jacques: Die Schweiz: Hilfe, Selbsthilfe und Solidaritaet entlang der Grenze (233-270). Vol. 2: "Ukraine, Frankreich, Boehmen und Maehren, Oesterreich, Lettland, Litauen, Estland": Golczewski, Frank: Die Revision eines Klischees: Die Rettung von verfolgten Juden im Zweiten Weltkrieg durch Ukrainer (9-82); Ammerschubert, Silke: Juden in Frankreich – Verfolgung und Rettung 1940-1944 (83-135); Rothkirchen, Livia: "...Gnade und Barmherzigkeit...": Die tschechischen Juden an der Schwelle des Todes (137-171); Koenigseder, Angelika: Oesterreich – ein Land der Taeter? (173-229); Vestermanis, Margers: Retter im Lande der Handlanger: Zur Geschichte der Hilfe fuer Juden in Lettland waehrend der "Endloesung" (231-272); Kibelka, Ruth: Litauen: Verfolgung und Rettung unter deutscher Besatzung (273-293); Gurin-Loov, Eugenia: Verfolgung der Juden in Estland (1941-1944): Rettungsversuche und Hilfe (295-307). Vol. 3: "Daenemark, Niederlande, Spanien, Portugal, Ungarn, Albanien, Weissrussland": Weiss, Hermann: Die Rettung der Juden in Daenemark waehrend der deutschen Besetzung 1940-1945 (11-86); Houwink ten Cate, Johannes Th.M.: Mangelnde Solidaritaet gegenueber Juden in den besetzten niederlaendischen Gebieten? (87-133); Rother, Bernd: Spanien – zwischen Hilfe und Restriktion (135-160); Hess, Renate: "Was Portugal getan hat, hat kein anderes Land getan" (161-206); Karsai, László: Tropfen im Meer:

Solidaritaet und Menschenrettung in Ungarn (207-245); Schmidt-Neke, Michael: Albanien – ein sicherer Zufluchtsort? (247-270); Chiari, Bernhard: Das Schicksal der weissrussischen Juden im "Generalkommissariat Weissruthenien": Eine Annaeherung an das Unbegreifliche (271-309). *ID*

0324 Berlin, Jeffrey B.: "Durch mich geht's ein zur Stadt der Schmerzerkorenen, durch mich geht's ein zum ewiglichen Schmerz, [...] Lasst, die ihr eingeht, alle Hoffnung fahren!" The Unpublished Correspondence of Heinrich Eduard Jacob in KZ Dachau and Buchenwald (1938-1939) (and Unpublished Letters with the German PEN-Club in London). *Germanisch-Romanische Monatsschrift* 49, 3 (1999) 307-331.

Depicts the horrors of the concentration camps, as described by authors such as Eugen Kogon, and publishes letters sent from Dachau and Buchenwald by Jacob (1889-1967), a prominent journalist and writer who was arrested in Vienna immediately after the Anschluss. Because of the harsh restrictions on correspondence, the letters do not openly mention camp conditions, but convey information by references to literary works. In his postwar writings, Jacob declared that he survived and helped others survive by telling stories and using his imagination to transport himself elsewhere, to a remembered or imagined place of wellbeing. *RW*

0325

ברנהיים-פרידמן, רחל: עגילים במרתף: לצמוח מעולמות נחרבים – סיפור אישי. תל-אביב: מורשת – בית עדות ע"ש מרדכי אנילביץ, 1999. 174 עמ'.

[Bernheim-Friedman, Rachel: *Earrings in the Cellar: To Grow from Ruined Worlds – a Personal Story.* Tel Aviv: Moreshet, 1999. 174 pp.]

Memoirs of a Jewish woman from Munkács, Hungary (now Mukachevo, Ukraine), born in 1922. Pp. 63-134 relate her experiences in the Holocaust. Describes the German occupation in March 1944 and her deportation, with her mother and two sisters, to Auschwitz; her mother and one of her sisters perished there. When Auschwitz was evacuated, Friedman went on the death march, from which she escaped with some friends. They entered a German village, posing as Hungarian fascists fleeing from the Russians. After the liberation, her friends were raped by Russian soldiers and then disappeared; Friedman managed to escape. She returned to Hungary, and soon emigrated to Israel. *RE*

0326

ברנהיים-פרידמן, רחל: האור בקצה המנהרה כבה. תל-אביב: משרד הבטחון – ההוצאה לאור, 1999. 89 עמ'.

[Bernheim-Friedman, Rachel: *The Light at the End of the Tunnel Went Out.* Tel-Aviv: Ministry of Defence, 1999. 89 pp.] An abridged version of her memoirs, published as "Agilim ba-martef" (Tel-Aviv: Moreshet, 1999).

0327 Bernstein, Michael André: Victims-in-Waiting: Backshadowing and the Representation of European Jewry. *New Literary History* 29, 4 (Fall 1998) 625-651.

Many historical and fictional works on the Holocaust, including those by Jews, are prone to "backshadowing" – a tendency to conceive the annihilation of European Jewry as inevitable, to blame the Jews of Europe for not foreseeing it, and to largely ignore the Jews of prewar Europe as individuals and members of communities of great variety and complexity. German and Austrian non-Jewish authors tend to standardize their Jewish characters, thereby perpetuating stereotypes (even if they are positive). Memoirs of Holocaust survivors serve as a corrective to other works on the Holocaust, because they depict the great diversity in Europe's prewar Jewish life, without stereotyping. *REK*

0328 Bernstein, Sara Tuvel: *The Seamstress: A Memoir of Survival.* New York: G.P. Putnam's Sons, 1997. 353 pp.

Memoirs of a Jewish woman, born Seren Tuvel, in Fulehaza, Transylvania in 1918. In 1931 she went to Bucharest to study. Returning to her hometown in 1940 (then part of Hungary), she and her father were arrested as alleged spies, tortured, and transferred to a prison in Budapest. Bernstein's father was later sent to an insane asylum, where he was eventually killed. Bernstein was released from prison and returned to Transylvania to help her mother and siblings, who had moved to Reghinal-Sasesc. She worked there as a seamstress, but lost her job and moved to Budapest to continue working, living with one of her sisters. In fall 1943 she was arrested by Hungarian policemen and sent to a labor camp; she was released just before the German occupation. She and her sister were deported in fall 1944 to Ravensbrueck and in January 1945 to Dachau. They were evacuated from Dachau on a train which was bombed, and were liberated by the American army. In 1949, Bernstein, with her husband and son, emigrated to Canada, and in 1960 to the U.S. *REK*

0329 Besenval-Amsallem, Daniela: Primo Levi: Un témoin pris dans les remous de l'histoire. *Revue d'Histoire de la Shoah* 162 (Jan-Apr 1998) 43-95.

Analyzes the writings of Primo Levi, especially "Se questo è un uomo" (1946) and "I sommersi e i salvati" (1986). Discusses changes in Levi's representation of the Auschwitz reality (based on his own experiences), and underlines the most significant messages in his works, such as the need to witness and to remember, and a consciousness that the Holocaust can be repeated. Reflects on aspects of his writings which provoked criticism and discussion (e.g. Wiesel considered that Levi relativized the victims as "possible executioners"). Deals with Levi's involvement in press discussions on the Holocaust and his responses to deniers, to articles by Nolte, etc. Concludes that Levi stressed the responsibility of the Nazis for the Holocaust, but refused to demonize the Germans as a people. *HV*

0330 Birger, Trudi; Green, Jeffrey M.: *Ho sognato la cioccolata per anni.* Trans.: Maria Luisa Cesa Bianchi. Casale Monferrato: Piemme, 1999. 223 pp. Translated from the English: "A Daughter's Gift of Love: A Holocaust Memoir" (Philadelphia: Jewish Publication Society, 1992). First published in German translation as "Im Angesicht des Feuers" (Muenchen: Piper, 1990).

0331 Birger, Zev: *Keine Zeit fuer Geduld: Mein Weg von Kaunas nach Jerusalem.* Muenchen: Luchterhand, 1997. 144 pp. On title page also: Mitarbeit: Elke Morlok.

Memoirs of a Jew born in 1926 in Kovno. In August 1941 he moved with his family (parents and brother) to the ghetto, where he was active in clandestine activities through the Irgun Brit Zion (e.g. printing the bulletin "Nitzotz," educational activities, construction of underground bunkers for hiding). When the liquidation of the ghetto began on 8 June 1944, Berger and his family hid in a bunker, but they were discovered and deported. His mother did not survive. The three men were taken to Stutthof and then to the Kaufering labor camp, where Birger's father died. His brother was killed after trying to escape from the labor camp. Birger was liberated at Kaufering in April 1945. For a time he worked as an interpreter in the American Army, and also became active in organizing illegal aliyah transports to Palestine. He and his wife emigrated to Palestine in November 1947. *ID*

0332 Birger, Zev: *No Time for Patience: My Road from Kaunas to Jerusalem: A Memoir of a Holocaust Survivor.* Trans.: Michele and Dan Sagir. New York: Newmarket Press, 1999. x, 150 pp. Originally published as "Keine Zeit fuer Geduld: Mein Weg von Kaunas nach Jerusalem" (Muenchen: Luchterhand, 1997).

0333

בשביל הזיכרון 31 (ינ'-פבר' 1999).

[*Bishvil ha-Zikaron* 31 (Jan-Feb 1999).]

חוברת זו מוקדשת לשוד הרכוש של יהודי אירופה בזמן מלחמת העולם השנייה, להשבת הרכוש של יהודי אירופה לאחר המלחמה ולהוראת הנושא בבתי-ספר תיכוניים בישראל.

התוכן: בקר, אבי: מבחנה של תודעה היסטורית: סוגיית הרכוש היהודי בשואה (4-11); פוזננסקי, רנה: צרפת: אריזציה קפדנית והשבת רכוש מהוססת (12-19); ראובני, שרי: הרצח והירושה: שוד הרכוש היהודי בימי השואה בהונגריה (20-24); ויינבאום, לורנס: הפשרת ההיסטוריה: השבת רכוש יהודי במזרח אירופה (25-35); לוין, איתמר: רכוש קורבנות השואה באוסטריה (36-41); לוין, איתמר: רכוש קורבנות השואה בבריטניה (42-44); שרייבר, רות: השבת הרכוש הקהילתי של יהודי גרמניה לאחר המלחמה (47-54); בלברג, אפרת: על האוצרות השדודים — כמה הערות דידקטיות (55-61).

This issue is dedicated to the plunder of the property of Europe's Jews during World War II, the restitution of that property after the war, and discussion of these subjects with high school students in Israel. Contents: [Beker, Avi: The Test of Historical Awareness: The Issue of Jewish Property in the Holocaust (4-11); Poznanski, Renée: France: Strict Aryanization and Hesitant Restoration of Property (12-19); Reuveni, Sári: The Murder and the Inheritance: The Plunder of Jewish Property during the Holocaust in Hungary (20-24); Weinbaum, Laurence: The Thawing of History: The Restitution of Jewish Property in Eastern Europe (25-35); Levin, Itamar: The Property of Victims of the Holocaust in Austria (36-41); Levin, Itamar: The Property of Victims of the Holocaust in Great Britain (42-46); Schreiber, Ruth: The Restoration of the Communal Property of the Jews of Germany after the War (47-54); Balberg, Efrat: On the Plundered Treasures – Some Didactic Comments (55-61).] *LFo*

0334

בשביל הזיכרון 33 (יוני-יולי 1999).

[*Bishvil ha-Zikaron* 33 (June-July 1999).]

חוברת זו מוקדשת לנושא "שואת יהודי הבלקן" ועוסק גם בהוראת הנושא לתלמידי תיכון בישראל. התוכן: באומן, סטיבן: שואות בלקניות (4-10); מוסק, משה: יהודי בולגריה — כפשע בינם לבין המוות (11-19); יושע, נסים: השועל הבלקני — על חלקה של בולגריה בהשמדת היהודים (20-24); יהודי יוון בשואה — סקירה היסטורית [מתוך "פנקס הקהילות — יוון" (1998).] (25-30); רפאל, שמואל: לשבור את השתיקה: מאבקם הספרותי של ניצולים מיוון לתיעוד קורותיהם בשואה (31-40); לבל, ג'ני: שואת יהודי יוגוסלביה (41-45); רפאל, שילהב: האי הנעזב: זיכרונותיה של נטה גטניו-אוסמו מקורפו (46-50); בלברג, אפרת: יהודי ארצות הבלקן בשואה — היחודיות והמשותף (51-59); יצחק כהן — עדות אישית [ראיינה: אפרת בלברג.] (59-62).

This issue deals with the Holocaust of the Jews in the Balkans, and how to teach the subject to high school students in Israel. Contents: [Bowman, Steven Barrie: Balkan Holocausts (4-10); Mossek, Moshe: The Jews of Bulgaria – a Narrow Escape (11-19); Yosha, Nissim: The Balkan Fox – on the Role of Bulgaria in the Extermination of the Jews (20-24); The Jews of Greece during the Holocaust – a Historical Survey [An extract from "Pinkas Hakehillot – Greece" (1998).] (25-30); Raphael, Shmuel: To Break the Silence: The Literary Struggle of Survivors from Greece to

Document Their Experiences of the Holocaust (31-40); Lebl, Ženi: The Holocaust of the Jews of Yugoslavia (41-45); Raphael, Shilhav: The Abandoned Island: The Memoirs of Nata Gattegno-Osmo from Corfu (46-50); Balberg, Efrat: The Jews of the Balkans in the Holocaust – the Uniqueness and That Which They Have in Common (51-59); Yitzhak Cohen – a Personal Testimony [Interview by Efrat Balberg.] (59-62).] *LFo*

0335

בשביל הזיכרון 34 (ספט'-אוקט' 1999).

[*Bishvil ha-Zikaron* 34 (Sept-Oct 1999).]

חוברת זו מוקדשת לנושא "השואה בהיסטוריוגרפיה" ועוסקת גם בהוראת הנושא לתלמידי תיכון בישראל. תוכן חלקי: נצר, שלמה: הכתיבה למען ההיסטוריה בימי השואה ובשנים הראשונות אחריה (13-4); כהן, בעז: ההיסטוריוגרפיה הישראלית של השואה — מגמות עיקריות (19-14); הרברט, אולריך: מדיניות השמדה: שאלות ותשובות חדשות בתולדות השואה (33-20); אלקסיון, נטליה: ההיסטוריוגרפיה הפולנית ושואת היהודים (41-34); ברונובסקי, יורם: הספרות הפולנית מול השואה (45-42); פוזננסקי, רנה: וישי והיהודים: משולי ההיסטוריה אל לב כתיבתה (51-46); ראובני, שרי: השואה וההצלה בהונגריה בהיסטוריוגרפיה ההונגרית (57-52); בלברג, אפרת: היסטוריוגרפיה, תודעה וזיכרון — השלכות חינוכיות (65-58).

This issue deals with the Holocaust in historiography, and how to teach the subject to high school students in Israel. Partial contents: [Netzer, Shlomo: Writing for the Sake of History during the Holocaust and in the First Postwar Years (4-13); Cohen, Boaz: Israeli Historiography of the Holocaust – Major Trends (14-19); Herbert, Ulrich: Extermination Policy: Questions and New Answers in Holocaust History [On historiography in West Germany and in unified Germany.] (20-33); Aleksiun, Natalia: Polish Historiography and the Holocaust of the Jews (34-41); Bronowski, Yoram: Polish Literature Facing the Holocaust (42-45); Poznanski, Renée: Vichy and the Jews: From the Margins of History to the Heart of Its Writing [Appeared in French in "Le Régime de Vichy et les Français" (1992).] (46-51); Reuveni, Sári: The Holocaust and Rescue in Hungary in Hungarian Historiography (52-57); Balberg, Efrat: Historiography, Consciousness and Memory – Educational Impacts (59-65).] *LFo*

0336 Boucher, Herbert: *Miracle of Survival: A Holocaust Memoir*. Berkeley, CA: Judah L. Magnes Museum, 1997. xi, 196 pp.

Memoirs of a Jew born in 1915 in Gevelsberg, Germany. At the end of 1933, Boucher (then Bouscher) moved to the Netherlands. His parents joined him in 1938. At the outbreak of the war, Boucher and his wife Hilde lived in Rotterdam; they were evicted from there and went to stay with relatives in Enschede. In September 1942 the Bouchers and their relatives went into hiding, mainly with Akke and Tjeerd van der Zee in the town of Sneek, Friesland, until the liberation. Boucher praises the many Righteous Gentiles who rescued him and his family. The Bouchers emigrated to the USA in 1947. Pp. 103-109 contain an interview with Tjeerd van der Zee conducted by Boucher in 1982. Pp. 113-136 present photographs and documents relating to the memoirs. Pp. 137-177 contain stories of the experiences of family and friends, some of whom survived the Holocaust while others perished. *DR*

0337 Boulanger, Gérard: La Shoah, d'Hitler à Papon. *La mémoire contre la nuit.* Bordeaux: Editions du Passant, 1997. Pp. 9-24.

Surveys the persecution of the Jews in Nazi Germany: the antisemitic legislation, the burning of "Jewish" and "Marxist" books, the November pogrom in 1938, the extermination policy which began in 1941 – the Einsatzgruppen, ghettos, and concentration and extermination camps. Maurice Papon was one of those responsible for the deportation of the French Jews. Concludes that he should be judged for "crimes against humanity." *HV*

0338 Brannigan, Augustine: Criminology and the Holocaust: Xenophobia, Evolution, and Genocide. *Crime & Delinquency* 44, 2 (Apr 1998) 257-276.

Criticizes Daniel Goldhagen's view, according to which the broad participation of Germans in the Holocaust is to be ascribed to the "exterminationist antisemitism" deeply imbedded in their psyches. Surveys other explanations (e.g. by Browning and Milgram) of the psychology of the Holocaust. These theories neglect the fact that the Nazis perpetrated many other kinds of genocide, or near-genocide, and the fervor of the "ordinary" German in perpetrating those crimes was no less. Based on a theory of the evolutionary psychologist Matt Ridley, contends that the psychology of the mass involvement with the genocide may be accredited to out-group animosities, or xenophobia, which are the reverse side of the coalitionist, "tribal" instinct lying at the core of human nature. *DR*

0339 Bretholz, Leo; Olesker, Michael: *Leap into Darkness: Seven Years on the Run in Wartime Europe*. Baltimore: Woodholme House, 1999. xviii, 274 pp.

Memoirs of Bretholz (as told to Olesker), born in Vienna in 1921, whose flight from the Nazis began in 1938 and continued throughout the war, taking him to Luxembourg, France, Belgium, and Switzerland, from where he was expelled to Vichy France. False papers helped him on several occasions. He was interned in the camps at Saint-Cyprien and

Rivesaltes. After being transported to Drancy, he was put on a train to Auschwitz, from which he jumped. In between escapes, he had several stays in prison and labor camps in France. Finally, he joined the resistance, helping others obtain false papers. In 1947 he emigrated to the USA; in 1962 he received confirmation of the death of his mother and sisters at the hands of the Nazis. *YC*

0340 Brockschmidt, Jens; Nadir, Shulamit: Zur Deportation der "Ostjuden": Der Erlebnisbericht von Susi Schmerler. *Sachor* 7 (1997) 60-66.

Pp. 63-66 contain the report by Schmerler (who later hebraized her name to Shulamit Nadir), written in Zbąszyń in February 1939, shortly before her 16th birthday. Because her family had emigrated to Germany from Belarus, they were deported from Bochum with the rest of the Polish Jews; she herself was deported from a "hachsharah" near Koeln. She describes the despair, the cold, the hunger, and the inadequate efforts of a local committee to help. The 24 members of her Hehalutz group made themselves responsible for a group of some 800 deportees. In his preface (pp. 60-62), Brockschmidt provides the historical background and describes the fate of her family (as told by her on a return visit to Bochum): in March 1939 she obtained a certificate for Palestine, but her parents and younger brother perished in the Holocaust. *RW*

0341 Broggini, Renata: *La frontiera della speranza: Gli ebrei dall'Italia verso la Svizzera, 1943-1945*. Milano: Mondadori, 1998. ix, 517 pp.

Based on diaries, letters, and interviews, discusses the fate of Jewish refugees from Italy, who, between 1943-45, came to the borders of Switzerland looking for a haven. Part of these refugees were refused entry into the country and were sent to certain death. Describes the difficult situation of Jews in Italy from 1938 on (racial laws, internment and persecution) and life in Swiss exile ("labor" camps, aid given to them, and refugee activities). Asserts that Swiss policy regarding refugees was inconsistent, influenced by differences in the criteria of the state and the cantons, by propaganda of Nazi circles, and by the hostility of the Swiss population toward foreigners. Pp. 451-517 contain a list of names of Jews (and their birth dates) who sought refuge in Switzerland. *MZ*

0342 Brown, Jean E.; Stephens, Elaine C.; Rubin, Janet E., eds.: *Images from the Holocaust: A Literature Anthology*. Lincolnwood, IL: NTC, 1997. xii, 579 pp.

Contains 100 selections, many of them excerpts; ca. three-quarters are historical accounts or biographical sketches, the rest are literary works. Each entry has a note about the author and an introduction, and all were published previously. Some are survivor memoirs, while some are by members of the second generation or others who want to preserve the history of the Holocaust. Chs. 1-6 (pp. 1-333) deal with the roots of antisemitism in Germany and Hitler's rise to power, Jews who hid from the Nazis and non-Jews who hid them, life on the run or inside Polish ghettos, and concentration camps. Chs. 7-10 (pp. 335-567) deal with the post-Holocaust period: the liberation of the camps, finding relatives and rebuilding lives, people who helped survivors, and the ongoing impact of the Holocaust. *YC*

0343 Bruchfeld, Stéphane; Levine, Paul Ansel: *"...om detta må ni berätta"* [..."Tell Ye Your Children"]. Stockholm: Regeringskansliet, 1998. [ca. 80] pp. See the abstract for the English translation: "Tell Ye Your Children...: A Book about the Holocaust in Europe, 1933-1945" (Stockholm: Regeringskansliet, 1998).

0344 Bruchfeld, Stéphane; Levine, Paul Ansel: *Tell Ye Your Children...: A Book about the Holocaust in Europe, 1933-1945.* Stockholm: Regeringskansliet, 1998. 84 pp. Originally published as "...om detta må ni berätta" (Stockholm: Regeringskansliet, 1998).

This booklet, sponsored by the Swedish government as part of an information program about the Holocaust, is aimed at helping adults teach their children about the events, and relates specifically to the experiences of children. Describes major aspects of the Holocaust, including German racism in the 1930s and various facets of the Final Solution (ghettos, camps). Includes illustrations (photos, drawings), chronological tables, poems, quotations (e.g. examples of racism), and a map of Auschwitz. Stress is placed on choices people and governments (including the Allies) made, physical and spiritual resistance, and the role of non-Jews as collaborators, rescuers, witnesses, or bystanders. Relates to Swedish aspects of the Holocaust as well. Includes a section on non-Jewish (especially Gypsy) victims of the Nazis. *YC*

0345 Cardosi, Giuliana; Cardosi, Marisa; Cardosi, Gabriella: *Sul confine: La questione dei "matrimoni misti" durante la persecuzione antiebraica in Italia e in Europa (1935-1945).* Torino: Silvio Zamorani editore, 1998. xxx, 297 pp.

Asserts that the attempt of the Nazi and fascist regimes to draw a clear line of separation between races resulted in a complex body of legislation and in contradictory policies about mixed marriages, influenced by political, historical, and legal conditions in each country. Pt. 1 (pp. 1-166), "Italia," presents testimonies and facsimiles of documents of nine Jews, partners in mixed marriages, who were deported between 1943-45. Pp. 143-166 describe aspects of the fascist legislation regarding mixed marriages. Pt. 2 (pp. 167-293), "Europa," deals with Nazi policy and legislation on mixed marriages in Germany and in occupied countries: France, Belgium, Netherlands, Luxembourg, Denmark, and Poland. *MZ*

0346 Carmelly, Felicia Steigman: *Shattered! 50 Years of Silence: History and Voices of the Tragedy in Romania and Transnistria.* Scarborough, Ont.: Abbeyfield, 1997. xxvi, 506 pp. On title page also: ...on behalf of the Transnistria Survivors Association, Toronto.

Pt. 1 (pp. 1-176), "Geographic and Historic Background," contains a historical essay on the Holocaust in Romania and Romanian-annexed Transnistria. Deals with the antisemitic policies of Antonescu's regime in 1940-44; pogroms in Romania before it entered the war (e.g. in Dorohoi and Bucharest), the pogrom in Iaşi in June-July 1941, and pogroms in reconquered Bessarabia and northern Bukovina; deportations of Jews to Transnistria; the Odessa massacre in October 1941 and the massacres in Bogdanovka, Domanyovka, and Akhmechetka; assistance given by Jews in Romania to the deportees in Transnistria; activities of the Zionist underground during the war. Pt. 2 (pp. 179-464), "Personal Testimonies," contains accounts of survival in Vapnyarka, Sharhorod, Mogilyov-Podolskii, Odessa, Bogdanovka, and other towns and camps in Transnistria. Pt. 3 (pp. 465-498) contains lists of those who perished in Transnistria, photographs of memorials, tables, documents, and maps. *DR*

0347 Chesnoff, Richard Z.: *Pack of Thieves: How Hitler and Europe Plundered the Jews and Committed the Greatest Theft in History.* New York: Doubleday, 1999. 325, [16] pp.

Based on published accounts and some interviews, discusses the theft of Jewish property (via expropriation, forced sales, discriminatory taxes, simple looting, etc.) by Nazis, collaborators, and neutrals during the Holocaust. The book is organized geographically: chs. 1-8 deal with Germany, Austria, Czechoslovakia, the Netherlands, Norway, France, Eastern Europe (Poland and Hungary), and Switzerland. Ch. 9, "The Other Neutrals," relates, briefly, to Spain and Portugal, Sweden, Argentina, Turkey, and the Vatican. The final chapter discusses the vagaries of restitution; the afterword updates the restitution struggle to July 1999. Among the topics discussed is the appropriation of art works owned by Jews, especially in Austria, and the failure to return them to their heirs after the war. *YC*

0348 Coppa, Frank J.: Pope Pius XI's "Encyclical" 'Humani Generis Unitas' against Racism and Anti-Semitism and the "Silence" of Pope Pius XII. *Journal of Church and State* 40, 4 (Fall 1998) 775-795.

Examines the history of the "hidden encyclical" prepared by Pius XI in 1938, but never issued by his successor Pius XII. The encyclical bore the title "The Unity of the Human Race" and was to denounce Nazi racism and antisemitism in explicit terms. Pius XI was known as a staunch anti-racist and had issued a number of anti-racist and anti-totalitarian pronouncements before. He was not afraid to provoke a breach between Germany and the Vatican. Pius XII, more cautious, was afraid that such a breach might be harmful for German Catholics. Moreover, a bluntly anti-racist encyclical might upset the racist USA and other countries. Pius XII was not an antisemite, but he regarded communism rather than Nazism as the primary danger for Christianity. His "silence" was deplored by many Catholic hierarchs. It is not clear whether the anti-racist encyclical, had it been issued, could have rescued many Jewish lives. *DR*

0349 Cornwell, John: *Hitler's Pope: The Secret History of Pius XII.* New York: Viking, 1999. xi, 430 pp. A paperback edition appeared in New York: Penguin Books, 1999 (xiv, 426 pp.).

A biography of Pius XII (b. Eugenio Pacelli, 1876-1958), focusing on his political activities vis-à-vis the Nazi regime in Germany. Dwells on the Pope's silence during the Holocaust, i.e. on the failure of the Vatican to deplore Nazi racism and to counteract the genocide of the Jews. The Church harbored an antisemitic spirit from early times, and at the time when Pacelli entered the Church it maintained that the Jews were to blame for their own misfortunes. Pacelli was instrumental in signing the Concordat with Nazi Germany in 1933, which gave the Nazis a free hand in solving the "Jewish problem." As Pope, he did not react to the Holocaust, although at least from 1942 he had ample information about it. He helped to rescue the Jews of Rome in fall 1943, but did not warn them of the impending danger. Believes that the Vatican could have opposed the Holocaust. The Pope's failure to respond was basically a failure of the papal office and of the prevailing culture of Catholicism. Personally, Pacelli was a staunch anti-communist, anti-liberal, and antisemite, which made him support totalitarian rightist regimes and rendered him indifferent to the plight of the Jews. *DR*

0350 Cornwell, John: *Le Pape et Hitler*. Trans.: Christoffe Beslon, Jeannie Carlier, Pierre-Emmanuel Dauzat. Paris: Albin Michel, 1999. 493 pp. Originally published as "Hitler's Pope: The Secret History of Pius XII" (New York: Viking, 1999).

0351 *Dachauer Hefte* 15 (Nov 1999).

This issue is entitled "KZ-Aussenlager – Geschichte und Erinnerung." Most of the articles are based on lectures held at a conference sponsored by the "Dachauer Hefte" in March 1999. Partial contents: Benz, Wolfgang: Die Allgegenwart des Konzentrationslagers: Aussenlager im nationalsozialistischen KZ-System (3-16); Hammermann, Gabriele: Die

Dachauer Aussenlager um Muehldorf [On Mettenheim and its subsidiary camps.] (77-98); Koenig, Ernest: Auschwitz III – Blechhammer: Erinnerungen (134-152); Schmidt, Alexander: Eine unauffaellige Geschichte: KZ-Aussenlager in der Region Nuernberg [Including a camp of Hungarian Jewish women attached to the Siemens-Schuckert-Werke.] (153-173); Schmidt, Monika: Zwangsarbeit und Lagerhaft als lebenslanges Trauma: Erfahrungen in Langenbielau und Peterswaldau [In Lower Silesia.] (174-195). *RW*

0352 Defonseca, Misha: *Misha: A Mémoire of the Holocaust Years*. Bluebell, PA: Mt. Ivy Press, 1997. 247 pp. Translated and edited from the French manuscript, which was co-authored by Vera Lee.

Memoirs of a Jewish woman from Belgium who, as a child, during the war, wandered through Germany, Poland, Ukraine, Romania, Italy, France, and back to Belgium, searching for her parents. She was seven years old when her parents left her with non-Jews in Brussels, before they were arrested in 1941. The negative attitude of her caretakers incited her to set out on her quest. During her wanderings she posed as a mute and moved through forests and along rivers, avoiding encountering people. For long periods she lived among wolves, who accepted her as one of their own. In spring 1945 she went back to Belgium, hoping her parents would return. As they did not, the authorities placed her with two sisters who took care of her; she never discovered what happened to her parents. Eventually she married and emigrated to the U.S. *LFo*

0353 Defonseca, Misha: *Survivre avec les loups: De la Belgique à l'Ukraïne, une enfant juive à travers l'Europe nazie 1941-1944*. Paris: Laffont, 1997. 238 pp. "Adaptation française de Marie-Therese Cuny." Unseen. See the abstract for the English translation: "Misha" (Bluebell, PA: Mt. Ivy Press, 1997).

0354

דויטש, יהודה: Bor: סחר עבדים במלחמת העולם השניה. נתניה: המחבר, 1998. 188 עמ'.

[Deutsch, Yehuda: *Bor: Slave Trade during World War II*. Netanya: The Author, 1998. 188 pp.]

A study of the Hungarian military labor service units which were leased by contract to the German Ministry of Defense to do slave labor under the direction of Organisation Todt in the copper mines around Bor, in Serbia. Discusses the terms of the contract signed by the Germans and the Hungarians in 1943, regarding which units were to be sent and what the Hungarians were supposed to get in return. Relates the harsh conditions at the work sites and the various camps where the units were based. 6,000

Hungarians, mostly Jews, were sent to Bor for forced labor (including the author) – 3,000 in August 1943 and another 3,000 in May 1944. In September 1944, 23 forced laborers were murdered by the Gestapo. On 7-8 October 1944, 680 forced laborers were murdered by the SS in Cservenka. During October, the town of Bor and the surrounding area was liberated by the Soviet army and Yugoslav partisans. Notes, throughout the book, the cruelty of the Hungarian commander of the labor service units from December 1943, Lieutenant-Colonel Ede Marányi. After the war, he disappeared and was never apprehended. *LFo*

0355 Eliach, Yaffa: Popular Jewish Religious Responses during the Holocaust and Its Aftermath. *Jewish Perspectives on the Experience of Suffering*, ed. Shalom Carmy. Northvale, NJ: Jason Aronson, 1999. Pp. 297-329.

For religious Jews during the Holocaust, beliefs and traditions became weapons for spiritual resistance. The marking of holy time was a major way that they expressed values totally opposed to those of the Nazis. Observance of tradition, to the degree possible (depending on whether one was in a ghetto, concentration camp, or elsewhere) testified to a bond with prewar reality and created a link with the future. Gender played a role, as men tended to focus on ritual and women on the communal aspect of holidays. Prayers, blessings, fasting, etc. were acts of spiritual survival. After the war, hasidic survivors tried to revive as much of the past as they could in new locations. Both secular and religious survivors appear to want religious ways of memorializing their dead. Museums are not seen to serve their purpose. Yet, if Orthodox Jews wish to keep the Holocaust from being de-Judaized, they need to make an effort to have more input into Holocaust museums and Holocaust studies. *YC*

0356 Favez, Jean-Claude: *The Red Cross and the Holocaust*. Ed. & trans.: John and Beryl Fletcher. Cambridge: Cambridge University Press, 1999. xxxiii, 353 pp. A slightly revised version of the original, published as "Une mission impossible? Le CICR, les déportations et les camps de concentration nazis" (Lausanne: Payot, 1988). Pp. 1-10 contain an introduction by John Fletcher.

0357 Fishman, Lala; Weingartner, Steven: *Lala's Story: A Memoir of the Holocaust*. Evanston, IL: Northwestern University Press, 1997. xxi, 355 pp.

Memoirs of Fishman, born in 1922 as Clara Weintraub; she grew up in Lvov. Describes the Soviet occupation of the city in 1939-41 and then the German occupation. Fishman's father was interned in the Janowska camp and was never heard from again. Her brother escaped to the USSR and joined the Soviet army. Fishman worked as a seamstress for the Luftwaffe, and therefore had working papers. In September 1942, when the Jews were forced into the ghetto, Lala fled with her mother, young sister, and a girlfriend. They had "Aryan papers," but were arrested by the Nazis in a small town and interned in prison. Fishman and her friend convinced the Nazis that they were Aryans, and were released. They then went to Kraków, found a flat and jobs and passed as Aryans. Later, Fishman moved to Katowice, where she was subsequently liberated. Her mother and sister were never heard from again. Fishman married Rabbi Morris Fishman, the JDC representative in the DP camps in southern Germany, and emigrated to the U.S. in 1947. Her brother emigrated to Palestine, and later to the U.S. *DR*

0358

פולמן-רבן, חווקה: לא נפרדתי מהם. קיבוץ לוחמי הגטאות: בית לוחמי הגטאות; קיבוץ דליה: מערכת, 1997. 226 עמ'.

[Folman-Raban, Havka: *I Did Not Say Goodbye to Them*. Kibbutz Lohamei Haghetaot: Beit Lohamei Haghetaot; Kibbutz Dalia: Ma`arekhet, 1997. 226 pp.]

Memoirs of a Jewish woman from Warsaw, born in 1924. Describes the German occupation and life in the ghetto. As members of the Dror youth movement, Folman and her two brothers, Wolf (Vevek) and Mordechai (Marc), were active in the resistance. The latter established underground schools and children's homes in various ghettos. Relates the activities of Dror, such as forging documents and smuggling weapons, and of its leaders, especially Yitzhak Zuckerman. Folman was a messenger for the organization. She was arrested with resistance fighters in Kraków in December 1942; the Nazis, who did not know that she was a Jew, sent her to Auschwitz as a political prisoner. Her brothers were killed by the Nazis in 1943. She was taken on a death march in January 1945 to Ravensbrueck, where she was liberated a few months later. Folman returned to Poland and worked for the Dror movement until late 1947 when she emigrated to Eretz-Israel. *RE*

0359 Fox, Tamar: The Death of Mala Zimetbaum: A Story of Resistance in Auschwitz. *The Jewish Quarterly* 174 (Sum 1999) 35-37. A Hebrew version appeared in "Barukh she-asani ishah?" (1999).

0360

פוקס, תמר: "אני תמיד חיה בנחת": מותה של מלה צימטבאום. ברוך שעשני אשה? האשה ביהדות – מהתנ"ך ועד ימינו, עורכים דוד יואל אריאל, מאיה ליבוביץ, יורם מזור. תל-אביב: ידיעות אחרונות, ספרי חמד, 1999. עמ' 27-33.

[Fox, Tamar: "Things Are Always Fine with Me": The Death of Mala Zimetbaum. *Barukh she-asani ishah? Ha-ishah be-yahadut – me-ha-tanach ve-ad yamenu*, eds. David Yoel Ariel, Maya Leibowitz, Yoram

Mazor. Tel-Aviv: Yedioth Ahronoth, Sifrei Hemed, 1999. Pp. 27-33.] An English version of pp. 30-33 appeared in "The Jewish Quarterly" 174 (1999).

Discusses current changes in the representation of women in studies on the Holocaust. Pp. 30-33 relate the case of Mala Zimetbaum (1918-1944). She was born in Poland but grew up in Antwerp. When the Nazis occupied Belgium, she joined a resistance group. In 1942, she was arrested and deported to Auschwitz along with her parents and sisters. There she was assigned the job of runner, carrying orders from the camp offices to the prisoners. In this capacity she managed to help many people and to save lives. She joined the underground movement in the camp, where she met and fell in love with Eduard (Edek) Galiński, a Polish political prisoner. In June 1944, they escaped from Auschwitz. After two weeks on the run they were captured and returned to the camp. Before her execution, Zimetbaum was asked by a friend how she was feeling; she replied: "Things are always fine with me." It is related that Galiński managed to hang himself on the gallows before his executors could do it. Zimetbaum, when led to the gallows, slashed her wrists with a razor; she was taken, bleeding, to the crematorium. *LFo*

0361 Fredj, Jacques, ed.: *Les archives de la Shoah*. Paris: L'Harmattan, 1998. 361 pp. On title-page also: Version française.

Proceedings of a conference held in Paris in 1996, organized by the Centre de Documentation Juive Contemporaine for archivists from 25 European countries, Israel, and the USA. They discuss sources available for research on the history of the Holocaust, and research works and institutions. Contents: Kaspi, André: Ecrire l'histoire de la Shoah (27-33); Klarsfeld, Serge: Ecrire l'histoire de la Shoah (34-45); Friedlaender, Saul: Ecrire l'histoire de la Shoah (45-53); Hilberg, Raul: Ecrire l'histoire de la Shoah (54-61); Henke, Josef: Utilisation des archives de la Shoah en Allemagne: Essai d'estimation (63-76); Steinberg, Maxime; Schram, Laurence: Les archives de la Shoah en Belgique (77-95); Klarsfeld, Serge: Les archives de la Shoah en France (96-104); Fredj, Jacques: Centre de Documentation Juive Contemporaine, Paris, France (105-118); Houwink ten Cate, Johannes Th.M.; Heidebrink, Iris: Evaluation des découvertes néerlandaises relatives à la Shoah (119-130); Fargion, Liliana: Les archives de l'histoire de la Shoah en Italie (131-156); Piper, Franciszek: Sources concernant l'histoire du camp de concentration d'Auschwitz (157-176); Kranz, Tomasz; Kielbon, Janina: Les sources d'archives et l'état de la recherche sur l'histoire du camp de Majdanek (177-196); Bergman, Eleonora: Treblinka: Etat de la documentation (197-200); Biedermann, Charles-Claude: Le Service International de Recherche (S.I.R.) – Arolsen (201-214); Ungar, Gerhard: Sources autrichiennes non exploitées: Les Archives des Autorités chargées des Affaires relatives aux Biens des

Victimes de Vienne (215-218); Evtimov, Bojidar: La question juive en Bulgarie 1940-1944 (219-224); Willemin, Georges: Les sources relatives à la Shoah dans les archives du CICR (225-234); Kolanovic, Josip: La Shoah – en Croatie: Documents et perspectives de recherche (235-254); Altman, Ilya A.: Centre Russe de Recherche et d'Education sur l'Holocauste (255-263); Iancu, Carol: Les Juifs de Roumanie pendant la Shoah dans les archives françaises (264-277); Lakos, János: La Shoah dans les archives publiques hongroises (278-282); Garofalo, Lucia; Procaccia, Micaela: Projet des archives de l'état italien relatif aux documents sur la persécution des Juifs d'Italie (283-287); Piliponis, Alfonsas: Perspectives d'études du génocide sur le territoire de Lituanie occupé par les Nazis (288-291); Basta, Jan: Les sources des archives sur l'Holocauste dans la Petite Pologne du Centre [On the area between the rivers Wisła and the San.] (292-305); Bergman, Eleonora: Utilisation future et sources d'archives inexploitées sur la Shoah en Pologne (306-310); Benjamin, Lya: Archives du Centre d'Histoire des Juifs de Roumanie (311-314); Koljanin, Milan: Sources historiques concernant la Shoah en Serbie (315-332); Kellerhals-Maeder, Andreas: La politique d'asile et de refoulement suisse durant la Seconde Guerre mondiale (sources et recherches historiques) (333-342); Gast, Uriel; Urner, Klaus: Sources sur la Shoah: Le centre de documentation "Histoire juive contemporaine", à l'Institut Fédéral Suisse de Technologie (ETH) de Zurich (343-347); Ioanid, Radu: L'Holocauste dans les archives de la Roumanie, de la Moldavie et de l'Ukraine (348-355). *TS*

0362 Friedlander, Albert Hoschander: Leo Baeck in Theresienstadt und sein Begriff des Leidens. *Juedische Denker im 20. Jahrhundert*, ed. Hanna Lehming. Hamburg: E.B.-Verlag, 1997. Pp. 13-28.

Discusses the role of Leo Baeck as the leader of German Jewry between 1933-45. When he was deported to Theresienstadt, ca. 47,000 Jews were already imprisoned there. Until the age of 70, Baeck had to do forced labor; later, he performed the duties of a rabbi and was involved with social services. Discusses his spiritual resistance, his writings on human suffering, and his teachings on history, philosophy, and Judaism. He tried to convince the prisoners that they were still thinking human beings, who could not be completely suppressed by the enemy. Discusses his book "Dieses Volk: Juedische Existenz," which he began to write in Theresienstadt. *ID*

0363 Frister, Roman: *The Cap or the Price of a Life*. Trans.: Hillel Halkin. London: Weidenfeld & Nicolson, 1999. 385 pp. Originally published as "Dyukan atsmi im tsaleket" (Tel-Aviv: Dvir, 1993).

0364 Gall, Matei: *Finsternis: Durch Gefaengnisse, KZ Wapniarka, Massaker und Kommunismus: Ein Lebenslauf in Rumaenien 1920-1990*. Konstanz:

Hartung-Gorre, 1999. 340 pp. Originally published as "Eclipsa" (Bucureşti: Du Style, 1997).

0365 Geissler, Frank, ed.: *Die Musik des osteuropaeischen Judentums – totalitaere Systeme – Nachklaenge: Kolloquium vom 02.-04.07.1994.* Leipzig: Dresdner Zentrum fuer Zeitgenoessische Musik; UniMedia, 1997. 136 pp.

Partial contents: Braun, Joachim: Die Bekaempfung der Moderne und der Antisemitismus (21-26); Diskussion (26-31); Vosteen, Annette: "Den Stoff durch die Form ueberwinden": Zu Viktor Ullmanns Oper "Der Kaiser von Atlantis" (54-57); Diskussion (58-60); Rienaecker, Gerd: Schostakowitschs Klage um juedisches Leid: Notate zum ersten Satz der "13. Symphonie" und zum Zyklus "Aus hebraeischer Volkspoesie" (87-91); Diskussion (91-92); Herrmann, Matthias: "Guenter Raphael ist Halbjude und hat als solcher keinen Platz im deutschen Musikleben": Bemerkungen ueber den Komponisten zwischen 1933 und 1945 (115-121); Diskussion (122-125). *RW*

0366 Gerlach, Christian: Konferencja w Wannsee: Los niemieckich Żydów a polityczna decyzja Hitlera o wymordowaniu wszystkich Żydów Europy [The Wannsee Conference: The Fate of German Jews and Hitler's Political Decision to Murder All of Europe's Jews]. *Biuletyn Żydowskiego Instytutu Historycznego* 185-186 (Jan-June 1998) 35-73. Appeared in German in "WerkstattGeschichte" 18 (1997) and in English in the "Journal of Modern History" 70 (1998).

0367 Gerlach, Christian: *Sur la conférence de Wannsee: De la décision d'exterminer les Juifs d'Europe.* Trans.: Julia Schmidt. Paris: Liana Levi, 1999. 141 pp. Translated from the German, which appeared as an article in "WerkstattGeschichte" 18 (1997). Appeared in English in the "Journal of Modern History" 70 (1998).

0368 Gilbert, Martin: *The Boys: Triumph over Adversity.* London: Weidenfeld & Nicolson, 1996. xv, 511 pp.

Relates the story of 732 ghetto and camp survivors, among them ca. 80 women, mainly from Poland and Czechoslovakia, but also from Germany and Romania, who as youngsters were transported to Britain in 1945-46, where they formed the '45 Aid Society. The book is based on letters sent to Gilbert and conversations with him, as well as on unpublished memoirs, and articles published in the "Journal of the '45 Aid Society" from 1976 on. Describes their prewar life in Eastern Europe, the beginning of World War II, the Holocaust, and their postwar life in Britain. Dealing with the Holocaust period, dwells on life in the ghettos (e.g. Łódź, Piotrków), work in labor camps (e.g. Skarżysko-Kamienna,

Płaszów, Rehmsdorf, Schlieben), the concentration camps Auschwitz and Buchenwald, Theresienstadt, death marches, and the liberation. *DR*

0369 Goepfert, Rebekka: *Der juedische Kindertransport von Deutschland nach England, 1938/39*. Frankfurt a.M.: Campus, 1999. 217 pp. Based on the author's diss. – Universitaet Muenster, 1997.

Examines the history of the "Kindertransport" rescue as part of the history of emigration from Germany during the period 1933-45, based on archival material and interviews with 26 of the children. Between December 1938 and September 1939, ca. 10,000 Jewish children were able to emigrate from the Third Reich to England. The "Kindertransport" rescue took place at a time when the borders of the whole world were closed to Jews. Discusses the situation in Germany in 1938, British immigration policy, the organization of the "Kindertransports," the children's lives in England, and the war years. The Jewish Refugee Committee in London, which was responsible for the children's expenses during their sojourn, had been founded in 1933 in order to assist refugees from Germany; in 1939 it became the Central Council for Jewish Refugees. With the outbreak of war, the "Kindertransports" ceased; the last ship arrived in England in May 1940 from the Netherlands. *ID*

0370 Goldstein, Kurt Julius: Wir sind die letzten – fragt uns: Spanienkaempfer, Auschwitz- und Buchenwald-Haeftling: *Reden und Schriften (1974-1999), mit einer autobiographischen Einfuehrung*. Ed.: Friedrich-Martin Balzer. Bonn: Pahl Rugenstein, 1999. 255 pp.

A collection of the writings and speeches of Goldstein, a German Jew born in 1914 in Dortmund. Pp. 12-58 contain his memoirs, "Die Geschichte meines Lebens." In 1928 Goldstein joined the communist youth movement and two years later became a member of the Communist Party. In 1933 he fled to Palestine. In 1936-39 he fought in the Spanish Civil War against Franco. After his escape from Spain, he was interned in Le Vernet in France between 1939-42, then handed over to the Gestapo and deported to Auschwitz. In 1945 he participated in a death march to Buchenwald, and in the self-liberation of the prisoners there in April 1945. After the war he settled in East Germany; from 1980-91 he lived in Vienna, and he now lives in Berlin. *ID*

0371 Grant, Reg G.: *The Holocaust*. Hove, East Sussex: Wayland, 1997. 64 pp.

A brief, illustrated history of the Holocaust, intended for young people. Discusses the persecution of the Jews in Nazi Germany, the implementation of the Final Solution, Jewish survival and resistance, and the postwar war crimes trials. *DR*

0372
גרייף, גדעון: "בכינו בלי דמעות...". ירושלים: יד ושם; תל-אביב: ידיעות אחרונות, ספרי חמד, 1999. 420 עמ'.

[Greif, Gideon: "*We Cried without Tears...*". Jerusalem: Yad Vashem; Tel-Aviv: Yediot Ahronot, Hemed Books, 1999. 420 pp.] First published in German (translated from the Hebrew manuscript) as "Wir weinten traenenlos...: Augenzeugenberichte der juedischen 'Sonderkommandos' in Auschwitz" (Koeln: Boehlau, 1995).

0373 Gross, Raphael; Lezzi, Eva; Richter, Marc R., eds.: *"Eine Welt, die ihre Wirklichkeit verloren hatte...": Juedische Ueberlebende des Holocaust in der Schweiz.* Zuerich: Limmat, 1999. 218 pp.

A collection of interviews with eleven Holocaust survivors living in Switzerland, including Jan Noach Trajster, Reine Seidlitz, Roland Kirilovsky, Fischl Rabinowicz, Judith Meyer-Glueck, Eduard Kornfeld, and Otto Klein (the others did not give their full names). Born between 1913-35, originally from Belgium, Germany, Hungary, Poland, Romania, and Slovakia, they grew up in different religious and cultural environments. Some of the interviewees focus on facts and events, relating their experiences in hiding and their escapes, or their daily life and suffering in ghettos and concentration camps. Others emphasize the influence of these events on their emotional life after the war. The different styles of narration vary according to the age of the person during the period of persecution. Those who experienced the Holocaust as children view the events differently than those who were adults at the time. Stresses the ambivalent role of Switzerland during World War II, and the need for breaking myths and coping with historical truths. *ID*

0374 Grynberg, Léon: *Mémoires de Léon Grynberg rescapé d'Auschwitz, 1903-1979.* Trans.: Lisa Wachtel. Orléans: CERCIL (Centre de Recherche et de Documentation sur les Camps d'Internement et la Déportation Juives dans le Loiret), 1998. 91 pp. Translated from the Yiddish.

Memoirs of a Polish Jew (1903-1979) from Minsk-Mazowieck relating his experiences before and during the Holocaust. He joined the Zionist Workers' Party but was imprisoned by the police from 1929-30 for illegal communist activities. In 1931, he emigrated to Paris, married, and became active in the French Communist Party. After the Nazi occupation, he was interned with his brother Moshe in Beaune-la-Rolande from May 1941 to June 1942, when they were deported to Auschwitz. Moshe was murdered a week later, but Léon survived numerous selections until his transfer to Warsaw in 1943 as part of the team in charge of the demolition of the ghetto. In May 1944, Grynberg was taken on a death march to Dachau and then to Muhldorf. In 1945 he was liberated at Innsbrueck and returned to Paris. His wife Guitele and his daughter Rosa (b. 1939) were rounded up at Vel d'Hiv in July 1942 and interned at Beaune-la-Rolande. Guitele was deported to Auschwitz. Rosa was placed in a children's home organized by the UGIF; she was deported to Auschwitz in July 1944. Both of them perished. *TS*

0375 Hahn-Beer, Edith; Dworkin, Susan: *The Nazi Officer's Wife: How One Jewish Woman Survived the Holocaust.* New York: Rob Weisbach Books, 1999. 305 pp.

Memoirs of a Jewish woman, born Edith Hahn in 1914 in Vienna. Her father died in 1936. Her two sisters emigrated to Palestine in the late 1930s. She remained in Vienna to study for a law degree, which she could not obtain due to the Anschluss. In 1941 she was sent for forced labor to Germany, working on a farm in Osterburg and a paper factory in Aschersleben. In June 1942 her mother was deported to Minsk and murdered. Hahn obtained "Aryan" papers with the help of non-Jewish friends, and went to Germany to work as a nurse's aide for the Red Cross. She met and married a member of the Nazi Party, Werner Vetter, a supervisor at an aircraft factory. Hahn lived with him in Brandenburg, and gave birth to a daughter. Vetter knew she was Jewish and protected her. In November 1944 he was drafted into the Wehrmacht, fought in Russia, and was taken prisoner. Later, as a judge in the Soviet sector of Berlin, Hahn obtained his release from the POW camp in Siberia. Upon his return they divorced. Hahn and her daughter emigrated to England in 1948, where she met and married Fred Beer. After his death, she emigrated to Israel. *SSC*

0376

הלבני, דוד וייס: עלה לא נדף: חיים של לימוד בצל המוות. תרגום: אבריאל בר-לבב. תל-אביב: הקיבוץ המאוחד, 1999. 142 עמ'.

[Halivni, David Weiss: *Not as a Driven Leaf: A Life of Learning in the Shadow of Destruction.* Trans.: Avriel Bar-Levav. Tel-Aviv: Hakibbutz Hameuchad, 1999. 142 pp.] Originally published as "The Book and the Sword: A Life of Learning in the Shadow of Destruction" (New York: Farrar, Straus and Giroux, 1996).

0377

היילפרין, אלכסנדר שמואל: אורות מאופל: קורות חיים וזכרונות מהדור הקודם. ירושלים: המחבר, [1999]. 182 עמ'.

[Halpern, Alexander Samuel: *Lights from the Darkness: A Life Story and Memoirs from a Past Generation*. Jerusalem: The Author, 1999. 182 pp.] An abridged English version appeared as "The Life Story of Rabbi Samuel Halpern" (Jerusalem: The Author, 1998).

Memoirs of an Orthodox Jew from Biecz, Poland (in Yiddish, Beitsh), born in 1922. Pp. 7-43 relate his experiences in the Holocaust, including the German occupation of Biecz in 1939. In summer 1942 he and four of his brothers were deported to the Prokocim labor camp, near Kraków. They escaped from the camp and joined their father and sister in the Kraków ghetto. In early 1943 the family was sent to the Płaszów labor camp, and in August 1944 they were transported to Mauthausen. Halpern, his father, and two brothers were sent to the labor camps of Melk and Ebensee, where they were liberated. Stresses, throughout, the efforts of Orthodox Jews to maintain religious practices in the labor camps. Halpern emigrated to Eretz-Israel in 1945 and then to the U.S., returning to Israel in 1987. *RE*

0378 Halpern, Alexander Samuel: *The Life Story of Rabbi Samuel Halpern, a Holocaust Survivor.* [Jerusalem]: The Author, [1998]. 50 pp. An abridged version of the author's Hebrew memoirs: "Orot me-ofel" (Jerusalem: The Author, 1999).

0379 Heer, Hannes, ed.: *The German Army and Genocide: Crimes against War Prisoners, Jews, and Other Civilians in the East, 1939-1944.* Trans.: Scott Abbott. New York: New Press, 1999. 223 pp. Based on the German edition: "Vernichtungskrieg: Verbrechen der Wehrmacht, 1941-1944" (Hamburg: Hamburger Edition, 1995).

Translation of a selection of the articles (accompanied by photographs) originally published in conjunction with an exhibition which opened in Hamburg in March 1995. The following articles deal with Wehrmacht units as spectators and participants in the massacres of Jews: Bartov, Omer: Preface [First appeared in "Memory & History" 9 (1997).] (10-17); Boll, Bernd; Heer, Hannes; Manoschek, Walter: Prelude to a Crime: The German Army in the National Socialist State, 1933-1939 (20-33); Manoschek, Walter: Serbia: The War against the Partisans, 1941 (34-72); Boll, Bernd; Safrian, Hans: The Sixth Army on the Way to Stalingrad, 1941-1942 (76-115); Heer, Hannes: Russia: Three Years of Occupation, 1941-1944 (116-171); Reemtsma, Jan Philipp: Afterword: On the Reception of the Exhibition in Germany and Austria (209-213). *RW*

0380 Heizner, Zmira H.: Yizkhor: "When Memory Comes" becomes "Memory Fields." *Jewish Studies Quarterly* 5, 1 (1998) 80-102.

Examines the Holocaust memoirs of Saul Friedlaender, "When Memory Comes" (1979) and Shlomo Breznitz, "Memory Fields" (1993), which have much in common. These two works reveal the conflict of Holocaust survivors, who desire to tell but at the same time need to conceal traumatic events. Discusses Friedlaender's identity crisis due to his separation from his parents at a young age and his experience of hiding in a monastery, as well as his use of historical research as a defense mechanism (e.g. he can relate his own experiences to those of the Jewish people) and his fear that the Holocaust will recur in Israel. Breznitz, who was also separated from his parents and hidden in a Catholic orphanage, uses similar defense mechanisms, integrating the past into the present and generalizing about human nature. Both authors distance themselves from their narratives, as though they were writing about someone else, and both analyze their stories rationally, without emotion. *RE*

0381 Herbert, Ulrich, ed.: *Nationalsozialistische Vernichtungspolitik, 1939-1945: Neue Forschungen und Kontroversen.* Frankfurt a.M.: Fischer Taschenbuch Verlag, 1998. 332 pp.

Based mainly on lectures given during the 1996/97 winter semester sponsored by the Historisches Seminar, Albert-Ludwigs-Universitaet, Freiburg. Partial contents: Herbert, Ulrich: Vernichtungspolitik: Neue Antworten und Fragen zur Geschichte des "Holocaust" [Traces the stages of German Holocaust historiography and draws conclusions from recent research on the interaction between local commanders and central authority, the motivation of the perpetrators, the extent of knowledge among German occupation personnel and the German public about the murder of the Jews, and the reactions of ordinary Germans.] (9-66); Aly, Goetz: "Judenumsiedlung": Ueberlegungen zur politischen Vorgeschichte des Holocaust (67-97); Pohl, Dieter: Die Ermordung der Juden im Generalgouvernement (98-121); Sandkuehler, Thomas: Judenpolitik und Judenmord im Distrikt Galizien, 1941-1942 (122-147); Browning, Christopher Robert: Die Debatte ueber die Taeter des Holocaust [On his debate with Goldhagen.] (148-169); Herbert, Ulrich: Die deutsche Militaerverwaltung in Paris und die Deportation der franzoesischen Juden (170-208); Manoschek, Walter: Die Vernichtung der Juden in Serbien (209-234); Gerlach, Christian: Deutsche Wirtschaftsinteressen, Besatzungspolitik und der Mord an den Juden in Weissrussland 1941-1943 (263-291); Dieckmann, Christoph: Der Krieg und die Ermordung der litauischen Juden (292-329). *RW*

0382 Hertling, Viktoria, ed.: *Mit den Augen eines Kindes: Children in the Holocaust – Children in Exile – Children under Fascism.* Amsterdam: Rodopi, 1998. 317 pp.

A collection of articles based on papers delivered at the conference "Children in the Holocaust – Children in Exile – Children under Fascism," held at the University of Nevada, Reno, in October 1996. Partial contents: Interview with Sonia Levitin: SNCAT-Reno TV-Station – October 2, 1996 [On her childhood experiences in Nazi Berlin until 1938, and on relatives who were killed in Belgium.] (18-25); Thunecke, Joerg; Mueller-Madej, Stella: "Wer leben will, stirbt und wer tot zu sein wuenscht, muss leben": Gespraech mit Stella Mueller-Madej ueber ihr Buch "Das Maedchen von der Schindler-Liste" (26-45); Bauer, Barbara: Drama – Maerchen – Gleichnis – Parabel: Wie Kinder den Holocaust erlebten und wie sie ihre Erfahrungen als Erwachsene darstellten (51-85); Haase, Donald: Overcoming the Present: Children and Fairy Tales in Exile, War, and the Holocaust (86-99); Viëtor-Englaender, Deborah: Education – Identity – Survival: Five Children's Formative Years in Third Reich Camps (100-110); Nussbaum, Laureen: Anne Frank, the Writer (111-121); Evans, Tamara S.: S.O.S. Schweiz: Emigranten schreiben Kinderbuecher [On children's books written between 1933-45 reflecting the political situation in Europe.] (164-178); Thielking, Sigrid: "Rosa Weiss wollte wissen...": Ansaetze und Strategien zur didaktischen Vermittlung des Themas Holocaust [Appeared in "Der Deutschunterricht" 4 (1997).] (179-191); Klimek, Julia: Art Spiegelman's "Maus": Drawing out the Implications of Witnessing in a Father/Son Relationship (192-206); Beckschulte, Klaus: "ich hasse die sprache die ich liebe": Leben und Werk von Robert L. Kahn [On his life (1923-1970) and work; Kahn was born in Nuernberg, emigrated to England with a Kindertransport in 1939, after the war settled in Canada and then the U.S., wrote poetry and taught German literature.] (207-219); Papanek, Hanna: Exile or Emigration: What Shall We Tell the Children? (220-236); Vogt, Jochen: Zwei, drei, viele Denkmaeler: Sieben Fragen an einen kleinen Text von Anna Seghers (237-251). *RW*

0383

הרצוג, הנרי ארמין: ...והשמים לא הזילו דמעה. תרגום: שלמה סוקול. קיבוץ דליה: מערכת, 1999. 340 עמ'.

[Herzog, Henry Armin: *...and Heaven Shed No Tears*. Trans.: Shlomo Sokol. Kibbutz Dalia: Ma`arekhet, 1999. 340 pp.] Originally published as "...and Heaven Shed No Tears" (New York: Shengold, 1995).

0384 *History of Photography* 23, 4 (Win 1999).

This issue, entitled "Photography and the Holocaust," edited by Sybil Milton and Genya Markon, contains articles discussing photographs taken during the war (and afterwards, in DP camps), and the significance of photographs as a form of historical data. Partial contents: Milton, Sybil H.: Photography as Evidence of the Holocaust (303-312); Rossino, Alexander B.: Eastern Europe through German Eyes: Soldiers' Photographs 1939-1942 (313-321); Tec, Nechama; Weiss, Daniel: The Heroine of Minsk: Eight Photographs of an Execution [On the Nazis' execution of Masha Bruskina and two other members of the resistance in Minsk in October 1941, and subsequent reactions in Belarus historiography. An earlier version, with just one photograph, appeared in "Holocaust and Genocide Studies" 11 (1997).] (322-330); Markon,

Genya: The Photo Archives of the United States Holocaust Memorial Museum (341-349); Pollin, Teresa Amiel: Photographs from Auschwitz [On a collection of 2,400 photos found in the archives of the Museum of Auschwitz-Birkenau, mainly family photos of Polish Jews.] (350-355); Milton, Sybil H.: Photography and the Holocaust: A Select Bibliography (379-382). *SSC*

0385 Humburg, Martin: Feldpostbriefe aus dem Zweiten Weltkrieg: Zur moeglichen Bedeutung im aktuellen Meinungsstreit unter besonderer Beruecksichtigung des Themas "Antisemitismus." *Militaergeschichtliche Mitteilungen* 58, 2 (1999) 321-343.

Analyzes letters sent home from the Soviet front by German soldiers. Only ca. 2% of these contain references to Jews, the great majority up to the end of September 1941 (15 out of 739 letters, by 10 out of 25 soldiers examined). Some of the references are neutral; the greater number find confirmation of previous prejudices about the Jews' physical and moral filthiness or about the dangerousness of Jewish Bolsheviks; a few express potential eliminationist antisemitism. Some of the letters mention mass shootings or remark on the disappearance of the Jews, but none criticize Nazi policy. Suggests possible explanations for the sparseness of references to Jews: indifference, acceptance of Nazi doctrine, or self-censorship, also out of consideration for the feelings of the recipient. The soldiers saw themselves more as victims than as perpetrators. Speculates also about the reactions of the families. *RW*

0386 Isacovici, Salomon; Rodríguez, Juan Manuel: *Man of Ashes*. Trans.: Dick Gerdes. Lincoln, NB: University of Nebraska Press, 1999. 236 pp. Originally published as "A7393, hombre de cenizas" (México, DF: Diana, 1990).

0387 Jackson, Livia Elvira Bitton: *I Have Lived a Thousand Years: Growing Up in the Holocaust*. New York: Simon & Schuster Books for Young Readers, 1997. 224 pp. A slightly revised version of her book "Elli: Coming of Age in the Holocaust" (New York: Times Books, 1980).

Memoirs of a Jew born in 1931 in Somorja, Hungary. After the Nazi occupation in March 1944, she was deported, with her parents and brother, to the Nagymagyar ghetto and then to Auschwitz. She and her mother were sent to the Płaszów labor camp, then back to Auschwitz, and from there to the Augsburg labor camp in Germany. In April 1945 they were sent to Muehldorf in the Dachau complex, where they found Jackson's brother, and to the Waldlager labor camp, from where they were evacuated by train. They were liberated by American forces at the Seeshaupt train station and returned to their hometown, now called

Samorin, in Slovakia. In 1951 they emigrated to the U.S. Jackson's father perished in the Holocaust. *REK*

0388 Jick, Leon A.: Method in Madness: An Examination of the Motivations for Nazi Mass Murder. *Modern Judaism* 18, 2 (May 1998) 153-172.

Rejects the conception that behind the Nazi resolution to exterminate the Jews in Europe there was "exterminationist antisemitism," shared by most of the Germans. From the beginning to the end, the Nazi policy toward the Jews was determined by expediency and utilitarianism. Racist ideology did play an important role in this policy, but only as a tool, underpinning the genuine intention of the Nazis – to create an agricultural empire in Europe, to reduce its non-German population, and to exploit the rest of the population as slave laborers. There was no place for Jews in such a society, thus they were to be removed from Europe. The mass murder of Jews began only after the German invasion of the USSR, which made perspectives for a "peaceful solution" of the Jewish problem dim. An important factor in the Nazi motives for the mass murder was that it was a profitable venture. The opportunities to exploit Jewish slave labor in the armament industry slowed down the murder from 1943 on. Suggests that the speedy deportation of the Hungarian Jews was motivated by a lack of workers for the production of V-2 rockets. *DR*

0389 *Judaism* 48, 1 (Win 1999).

The section entitled "Primo Levi, Witness," contains four articles on Levi's writings, which serve as his witness to the Holocaust. Contents: Gordon, Robert S.C.: Primo Levi's "If This Is a Man" and Responses to the Lager in Italy 1945-47 [On Holocaust memoirs published in Italy just after the war.] (49-57); Cheyette, Bryan H.: The Ethical Uncertainty of Primo Levi [Excerpted from "Modernity, Culture and 'the Jew'" (1998).] (57-67); Wilson, Jonathan: Primo Levi's Hybrid Texts (67-72); Boone, Susan Livingston: Unvarnished Truth: The Chemistry of Shame in Primo Levi (72-83). *SSC*

0390 Kalter, Joachim: *Eine juedische Odyssee: Von Leipzig nach Polen abgeschoben und deutsche Lager ueberlebt: Ein Bericht 1938-1946 = A Jewish Odyssey: Deportation from Leipzig to Poland and Survival in German Camps: A Report 1938-1946.* Konstanz: Hartung-Gorre, 1997. 142 pp. In German and English.

Memoirs of a Jew born in 1923 in Leipzig, who was expelled from Germany to Poland, with his parents and two brothers, in October 1938. The family settled in Tarnów, the father's native town. In November 1939, Kalter was taken for forced labor; in April 1940 he was transferred to the Dębica labor camp. In December 1940 he managed to escape, returned to

Tarnów, and hid there, but several months later he was captured and sent back to the camp. His mother and elder brother were deported and killed in 1942, his father in 1943. In 1944 Kalter was sent to Auschwitz, and in November joined his younger brother in Monowitz; together they were taken in January 1945 on a death march to Gleiwitz, and a transport to Buchenwald and then Langenstein, where they were liberated. After the war Kalter emigrated to the USA; these memoirs were written during his journey there in summer 1946. Pp. 9-13, 77-81 contain an introduction by Erhard Roy Wiehn, in German and English. Pp. 67-76 present a reaction to the book by James Stuart Brice. *ID*

0391 Katzenstein, Petra et al., eds.: *Auschwitz.* Amsterdam: Nederlands Auschwitz Comité, 1997. 23 pp.

A booklet relating briefly the history of concentration camps in general, and Auschwitz in particular (pp. 5-15). Describes, also, the set-up of the camp, the present-day museum at the camp site, and the activities of the Nederlands Auschwitz Comité (pp. 16-23). *MF*

0392 Kerbel, Anja Jacobsohn: "Bien-aimée Anja": Recueil de lettres d'un temps difficile. *Revue d'Histoire de la Shoah* 166 (May-Aug 1999) 195-237. Translated and presented by Nadine Ganivel.

Contains the correspondence of Anja Jacobsohn with her family and other persons in the period 1939-43 and her notes on their fate during the war. Anja was born in 1930 in Breslau. In 1939 she was sent by her parents (along with 500 other German-Jewish children) to Sweden to an adoptive family, where she survived the war. Her older sister survived in the Netherlands on false papers. Her parents escaped from Germany to Belgium, but were deported from there to France, and in 1942 to a concentration camp; they did not survive. *HV*

0393 Kertész, Lilly: *Von den Flammen verzehrt: Erinnerungen einer ungarischen Juedin.* Trans.: Zsuzsanna Szekelyne Egri. Bremen: Donat, 1999. 266 pp. Originally published in Hungarian as "Mindent felfaltak a lángok" (Tel-Aviv: The Author, 1994).

0394 Kirsch, Jan-Holger: Kommandant in Auschwitz: Autobiographische Aufzeichnungen von Rudolf Hoess. *Geschichte in Wissenschaft und Unterricht* 49, 7-8 (July-Aug 1998) 421-439.

Discusses the autobiography of Hoess, commandant of Auschwitz between 1940-43. Deals with Hoess's background and career until 1940, Hoess's own views concerning his acts, and various interpretations of his views. From December 1934 to May 1938 Hoess held various administrative positions at Dachau. In 1938 he was transferred to

Sachsenhausen. In 1939, when the decision was made to establish a concentration camp at Auschwitz, Hoess was appointed its commandant; his task was to prepare a site for mass murder. By 1943 Hoess was responsible for the murder of one million victims, most of them Jews. He was arrested and imprisoned in March 1946 in Kraków, where he wrote his autobiography. In April 1947 he was convicted, and executed in the area of Auschwitz. *ID*

0395 Kisielewicz, Danuta: Żydzi polscy w obozach jenieckich Rzeszy Niemieckiej w czasie II wojny światowej [Polish Jews in the German Camps for Prisoners of War during World War II]. *Biuletyn Żydowskiego Instytutu Historycznego* 191 (Sept 1999) 3-11.

Describes the fate of Jewish soldiers in the Polish army who were taken prisoner in September 1939 and sent to Nazi POW camps. There were about 50,000 Jews among the 500,000 Polish prisoners. They were separated from the non-Jews, in violation of the international law for the protection of prisoners of war. The Jews were forced to build camps; they were prohibited from maintaining contact with the Polish prisoners. In 1940, the Jewish POWs were transported to the Generalgouvernement, handed over to the SS and Gestapo, and sent either to labor camps or extermination camps. The fate of Jewish army officers was different. Most of them survived the war. The Wehrmacht tried to isolate them in special ghettos in the "oflags," but the Polish officers took a firm stance against segregation of the Jews. Concludes that the Nazis did not murder the Jewish officers for fear that Nazi officer-POWs would be murdered in retaliation. *HV*

0396 Klein, Mordehay: *My Life – Our History*. Montreux: Minerva Press, 1997. 140 pp.

Memoirs of a Jew born in Oradea, Transylvania, in 1929. Pp. 21-29, "Auschwitz," and pp. 30-54, "Camp Kaufering 1945," relate his experiences in the Holocaust. He was deported to Auschwitz with his mother and younger brother, who were immediately killed. Since he was considered fit for work, he was transferred to Kaufering, where he was liberated by the Americans. In 1946 he emigrated to Eretz-Israel. In 1983 he and his family emigrated to the U.S. *YC*

0397 Lagrange, Simone: *Coupable d'être née: Adolescente à Auschwitz*. Paris: L'Harmattan, 1997. 203 pp.

Memoirs of a French Jewish woman, born Simone Kadoshe (1931), relating her and her family's experiences in the Holocaust. In 1943 her sister Phiby was arrested by the Gestapo and disappeared, leaving a husband and three children. The husband died and two of the children were arrested by Klaus Barbie and deported to Auschwitz, where they perished. From 1942-44 Simone acted as a messenger for the Résistance in Lyon until she and her parents were betrayed by a neighbor. She and her mother were interrogated and tortured by Barbie and deported to Auschwitz, where her mother died. In 1945 she joined the death march to Ravensbrueck. On the way, she was reunited with her father, who was, however, immediately assassinated in her presence. She escaped a mass murder of survivors in a Polish forest and was liberated by Russians. She returned to Paris in 1945; a sister and brother of hers also survived. Simone was one of the plaintiffs and principal witnesses in the indictment of Barbie in 1987 and helped found the museum of the deported Jewish children at Izieu. *TS*

0398
לנצמן, אברהם: שנים ללא שמש: שלטון האימים של הרייך הגרמני השלישי: שישים שנה להקמת מחנה הריכוז זאקסנהוזן. תל-אביב: בן-יוסף, 1996. 168 עמ'.

[Lancman, Abram: *Years without Sunshine: The German Third Reich's Regime of Terror: 60 Years since the Establishment of the Sachsenhausen Concentration Camp*. Tel-Aviv: Ben-Yosef, 1996. 168 pp.]

Sachsenhausen was established in 1936 as a concentration camp for political prisoners. The first 1,800 Jews were brought there in November 1938, after the "Kristallnacht" pogrom. Ca. 200,000 people passed through the camp, many of them Jews, and ca. 100,000 perished there. The prisoners were forced to work in various factories in the area, mostly in the production of tanks, planes, and ammunition. Discusses the participation of the Wehrmacht in the murder of Jews, the importance of remembering the Holocaust, the Nuremberg Trials, Hitler's ideology, and the German army's support of Hitler in spite of the fact that many officers did not completely agree with him. Pp. 105-163 contain memoirs of Lancman, from his book "Ne`urim ba-drakhim" (1992), relating his experiences at the camps of Pionki, Małusy Wielkie, and Sachsenhausen. *RE*

0399 Lanzmann, Claude: *Un vivant qui passe: Auschwitz 1943-Theresienstadt 1944*. Paris: Mille et Une Nuits; Arte, 1997. 62 pp.

An interview with Maurice Rossel, a delegate of the International Committee of the Red Cross in Germany between 1942-45. The interview was conducted by Lanzmann in 1979, during the preparation of his film "Shoah." Rossel mentions his visit to Auschwitz in 1943, but talks more about his one day visit to Theresienstadt on 23 June 1944. Lanzmann evokes a report written by Rossel after the visit to Theresienstadt, and they discuss the image of the camp as Rossel then saw it and the camp reality which was very different and totally hidden from the eyes of the

Red Cross visitor. This interview was screened on the Arte TV station on 12 November 1997. *HV*

0400 *The Last Days*. Introd.: David Cesarani. London: Weidenfeld & Nicolson, 1999. 240 pp. On title-page also: Steven Spielberg and Survivors of the Shoah Visual History Foundation.

Based on the documentary film "The Last Days" by James Moll, presents six accounts by survivors of the Holocaust in Hungary: Alice Lok Cahana, Irene Zisblatt, Bill Basch, Renée Firestone, Dario Gabbai, and Tom Lantos. The accounts are from videotaped testimonies recorded by the Survivors of the Shoah Visual History Foundation. The text is interspersed with photographic sections: Roundups, Ghettoization, Deportation and Selection, In the Camps, Disposal of the Bodies, and Liberation. The accounts are supplemented by those of three American soldiers who liberated Dachau. *YC*

0401

לזר, דב: דור ראשון בארץ: זכרונות. נהריה: [המחבר], 1996. 242 עמ'.

[Lazar, Dov: *First Generation in Israel: Memoirs*. Nahariya: The Author, 1996. 242 pp.]

Memoirs of a Jew from Budapest, born in 1914. Pp. 23-54 describe Lazar's experiences in the Holocaust. He married in January 1944, while serving in a forced labor battalion. On 19 March the Nazis occupied Hungary, and he was sent with the labor battalion to Galicia and then to Mauthausen and Gunskirchen, where he was liberated on 4 May 1945. Lazar and his wife emigrated to Palestine in 1947. *RE*

0402 Lee, Carol Ann: *Rosas da terra: A biografia de Anne Frank* [Roses from the Earth: The Biography of Anne Frank]. Trans.: Elsa T.S. Vieira. Lisboa: "Livros Brasil," 1999. 367 pp. Originally published as "Roses from the Earth: The Biography of Anne Frank" (London: Viking, 1999).

0403 Lee, Carol Ann: *Roses from the Earth: The Biography of Anne Frank*. London: Viking, 1999. xxii, 297 pp.

Based on documentary research and interviews, presents an account of the life of Anne Frank, from her birth in Frankfurt in 1929 to her death in Bergen-Belsen in 1945. Presents new information about her family and those who helped hide them in Amsterdam. Discusses their betrayal to the Nazis and the probable betrayer. Supplements Anne's diary, and comments on it. Fills in background about the fate of the Dutch Jews, many of whom, like the Franks, were sent to Auschwitz (where Anne's mother perished). Anne and her sister Margot were sent from Auschwitz to Bergen-Belsen; both of them died just before the liberation. Includes information on the fate of Anne's diary and its publication, including the role of her father, Otto Frank, the only survivor of the family. *YC*

0404

לעווינסאן, לעא: מיין פארניכטעטע וועלט. תל-אביב: פארלאג י.ל. פרץ, 1999, 158 עמ'.

[Levinson, Leo: *My Annihilated World.* Tel-Aviv: I.L. Peretz Pub. House, 1999. 158 pp.]

Memoirs of a Jew from Kovno, Lithuania, who also lived in the nearby villages, as a student and as a teacher. Pp. 73-147 describe his experiences in the Holocaust, including the German occupation of Kovno in June 1941, life in the ghetto, forced labor, and the deportations (particularly the deportation of children and the elderly in March 1944). Levinson was sent to the nearby Šančiai labor camp, deported to Dachau, and liberated during a death march. *RE*

0405 Lewy, Guenter: Gypsies and Jews under the Nazis. *Holocaust and Genocide Studies* 13, 3 (Win 1999) 383-404.

Traces briefly the Nazi policy toward Gypsies in 1933-45 in Germany, Austria, Czechoslovakia, and some areas of Poland, the USSR, and Yugoslavia. Dwells briefly on Gypsy-Jewish relations in the ghetto of Łódź. Compares Nazi policy and the genocide of the Gypsies to that of the Jews. Argues that, despite some parallels, the treatment of each group differed fundamentally. The Nazis never formulated a plan for a "final solution of the Gypsy problem" analogous to that of the Jews. Large numbers of "racially pure" and "socially adjusted" Gypsies were exempted from deportation and murder. During the last stage of the war some deported Gypsies were even permitted to return to Germany. *DR*

0406 Linder, Bert: *Verdammt ohne Urteil: Holocaust-Erinnerungen eines Ueberlebenden.* Ed. of the German version: Gerhard Landauf. Graz: Verlag Styria, 1997. 328 pp. Originally published as "Condemned without Judgment" (New York: Shapolsky, 1995).

Memoirs of a Jew born in 1911 in Vienna. He and his brother fled to Belgium in March 1938. Linder's father was deported to Dachau and, after his release, emigrated with his wife to England. In May 1940, with the Germans approaching Belgium, Linder was arrested and expelled to France. He was interned in Saint-Cyprien, but escaped, joined the resistance, found refuge on farms in southern France, and married the daughter of German Jewish refugees. In autumn 1943 he fled with his wife and son, and hundreds of Jewish refugees, toward Italy. They were caught by the SS and sent to Drancy; from there they were deported to

Auschwitz. His wife and son were murdered. Linder worked in the Buna factory; in January 1945 he was evacuated to Dora and then to Bergen-Belsen, where he was liberated. He married again, and in 1951 emigrated to the USA. *ID*

0407 Lozowick, Yaacov: Malice in Action. *Yad Vashem Studies* 27 (1999) 287-330. Appeared in Hebrew in "Yad Vashem: Kovetz mehkarim" 27 (1999).

Examines the activities of the Department of Jewish Affairs IVB4 in the RSHA, headed by Eichmann, in its capacity as the main organizer of the deportation of Jews from the greater Reich (including Austria and the Czech Lands). Dwells on the operational procedures of this department, and on some special problems confronted by it (e.g. the treatment of various categories of Jews "protected" from deportation). Concludes that the Nazi officers in charge fully understood the gravity of what they were doing, and in persisting in their activities they demonstrated a greater degree of evil than if they had not known. Their participation in the deportation (in fact, genocide) of Jews is an example of the active malice and antisemitism of the officials in department IVB4. *DR*

0408 לזוביק, יעקב: אנטישמים בפעולה. יד ושם 27 (1999) 227-257.

[Lozowick, Yaacov: Antisemites in Action. *Yad Vashem* 27 (1999) 227-257.] Appeared in English in "Yad Vashem Studies" 27 (1999).

0409 Mannarino, Damien: La mémoire déportée. *Revue d'Histoire de la Shoah* 162 (Jan-Apr 1998) 12-42. Part of the author's M.A. thesis – Sorbonne, 1995.

Surveys works about the concentration camps published in France between 1944-92. Distinguishes three periods, and characterizes them: 1. the return (1944-51), in which a relatively important number of memoirs and diaries appeared, mainly about the French camps; 2. repression (1951-80), a period when memory was repressed and relatively few books appeared; 3. the awakening (1981 to the present), a period in which many books and articles appeared, and many books were translated into French from other languages. Deals with the major themes of the publications, and with the evolution of ways of representation of concentration and extermination camps. *HV*

0410 Marcus, Paul: The Religious Believer, the Psychoanalytic Intellectual, and the Challenge of Sustaining the Self in the Concentration Camps. *JPCS – Journal for the Psychoanalysis of Culture & Society* 3, 1 (Spr 1998) 61-75.

Asserts that concentration camp prisoners who were intellectuals – writers, scientists, and especially psychoanalysts – were less capable of preserving their humanity and more vulnerable to Nazi assaults than those with strong religious, political, or moral convictions. Compares the experiences of psychoanalysts and devout Jews in the Holocaust. Bruno Bettelheim, who was interned for a short time in a concentration camp, claimed that psychoanalysis helped him in certain instances, but it could not explain the influence of the camp environment on its victims and could not be used to guide inmates in their struggle for survival. On the other hand, religious inmates were able to transcend painful realities through their faith and the rituals which they practiced, often in a communal context. While many intellectuals were tormented by feelings of humiliation because their intellectual powers were useless in the camps, devout Jews were more able to maintain and express their values. *RE*

0411

מרכוס, צבי: אוד מוצל מאש. רמת השרון: המחבר, 1996. 139 עמ'.

[Markus, Zwi: *An Ember Saved from the Fire.* Ramat Hasharon: The Author, 1996. 139 pp.]

Memoirs of a Jew from Pomoryany, Ukraine, born in 1923. Pp. 28-42 describe his experiences in the Holocaust. After the German occupation in June 1941, some members of his family were killed by the Nazis. Markus was taken for forced labor, but escaped and hid in a forest with two of his brothers and other people from his town. Some of the group were killed during the war; those who survived were liberated in June 1944 by Russian forces. One of Markus' brothers was killed by Ukrainian bandits soon after the liberation. Pp. 86-100 contain the memoirs of his wife, Dunia Kotler Markus (1931-1996), born in Kovno. Describes her life in the ghetto until its liquidation in June 1944, when she and her mother were deported to Stutthof. They were liberated from a death march by Russian forces. Zwi and Dunia Markus married in 1951 and emigrated to Israel. *RE*

0412 Marton, Alexandru: *Speranţa.* [Tel-Aviv]: Minimum, 1998. 111, [32] pp.

Memoirs of a Jew born in 1927 in Topliţa, Transylvania. He was deported by the Hungarians, together with his family, from Miercurea Ciuc to Cluj. There they were interned in the ghetto, and then deported to Auschwitz. Marton describes his experiences as a forced laborer in the concentration and labor camps of Auschwitz, Mauthausen, Melk and Ebensee. After the war he returned to Romania. His parents and brother did not survive the war. In 1961 Marton emigrated to Israel. *LFo*

0413
מרטון, אלכסנדר: 186 המדרגות: קורותיו של נער יהודי במחנות אושוויץ, מאוטהאוזן, מלק ואבנזה. תרגום: דב שנער. ירושלים: יד ושם, 1999. 130 עמ'.

[Marton, Alexandru: *The 186 Steps: The Story of a Jewish Boy in the Camps Auschwitz, Mauthausen, Melk and Ebensee*. Trans.: Dov Shinar. Jerusalem: Yad Vashem, 1999. 130 pp.] Translated from the Hungarian manuscript. Appeared in Romanian as "Speranţa" (Tel Aviv: Minimum, 1998).

0414
מרטון, דוד ינו: אלבום משפחתי, לאור האירועים ההסטוריים (1900-1995). תרגום: דב שנער. ירושלים: המחבר, 1996. 336 עמ'.

[Márton, Dávid Jenő: *A Family Album, in Light of the Historic Events (1900-1995)*. Trans.: Dov Shinar. Jerusalem: The Author, 1996. 336 pp.] Originally published in Hungarian as "Családi Album" (Jeruzsálem: The Author, 1995).

0415 Maxwell, Leonard: *Heroes at the Crossroads of Jewish History*. Banham, Norfolk: Erskine Press, 1997. 171 pp.

Discusses eight cases of heroism in Jewish history, from Joshua ben Nun to Yitzhak Rabin. Ch. 7 (pp. 121-143), "Zivia Lubetkin: The Warsaw Freedom Fighters and the Jewish Martyrs of Europe," relates the history of the Holocaust briefly (pp. 121-131) and the biography of Lubetkin, a leader of the Warsaw ghetto uprising. *DR*

0416 Melton, Judith M.: *The Face of Exile: Autobiographical Journeys*. Iowa City: University of Iowa Press, 1998. xx, 225 pp.

Discusses a wide range of autobiographical writings by Jews and non-Jews that express various aspects (especially psychological ones) of the experience of exile from fascist-dominated Europe of the 1930s-40s. Ch. 3 (pp. 28-49), "The Persecution and Flight of the Jews: Narratives of Survival," briefly discusses narratives of the "Kindertransports" and of Nazi brutality experienced by children. Cites an account of the "Kristallnacht" pogrom in Berlin by Frederic Zeller and an incident in the town of Žilina, Slovakia, told by Shlomo Breznitz. The danger and anxiety of hiding in Nazi-occupied Poland is illustrated by an excerpt from the autobiography of Yehuda Nir, and survival through dissociation from other Jews by Jacov Lind's "Counting My Steps" (1968). Youthful escapes from Germany and Austria, as described in works by George Clare, by Zeller, and by Lilian Furst (and her father Desider), illustrate the terror experienced by some of those who attempted to flee Hitler's Europe. *YC*

0417 Mendel, David: Italy, Great Britain and Jewish Identity: A Personal Comment on Primo Levi. *European Judaism* 31, 2 (Fall 1998) 81-90.

Levi often said that before the introduction of the racial laws in Mussolini's Italy in 1938 he had been unaware that he was a Jew and regarded Jews in Italy as "Italians without Christmas trees." Expresses doubts that any Jew of Levi's generation, Italian or British, could fail to note his Jewishness, especially in the antisemitic atmosphere of the 1930s. Italy and Britain never were, and still are not, free from antisemitism. Surmises that the feeling common to many Italian Jews that they were Italians and that there was no antisemitism before 1938 may be explained by the fact that Italians, both Catholic and Jewish, began to think of themselves as Italians only after 1861, whereas the British always felt themselves members of one great nation, opposed to all "foreigners." *DR*

0418

מישל, ארנסט: הייתי מספר 104995. תרגום: מאיר סלומון. [קיבוץ עין הנצי"ב: המחבר, 1996]. [6], 128 עמ'.

[Michel, Ernst: *I Was No. 104995*. Trans.: Meir Salomon. Kibbutz Ein ha-Natziv: The Author, 1996. 6, 128 pp.] Translated from the German manuscript (written in 1947).

Memoirs of a Holocaust survivor, born in 1923 in Mannheim, Germany, relating his experiences in Auschwitz in 1943-45. Describes everyday life in the camp, the forced labor, the hospital in Auschwitz III, and his experiences after the evacuation in January 1945, when he was transferred to Gleiwitz, Buchenwald, and the Berga-Elster labor camp. During the death march from Berga-Elster in April 1945, he escaped with some friends; they wandered around Saxony until the liberation. *LFo*

0419 Minco, Bill: *Koude voeten – begenadigd tot levenslang: Het relaas van een joodse scholier uit het Geuzenverzet: Oranjehotel – Untermassfeld – Mauthausen – Auschwitz – Dachau* [Cold Feet – Sentenced to Life Imprisonment: The Story of a Jewish Student from the Geuzen Resistance: Oranjehotel – Untermassfeld – Mauthausen – Auschwitz – Dachau]. Nijmegen: SUN, 1997. 203 pp.

Memoirs of a Jew, born Sebil Minco, in Almelo, Netherlands in 1922. His family (himself, parents and brother) first lived in Groningen and afterwards in the southern part of the Netherlands. In February 1942 he was arrested, accused of espionage, and interned in the prison of Untermassfeld. In May 1943 he was transferred to Mauthausen, and in August of that year to Auschwitz. With the approach of the Russian army, he was taken on a death march to Gross-Rosen, and transferred from there to Dachau, where he was finally liberated. His parents and brother survived, but his parents were unable to cope after the war and his mother was placed in a home for the mentally ill. His brother emigrated to Palestine, but was killed in Jerusalem on Independence Day, 1948. Minco now lives in Hilversum. *MF*

0420 Muehlstein, Verena, ed.: *Leo Baeck – zwischen Geheimnis und Gebot: Auf dem Weg zu einem progressiven Judentum der Moderne*. Karlsruhe: Evangelische Presseverb. fuer Baden; Osnabrueck: secolo, 1997. 279 pp. Opposite title page: Beitraege von Tagungen der Bertelsmann Buch AG, der Evangelischen Akademie Baden und der Evangelischen Akademie Berlin-Brandenburg 1996 und 1997.

Partial contents: Friedlander, Albert Hoschander: Ueberleben in Theresienstadt und Leben mit der Schoa (52-65); Opfermann, Charlotte Guthmann: Erinnerungen an Rabbiner Dr. Leo Baeck in Theresienstadt (66-71); Klappert, Bertold: Der Midrasch aus Theresienstadt und das Testament Leo Baecks: Eine Einfuehrung in "Dieses Volk: Juedische Existenz" [Guetersloher Verlagshaus, 1996] (88-116). *ID*

0421 Mueller, Melissa: *Anne Frank (the Biography)*. Trans.: Rita & Robert Kimber. New York: Metropolitan Books, Henry Holt & Co., 1998. xvii, 330 pp. Originally published as "Das Maedchen Anne Frank: Die Biographie" (Muenchen: Claassen, 1998).

0422 Mueller, Melissa: *Das Maedchen Anne Frank: Die Biographie*. Muenchen: Claassen, 1998. 447 pp.

A biography of Anne Frank, based on her diary and on information received from survivors of her family (cousins of Anne's mother) and others who knew her, intended to show how she became the person she was. Traces the living conditions of Frank's extended family (including grandmothers, uncles) from the rise of the Nazi regime, first in Germany and later in the Netherlands. Discusses the difficulties of daily life for the Franks and other German Jewish families in the Netherlands. Examines Anne's life and those of her friends and family, illustrating the process of gradual exclusion, humiliation, deprivation of rights, and persecution. Deals, also, with the betrayal which led to the discovery of the Franks' hiding-place and to their deportation. *ID*

0423 Nemtsov, Jascha; Schroeder-Nauenburg, Beate: Musik im Inferno des Nazi-Terrors: Juedische Komponisten im "Dritten Reich." *Acta Musicologica* 70, 1 (Jan-June 1998) 22-44.

From November 1938 non-Aryan musicians – among them soloists, conductors, composers, and professors of music departments in universities – were excluded from German musical life. They were allowed to practice their professions only in the framework of the Juedischer Kulturbund, which was controlled by the Nazis and also served for purposes of Nazi propaganda. Many composers emigrated. The remaining Jewish population eventually could no longer pay the expenses for performances of the Kulturbund; these were limited gradually until the final prohibition in September 1941. Some well-known composers were deported to Theresienstadt – among them Viktor Ullmann, Hans Krása, Pavel Haas, Gideon Klein, Sigmund Schul – where they participated in the organization of musical activities. Discusses the works of these composers, and relates their fate. *ID*

0424 Obenaus, Herbert: Vom SA-Mann zum juedischen Ghettoaeltesten in Riga: Zur Biographie von Guenther Fleischel. *Jahrbuch fuer Antisemitismusforschung* 8 (1999) 278-299.

Fleischel (b. 1903 in Berlin), who grew up as a Protestant and converted to Catholicism in 1919, discovered only in 1936 that both his parents were Jewish (they converted in 1897). He then tried to resign from his membership in Nazi organizations, including the SA, but concealed his Jewish descent from his employer and associates and tried to ignore it himself. This denial led him to flout the Nuremberg sexual laws and, when his identity came to light, he was imprisoned for "Blutschande." In December 1941 he was deported with a transport from Hannover to the Riga ghetto. There the SS commander appointed him head of the Hannover, and later, also, of the Hamburg and Vienna transports. Because of his flair for organization, he obtained some benefits for the Jews; but he remained at a distance from them and terrorized them by his relentless enforcement of SS rules. He died of cancer in 1943. *RW*

0425 Ofer, Dalia: Cohesion and Rupture: The Jewish Family in East European Ghettos during the Holocaust. *Studies in Contemporary Jewry* 14 (1998) 143-165.

A study based mainly on Jewish wartime and, to a lesser extent, postwar records (diaries, personal archives, memoirs, Jewish Council records, etc.) in three large ghettos: Warsaw, Łódź and Kovno. Focusing on these three cities, examines the impact of ghetto existence on the Jewish family in Eastern Europe. The Holocaust challenged the very basis of family cohesion. Ghettoization exacerbated the material disparity between Jews, causing a deterioration of moral standards in the ghettos. The ghetto saw a reduction in the position of men as chief providers and raised the position of women; sometimes, children became the main providers. Some families were disrupted, while many others tried to retain some form of "normal" family life. Starvation and forced labor greatly aggravated family relations. The most crucial factors affecting family life were the deportations to the death camps and the mass murders. In an atmosphere of dread, the family could be either a source of support or a burden. Concludes that the impact of the Holocaust on family relations was ambivalent and contradictory. *DR*

0426 Ofer, Dalia: Holocaust Historiography: The Return of Antisemitism and Ethnic Stereotypes as Major Themes. *Patterns of Prejudice* 33, 4 (Oct 1999) 87-106.

The first comprehensive accounts of Nazism (e.g. by Ernst Fraenkel and Franz Neumann, 1941-42) did not highlight antisemitism as a cornerstone of Nazi ideology and policy. The historians of the 1950s-60s (Poliakov, Reitlinger, Hilberg) historicized the Holocaust and depicted Nazi antisemitism as a continuation of pre-modern German antisemitism. Later, historians drew a line between pre-modern antisemitism and that of Hitler; however, functionalists downplayed antisemitism as a factor in the genesis of the Holocaust. In the mid-1990s, antisemitism has returned to Holocaust historiography as a major factor explaining the events. Three recent books, Goldhagen's "Hitler's Willing Executioners," Saul Friedlaender's "Nazi Germany and the Jews," and John Weiss's "Ideology of Death" may be regarded as a reaction to the bureaucratic conception of the Holocaust. Goldhagen rejects completely the conclusions of the previous historiographic school, while the two others combine their ideological approach with some of the findings of the functionalists. *DR*

0427

אופיר, אפרים: נרדפים ללא מוצא: סיפורה של "סטרומה". תל-אביב: א.ק.מ.או.ר; קיבוץ דליה: מערכת, 1999. 232 עמ'.

[Ofir, Efraim: *Hounded with No Way Out: The Story of "Struma."* Tel-Aviv: A.C.M.E.O.R; Kibbutz Dalia: Ma`arekhet, 1999. 232 pp.]

Discusses the sinking of the "Struma" refugee boat off the Turkish coast in February 1942. 769 Jews who had embarked in Romania lost their lives in the explosion. Investigation of the incident is difficult, since some of the Romanian documents seem to have disappeared and the Turkish documents are not available. Surveys background information, including Romanian anti-Jewish policies during World War II, British restrictions on Jewish immigration to Palestine, the situation of the refugees, the condition of the boat, and Turkish indifference to the plight of the Jews. Pp. 183-232 contain a list of the "Struma"'s passengers, as well as photographs and personal information. *RE*

0428 Oliner, Samuel P.; Lee, Kathleen M.: *Wilhelm Bachner – ein Jude bei der Reichsbahn.* Trans.: Anna Kaiser. Gerlingen: Bleicher Verlag, 1999. 301 pp. Originally published as "Who Shall Live: The Wilhelm Bachner Story" (Chicago: Academy Chicago Publishers, 1996).

0429 O'Neil, Robin: Bełżec – the "Forgotten" Death Camp. *East European Jewish Affairs* 28, 2 (1998-1999) 49-62.

In 1997-98, two archaeological investigations were carried out at the site of the Nazi extermination camp Bełżec. Over 300 artefacts considered "interesting" and 33 mass graves were found at the site. Maps were drawn reconstructing the layout of this extermination camp. This research revealed evidence of the true nature and purpose of the death camp. Pp. 59-62 contain maps and photographs. *DR*

0430 O'Neil, Robin: Bełżec: A Reassessment of the Number of Victims. *East European Jewish Affairs* 29, 1-2 (Sum-Win 1999) 85-118.

Based on a comparison of several sources along with original research, concludes that the number of Jewish victims at Bełżec was far greater than the generally accepted figure of 600,000. Combining monthly figures of transportations to Bełżec between February-December 1942, when this death camp operated, with the figures for 1943 for the extermination of Jews in the Galician District at Auschwitz, Sobibór, and Treblinka (listed on pp. 89-104), arrives at a total number of victims of 1,012,342 (of whom 800,555 were killed at Bełżec). *YC*

0431 Opp, Karl-Dieter: Can Identity Theory Better Explain the Rescue of Jews in Nazi Europe than Rational Actor Theory? *Research in Social Movements, Conflicts and Change* 20 (1997) 223-253.

Sociological rational choice theory (RCT), i.e. cost and benefits scheme, taken in its simplest form, cannot explain rescue behavior during the Holocaust. As a high-cost and low-benefit situation, the rescue of Jews would be improbable. Therefore, the "identity theory" was proposed by Kristen R. Monroe as a tool to explain this behavior. Monroe holds that the sense or definition of self of the rescuers of Jews is the strongest factor determining their rescue behavior; the rescuer neglects the risk involved and other constraints. Criticizes Monroe's theory; argues that the more sophisticated version of RCT, namely that which assumes that rescuers do not have full and objective knowledge of the cost of their acts, and that the benefits need not necessarily be material, can be applied to the rescue situation. Empirical data, collected by Klingemann and Falter in 1993 in Germany, as well as those collected by Monroe herself in 1991, seem to confirm the broader version of RCT rather than identity theory. *DR*

0432 Peleg-Mariańska, Miriam; Gruess, Noe, eds.: *The Children Accuse*. Trans.: Bill Johnston. London: Vallentine Mitchell, 1996. xxxv, 279 pp. Originally published as "Dzieci Żydowskie oskarżaja" (Kraków: Jewish Historical Commission, 1946). Another version appeared in 1993.

Presents 70 eyewitness accounts collected by the author in 1945 from Jewish children and teenage survivors, as well as from adults who worked alongside Jewish children in wartime Poland. The narratives cover life in the ghettos of Warsaw, Przemyśl, Będzin, Lvov, Boryslav, Horodenka and Pruzhany; mass killings in Białystok and other places; the labor camp Płaszów; life on the "Aryan side," in hiding and in prison; and partisan warfare in Ukraine and Belarus. Relates, as well, the activities of orphanages and relief organizations (Centos, TOZ) which operated in the ghettos of Warsaw, Kraków, and Częstochowa. *DR*

0433 Pelinka, Anton: Schoah – Christentum – Oesterreich. *Harmonie und Gewalt: Oesterreich, Europa und die Zukunft der Vergangenheit*, eds. Wilfried Graf, Roswitha Riepl. Muenster: agenda, 1997. Pp. 87-96.

Argues that as Jews are defined by religion, so antisemitism is a religious, Christian phenomenon, and the Holocaust is largely the product of Christianity. Both the Catholic and the Protestant Churches in Austria and Germany collaborated with the Nazis or at least remained silent in the face of the persecution of the Jews; Christians who resisted were not backed by the Church. Notes the Austrian roots of Nazism and the massive participation of Austrians in the Holocaust, and Austria's refusal after 1945 to admit its share in the guilt or respect the sensitivities of the survivors. Asserts that Nazism was in essence eliminationist antisemitism as an end in itself, and the only even partially satisfactory causal explanation for the Holocaust is the antisemitism of a secularized Christianity. *RW*

0434 Perl, Lila; Lazan, Marion Blumenthal: *Four Perfect Pebbles: A Holocaust Story*. New York: Greenwillow Books, 1996. xi, 130 pp.

Memoirs of Lazan, born in 1935 in Hoya, northwestern Germany, based on Lazan's own memories and those of her mother, Ruth Blumenthal Meyberg. In spring 1938 the family moved to Hannover; there, they experienced the "Kristallnacht" pogrom, during which Lazan's father was arrested and sent to Buchenwald. He returned soon after and the family emigrated to the Netherlands, where they were interned in the Westerbork refugee camp. Conditions worsened in the camp from July 1942, when the Germans took over and it became a transit camp for deportations. In February 1944 the family was deported to Bergen-Belsen; in April 1945 they were sent on a death transport, and were liberated by Soviet troops. Lazan's father died of typhus soon after liberation. Lazan, with her mother and brother, returned to the Netherlands and emigrated to the U.S. in 1948. *RE*

0435 Perz, Bertrand; Sandkuehler, Thomas: Auschwitz und die "Aktion Reinhard" 1942-45: Judenmord und Raubpraxis in neuer Sicht. *Zeitgeschichte* 26, 5 (Sept-Oct 1999) 283-316.

Observes that historians have applied the name "Aktion Reinhard" only to the extermination of the Jews of the Generalgouvernement in local massacres and in the death camps Bełżec, Sobibór, and Treblinka, and to the systematic seizure of their property. Auschwitz was either excluded from this definition or included only in regard to the seizure of property, not the mass killings; this because the name is mentioned in contemporary documents and postwar testimony in connection with the former, but not the latter. Describes the administration and procedures for the disposal of Jewish property, both in Lublin (the headquarters of "Aktion Reinhard") and in Auschwitz, showing that they followed the same guidelines; moreover, there was no clear division between the personnel engaged in mass murder and those engaged in the disposal of property. Concludes that Auschwitz was also part of "Aktion Reinhard," and that the Aktion was organized centrally rather than locally. The Central Economic Office of the SS played its part in the mass killings no less than the RSHA. Pp. 301-307 contain testimony by Karl Ernst Moeckel, head of the administration in Auschwitz from April 1943-January 1945, describing the procedures for utilization of the property "from the Aktion Reinhard." The testimony was written in 1947 in Kraków, where he was sentenced to death along with other war criminals. *RW*

0436 Platovsky Stein, Milan: *Sobre vivir: Memorias*. Barcelona: Editorial Andres Bello, 1997. 493 pp.

Memoirs of a Jew born in Prague in 1922. Pp. 9-176 focus on his experiences in the Holocaust. Describes the situation of the Jews in Prague under Nazi occupation. His brother was taken prisoner in 1941 and killed during a transport to the Minsk ghetto in 1942. His mother was deported to Theresienstadt; she perished in Auschwitz in September 1944. In 1942 Milan was assigned to work on a road in Karlstein, near Prague, and was subsequently employed in other building projects. In April-October 1944, he worked in Theresienstadt; then he was transferred to Auschwitz and from there to Golleschau, where he worked in a cement factory. When the Soviet army approached, the camp was evacuated. He survived a death march to Sachsenhausen. There he worked in a factory adjacent to the camp until April 1945, when the camp was evacuated and the prisoners marched toward Luebeck. On the way he was liberated and returned to Prague. In 1948 he escaped and spent two years in France. In 1950 he and his wife emigrated to Chile. *MZ*

0437 Polen, Nehemia (Nathan): Theological Responses to the Hurban from within the Hurban. *Jewish Perspectives on the Experience of Suffering,* ed. Shalom Carmy. Northvale, NJ: Jason Aronson, 1999. Pp. 277-295.

Compares the only three known theological works written in response to the Holocaust by ultra-Orthodox rabbis who were also victims. "Siftei Shlomo," by Shlomo Zalman Unsdorfer (1887-1944), a collection of sermons delivered in Bratislava during the war and published in 1972, reflects a traditional response to unprecedented Jewish suffering; it represents unshaken faith. "Em ha-banim semehah," by Yissakhar Shlomo Teichthal (1884?-1945), published in Budapest in 1943, reflects a transformation of both his own personal and communal views. He espoused Zionism as a human, rather than Divine, initiative to save the surviving Jews, and blamed traditional religious leaders for failing to appreciate this possibility earlier when myriads might still have been saved. "Esh kodesh," by Kalonymus Klemish Shapira (1889-1943), written in Warsaw in 1939-42 and published in 1960, expressed the most radical theological shift. Lack of Divine intervention did not mean lack of involvement; God was conceived as weeping inconsolably in communion with his people, and their suffering was seen as an offering to Him. *YC*

0438

פולק, אורליה: שלוש שנות גירוש. עריכה ותרגום: טובה ינאי. קיבוץ דליה: מערכת, 1998. 172 עמ'.

[Pollak, Aurelia: *Three Years of Deportation*. Ed. & trans.: Tova Yanay. Kibbutz Dalia: Ma`arekhet, 1998. 172 pp.] Translated from the German manuscript, "Drei Jahre Deportation."

Memoirs of a Holocaust survivor (Trnava, 1896-Israel, 1985) who was deported in April 1942 from Slovakia – together with her husband, her parents, and some of her sisters – to Auschwitz. After three months of hard labor she was sent to work as a clerk in the political division of the SS administration of Auschwitz, the department which also controlled a large part of the Nazi camps in Silesia. Her work enabled her to closely follow aspects of the Nazi extermination and forced labor systems and testify about them at the Eichmann trial. After the evacuation of Auschwitz she was transferred to Ravensbrueck; from there she was taken on a death march to the camp Neustadt-Glewe, where she was liberated by the Soviets in May 1945. None of her family survived the war. After the war, Pollak returned to Trnava and eventually emigrated to Israel. Pp. 131-172 present additional information related to Pollak's experiences written by her niece, Tova Yanay. *LFo*

0439 Pollak, Aurelia: *Three Years of Deportation*. Ed. & trans.: Tova Yanay. Ra`anana, Israel: DocoStory, 1999. 200 pp. Translated from the German manuscript, "Drei Jahre Deportation." See the abstract for the Hebrew translation: "Shalosh shnot gerush" (Kibbutz Dalia: Ma`arekhet, 1998).

0440

רפל, יואל, עורך: לקראת מסע: במחוזות הזכרון במזרח אירופה: קובץ מאמרים לזכרו של משה קול. משואה 27 (1999).

[Rappel, Yoel, ed.: Landscapes of Memory: A Companion to a Voyage to Eastern Europe: A Collection of Articles in Memory of Moshe Kol. *Massuah* 27 (1999).]

קובץ מאמרים המעניק מבט אל העולם היהודי שהתקיים במזרח אירופה לפני מלחמת העולם השנייה. המאמרים מתארים את גורלם של הקיבוצים היהודיים בתקופת מלחמת העולם השנייה, עם דגש על פולין. נועד לסייע בהכנת משלחות הנוער לפולין. רוב המאמרים הופיעו קודם בבמות שונות. תוכן חלקי: האוזנר, גדעון: ההשמדה בפולין (211-223); גוטמן, ישראל: בגטו ורשה (225-232); גרייף, גדעון; קיליאן, אנדריאס: מיתקני ההמתה ההמונית במחנה אושוויץ-בירקנאו (237-258); קרן, נילי: "אי-הילדים" בבירקנאו: צריף הילדים במחנה המשפחות הצ'כיות — B2B (259-271); לפידות, אורי: מחנה מאידנק: מקומו ותפקידיו במסגרת מחוז לובלין (273-282); בן-נפתלי, אילה (איה): "הפורט התשיעי"— קורות אתר ההרג (283-301); קרן, נילי: טרזיאנשטאט — גן עדן לילדים (303-312); פולמן-רבן, חווקה : הבית ברחוב דז'לנה 34 (313-319); גוטמן, ישראל: מסע השיטנה והרצח (355-368); יונג, ג'יימס אדוארד: שרידי הזיכרון (387-400).

A collection of articles related to Jewish life in Eastern Europe before World War II and during the war, focusing on Poland. Intended for use in preparing the Israeli organized youth group tours to Poland. Most of the articles were published previously. Partial contents: [Hausner, Gideon: The Extermination in Poland [An extract from the opening speech of the prosecutor at the Eichmann trial; appeared in "Ha-yo`etz ha-mishpati la-memshalah neged Adolf Eichmann" (Jerusalem: Ministry of Information, 1961).] (211-223); Gutman, Yisrael: In the Warsaw Ghetto [An extract from his book "Mered ha-netsurim" (Merhavia: Sifriat Poalim, 1963).] (225-232); Greif, Gideon; Kilian, Andreas: The Mass Killing Installations in the Auschwitz-Birkenau Camp (237-258); Keren, Nili: "The Children's Island" in Birkenau: The Children's Barracks in the Czech Families' Camp – B2b [A revised version of a chapter from her book "Resisei yaldut" (1993).] (259-271); Lapidot, Uri: The Majdanek Camp: Its Position and Functions in the Framework of the Lublin District (273-282); Ben Naftali, Ayala (Aya): "The Ninth Fort" – a Chronicle of the Murder Site [In Kovno.] (283-301); Keren, Nili: Theresienstadt – a Paradise for Children [From her book "Resisei yaldut" (1993).] (303-312); Folman-Raban, Havka: The House at 34 Dzielna Street [In Warsaw. From her book "Lo nifradeti mehem" (1997).] (313-319); Gutman, Yisrael: The Campaign of Hatred and Murder [On antisemitism in postwar Poland, and the Kielce pogrom. From his book "Ha-yehudim be-Polin aharei Milhemet ha-Olam ha-Shniah" (1985).] (355-368); Young, James Edward: The Ruins of Memory [Translated from "The Texture of Memory" (1993).] (387-400).] *LFo*

0441 Reading, Anna: Scarlet Lips in Belsen: Culture, Gender and Ethnicity in the Policies of the Holocaust. *Media, Culture & Society* 21, 4 (July 1999) 481-501.

Gender is a factor in Nazi policy toward the Jews that has not yet been understood. Some aspects relate to biology, e.g. the sterilization of Jewish women or forcing them to commit infanticide when pregnant. Others relate to culture, e.g. eliminating possibilities of feminism. Neither the process of selection for death at extermination camps (half of the men vs. three-quarters of the women) nor methods of extermination were gender neutral. Specifically feminine experiences are sometimes discussed in Holocaust memoirs. In this context, provision of lipstick to survivors of Belsen by the Allies signified a "return of gender... of Jewish women's femininity, an aspect of humanity, along with Jewish men and women, that Nazi policies attempted to destroy." *YC*

0442 Rodgers, Silvia: *Red Saint, Pink Daughter: A Communist Childhood in Berlin and London.* London: André Deutsch, 1996. x, 282 pp.

Memoirs of Rodgers (née Szulman), born in 1928 in Berlin, the daughter of Polish Jewish immigrants who were dedicated communists. Pp. 1-183 describe her childhood in Berlin, relating also to antisemitism in interwar Poland and in Nazi Germany. Her family emigrated to England in March 1939. Relates, also, to antisemitic prejudice she encountered in England. After the war she married Bill Rodgers, who became a British cabinet minister. *REK*

0443 Rosenberg, Blanca Nebenzahl: *Versuch zu ueberleben: Polen 1941-1945.* Trans.: Herbert Uhl, Fritz Kunle. Frankfurt a.M.: Juedischer Verlag, 1996. 262 pp. Originally published as "To Tell at Last: Survival under False Identity, 1941-45" (Urbana: University of Illinois Press, 1993).

0444 Rosengarten, Israel J.: *Survival: The Story of a Sixteen-Year-Old Jewish Boy.* Syracuse, NY: Syracuse University Press, 1999. xviii, 218 pp. Originally published as "Overleven: Relaas van een zestienjarige joodse Antwerpenaar" (Antwerpen: Uitgeverij C. de Vries-Brouwers, 1996).

0445

רוזנטל, אליהו: קדיש לא קיים...: פרשת השואה של המשפחה, 1939-1944: כרוניקה, עדויות ואסופת מסמכים. ירושלים: [המחבר]. 1996. 48, [17] עמ'.

[Rosenthal, Eliahu: *There Is No Kaddish...: The Chapter of the Shoah of the Family, 1939-1944: Chronicle, Testimonies, and a Collection of Documents.* Jerusalem: The Author, 1996. 48, 17 pp.]

Chronicles Rosenthal's efforts to determine the fate of various members of his family, from Łódź, Sulejów, and Brest-Litovsk in Poland. 36 of his relatives perished in the Holocaust. *RE*

0446

רוזנבלום, דוד: והילד דוד.... תל-אביב: [המחבר], 1996. 93 עמ'.

[Rozenblum, David: *And the Boy David....* Tel-Aviv: The Author, 1996. 93 pp.]

Memoirs of a Jew from Łódź, born in 1930. Describes the German occupation of Poland and the hardships of life in the Łódź ghetto. Rozenblum and his parents were deported to Auschwitz in August 1944; from there he and his father were sent to various labor camps, satellites of Gross-Rosen. His mother was sent to Stutthof, where she was killed. After their liberation by the Russian army in May 1945, Rozenblum and his father returned to Łódź and then emigrated to Palestine in spring 1946. *RE*

0447

רובינשטיין, שמעון: המארה שהתגשמה: על כמה טרגדיות קטנות בתוך הטרגדיה הגדולה ששמה "סטרומה". ירושלים: המחבר, 1997. 13, [14] עמ'.

[Rubinstein, Shimon: *The Curse That Came True: On Some Small Tragedies within the Great Tragedy Called "Struma."* Jerusalem: The Author, 1997. 13, 14 pp.]

Discusses the sinking of the "Struma" off the Turkish coast in February 1942. The ship was carrying 769 Jewish refugees from Romania. It was supposedly torpedoed by mistake by a Russian submarine. Mentions the possibility of cooperation between the USSR and Great Britain in sinking the ship, since the latter refused to allow the passengers to enter Palestine. Includes short biographies of some of the passengers who perished, from Bárlad, Romania. Discusses the testimony of the sole survivor, David Stolear. Relates the story of Medea Salmovici (Marcovici), who was saved because she was taken off the ship and sent to a hospital in Istanbul; her husband, who was not allowed to debark, drowned with the other victims. The last 14 pp. contain photographs and documents. *RE*

0448 Ruch, Martin: *In staendigem Einsatz: Das Leben Siegfried Schnurmanns: Juedische Schicksale aus Offenburg und Suedbaden 1907-1997.* Konstanz: Hartung-Gorre, 1997. 112 pp.

Pp. 21-42 present the text of an interview with Schnurmann, a German Jew born in 1907 in Offenburg. He emigrated to Sweden in 1937 and joined a Hachsharah program in order to prepare for life in Palestine. Pp. 43-71 contain correspondence of Schnurmann with his parents, Elias and Rosa, and two sisters, Elsa and Berta, between 1937-44. His father was interned in Dachau briefly in November 1938. In 1939 his parents emigrated to Luxembourg. In 1943 they were deported to Theresienstadt, where they died. Elsa emigrated to Paris in 1933; in 1939 she was interned in Gurs, escaped and was caught several times, and finally went into hiding. After the war she returned to Paris. Berta could not obtain a certificate to emigrate to Palestine because of her age (she was born in 1903), but joined the last transport in 1940 and entered Palestine illegally. Schnurmann remained in Sweden during the war, emigrated to Palestine in 1945, and returned to Germany in 1951. *ID*

0449 Salamon, Julie: *The Net of Dreams: A Family's Search for a Rightful Place.* New York: Random House, 1996. 337 pp.

Salamon, a writer and film critic, describes a trip she took with her mother and stepfather in 1993 to Poland, Slovakia, and Ukraine, including a visit to the set where Spielberg was filming "Schindler's List." Pp. 89-175 recount the experiences of her parents, Lilly (née Rapaport) and Alexander (Sándor) Salamon in the Holocaust, and their life in the U.S. after the war. Her mother, born Szimi Rapaport, in 1922 in Huszt (now Ukraine), was deported to Auschwitz in 1944, and from there to the Zittau labor camp in Germany. Her father, a physician, was deported from Huszt to Dachau, escaped from a train leaving the camp, crossed the border into Czechoslovakia, and joined a group of partisans until the end of the war. His wife and daughter were murdered by the Nazis. Szimi and Sanyi met after the war in Prague, married, and emigrated to the U.S. *REK*

0450 Sale, Giovanni: Pio XII, la guerra e gli ebrei. *La Civiltà Cattolica* 3586 (20 Nov 1999) 359-372.

Defends Pius XII against accusations which arose after his death, that he was silent in the face of Nazi crimes and did not do enough to save Jews during the Holocaust. Focuses on the book by John Cornwell, "Hitler's Pope: The Secret History of Pius XII" (1999). Cornwell tries to prove that Pacelli was an antisemite based on documents from the years 1917-20, when he was the apostolic nuncio in Munich. Asserts that the documents prove only that Pius XII was a preconciliar Christian and consequently could not have been a philosemite, but as Pope he followed the policy of his predecessor, Pius XI. After 1942 he avoided useless declarations against the Nazis, preferring diplomatic action, local involvement of bishops, and, when necessary, his own intervention in order to save Jews. Relates to the writings of Pinchas E. Lapide to support his views. Lapide maintains that 800,000 Jews were saved as a consequence of Pius XII's intervention. *MZ*

0451

שצקר, חיים: פרשת קלאדובו — שאבאץ (שני ספרים והרהור שלישי). הציונות 21 (1998) 177-193.

[Schatzker, Chaim: The Kladovo-Šabac Affair (Two Books and a Third Reflection). *Ha-Tzionut* 21 (1998) 177-193.]

Compares the presentation of the "Kladovo-Šabac affair" – when a group of about 1,000 refugees from Austria and Germany on its way to Palestine was trapped in Yugoslavia by the Nazi invasion – in two research works by Israeli and Austrian scholars: "Pareshat Kladovo-Šabac" ("The Kladovo-Šabac Affair"), by Dalia Ofer and Hannah Weiner (1992), and "Gescheiterte Flucht: Der juedische 'Kladovo-Transport' auf dem Weg nach Palaestina, 1939-42," by Gabriele Anderl and Walter Manoschek (1993). Questions the attempt of the Israeli work to refrain from passing judgment on the conduct of the Zionist establishment in Palestine toward the group, whose members probably could have been rescued, as were other groups of refugees who reached Palestine safely at that time. *LFo*

0452

שילר, יוז'ף: מפעל ההצלה בשטרסהוף על רקע שואת יהודי הונגריה 1944-1945. תרגום: אלכסנדר ברזל. קיבוץ דליה: מערכת; קיבוץ לוחמי הגיטאות, בית לוחמי הגיטאות, 1999. 108 עמ'.

[Schiller, József: *The Strasshof Rescue Operation on the Background of the Holocaust of Hungarian Jewry 1944-1945*. Trans.: Alexander Barzel. Kibbutz Dalia: Ma`arekhet; Kibbutz Lohamei Haghetaot: Beit Lohamei Haghetaot, 1999. 108 pp.] Originally published as "A strasshofi mentőakció és előzményei (1944-1945)" (Budapest: Gordius, 1996).

0453 Schilling, Donald G., ed.: *Lessons and Legacies II: Teaching the Holocaust in a Changing World.* Evanston, IL: Northwestern University Press, 1998. xii, 233 pp.

Based on papers presented at the second "Lessons and Legacies" conference, held in 1992 in Northbrook, Illinois. Contents: Marrus, Michael Robert: Good History and Teaching the Holocaust [Appeared in "The Holocaust Fifty Years After" (1993?).] (13-25); Weinberg, Gerhard L.: The Holocaust and World War II: A Dilemma in Teaching (26-40); Browning, Christopher Robert: Ordinary Germans or Ordinary Men? Another Look at the Perpetrators [An earlier version appeared in "History & Memory" 8 (1996).] (41-54); Fenigstein, Allan: Reconceptualizing the Psychology of the Perpetrators (55-84); Porat, Dina: Jewish Decision-

Making during the Holocaust [Compares decision-making by the Kovno Judenrat, the youth movement underground in Vilna, the Jewish Agency in wartime Palestine, and Stephen Wise.] (85-102); Baumel, Judith Tydor: Gender and Family Studies of the Holocaust: A Historiographical Overview [Appeared in "Women" 7 (1996).] (105-117); Laor, Dan: The Legacy of the Survivors: Holocaust Literature in Israel (118-130); Baron, Lawrence: Teaching about the New Psychosocial Research on Rescuers in Holocaust Courses [Appeared in "Shofar" 10 (1992).] (131-139); Doneson, Judith E.: Why Film? [Dwells, especially, on the film "Europa, Europa."] (140-156); Lee, Marshall; Steele, Michael R.: The Affective Approach in the Interdisciplinary Holocaust Classroom (159-174); Hammer, Reuven: Commemorations and the Holocaust (175-191). *DR*

0454 Schmetterling, Astrid: "Life? or Theatre?": The "Wildly Unusual" Work of Charlotte Salomon. *The Jewish Quarterly* 45, 3 [171] (Fall 1998) 48-54.

Describes the life and creative work of Charlotte Salomon (1917-1943), a Jew from Berlin, who began to experience Nazi persecution in 1933 when she was forced to leave school. In February 1936 she was permitted to attend the Berlin Academy of Arts, where she suffered from antisemitism. After the "Kristallnacht" pogrom she joined her grandparents in southern France. There, between 1940-42, she wrote the autobiographical work "Life? or Theatre?," which is composed of various types of texts, drawings, and music, and includes elements of biography and fiction – a description of her own life as theater. In 1943 Salomon married Alexander Nagler, but the two were deported soon after and killed in Auschwitz. Just before her deportation, she gave her work to friends for safekeeping. *REK*

0455 Schwemer, Ulrich; Vorndran, Hans-Georg, eds.: *"Wer haette das geglaubt!" Erinnerungen an die Hachschara und die Konzentrationslager*. Heppenheim: Evangelischer Arbeitskreis Kirche und Israel in Hessen und Nassau, [1998]. 79 pp. First published in Hebrew in 1985.

A compilation of brief memoirs by 25 members of Zionist hakhsharot in Nazi Germany who survived the concentration camps. In 1941 the hakhsharah centers were turned into forced-labor camps under the direct control of the SS; in April 1943 the members were deported to Auschwitz. Most of them, young and able-bodied, were sent to Monowitz or other labor camps. Some groups, or parts of them, were allowed to remain together; even where this was not the case, they tried to remain in contact with each other and with members of other Zionist youth groups from Germany and Poland. Their often heroic mutual support enabled relatively many of them to survive. After the war they reorganized and eventually reached Israel. Based on a gathering of the survivors at Netzer Sireni, Israel, in 1985. *RW*

0456 Scorin, Jérôme: *L'itinéraire d'un adolescent juif de 1939 à 1945: Drancy – Bordeaux – La Lande – Nancy – Lyon – Drancy – Auschwitz – Stutthof – Vaihingen – Ohrdruf – Erfurt – Buchenwald – Crossen – Nancy.* [Nancy: The Author], 1997. 205 pp.

Memoirs of a Jew born in France in 1924 to Polish immigrants. He, his parents, and two brothers fled from Nancy to Bordeaux in 1940, but were arrested and interned in La Lande. He escaped and returned to Nancy, but was warned of impending raids and, with false identity papers, fled to join his sister Régine in Lyon. She was able to hide three cousins but her parents and brother Léon were arrested and perished in Auschwitz. Following the Nazi occupation of Lyon, Jérôme was ordered to join the labor force for youth in Savoie but deserted and returned to Lyon where he joined the Jewish forces of the Résistance. In 1944 he was arrested and deported to Auschwitz. From there he was sent to other camps. In 1945 he was taken on a death march to Erfurt. He returned to Nancy after the war. Régine was sent to Auschwitz but in 1944 was transferred to a munitions factory in Czechoslovakia, where she survived the war. Pp. 35-46 contain a list of Jews interned at La Lande. *TS*

0457 Seltzer, William: Population Statistics, the Holocaust, and the Nuremberg Trials. *Population and Development Review* 24, 3 (Sept 1998) 511-552.

Examines the role of population statistics in the Holocaust and at the Nuremberg Trials, based on a study of national data systems and of the activities of several statisticians and demographers. Points out that although population registers are useful to governments, they can also be used as an "operational tool of genocide." The murder of the Jews was an administrative process, involving planning, organization, and coordination, all of which made use of statistical systems. Examines the statistical processes in five countries – Germany, Poland, France, the Netherlands, and Norway – where Jewish population registers were ordered by the Nazis in order to facilitate the roundup and deportation of Jews. Discusses the activities of SS statistician Richard Korherr during the war, and the use of population data in the prosecution of war criminals at the Nuremberg Trials. *RE*

0458 Skakun, Michael: *On Burning Ground: A Son's Memoir*. New York: St. Martin's Press, 1999. 235 pp.

A biography of the author's father, Joseph Skakun, who was a yeshiva student in Novogrudok, Belarus, before the German occupation. Skakun survived a Nazi massacre in December 1941, in which his mother and other relatives were killed. After two escapes from the Novogrudok ghetto, he ended up in the ghetto of Vilna, from which he was sent to a labor camp in Sorok Tatar. There, he borrowed the identity papers of a

Polish Tatar, Stefan Osmanov, and volunteered for work in Germany. In the village of Stensch he worked for a Nazi farmer, in constant fear of being found out. Ironically, the Gestapo "promoted" him from a Pole to a Lithuanian. Thus, he became eligible to join the Waffen SS, but liberation by the Red Army saved him from actual collaboration. After living in a DP camp, he emigrated to Paris in 1948 and to New York in 1959. *YC*

0459 Solchany, Jean: De la régression analytique à la célébration médiatique: Le phénomène Goldhagen. *Revue d'Histoire Moderne et Contemporaine* 44, 3 (July-Sept 1997) 514-529.

A critique of Goldhagen's "Hitler's Willing Executioners" (published in French in 1997). Ridicules Goldhagen's presentation of his book as a "radical revision" of everything written up to now on the causes of the Holocaust. Regarding the "eliminationist antisemitism" of the Germans from the 19th century on, disagrees that antisemitism was so widespread amongst the German population. Argues that Goldhagen attributes to the majority of Germans something that was characteristic of the minority. In Goldhagen's view, antisemitism was the most important factor in the implementation of the Final Solution, which is a huge simplification. States that 19th-century Germany had much more than one political culture. Criticizes Goldhagen, also, for his not always scientific language, in which moral indignation is present. A positive aspect of the book is the description of the behavior of ordinary people, but lack of description of the elites' behavior is another example of simplification of the Holocaust's complexity. *HV*

0460 Spitzer, Federica; Weisz, Ruth: *Theresienstadt: Aufzeichnungen.* Mit einem Beitrag von Wolfgang Benz. Berlin: Metropol, 1997. 171 pp.

Memoirs of two Jewish women – Federica Spitzer, born in 1911 in Vienna (pp. 9-97), and Ruth Weisz, born in 1922 in Goeding, Moravia (pp. 98-157) – on their internment in Theresienstadt. Federica and her parents arrived in Theresienstadt in October 1942, suffered from starvation and epidemics, and narrowly escaped deportation to an extermination camp. Weisz and her family moved to Prague in 1938; from there they were deported to Theresienstadt in 1943, and she was sent to Auschwitz in 1944. The essay by Wolfgang Benz (pp. 158-171), "Theresienstadt: Ort der deutschen Geschichte," deals with the role and function of Theresienstadt as a concentration camp. *ID*

0461 Sprenger, Isabell: *Gross-Rosen: Ein Konzentrationslager in Schlesien.* Wien: Boehlau, 1996. viii, 425 pp.

Surveys the establishment of a labor camp at Gross-Rosen in 1940 and its development and transformation into a concentration camp in 1942. Pp.

121-137, "Juedische Haeftlinge in Gross-Rosen," report on Jews who were deported to the camp at the end of 1941. They were separated from other prisoners, they had to work harder for longer hours, they were beaten and maltreated, and in case of sickness had no right to medical treatment. After only a short period in the camp, they became weak and had little chance for survival. There is no data on the number of Jews interned between August 1940-October 1942, but 218 Jews died and the 37 Jews who remained were deported to Auschwitz. An expansion took place in 1944, when 28 labor camps in the area were taken over by Gross-Rosen; the labor camp complex held 60,000 Jewish prisoners, 26,000 of them women. There are no data on the victims. *ID*

0462 Staudte-Lauber, Annalena: *Stichwort: Holocaust*. Muenchen: Wilhelm Heyne Verlag, 1997. 95 pp.

Surveys the history of the Holocaust, including the rise of the Nazi regime, anti-Jewish legislation, the "Kristallnacht" pogrom, and the exclusion, persecution, and deportation of Jews to concentration and extermination camps. Deals, also, with the aftermath of the Holocaust: commemoration and memorials, reparation payments, memoirs of survivors, Holocaust denial, and the "Historikerstreit." *ID*

0463 Steinmacher, Gabriele: Juedische Bibliotheken unter dem Nationalsozialismus: Buecherdiebstahl waehrend des Zweiten Weltkrieges. *Myosotis* 1 (1998) 3-12.

Traces the fate of Jewish seminary, community, and private libraries and archives which were confiscated by the Nazis, first in Germany and later in the occupied countries. The confiscations were coordinated by Rosenberg's "Einsatzstab" and the books were sent to German libraries. Judaica and Hebraica were concentrated in the libraries of the Institut zur Erforschung der Judenfrage in Frankfurt am Main and of the RSHA in Berlin; they were to serve Nazi research on the "Jewish question" and antisemitic propaganda. Enumerates the main libraries confiscated. After the war, ca. two million of the four million stolen books were recovered by the Commission on Jewish Cultural Reconstruction. Where possible, they were restored to their original owners; the rest were distributed to Jewish libraries, mainly in Israel. *RW*

0464 Stone, Dan: Modernity and Violence: Theoretical Reflections on the Einsatzgruppen. *Journal of Genocide Research* 1, 3 (Nov 1999) 367-378.

Criticizes the recent historiography of the Holocaust which tends to explain the Nazi genocide as an outcome of modernity. These historians focus on "industrial killing" in the death camps, and present Nazi Germany as a civilized country which eradicated violence and in which mass murder served some purpose. In fact, society can combine civilization and barbarity without contradiction. Antisemitism, sadism, wanton brutality, the desire to break free of a rationalized world were no less driving forces in the mass murder. Einsatzgruppen reports and postwar testimonies of their members can produce the impression that these mass murderers killed out of necessity and in an orderly fashion. In fact, the motives that they present are but a post factum rationalization of their violence. Historians must treat the murderers' reports with caution and give voice to their victims. *DR*

0465 Straeten, Herbert: *Andere Deutsche unter Hitler: Zeitberichte ueber Retter vor dem Holocaust.* Mainz: v. Hase & Koehler, 1997. x, 192 pp.

Presents 26 stories of Germans who risked their lives to save Jews, in Germany itself and as soldiers or officials in the occupied countries, and who were honored by Yad Vashem as Righteous Gentiles. The introduction (pp. 1-15) discusses the newly-awakened need of Germans to confront responsibility and shame for the Holocaust, but also warns against attribution of collective guilt – there were other Germans. Pp. 181-187 contain the names of ca. 300 Germans recognized as Righteous Gentiles by Yad Vashem. *RW*

0466 Stránský, Pavel: *Als Boten der Opfer: Von Prag durch Theresienstadt, Auschwitz, Schwarzheide und zurueck: Tschechisch-juedische Schicksale 1939-1997.* Konstanz: Hartung-Gorre, 1997. 90 pp.

Pp. 13-48 contain the memoirs of Stránský, a Czech Jew (b. 1921). Describes the children's block of the "Czech family camp" in Auschwitz (he was in charge of the house of the 3-6 year olds), including visits by Mengele, who took children on his lap and played with them; the liquidation of his transport after half a year; the labor camp at Schwarzheide; and the death march to Theresienstadt in April 1945. Stresses SS tactics to dehumanize the prisoners. Also describes his dismissal from his job in a Czech government office during the Slánský trial and the silence about the Holocaust imposed by the communist regime. *RW*

0467 Strobl, Ingrid: *Die Angst kam erst danach: Juedische Frauen im Widerstand in Europa 1939-1945.* Frankfurt a.M.: Fischer Taschenbuch Verlag, 1998. 479 pp.

Discusses the role of Jewish women in resistance activities in Belgium, France, the Netherlands, Hungary, and Poland during World War II, based on archival material, private documents, letters, and diaries, as well as interviews with 43 former resistance fighters. Most of the interviewees had an academic or professional education; 90% were immigrants or children of immigrants. Their activities focused on rescue, especially of children, and on armed resistance, for both of which the falsification of documents was a very important function. In France, Belgium, and the Netherlands, armed resistance was organized primarily by communist groups. The Jewish resistance in Western Europe, apart from the Armée Juive in France, arose from the youth movements, which were active especially in organizing soup kitchens and homes for children. In Hungary, the development of events after the German occupation was so rapid that organization of an armed resistance was not possible; resistance dealt with rescue. In all of these groups, women held leading positions, but especially in the left-wing youth movements in Poland, and in social and educational institutions in France. *ID*

0468 Teichman, Milton: How Writers Fought Back: Literature from the Nazi Ghettos and Camps. *Judaism* 47, 3 (Sum 1998) 338-350.

Writing in the ghetto was a form of spiritual resistance. Focuses on writers in ghettos, and even in camps, who intended to leave a record of Nazi crimes and Jewish sufferings in the hope that after the war justice would be done. Dwells on the history of the Ringelblum Archives, compiled in the Warsaw ghetto, which served as a pattern for similar ventures in Poland and Lithuania, e.g. the archives of the Białystok, Vilna, and Kovno ghettos. Less daring, but also significant, was Rumkowski's archive in the ghetto of Łódź. Besides such collective efforts to document the horrors of the Holocaust, there were individuals who recorded events: Isaiah Spiegel, Jozef Zelkowich, and Simcha Bunim Shayevich in Łódź; Yitzhak Katzenelson, Abraham Lewin, and Chaim Kaplan in Warsaw; Abraham Tory in Kovno; Zalman Gradowski, who gave an account of life in the Sonderkommando in Auschwitz; and others. Dwells on the psychological problems of writing in ghettos and camps. *DR*

0469 *Terezínské Studie a Dokumenty* (1998).

Most of the articles listed below appeared in German in "Theresienstaedter Studien und Dokumente" 1995, 1998, and 1999: Kárný, Miroslav: Vyřazení židů z veřejného života protektorátu a historie "čestného árijství" (11-41); Milotová, Jaroslava: Okupační aparát a příprava transportů do Lodže (42-66); Seemann, Richard: Čeští Židé v Lodži [Czech Jews in Łódź] (67-85); Hampel, Jens: Osud židovského obyvatelstva města Jihlavy 1938-1942 (86-113); Franková, Anita: Terezínská výchova: Zprávy k prvnímu výročí terezínských domovů v L 417 [Pp. 121-152 contain the texts or translations of the reports by Otto Zucker, Albert Fischer, Egon Redlich, Franz Kahn, Ota Klein, G. Baeuml, Maximilian Adler, Berta Freund, Walter Eisinger, Rosa Engländerová, Jan Brammer, Sigi Kwasniewski, Friederike Brandeis, and Hans Krása.] (114-153); Tarsi, Anita: Staré ženy z Německa v ghettu Terezín (154-

178); Chládková, Ludmila: O jednom domě v terezínském ghettu (179-186); Bondy, Ruth: Prominenti až do odvolání (187-204); Margry, Karel: Zajímavý předchůdce – první terezínský film (1942) (205-231); Makarová, Elena: Akademie přežití (232-258); Kukáňová, Zlatuše; Matušíková, Lenka: K evidenci zemřelých v terezínském ghettu (259-272); Franc, Miroslav: Pracovní komando Wulkow (273-288); Sladký, Hans H.: Epizody ze života přeživšího (289-305); Čapková, Kateřina: Svědectví Salmena Gradowského o rodinném táboře v Birkenau (306-310); Gradowski, Zalman: V srdci pekla (311-330); Munk, Jan: Rozvojové koncepce Památníku Terezín a motivační struktura jeho návštěvníků [In the 1990s.] (331-340); Kopold, Bedřich: Osvobození vězňové v československé východní armádě [Freed Prisoners in the Czech Army in the East. On prisoners of the Nazis (mostly Jews) who, after their liberation, joined the Czech army formed in the USSR to fight the Nazis.] (341-354); Frankl, Michal: Případ Jeanette Jennyové (355-363); Škorpil, Pavel: Případ Bedřicha Laufera [The Case of Bedřich Laufer] (364-374). *SSC*

0470 Ternon, Yves: La conscience perdue du docteur Muench, médecin SS à Auschwitz. *Revue d'Histoire de la Shoah* 165 (Jan-Apr 1999) 163-184.

Traces the figure of Hans Muench, who was an SS physician in Auschwitz and the only one acquitted in the Kraków trial in 1947. He refused to participate in selections, and saved the lives of ten women on whom he had performed medical experiments. For these reasons, a legend arose that he was the only "human" doctor in Auschwitz. A recent interview (in "Der Spiegel," 1998), and other research, shows that he was and has remained a Nazi ideologue up to the present. His refusal to participate in selections was only a technical problem, and not a moral one. His Nazi credo has not changed, and in a very cynical and perverted way he now banalizes and relativizes Auschwitz. He saved some Jewish women because he needed them for medical experiments, and he was pleased that he could perform experiments that were impossible in other times. *HV*

0471 *Theresienstaedter Studien und Dokumente* (1998).

Contents: Kárný, Miroslav: Die Ausschaltung der Juden aus dem oeffentlichen Leben des Protektorats und die Geschichte des "Ehrenariertums" (7-39); Milotová, Jaroslava: Der Okkupationsapparat und die Vorbereitung der Transporte nach Lodz (40-69); Hampel, Jens: Das Schicksal der juedischen Bevoelkerung der Stadt Iglau 1938-1942 (70-99); Tarsi, Anita: Das Schicksal der alten Frauen aus Deutschland im Ghetto Theresienstadt [Also relates to their psychological reactions.] (100-130); Chládková, Ludmila: Von einem Haus im Theresienstaedter Ghetto [On the basis of a register kept by Margarete Pedde from

September 1942-February 1943, analyzes statistically the ages and national origins of the 679 residents who lived in House L415 or passed through it during this period, and traces their subsequent fate.] (131-141); Franková, Anita: Theresienstaedter Erziehung: Berichte zum ersten Jahrestag der Theresienstaedter Heime in L417 [Pp. 149-180 contain the texts or translations of the reports by Otto Zucker, Albert Fischer, Egon Redlich, Franz Kahn, Ota Klein, G. Baeuml, Maximilian Adler, Berta Freund, Walter Eisinger, Rosa Engländerová, Jan Brammer, Sigi Kwasniewski, Friederike Brandeis, and Hans Krása.] (142-148); Margry, Karel: Ein interessanter Vorgaenger: Der erste Theresienstadt-Film (1942) (181-212); Makarová, Elena: Die Akademie des Ueberlebens [On the lectures organized in Theresienstadt by the Jewish internees.] (213-238); Franc, Miroslav: Arbeitskommando Wulkow (239-256); Sladký, Hans H.: Episoden aus dem Leben eines Ueberlebenden (257-276); Bondy, Ruth: Ueberleben [In Theresienstadt, Auschwitz, forced labor in Hamburg, and Bergen Belsen.] (277-287); Osers, Jan: Flucht aus Zamość, herausgegeben und kommentiert von Raimund Kemper [The memoirs of two Czech Jews, Osers and Jan Bachrich, deported from Theresienstadt to the Zamość ghetto. They fled from Zamość, independently of each other, changing their identities frequently. Both were repeatedly arrested in the countries in which they sought refuge, Hungary and Slovakia. Bachrich was sent to Auschwitz, volunteered for a labor camp in Warsaw, and survived a death march and forced labor in Kaufering. Osers experienced the end of the war in Hungary.] (288-321); Pick, Miloš: Erinnerungen an die Widerstandsbewegung [In Liban, Czechoslovakia, Auschwitz, Meuselwitz (a labor camp affiliated with Buchenwald), and on a death march through the Sudeten.] (322-341); Munk, Jan: Entwicklungskonzeptionen der Gedenkstaette Theresienstadt und die Motivationsstruktur ihrer Besucher [In the 1990s.] (342-355); Kukáňová, Zlatuše; Matušíková, Lenka: Zur Erfassung der Verstorbenen im Theresienstaedter Ghetto [In Czech archives.] (356-367); Frankl, Michal: Der Fall der Jeanette Jenny [Jenny, a Swiss woman working at the Swiss consulate in Prague, was imprisoned by the Gestapo for trying to help a Jewish friend avoid deportation and for denouncing the Nazis' persecution of Jews.] (368-378). *RW*

0472 *Theresienstaedter Studien und Dokumente* (1999).

Partial contents: Kárný, Miroslav: Fragen zum 8. Maerz 1944 [On the gassing of the Jews of the September 1943 transport from Theresienstadt, who had spent six months in the Auschwitz "family camp."] (9-42); Flusser, Jindřich: Ein Rueckblick [A memoir on Auschwitz and Blechhammer, where the author worked part of the time as a medical assistant.] (43-75); Bondy, Ruth: Ein Brotwunder: Das Hilfswerk von Heinz Prossnitz (76-104); Čapková, Kateřina: Das Zeugnis von Salmen Gradowski (105-111); Gradowski, Zalman: Im Herzen der Hoelle [A member of the Sonderkommando describes the gassing of the Jews from the Czech "family camp" in Auschwitz on 8 March 1944.] (112-140); Lorencová, Anna; Hyndráková, Anna: Die tschechische Gesellschaft und die Juden in den Erinnerungen von Zeitzeugen [Based on reminiscences of Jews on the pre-Holocaust and Holocaust periods, recorded by the Holocaust Section of the Jewish Museum in Prague.] (141-167); Tučková, Anna: Ich war Mischling ersten Grades [In Brno.] (168-179); Zimmermann, Volker: Taeter und Zuschauer: Die Judenverfolgung im "Sudetengau" 1938-1945 (180-203); Sever, Max: Die technischen Einrichtungen Theresienstadts (204-216); Peduzzi, Lubomír: Aus dem Theresienstaedter Musikleben (217-225); Fatran, Gila: Franz Kahn [Appeared in Hebrew in "Yalkut Moreshet" 51 (1991).] (226-241); Poloncarz, Marek: Die Evakuierungstransporte nach Theresienstadt (April-Mai 1945) (242-262); Brenner, Hans: Frauen in den Aussenlagern von Flossenbuerg und Gross-Rosen in Boehmen und Maehren [On women, including Jews, in many small satellite camps.] (263-293); Pospíšil, Leopold: Zwei Tagebuecher: Anne Frank und Otto Wolf [Wolf, b. 1927, hid with his family in and around Tršice, Czechoslovakia. They were discovered on 18 April 1945 and Otto was shot.] (294-298). *RW*

0473 Touboul, Camille: *Le plus long des chemins: De Marseille à Auschwitz.* Nice: Editions du Losange, 1997. 196 pp.

Memoirs of a Jewish woman born in Casablanca in 1921. Her family moved to Marseille when she was a child. In November 1943 she was arrested by the Gestapo and deported via Drancy to Auschwitz. Describes life and conditions in the camp. In January 1945 she was taken from Auschwitz to Wrocław in a death march, then to Ravensbrueck and to Chemnitz where she was liberated by American troops. After the war she returned to Marseille. Both of her parents survived the war. *HV*

0474

אונסדארפער, שמחה בונם: דער געלער שטערן: לעבנס געשיכטע אין די נאצישע טויטן לאגערן. תרגום: ל. הלפרין. ירושלים?: חסר מו"ל, 1996? 162 עמ'.

[Unsdorfer, Simha Bunem: *The Yellow Star: A Life Story in the Nazi Extermination Camps.* Trans.: L. Halperin. Jerusalem?: no pub., 1996? 162 pp.] Originally published as "The Yellow Star" (New York: T. Yoseloff, 1961). Appeared in Hebrew as "Tlai Tsahov" (Jerusalem?: no pub., 1995?).

0475 Varon, Laura: *The Juderia: A Holocaust Survivor's Tribute to the Jewish Community of Rhodes.* Westport, CT: Praeger, 1999. x, 167 pp.

Memoirs of a Jew born in 1926 in Rhodes. Relates the introduction of the Italian racial laws in 1938-39, and how the community adjusted to the restrictions. In spring 1944 the Nazis occupied the island. The Jews were transported by ship to Athens, where they encountered Nazi brutality in the Haidary detention camp. From there they were deported to Auschwitz, where most of her relatives (including her parents) were killed. Varon remained in Birkenau with her sister Stella. Gives details about life in the women's camp. When Stella became ill, Varon took her to the camp infirmary, thereby saving her from death in a selection. Varon was transferred to Dachau, and then sent on a death march to Bergen-Belsen, where she was liberated. After the war she was taken to Sweden to recover from severe illness. In 1947 she was reunited with her uncle, two brothers, and her sister in Italy. She emigrated to the U.S. in 1961. Of the 2,000 Jews who were deported from Rhodes in 1944, only 104 survived. *YC*

0476 Viviani, Aglaia: La guerra con occhi di bambina: Autobiografie d'infanzia: Janina David, Giacoma Limentani e Hannele Zuerndorfer. *La Rassegna Mensile di Israel* 64, 1 (Jan-Apr 1998) 145-167.

Deals with the issue of maintaining or recovering Jewish identity during and after the trauma of the Holocaust. Refers to the published Holocaust diaries of three young Jewish girls: Janina and Giacoma, who were hidden in Catholic convents in Poland and Italy respectively, and Hannele, who found asylum as a refugee in England with Christian families, but was seen as an "enemy alien" during the "Blitz." Outlines the children's trauma as a result of the loss of their families and their identities in hiding. *AA*

0477 Vries, Willem de: *Sonderstab Musik: Music Confiscations by the Einsatzstab Reichsleiter Rosenberg under the Nazi Occupation of Western Europe*. Trans. of texts: Lee K. Mitzman. Amsterdam: Amsterdam University Press, 1996. 271 pp.

Against the background of the Nazi purge of Jews and "Jewish elements" from German culture, under the leadership of Alfred Rosenberg, describes the activities of the "Sonderstab Musik," whose aim was to locate music manuscripts, books, and instruments. This enterprise, headed by the musicologist Herbert Gerigk, was extended to Nazi-occupied Europe, where it carried out extensive confiscations of Jewish property related to music. Particular attention is given to confiscations in France, Belgium, and the Netherlands. Among the Jews who lost possessions were the composer Darius Milhaud and the harpsichordist Wanda Landowska. Most Jewish musical property remained in the hands of German citizens after the war. *YC*

0478 Wasserstein, Bernard: Polish Influences on British Policy Regarding Jewish Rescue Efforts in Poland 1939-1945. *Polin* 11 (1998) 183-191.

Analyzes the effects of the Polish government-in-exile's attitudes on the policies of the British government towards the "Jewish question" and the issue of the rescue of Polish Jews. States that the Polish desire to reduce the number of Jews in Poland influenced all the discussions between the two governments. The Poles wished to stimulate a large-scale Jewish emigration to Palestine and the British sought to implement the policy of the White Paper of 1939 which limited immigration. British officials knew about the antisemitism in interwar Poland and its persistence during the war. However, in 1941 they advised Poland to assimilate the Jews and not to try to get rid of them. This is why the issue of rescue of Jews from Poland was not raised in discussions between the British and Polish governments at any stage in the war. Concludes that the British attitude to the problem of the rescue of the Jews showed a failure of elementary human solidarity and shows how this attitude was connected with the attitude of Poles towards the Jewish population. *HV*

0479 Waysand, Georges: *Estoucha: Récit.* Paris: Denoël, 1997. 438 pp.

A biography of the life of the author's mother, Esther (Estoucha) Zilberberg. She was born in Kalisz, Poland; as a young woman she emigrated to Belgium, where she studied medicine. An active communist, she went to Spain in 1936 to serve as a sanitary aide in the Civil War. After the war, she entered France illegally. Following the German occupation of France, she fled to the south and later returned to Belgium with her friend, Jean Waysand, a Jew from Poland. In 1942 their son Georges was born. His parents placed him with a Christian family and joined the communist resistance, living on "Aryan" documents. Jean was arrested and executed in December 1942. Estoucha was arrested in Lille in February 1943 as a resistant communist and interned in various prisons. In December 1943 she was transferred to Ravensbrueck, and from there to Mauthausen where she was liberated. In 1955 she finished her medical studies in France and became a doctor. *HV*

0480 Weiss, Yfaat: Polish and German Jews between Hitler's Rise to Power and the Outbreak of the Second World War. *Leo Baeck Institute Year Book* 44 (1999) 205-223.

German Jews, who had attained civil equality as early as the late 19th century and whose economic condition was relatively good, had rendered help to Polish Jews suffering from antisemitism and had waged a relentless struggle for their rights. With the Nazi rise to power in 1933, the roles of the two communities were reversed; now the Polish Jews, who had gained civil rights in 1919, had more freedom of action than their German brethren, and could wage a struggle both in Poland against growing antisemitism there, and in the world arena on behalf of German Jews. Dwells on the controversy over the distribution of immigration certificates by the Jewish Agency, since the demand of German Jews for entry permits to Palestine grew drastically after 1933. Dwells, also, on the key role of Polish Jews in the anti-Nazi economic boycott in 1933-35. The boycott brought about a clash of interests between Polish Jews and German Jews who feared a Nazi backlash, and a controversy with the Zionist movement concerning the Transfer Agreement. *DR*

0481 Wilson, Nelly: Back to the Monsters: A Holocaust Survivor Reads Goldhagen's "Hitler's Willing Executioners." *The Jewish Quarterly* 168 (Win 1997-1998) 62-64.

Wilson was deported as a child from Vienna to Theresienstadt, and from there to Auschwitz and a labor camp in Silesia. She agrees that her compatriots in Austria were willing collaborators, but states that Goldhagen weakens his case by insisting that all the people were collaborators, and that they all knew about the mass killings. In order to inculpate the "Volk," he endows it with unusual powers of perception and understanding, as well as an incredible capacity for protest. He does not take the element of fear into account in explaining the actions of the ordinary person, nor the fact that many people did not know about the killings; even the victims did not know, or, if they were told, did not believe it. Much more to the point is the knowledge a good many German citizens must have had of the forced labor camps. Relates incidents from her own experience. To state that anybody is capable of doing what the Germans did shows up the dark side of humanity. To blame all the Germans is too simplistic. However, it is true that the extermination policy was conceived, planned, and organized in Germany. *SSC*

0482

יחזקאל (פרידלר), אילה: לארוג את סיפור החיים: רה-ביוגרפיה של ניצולי שואה. קיבוץ לוחמי הגיטאות: בית לוחמי הגיטאות; תל-אביב: הקיבוץ המאוחד, 1999. 272 עמ'.

[Yeheskel (Fridler), Ayala: *Weaving the Story of One's Life: Re-Biography of Holocaust Survivors.* Kibbutz Lohamei Haghetaot: Beit Lohamei Haghetaot; Tel-Aviv: Hakibbutz Hameuchad, 1999. 272 pp.] Based on the author's diss. – Hebrew University of Jerusalem, 1997.

A study based on interviews with eleven Holocaust survivors who live in Beer-Sheva, Israel: Shoshana Adler, Miriam Yahav, Asher Zohar, David Efrati, Ida Buchsbaum, Miriam Livni, Binyamin Livnat, Ya`akov Belogokovsky, and three others who are not identified. The interviewees were born between 1920-27 – seven in Poland, three in Hungary, and one in Czechoslovakia. They relate their wartime experiences in ghettos, labor camps, and concentration camps. The biography of each interviewee is followed by an analysis of the mechanism used by that person for rehabilitation after the Holocaust and for creating a re-biography, in order to establish a “continuous identity” and achieve a “tikkun,” a correction of evil. *RE*

0483

יחזקאל (פרידלר), אילה: סיפור חיים של ניצולי שואה: ניסיון ליישום המושג רה-ביוגרפיה. דיס' — האוניברסיטה העברית בירושלים, 1997. 275, 14 עמ'.

[Yeheskel (Fridler), Ayala: *Life Stories of Holocaust Survivors: An Attempted Application of the Concept of Re-Biography.* Diss. – Hebrew University of Jerusalem, 1997. 275, 14 pp.] With an English summary.

See the abstract for the author's book, “Le'erog et sipur ha-hayim: Re-biografia shel nitzolei Shoah” (1999), based on this dissertation. The book gives the real names of the subjects interviewed.

0484

דאס אידישעווארט 336 (אדר א' תשנז).

[*Dos Yiddishe Vort* 336 (Adar A 5757 = Feb 1997).]

”דער חורבן – א געשיכטליכער איבערבליק“: אחי-שמשון רפאל (פסויד'): די ערשטע יארן פון רדיפות, גירושים, פיין און רציחות (35-29); אדלער, דוד: די דייטשע טויט-פאבריקן (42-36); גאטעסמאן, ישעיה: די אייגנארטיקייט פון חורבן אונגארן (45-43); דער קידוש השם פון לעצטן אסטראוועצער רבי'ן הי"ד (46); זשולטי, בעריש: יום-כיפור ביי דער תליה אין לאגער 'בלעכהאמער': זכרונות פון א קאצעטלער (48-47); שטשעגאוווסקי, אברהם דוד: מיט הקדוש ר' יהודה לייב ארלעאן און ראדאמסקער בני-תורה הי"ד אין קראקאווער געטא (49); בוים (בוימעלגרין), אפרים דוד: תפילות און תקיעת שופר אין בונא (50); לייזערזאן, גרשון: א צוואה פון א קדוש [ר' שמואל יהודא גריסגאטט, סלובקיה, טבת תש"ה] (51); בריוו פון ”יענער-זייט“ [נאספו ע"י גוטא שטרנבוך; כולל מכתבים שנשלחו אליה בזמן המלחמה ולמשפחת שטרנבוך בשווייץ.] (55-52); בעסער, יחזקאל: דער ארץ-ישראל ישוב און דער חורבן (58-56); טאפעל, מרדכי: דער ענדלאזער וועג: די ”טויט-מארשן“ ערב דער באפרייאונג (60-59); אדלער, דוד: די געשיכטע וואס איז נאך נישט גענוג דערציילט געווארן: די גייסטיקע גבורה פון אידן אין די געטאס און קאנצענטראציע-לאגערן (62-61); גלאטשטיין, מרדכי לייב: חורבן און תקומה: פון ווארשעווער געטא, איבער בודזין, אוישוויץ און דאכאו, ביז צו פעלדאפינג און מינכען צו א באנייט לעבן (71-69); פרוינדליך, שמחה זימל: רעטונג אין דער לעצטער מינוט (73-72); ביטערספעלד, יקותיאל: הצלה אין באהעלטעניש (75-74); בלוי, צ.פ.: הצלה פון סקווערער רבי'ן זצ"ל (81-76); אחי שמשון רפאל (פסויד'): אין אנדענק פון די הצלה-העלדן זכרונם לברכה [בעיקר על רבנים בארה"ב ובאירופה שעסקו בהצלה.] (86-93); געניסלאוו, מרדכי: צום אייביקן אנדענק פון מיין טראגיש אומגעקומענער משפחה (151-150); פישהאף, יחיאל בן-ציון: באגלייט פון מלאכים...: די רעטונגס-געשיכטע פון 72 פליטים אויפן וויטען מזרח [מליטא לשנחיי.] (159-157).

This issue is devoted to the “She'erith ha-Pleitah” – the Jews who survived the Holocaust, and how religious Jewish life and learning was renewed after the war. The following articles describe the Holocaust period: [Achi Shimshon Raphael (pseud.): The First Years of Persecution, Expulsions, Torment and Murder (29-35); Adler, David: The German Death-Factories (36-42); Gottesman, Yeshaya: The Uniqueness of the Hungarian Destruction (43-45); The Martyrdom of the Last Ostrowiecer Rebbe (46); Zolty, Berish: Yom Kippur at the Hanging in Camp "Blechhammer": Memoirs of a Concentration Camp Inmate (47-48); Shchegovskii, Avraham Dovid: With the Martyr R. Yehudah Leib Orlean and Torah Scholars from Radom in the Kraków Ghetto (49); Boim (Boimelgreen), Efraim Dovid: Prayers and Blowing the Shofar in Buna (50); Leisersohn, Gershon: The Last Will of a Martyr [R. Shmuel Yehudah Grussgott, Slovakia, Teveth 5705 = January 1945.] (51); Besser, Yehezkel: The Yishuv in Eretz-Israel and the Holocaust (56-58); Toppel, Mordechai: The Final Way: The "Death Marches" on the Eve of Liberation (59-60); Adler, David: The Story Which Has Not Yet Been Told Enough: The Spiritual Heroism of Jews in the Ghettos and Concentration Camps (61-62); Glatstein, Mordechai Leib: Destruction and Resurrection: From the Warsaw Ghetto, through Będzin, Auschwitz and Dachau, to Feldafing and Munich and a New Life (69-71); Freundlich, Simcha Zimmel: Rescue in the Last Minute (72-73); Bittersfeld, Yekutiel: Rescue in Hiding (74-75); [Blau, Ts.P.: Rescue of the Skverer Rebbe zts"l (76-81); Achi Shimshon Raphael (pseud.): In Memory of the Rescue Heroes, May Their Memory Be for a Blessing [Especially on rabbis in the USA and Europe who engaged in rescue.] (86-93); Genislav, Mordechai: In Eternal Memory of My Tragically Destroyed Family (150-151); Fishhof, Yehiel Ben-Zion: Accompanied by Angels...: The Rescue Story of 72 Refugees to the Far East [From Lithuania to Shanghai.] (157-159).] *SSC*

0485

צימרמן, משה, עורך: הרייך השלישי: מאזן היסטורי. ירושלים: הוצאת מאגנס, האוניברסיטה העברית, 1999. 211 עמ'.

[Zimmermann, Moshe, ed.: *The Third Reich: A Historical Reassessment.* Jerusalem: Magnes Press, Hebrew University, 1999. 271 pp.]

תוכן חלקי: צימרמן, משה: הרייך השלישי: אפיזודה היסטורית? (15-9); דינר, דן: מאורע וזיכרון: על וריאציות בזיכרון ההיסטורי (51-4) ;אשהיים, סטיבן אדוארד: הנאציזם והשואה בשיח המערבי והישראלי (65-52); יאקובייט, זיגריד: הלבוש בגרמניה הנאצית (157-142); ברקאי, אברהם: חיסולם הכלכלי של יהודי גרמניה: כוונה אידאולוגית או דינמיקה פונקציונליסטית? (185-178); מרגלית, גלעד: השיח הגרמני על רדיפת הצוענים בתקופת הרייך השלישי (196-186); היים, סוזאנה: רציונליזציה כלכלית ומדיניות דמוגרפית במזרח אירופה בתקופת הכיבוש הגרמני (223-215).

A collection based on papers delivered at a conference organized by the Koebner Center of German History at the Hebrew University of Jerusalem, 1995. Partial contents: [Zimmermann, Moshe: The Third

Reich: A Historical Episode? (9-15); Diner, Dan: Event and Memory: On Variations in Historical Memory [Appeared in German in "Shoah, Formen der Erinnerung: Geschichte, Philosophie, Literatur, Kunst" (1996).] (40-51); Aschheim, Steven Edward: Nazism and the Holocaust in Western and Israeli Dialogue [On responses to the Holocaust, mainly in Germany and Israel.] (52-65); Jacobeit, Sigrid: Dress in Nazi Germany [Also on the anti-Jewish boycott and restrictions against Jews.] (142-157); Barkai, Avraham: The Economic Liquidation of German Jews: Ideological Aim or Functional Dynamics? (178-185); Margalit, Gilad: The German Dialogue on the Persecution of Gypsies in the Third Reich [Compares their persecution to that of the Jews.] (186-196); Heim, Susanne: Economic Rationalization and Demographic Policy in Eastern Europe during the German Occupation [Also contends that Nazi anti-Jewish measures were meant to solve economic problems in the occupied territories.] (215-223).] *RE*

Austria

0486 John, Michael: Aggressiver Antisemitismus im oesterreichischen Sportgeschehen der Zwischenkriegszeit: Manifestationen und Reaktionen anhand ausgewaehlter Beispiele. *Zeitgeschichte* 26, 3 (May-June 1999) 203-223.

Chronicles violent antisemitic incidents in Austria at competitive sports events involving teams of the Jewish sports organization Hakoah or other teams with large numbers of Jews. Young bullies and street children attacked Jewish spectators; sometimes members of the rival team intentionally injured Jewish players. At the universtities, Jewish fraternities were not considered worthy of defending their honor in saber duels; instead there were bloody fights with fists, canes, and rubber truncheons. More and more young Jews defended themselves; at times they were the attackers. Hakoah formed a defense squad to protect the Leopoldstadt from antisemitic rioters. The anti-Nazi dictatorship of the "Staendestaat" suppressed such violence. After the Anschluss, some top Jewish athletes were aided by their "Aryan" rivals. Argues that the physical antisemitism of the 1920s contradicts the thesis that until the Nazi period there existed only "cultural antisemitism," which, however vicious, stopped short of aggressive action. *RW*

0487 *Juedische Brigittenau: Auf den Spuren einer verschwundenen Kultur.* Wien: Gebietsbetreuung Brigittenau, 1997. 87 pp. On title page also: Katalog zur gleichnamigen Ausstellung der Gebietsbetreuung Brigittenau, 5.-28. November 1997.

The catalogue of an exhibition on Jewish life in Brigittenau, a district of Vienna, from 1900 to the present. Pp. 25-48 deal with the Holocaust period: the "Anschluss" in March 1938, the anti-Jewish measures, the "Kristallnacht" pogrom, Aryanization, persecution, and deportation. Pp. 49-60 include a list of 158 names of former Jewish residents of Brigittenau who were deported to concentration camps. *ID*

0488 Kratz, Kaethe et al., eds.: *Verlorene Nachbarschaft: Die Wiener Synagoge in der Neudeggergasse: Ein Mikrokosmos und seine Geschichte.* Wien: Mandelbaum, 1999. 303 pp.

Presents memoirs of 17 Holocaust survivors (pp. 35-140) who lived in Vienna until the Nazi takeover; after the Anschluss and the "Kristallnacht" pogrom they left the country. Includes essays (pp. 181-253) which discuss the experiences of survivors, the "Kristallnacht" pogrom, antisemitism in Austria, and the synagogue. Partial contents: Brown, Curtis L.: Mit zwanzig Dollar in den Wilden Westen (35-41); Bar-On, Dan: "Dass ich Jude bin und einen Buckel hab..." (43-47); Segal, Lore: Das ganz kleine Dorf Fischamend (49-53); Edelstein, Elieser L.: "Dieser Ghettogeist, der da herrschte..." (55-60); Lederer, Gerda: "Ich war froh, Opfer und nicht Angreifer zu sein" (61-66); Eres, Konrad: "Wir sind Spiessruten gelaufen" (67-72); Kelemen, Anne: Die Finesse der Erniedrigungen (73-78); Reiss, Abraham: "...dass unseres Bleibens nicht ist" (79-84); Prochnik, Wilhelm Guillermo: "Ich war kein Kaempfer" (95-101); Slutzkaja-Brand, Minna: "Am Suedbahnhof hab ich das Pflaster gekuesst" (113-119); Schweitzer, Sabine: Erinnerungen im Schatten der Shoah: Ehemalige Wiener Juden/Juedinnen in Argentinien, Israel und den USA erzaehlen ihr Leben (162-180); Weinzierl, Erika: Antisemitismus in der Ersten Republik (181-186); Streibel, Robert: Die Zerstoerung der Tempel und was wir alles nicht wissen (187-198). *ID*

0489 Schwarz, Peter: *Tulln ist judenrein! Die Geschichte der Tullner Juden und ihr Schicksal von 1938 bis 1945: Verfolgung – Vertreibung – Vernichtung.* Wien: Loecker Verlag, 1997. 376 pp.

Discusses the history and fate of the Jewish community of Tulln, Austria, between 1938-45. Emphasizes the existence of latent political antisemitism in Austria before 1938, and its economic and racial aspects. Describes the Anschluss in 1938, the "Kristallnacht" pogrom in Tulln, antisemitic propaganda, and the expropriation and Aryanization of Jewish property. 71 Jews were registered in Tulln in 1934; in 1940 only six remained, who had no chance of emigrating because of their age or lack of funds. Deals with the deportation of former Jewish residents of Tulln from Vienna to concentration and extermination camps in the district of Lublin in 1941-42. The town was declared "judenrein" in 1940. 36 Jews from Tulln were murdered in the Holocaust. *ID*

0490 Szita, Szabolcs: *Verschleppt, verhungert, vernichtet: Die Deportation von ungarischen Juden auf das Gebiet des annektierten Oesterreich 1944-1945.* Trans.: Schmidtné Tasnádi Ágnes, Winfried Schmidt. Wien: Werner Eichbauer Verlag, 1999. 279 pp. Originally published as "Utak a pokolból: Magyar deportáltak az annektált Ausztriában, 1944-1945" (Budapest: Metalon, 1991).

Baltic States

(see also Estonia; Latvia; Lithuania)

0491 Anolik, Binyamin: *Missiya pamyati* [The Mission of Memory]. Trans.: Olga Medvedeva. Moskva: Rossiiskoye obshchestvo Yanusha Korchaka, 1997. 135 pp. Translated from the Polish (1996). Originally published as "Bi-shelihut ha-zikaron" (Kibbutz Lohamei Haghetaot: Beit Lohamei Haghetaot, 1990).

Memoirs of a Jew born in 1926 in Vilna, describing his experiences in the Holocaust. First he was interned in the Vilna ghetto; in 1943 he was transferred to Estonia where he went through various labor camps (Viivikonna, Klooga, Lagedi, etc.). In September 1944, when the Nazis began to liquidate all the remaining camp inmates, Anolik succeeded to flee eastward where he met with the Soviet Army. In 1963-69 he testified at trials against Nazi war criminals held in Germany, in particular against Helmut Schnabel who was a Lagerfuehrer in Estonian camps. In 1985 he participated in the first conference of Righteous Gentiles in Poland. Describes Polish rescuers of Jews whom he knew personally. Contains, also, recollections on Janusz Korczak, recorded from other Holocaust witnesses. *DR*

0492 Anolik, Binyamin: *Pamięć przywołana* [Recalled Memory]. Kraków: Oficyna Cracovia, 1996. 179 pp. Unseen. Originally published as "Bi-shelihut ha-zikaron" (Kibbutz Lohamei Haghetaot: Beit Lohamei Haghetaot, 1990). See the abstract for the Russian translation: "Missiya pamyati" (1997).

0493 Friedman, Tuviah, ed.: *Die drei SS- und Polizeifuehrer im Ostland – in Lettland-Riga: SS-Brigadefuehrer Schroeder, in Litau-Kowno: SS-Brigadefuehrer Wysocki, in Estonien-Reval: SS-Brigadefuehrer Moeller – die verantwortlich waren fuer die Ermordung der Juden im Ostland 1941-1944: Dokumentensammlung.* Haifa: Institute of Documentation in Israel for the Investigation of Nazi War Crimes, 1998. Various pagings.

A collection of documents, correspondence, and personal papers concerning the identity, role, and activities of three SS-Brigadefuehrers – Schroeder, Wysocki, and Moeller – who were responsible for the annihilation of Estonian, Latvian, and Lithuanian Jewry. Ca. 180,000 Jews were murdered in the Baltic States between December 1940-December 1941. Contains, also, a list of war criminals from the area and a report on the Kovno ghetto. *ID*

0494 Friedman, Tuviah, ed.: *Die Vernichtung der Juden in Estland, Lettland und Litauen durch die SS-Einsatzgruppen und die Mithilfe der lettischen und litauischen Hilfspolizei: Dokumentensammlung.* Haifa: Institute of Documentation in Israel for the Investigation of Nazi War Crimes, 1999. Various pagings. An earlier version appeared as "SS-Obergruppenfuehrer Friedrich Jeckeln verantwortlich fuer die Ermordung der Juden in Litauen, Lettland und Estland 1941-1944" (1997).

Includes two lists of Lithuanian collaborators with the Nazis compiled by the Organization of Lithuanian Jews after the war.

0495 Stranga, Aivars: *Ebreji un diktatūras Baltijā (1926.-1940. gads)* [Jews and Baltic Dictatorships (1926-1940)]. Riga: [The Author], 1997. 171 pp.

Short periods of liberal democracy in the Baltic States during the interwar period were followed by dictatorial regimes: in Lithuania in 1926, in Latvia in 1934, and in Estonia in 1938. These regimes conducted a policy of "national modernization" which became increasingly antisemitic. These countries already had an antisemitic tradition. Extreme right organizations existed in all three countries; the most radical was the Latvian Perkonkrusts. Compares relations between the Jews and the governments in the Baltic States to those in Poland, Romania, and other Eastern European countries. There was no "eliminationist antisemitism" in the Baltics, although in Lithuanian culture its first offshoots could be discerned. The wartime attitudes of Latvians and Estonians toward the Jews were marked with outright indifference. *DR*

Belarus

0496

בונה, נחום: יהודי פינסק בגיטו — תמונת מצב. ילקוט מורשת 64 (נוב' 1997) 57-65.

[Bone, Nahum: The Jews of Pinsk in the Ghetto – Situation Report. *Yalkut Moreshet* 64 (Nov 1997) 57-65.]

Analyzes a Nazi document from 1942 (discovered in Brest in 1994) which lists Jews in the Pinsk ghetto. The list includes 17,344 names and gives date of birth, address in the ghetto, and profession and place of work in the ghetto. Relates, briefly, the history of the ghetto and compares data previously available on the population (compiled mainly from testimonies) with the data provided by the list. *LFo*

0497

פתל-כנעני, תקוה: גרודנה — ניסיונות התנגדות, בריחה והצלה. דפים לחקר תקופת השואה 14 (1997) 51-75.

[Fatal-Knaani, Tikva: Grodno – Resistance Attempts, Escape and Rescue. *Dapim le-Heker Tekufat ha-Shoah* 14 (1997) 51-75.]

Describes the Nazi occupation of Grodno (now Hrodna, Belarus) and the establishment of the ghetto and the Judenrat, including the latter's attitude toward resistance attempts initiated by the youth movements; the links between the local Zionist youth movements and the movements in other ghettos, such as Warsaw and Vilna; resistance and escape to the partisans; and rescue episodes, including arrangement of false identities, placing children with non-Jews, and receiving aid from non-Jews. *LFo*

0498 Garrard, John Gordon; Garrard, Carol: Barbarossa's First Victims: The Jews of Brest. *East European Jewish Affairs* 28, 2 (1998-1999) 3-48.

During the 500 years of its existence, the Jewish community of Brest suffered pogroms and decimations many times. The city was the first one captured by the Germans in the course of their invasion of the USSR in World War II. States that Brest was the site of the first mass murder of the Jews in the Holocaust. The mass murder in Brest took place in July 1941; the ghetto was established in December 1941, and forced labor, hunger, and epidemics ensued. In summer 1942, the Nazis began to kill the remaining Jews of Brest (as well as those of nearby towns) in Bronnaya Gora, and in October the ghetto was liquidated. Presents lists of Jewish survivors and non-Jewish rescuers of Jews. The Soviet authorities kept silent about the genocide of Jews in Brest for decades. Now, the names of Jews who perished in Brest in 1941-44 are being entered in a computerized database created in the framework of the PHOENIX Project. *DR*

0499 Gerlach, Christian: *Kalkulierte Morde: Die deutsche Wirtschafts- und Vernichtungspolitik in Weissrussland 1941 bis 1944*. Hamburg: Hamburger Edition, 1999. 1231 pp.

Ch. 7 (pp. 503-774), "Die Ermordung der weissrussischen Juden," gives a chronologically-organized description of the Holocaust in Belarus. Contends that the genocide of Jews here, as in all the Nazi-occupied Soviet territories, was part of the Nazi plan for economic exploitation and de-industrialization of the conquered Soviet East. The Nazi plans envisaged mass extermination, mainly through starvation, of several tens of millions of people. The Jews, seen as part of the economically useless urban population, and as a harmful element, prone to resist Nazi measures, were to be murdered first. Only a handful of Jews were considered useful as "specialists." Rejects the view that there was a conflict between the Army and the SS and police over the murder of Jews and the exploitation of Jewish labor. It was the military that launched the first anti-Jewish measures, including ghettoization, and that was later instrumental in the mass murder of Jews; the motives were mostly economic. Jews in eastern Belarus were murdered with greater speed than those in the western areas (formerly Polish) – not only because they were regarded as more pro-Soviet, but also because the east suffered greater devastation in the battles of 1941. *DR*

0500 Ioffe, Emanuil G.: K voprosu o kollaboratsionizme v Belarusi v 1941-1944 gg. [On the Question of Collaboration in Belarus in 1941-1944]. *Yevrei Belarusi* 3-4 (1998) 164-178.

Relates numerous cases when collaborators from the local population serving in the pro-Nazi police and military units, as well as Nazi collaborators from Lithuania and other countries, participated in the genocide of Jews in Belarus. Considers events in Slutsk, Borisov, Minsk, Narovlya, labor camp Koldychevo, and other places. *DR*

0501 Ioffe, Emanuil G.: Kolki zh yawreyaw zahinula na belaruskai zyamli w 1941-1945 hh.? [So, How Many Jews Perished on Belorussian Soil in 1941-45?]. *Belaruski Histarychny Chasopis* 4 (Oct-Dec 1997) 49-52.

Estimates the number of Jews who perished during the German occupation of Belarus as ca. 811,000; this number includes 48,000 Jews brought to Belarus from other countries. *DR*

0502 Kagan, Jack; Cohen, Dov: *Surviving the Holocaust with the Russian Jewish Partisans.* London: Vallentine Mitchell, 1998. xi, 275 pp.

Pt. 1 (pp. 1-118), "My Life," contains memoirs by Dov Cohen (b. Berl Kagan, 1922); pt. 2 (pp. 119-254), "How I Survived," contains memoirs of his cousin, Jack (Idel) Kagan (b. 1929), both of them from Novogrudok, Belarus. They describe the Holocaust in Novogrudok: the execution of 52 male Jews on 26 July 1941, the massacre of 5,000 Jews on 5-8 December 1941, the establishment of the ghetto, and massacres on 7 August 1942 and 7 May 1943. Before the last massacre of ca. 250 "specialists" who worked in a labor camp in Novogrudok, Kagan and some others dug a tunnel and fled to the partisans. Cohen managed to leave the camp and join the partisans at the end of 1942. They relate the activities of the Jewish partisan group headed by the Bielski brothers, of which they were members. After the war, Cohen settled in Israel and

Kagan in England. Pt. 2 includes facsimiles of documents pertaining to the Novogrudok ghetto and to partisan activities in the area. *DR*

0503 Kagan, Jack; Cohen, Dov: *Yak my perazhyli Khalakost z yawreiskim partyzanskim atradam u Belarusi: Dakumentalnaya apovests* [How We Survived the Holocaust with a Jewish Partisan Unit in Belarus: A Documentary Story]. Minsk: Medysont, 1999. 308 pp. Unseen. Appeared in English as "Surviving the Holocaust with the Russian Jewish Partisans" (London: Vallentine-Mitchell, 1998).

0504

קפלן, אידה: נדדתי עם ילדי ביערות ביילורוסיה. ילקוט מורשת 64 (נוב' 1997) 107-135.

[Kaplan, Ida: I Wandered with My Child in the Forests of Belarus. *Yalkut Moreshet* 64 (Nov 1997) 107-135.]

Memoirs edited and annotated by Shmuel Maayan, based on interviews conducted with Kaplan in 1978 and 1983. Kaplan was born in 1911 in Kalisz; she lived in Łódź and then in Białystok until the Nazi occupation. In July 1941 she fled from Białystok, along with her son (b. 1937), to Voronovo, Belarus, where she had relatives. They were interned in the ghetto in Voronovo, and escaped before the ghetto's liquidation in May 1942. While in hiding, Kaplan gave her son to a non-Jewish woman and returned to Voronovo. She was taken, together with other Jews who survived, to the ghetto in Lida. There, her son was brought back to her. In September 1942 they escaped to the forests and joined the family camp of the Bielski brothers, where they remained until the liberation. *LFo*

0505 Kruglov, Aleksandr I.: Istrebleniye inostrannykh yevreyev v raione Minska v 1941-43 gg. [The Extermination of Foreign Jews in the Vicinity of Minsk in 1941-43]. *Yevrei Belarusi* 3-4 (1998) 155-163.

Using mainly German wartime documents, describes the mass murder of the Jews from Germany, Austria, and Czechoslovakia who were deported to Minsk from November 1941 until the last transport which arrived on 9 October 1942. Part of them were murdered immediately after their arrival in Maly Trostinets, others – later; the last group, ca. 1,000 Western Jews, were deported to Majdanek in September 1943. *DR*

0506

מאירוביץ, אהרן: ימים יגידו: סיפורים, רשימות, מחזה. תל-אביב: רשפים, 1996. 348 עמ'.

[Meirovitch, Aaron: *Time Will Tell: Stories, Chronicles, a Play*. Tel-Aviv: Reshafim, 1996. 348 pp.]

Pp. 260-296, "Yankele Orchik's," is an expanded version of an article published previously in "Massuah" 21 (1993). *LFo*

0507 Migunova, Yelena: Antisemitskaya propaganda na okkupirovannykh territoriyakh Belarusi v 1941-1944 gg. [Antisemitic Propaganda in the Occupied Territories of Belarus in 1941-1944]. *Yevrei Belarusi* 3-4 (1998) 101-107.

The "Jewish theme" was a major one in the Nazi and pro-Nazi propaganda in Belarus from the first to the last days of the Nazi occupation. Its main topics were: Jews as the main impediment on the way to Belorussian national revival; Jews as the instigators of the war; Jews as the common enemy of humanity; Germany is at war only against Jews, not against the Slavs, etc. In the last stage of Nazi rule, the main motif became that, in the event of a Bolshevik victory, the "Jewish NKVD" would severely punish all those who lived under German rule. The Nazis also tried to infect the partisan movement with antisemitism in order to corrupt it from within. *DR*

0508 Milyakova, L.V.: Zabytyi pogrom: Lida, aprel 1919 g. [The Forgotten Pogrom: Lida, April 1919]. *Berastseiski Khranohraf* 2 (1999) 335-349.

On 16-17 April 1919, the Polish army units which captured the town of Lida (Belarus) carried out a brutal pogrom. The Jews, largely neutral in the course of the Polish-Soviet war, were accused of pro-Soviet sympathies and of committing atrocities against Poles; dozens of Jews were killed. Presents a report on this event, originally written in Polish, translated into Russian and slightly abridged. The report, compiled ten days after the pogrom, was found by Milyakova in an archive in Moscow. Pp. 335-339 contain an historical introduction. *DR*

0509 Rozenblat, Yevgenii S.: Kholokost yevreyev Belarusi v otechestvennoi i zarubezhnoi istoriografii: Sotsialno-ekonomicheskiye aspekty [The Holocaust of Belorussian Jewry in Domestic and Foreign Historiography: Socio-Economic Aspects]. *Berastseiski Khranohraf* 2 (1999) 205-226.

Discusses the main tendencies in Soviet and post-Soviet historiography concerning the Holocaust in Belarus, as well as the main topics dealt with in Western historiography on that subject. Focuses on the social aspects: conditions in the ghetto, the Jewish Councils, etc. Pp. 219-226 contain bibliographical notes. *DR*

0510 Rozenblat, Yevgenii S.: Yevreiskoye naseleniye Zapadnoi Belarusi v khozyaistvennykh strukturakh Vermakhta (1941-1943 gg.) [The Jewish Population of Western Belarus in the Wehrmacht's Economic Structures (1941-1943)]. *Yevrei Belarusi* 3-4 (1998) 108-127.

Examines the exploitation of Jewish labor resources in Western Belarus (divided in the occupation period between the General Commissariate White Ruthenia, Reich Commissariate Ukraine and General District Białystok) by the Nazi occupation authorities. Only a small part of the prewar Jewish artisans (who survived the first massacres in summer 1941) continued to work as artisans; the majority were employed as factory workers. The Jewish workers were denied all the meager advantages that were allotted to "Aryan" workers; 20% of their wages were deducted. There were other deductions from the wages of Jews. Examination of the wartime economic records in the archives of Brest, Pinsk, and Grodno shows that the Jews in these cities were employed in the most difficult tasks, and that large fines were often levied on them. The massacres and deportations of Jews which resumed in the fall of 1942 heavily damaged the Nazi economy in the region. *DR*

0511 Rubenchik, Abram: *Pravda o Minskom getto* [The Truth about the Minsk Ghetto]. Ed.: Grigorii Rozinskii. Tel-Aviv: Krugozor, 1999. 192 pp.

Memoirs of a survivor of the Minsk ghetto, who was a teenager at the time. Describes the first days of the Nazi occupation, the establishment of the ghetto in Minsk, the Jewish Council and police, how children provided the ghetto with food, and the mass executions in 1941-42. During the "action" of 18 July 1942, Rubenchik lost five of his six brothers and sisters. In summer 1942, his sister Yokha fled the ghetto and joined the partisans; later she brought over many other Jews, including the rest of her family, to the partisans. Rubenchik's parents were received by the family camp headed by Sholom Zorin. Describes resistance activities in which he and Yokha participated. Dwells on cases of antisemitism amongst the partisan units. After the war he and his sister settled in Israel. The book is interspersed with excerpts from documents and from historical literature, and photographs. *DR*

0512 Ryzhik, Raisa N.: *Spasi i pomilui: Ocherk moyei zhizni* [Save and Give Mercy: A Sketch of My Life]. Vitebsk: Belorusskoye obyedineniye yevreiskikh obshchin i organizatsii, 1997. 57 pp.

Memoirs of a Jew born in 1927 in Ostrovno, Belarus (near Vitebsk). When the war broke out, her family failed to evacuate and, together with other Jews of Ostrovno, was interned in a ghetto. On 30 September 1941 all of the inmates of this small ghetto were murdered by the Nazis. Ryzhik fled, wandered across northeastern Belarus posing as a non-Jew, and saw many cases of both betrayal and generous help on the part of peasants. For a short time she fought in a partisan unit whose members knew she was Jewish. In mid-1942 she was conscripted, as a non-Jew, for work in Germany. In the USSR after the war, because she was a former "prisoner of war" she suffered due to limitations on her rights. *DR*

0513
סמילוביצקי, ליאוניד: גורל הילדים היהודים בביילורוסיה בשנות השואה, 1944-1941.
ילקוט מורשת 68 (אוקט' 1999) 144-121.

[Smilovitskii, Leonid L.: The Fate of the Jewish Children in Belorussia during the Holocaust, 1941-1944. *Yalkut Moreshet* 68 (Oct 1999) 121-144.]

Examines the fate of Jewish children, and children born to mixed-marriage couples, during the invasion of Belarus and the evacuation of Belorussians to the Soviet hinterland, and in the ghettos and camps. Relates, also, to rescue of Jewish children from Belarus and to the children's reactions to their Holocaust experiences. Includes statistical data on wartime losses in Belarus and numbers of victims in particular locales, but there is not enough information to quote an overall number of Jewish child victims. *LFo*

0514 Smilovitskii, Leonid L.: Historiography of the Holocaust in Belorussia, 1996-1998. *Shvut* 7 [23] (1998) 189-203.

A continuation of the author's article on the historiography of Belorussian Jewry ["Shvut" 3 (1996)]. The latter half of the 1990s have been marked by increasing interest in the history of the Holocaust in Belarus. Works on this topic have appeared in Belarus, Russia, Israel, the USA, and Britain. Mentions and describes briefly the contents of many articles, memoirs, collections of documents, surveys, and monographs. The academic level of these publications varies greatly, from "dilletante" to serious research. Most of the works are of a descriptive nature. Despite the proliferation of literature dealing with the Holocaust in Belarus in recent years, work in this field is only beginning. *DR*

0515 Smilovitskii, Leonid L.: Ilya Ehrenburg and the Holocaust in Byelorussia. *East European Jewish Affairs* 29, 1-2 (Sum-Win 1999) 61-73.

Presents a three-page extract from a report by Soviet Jewish writer Ilya Ehrenburg (1891-1967) to a meeting of the Jewish Anti-Fascist Committee in Moscow on 26 July 1944 about Nazi atrocities, the results of which he had just witnessed in Belarus. Ehrenburg also collected eyewitness testimonies there. The author presents information from archival sources about places mentioned by Ehrenburg. Trostinets (where the Soviet writer's account of wagons for gassing victims was confirmed, as was the German destruction of evidence of their mass murder) was the death camp with the fourth highest number of victims: the 55,000 Jews murdered there comprised one-tenth of the total number of victims in Belarus. Other locations in Belarus about which information is provided are Borisov, Rakov, Ivenets, and Smorgon. *YC*

0516 Vinnitsa, Gennadii R.: Tragediya yevreyev Shklova (po vospominaniyam ochevidtsev) [The Tragedy of the Jews of Shklov (According to the Recollections of Eyewitnesses)]. *Yevrei Belarusi* 3-4 (1998) 128-136.

Relates the events of the Holocaust in Shklov (eastern Belarus), based on interviews with survivors and non-Jewish eyewitnesses as well as on a Soviet document compiled after the liberation of the town. Part of the Jews of Shklov and its vicinity were collected in a ghetto in the town, and part in the nearby village Ryzhkovichi. In October 1941 most of them were murdered by the Nazis and their accomplices. Some of the Jews were hidden by non-Jews. In all, ca. 6,000 Jews of Shklov were murdered. *DR*

Belgium

0517 *Les Cahiers de la Mémoire Contemporaine* 1 (1999). In French or Flemish.

The Fondation de la Mémoire Contemporaine was founded in 1994. It deals mainly with the memory of the Holocaust, but also covers themes such as wartime resistance, immigration, and postwar reconstruction of the Jewish commmunities. Partial contents: Goldberg, Martine: Méthodologie et structures [About the activities of the Foundation and its methods of research.] (45-54); Klein, Maya; Errera, Milantia: Fragments de vies: Témoignages [Presents two interviews, with a Jewish man and woman (their names are withheld), both born in Poland in the interwar period. The woman emigrated to Belgium with her mother. During the Nazi period, as she was a social worker and joined the Association des Juifs de Belgique, she was not deported in 1943 to Auschwitz. After the war she worked for the AIVG (L'Aide aux Israélites Victimes de la Guerre) which, among others, helped children orphaned in the Holocaust. The man talks about his life in Poland in the interwar period and his deportation during the war to labor camps in Germany. After the war he emigrated to Belgium.] (65-110); Bloch, Jean: La communauté juive de Belgique et la Deuxième Guerre mondiale: A propos du Musée Juif de la Déportation et de la Résistance [Appeared previously in the "Revue d'Histoire de la Shoah" 161 (1997).] (111-121; on pp. 123-132 in Flemish); Denhaene, Godelieve: Les Juifs dans certains documents communaux de Schaerbeek pendant la Deuxième Guerre mondiale [Comments on documents which illustrate the persecution of the Jews, from an exhibition held in Schaerbeek in 1994. Facsimiles of the documents are appended on pp. 140-150.] (133-150); Wulliger, Elisabeth: L'Institut "Mes Enfants" à Ixelles (1920-1945) [Traces the figure of Berthe Vanderstock, the founder of an educational institution "Mes

Enfants," who saved some Jewish children from deportation during the war.] (151-155). *HV*

0518 Schreiber, Jean-Philippe, ed.: *Hertz Jospa – Juif, résistant, communiste.* Bruxelles: EVO (Editions Vie Ouvrière), 1997. 160 pp.

A collection of articles in memory of Hertz Jospa (1905-1966). The following discuss his activities during the Holocaust period: Faux, Jean-Marie: Hertz et Hava Jospa [Recounts the lives of the founders of the Comité de Défense des Juifs, a Belgian organization dedicated to the rescue of Jewish children during the Shoah.] (13-37); Steinberg, Maxime: Juifs et communistes dans la guerre: De la mémoire à l'histoire (53-99); Schreiber, Jean-Philippe: En guise de conclusion: Le débat de la mémoire (123-134). *TS*

Czechoslovakia

(see also Slovakia)

0519 Auerbach, Edgar: Meine Eltern Herbert und Manja Auerbach: Ein Lebensabschnitt. *Mnemosyne* 22 (1997) 14-31.

Herbert Auerbach, an industrial chemist from Berlin, had moved in 1930 to Aussig in the Sudetenland to become director of an internationally-known chemical concern. Just before the Nazis occupied the Sudetenland, the author emigrated to Palestine; his parents fled to Prague. After the Nazi occupation, they attempted to obtain exit visas in order to reach England, for which they had visas and an offer of work. The Nazis heaped difficulties in their way; one let slip that they did not want Auerbach's talents to benefit the enemy. Describes many frustrating, humiliating, and sometimes dangerous encounters, in which they were saved by luck, their own presence of mind, and twice by a humane gesture by a Nazi official. They reached England days before the outbreak of the war. Also relates the story of a cousin who survived in Berlin with the aid of his non-Jewish wife and friends. *RW*

0520 Ludwig, Mark: Tales of Terezín. *Index on Censorship* 27, 6 (Nov-Dec 1998) 156-165.

The Nazis applied the term "degenerate" to certain forms of music, e.g. jazz, atonal music, music written by Jews, and music written by Germans experimenting with new musical forms. A number of Jewish composers were deported to Theresienstadt, where four of them emerged as the main creative forces: Gideon Klein, Pavel Haas, Hans Krása, and Viktor Ullmann. Instruments were initially smuggled into the ghetto and performances were clandestine, but from 1942 the Nazis allowed the performances and other cultural activities to take place; later they were used as propaganda to counter reports of Nazi atrocities. *REK*

0521 Mayers, Robin: Jewish Composers in Theresienstadt. *Jewish Affairs* 54, 2 (Win 1999) 18-20.

From November 1941 several Jewish composers were incarcerated in the Theresienstadt ghetto, including Erwin Schulhoff, Pavel Haas, Hans Krása, Viktor Ullman, and Gideon Klein. New music was created in the ghetto, and there was a wealth of performances due to the composers and instrumentalists imprisoned there and to the musically sophisticated audience. Krása's children's opera "Brundibar" (composed in 1938) was reconstructed for this unique environment and performed 55 times. In Theresienstadt, Ullman composed the opera "Der Kaiser von Atlantis," a satire on Nazi tyranny. The opera reached full stage rehearsal before the Nazi authorities banned it in October 1944, when the deportations to Auschwitz and other death camps were stepped up. Ullman and his wife died in Auschwitz, but many manuscripts, entrusted to friends, survived. *YC*

0522 Rothkirchen, Livia: The Protectorate Government and the "Jewish Question," 1939-1941. *Yad Vashem Studies* 27 (1999) 331-362. An expanded version of a lecture delivered in Czech at a conference in Ostrava in 1994, published in "Nisko 1939/1994" (1995). Appeared in Hebrew in "Yad Vashem: Kovetz mehkarim" 27 (1999).

Examines the ambivalent politics in regard to the "Jewish question" in the Czech puppet government headed by Alois Eliáš (April 1939-September 1941). Eliáš was most probably not antisemitic, and his laws excluding Jews from public and economic life were dictated by utilitarian motives. Eliáš did not succumb to Nazi pressure and did not introduce a racial definition of the Jew. His attempt to exempt some Jews from the anti-Jewish laws failed because of the German counteraction. On 28 September 1941 Eliáš was arrested for his contacts with the Czech government-in-exile and later executed; immediately after his arrest the deportations of Czech Jews began. Notes that the passivity of Czech governments in regard to Jewish issues led to the indifference of the Czech population toward the fate of the Jews. *DR*

0523

רוטקירכן, ליוויה: ממשלת הפרוטקטורט ו"השאלה היהודית", 1939-1941. יד ושם 27 (1999) 259-285.

[Rothkirchen, Livia: The Protectorate Government and the "Jewish Question," 1939-1941. *Yad Vashem* 27 (1999) 259-285.] Appeared in English in "Yad Vashem Studies" 27 (1999).

0524 Stargardt, Nicholas: Children's Art of the Holocaust. *Past & Present* 161 (Nov 1998) 191-235.

Ca. 4,000 drawings and paintings by children interned in Theresienstadt, and subsequently killed, were recovered when the camp was liberated in 1945. The works were created mostly by girls, under the tutelage of Friedl Dicker-Brandeis, who saw art as a form of creative release. Only a small fraction of the pictures depict life in the ghetto; many depict images of home and the countryside, and convey optimism. Queries whether the optimistic paintings actually reflect the children's attitudes. Describes the children's homes – including cultural activities and warm relationships – as a largely positive experience of communal life. Observes that the works depicting life in the ghetto tend to compartmentalize the “good” of the children's homes and the “bad” of the adult barracks. Although the children were aware of the suffering of the adults, they seemingly concentrated on their own experiences, memories, and hopes for the future. The question of whether the reticence of the children to depict suffering should be attributed to trauma or was itself part of a coping strategy cannot be answered. *REK*

0525

טרסי, אניטה: יומנים וזיכרונות מאוחרים: נשים קשישות מן הרייך בגטו טרזין. דפים לחקר תקופת השואה 14 (1997) 149-186.

[Tarsi, Anita: Later Diaries and Memoirs: Elderly Women from the Reich in the Theresienstadt Ghetto. *Dapim le-Heker Tekufat ha-Shoah* 14 (1997) 149-186.]

Describes the experiences of middle-aged and elderly women from Nazi Germany in Theresienstadt, as reflected mainly in diaries, testimonies, and memoirs found at Terezin House, Israel. Also relates to the presentation of the elderly in writings and works of art created in the ghetto. Focuses on changes in the attitudes of older women toward life and death, reactions to imprisonment and everyday life in the ghetto, attitudes of Czech deportees toward women from the Reich, and postwar reflections on Holocaust experiences and survival. *LFo*

0526 *Terezínské Listy* 27 (1999). With English and German summaries.

Partial contents: Lipková, Renáta: Právo v nuceném společenství [The Application of Law in Involuntary Communities. Excerpts from the author's dissertation on laws and their application in the Terezín ghetto.] (20-37); Kárná, Margita: Léta našeho života [Years of Our Lives. Memoirs of the author's (b. Margita Krausová, 1923) experiences in Terezín, Auschwitz, and Kudowa (in Poland).] (100-110); Kryl, Miroslav: Terezínský deník Miloslava Kuglera [The Terezín Diary of Miloslav

Kugler. Excerpts covering the period 1939-1942, also mentioning Jewish transports and Jews in the ghetto.] (125-134). *SSC*

0527 Weil, Jiří: *Elegie fuer 77 297 Opfer: Juedische Schicksale in Boehmen und Maehren, 1939-1945.* Trans.: Avri Salamon. Konstanz: Hartung-Gorre, 1999. 36 pp. Originally published as "Žalozpěv za 77297 obětí" (Praha: Československý Spisovatel, 1958).

Commemorates the 77,297 Czech Jews murdered in the Holocaust, whose names are carved on 18 plaques in the Pinkas Synagogue in Prague. Pp. 13-32 present, on each page, an aspect of the Nazis' cruel measures, examples of individual suffering from those measures, and an appropriate biblical verse. Weil himself was persecuted during the Nazi occupation as a Jew, and also for his political activities. Aided by a Czech policeman, he escaped prison, faked suicide, and went into hiding. *ID*

Denmark

0528 Buckser, Andrew: Group Identities and the Construction of the 1943 Rescue of the Danish Jews. *Ethnology* 37, 3 (Sum 1998) 209-226.

Argues that the traditional account of the rescue of the Danish Jews oversimplifies the motivations involved and presents only a partial story. Rescue accounts tend to see Danes, Jews, and Germans as homogeneous groups, and to minimize divisions within each group which sometimes affected the course of the rescue. Many Danes, including the police, initially viewed resistance with ambivalence, and some Danes were hostile toward Jews. Notes that Denmark agreed to protectorate status for the first three years of the war, and supplied Germany with important provisions. After the war, ca. 15,000 Danes were arrested for collaboration. Danes were concerned about their image; therefore, the group uniformity of the rescue story reassured them. Motivation for the rescue is generally attributed to Danish ethical responsibility. This view is also beneficial for the Jews. Although Jews were sometimes subject to prejudice and maintained a low profile before the war, they became emblems of Danish tolerance after the rescue and felt they could be "publicly happy about being Jewish." The benefits of the traditional story explain its popularity. *REK*

0529 Buckser, Andrew: Modern Identities and the Creation of History: Stories of Rescue among the Jews of Denmark. *Anthropological Quarterly* 72, 1 (Jan 1999) 1-17.

Anthropological analysis of the construction of history has a specific interest in a case of poorly-defined group identities such as the highly assimilated Jews of Copenhagen. Examines the ways in which Copenhagen Jews tell the story of the rescue of Danish Jews in 1943. The study is based on interviews conducted by Buckser in 1998 with five members of the Jewish community in Copenhagen, of whom four are Holocaust survivors and one was born after the war. Analysis of their stories shows that there is no common Jewish narrative on the event which has such importance for both Jewish and Danish historical memories. What is common to all the narratives is that they address problems of sameness and difference endemic to Danish Jewish life. *DR*

0530 Neuman, Joshua: Representing Righteousness: Constructing the Rescue of the Jews of Denmark. *The Harvard Mosaic* 24 (Sum 1999) 16-23.

The rescue of Danish Jews in 1943 is shrouded by myths. The truth is more prosaic: Danish Jews were more assimilated than Jews in any other country; there was antisemitism in Denmark; many of the Danes who participated in the rescue profited from it; Danish administrative affairs ranked low on the German list of priorities. In spite of all this, the rescue of Jews in Denmark is a compelling story worth telling. *DR*

0531 Schoenborn, Paul Gerhard: Kaj Munk: Daenischer Pfarrer, Dichter und Maertyrer (1898-1944). *Freiburger Rundbrief* 6, 4 (1999) 250-258. A slightly abridged version of an article first published in "Impressum" 7 (1997).

Describes the opposition of the Danish Lutheran Church to Nazi plans to deport the Danish Jews in October 1943; a pastoral letter to this effect was read from all the pulpits on October 3rd. It was followed by a proclamation of the Danish resistance calling on Danes to cease all cooperation with the Germans. The Danes succeeded in saving most of the Jews. Munk, pastor of the rural community Hørbylunde and a writer, had condemned Nazi racism in a drama – "Han sidder ved Smeltediglen" ("He Sits at the Melting Pot") (1938) – and in newspaper columns and sermons, all stressing that Jesus was a Jew. In a sermon preached at a protest service in the Copenhagen cathedral on 5 December 1943, he affirmed the duty of the Church to oppose persecution. A month later the SS arrested and murdered him. *RW*

Estonia

0532 Weiss-Wendt, Anton: Sovetskaya okkupatsiya Estonii 1940-1941 godov i yevrei [The Soviet Occupation of Estonia in 1940-1941 and the Jews]. *Yevrei v Rossii: Istoriya i kultura: Sbornik nauchnykh trudov, Vol. 5*, ed. Dmitri Elyashevitch. Sankt-Peterburg: Peterburgskii Yevreiskii Universitet, 1998. Pp. 219-245. An English version appeared in "Holocaust and Genocide Studies" 12 (1998).

France

0533 Afonso, Rui: Le "Wallenberg portugais": Aristides de Sousa Mendes. *Revue d'Histoire de la Shoah* 165 (Jan-Apr 1999) 7-28.

Mendes was Portugal's consul general in Bordeaux at the outbreak of World War II. Describes his activities to rescue thousands of people (Jews and others) who were in danger of being deported after France was occupied by the Nazis. In June 1940, he gave out thousands of Portuguese visas without receiving confirmation for them by the authorities in Lisbon, as he should have done according to a decree by Salazar. For this insubordination, Mendes was recalled to Lisbon in July 1940; his diplomatic career was over. After the war, throughout Salazar's regime, Mendes and his family lived in poverty. He died in 1954. In 1967 Israel posthumously awarded him the title "Righteous among the Nations." In Portugal, Mendes was rehabilitated in 1987. *HV*

0534 Angel, Jacques: Drancy – les premiers mois: Témoignage. *Revue d'Histoire de la Shoah* 165 (Jan-Apr 1999) 185-207.

Memoirs of a French Jew (born ca. 1922), relating his and his father's arrest in Paris in August 1941 and their internment in Drancy. Describes their life in the camp, the living conditions, the other prisoners, the hunger. Angel fell ill (with edema), and due to fear of an epidemic in the camp, the authorities liberated him after a few weeks. *HV*

0535 Badinter, Robert: *Un antisémitisme ordinaire: Vichy et les avocats juifs (1940-1944).* Paris: Fayard, 1997. 257 pp.

Describes the interwar xenophobia and antisemitism of the French people (including French Jews) which made possible the Vichy anti-Jewish measures. In Paris there were ca. four hundred Jewish lawyers (out of 2,500); there was a high degree of hostility towards them in the traditional, conservative milieu. Surveys Vichy policy towards Jewish lawyers, the enactment of the statutes of the Jews, and their consequences for Jews in the legal profession. Asserts that members of the Parisian bar approved of the introduction of a numerus clausus, and later of the near total exclusion of Jews from the bar. Discusses the creation of the Commissariat Général aux Questions Juives, and the role of its chief, Vallat, and later, Darquier de Pellepoix, in the implementation of the Vichy laws. Concludes that the anti-Jewish measures imposed by the Vichy regime did not provoke protests among French lawyers. The Jews, in this profession as in the general population, were considered as unassimilable strangers, which was one of the reasons for the lack of public protest. *HV*

0536 Bloch, Lydie: Car, ils ont une histoire.... *La mémoire contre la nuit.* Bordeaux: Editions du Passant, 1997. Pp. 37-49.

Traces the history of her family; Bloch's parents were Jewish immigrants from Eastern Europe. Their three children were born in France; Lydie was born in 1941. The family survived the war in a small village, 60 km from Paris. Bloch's father was involved in the resistance movement. *HV*

0537 Bruno, Jean; Monicault, Frédéric de: *L'affaire Papon: Bordeaux, 1942-1944.* Paris: Tallandier, 1997. 180 pp.

An investigation by two journalists on the period when Maurice Papon was secretary general of the police in the Gironde region, in charge of the Jewish question. Traces the deterioration of French-Jewish relations back to the 1930s, much before the Jews' official relegation to inferior status in 1941. The authors interviewed various French personalities who were closely connected with the occupation government. Emphasizes the role of Michel Slitinsky who, in 1981, first revealed evidence to the press of Papon's signature on antisemitic decrees in Bordeaux. Concludes that doubts regarding Papon's role in the Résistance (an argument put forward by the defense) dominated the trial, rather than confrontation with the real issue – Gaullist disregard of his wartime past. *TS*

0538

כהן אשר: מגוון של מצילים בצרפת. דפים לחקר תקופת השואה 14 (1997) 247-272.

[Cohen, Asher: A Variety of Rescuers in France. *Dapim le-Heker Tekufat ha-Shoah* 14 (1997) 247-272.]

Contends that the policies of French authorities during 1942-44, French public opinion, and rescue operations initiated by Jews impeded the implementation of the Final Solution in France. 75% of French Jewry survived the Holocaust. Highlights rescues carried out by individuals in various parts of France. Also deals with operations initiated by Jewish organizations and their involvement in the activities carried out by non-Jews. *LFo*

0539

כהן, ירחמיאל (ריצ'רד): לדרכו של אשר כהן בחקר יהדות צרפת בתקופת השואה. דפים לחקר תקופת השואה 14 (1997) 13-19.

[Cohen, Richard I. (Yerachmiel): On Asher Cohen's Research on the Jews of France during the Holocaust. *Dapim le-Heker Tekufat ha-Shoah* 14 (1997) 13-19.]

A lecture presented at a meeting convened at Yad Vashem in memory of the historian Asher Cohen (1936-1996). Surveys Cohen's work and summarizes his conclusions regarding the attitude of France and the French people toward Jews and the anti-Jewish measures during the Holocaust, and French involvement in rescue activities. *LFo*

0540 Cointet, Michèle: *L'Eglise sous Vichy, 1940-1945: La repentance en question.* [Paris]: Perrin, 1998. 404 pp.

Examines the impact of the Catholic Church on government policy in France during the Holocaust period. Notes that antisemitism had existed in the Church since antiquity. Although Christian-Jewish relations improved during the interwar period, philosemitism was perceptible mostly in academic circles. Contends that despite antisemitic legislation inspired and abetted by certain members of the clergy, the protests against the roundup of Jews originated with a group of bishops. Ch. 5 (pp. 162-275), "L'Eglise et l'antisémitisme," evaluates the extent of complicity of the Church with the Vichy regime by analyzing the type of response at each stage of antisemitic legislation, beginning in 1940. Traces the roles of Cardinal Gerlier and Mgr. Saliège, archbishop of Toulouse, in the defense of French Jews, particularly after the Vel d'Hiv roundup in 1942. Describes efforts by the clergy to aid Jews, including clandestine placement on farms and convents. Pp. 265-275 recount the life of Père Jacques de Jésus, immortalized by Louis Malle in the film "Au revoir les enfants," who hid Jewish children in his school. They were denounced to the Gestapo, the children were deported to Auschwitz, and the priest died in Mauthausen in 1945. *TS*

0541 Copfermann, Emile: *Des les premiers jours de l'automne.* Paris: Gallimard, 1997. 214 pp.

Reflects on his childhood before the Second World War and during the war. His parents emigrated from Romania to France; their three sons were born in Paris. The parents were arrested and deported in September 1942. The boys survived the war kept by a Catholic family in Boutigny. Focuses on issues of everyday life and the feelings of a child in this period. *HV*

0542 Daumas, Jean-Claude: Une entreprise juive sous l'Occupation: Blin et Blin, 1940-1944. *Archives Juives* 31, 1 (1998) 78-94.

Traces the history of legislation to Aryanize Jewish property in France, from 1940 to 1944. Ernest, Maurice, and André Blin, Jewish owners of a textile factory in Elbeuf, decided to transfer ownership to a non-Jew. Charles Bedaux, an associate, took over management of the company until the return of the Blin brothers in 1944. Justifies the company's cooperation with the Germans as a measure to save the company from bankruptcy. Studies the question of the employees' support for the Résistance, although they did not actively join. The Blin brothers and their families survived the war in hiding, aided by friends, in the Corrèze region. After the war, their company's ownership reverted back to them, as agreed. Argues that the dichotomy collaboration/resistance, the model referred to by many historians, was not applicable to the case of Blin et Blin. *TS*

0543 Epstein, Eric: Fit to Be Tried: Maurice Papon and the Vichy Syndrome: Defeat and Collaboration. *Journal of Genocide Research* 1, 1 (Mar 1999) 115-121.

Outlines briefly the history of the Holocaust under the Vichy regime in France. Until the 1980s, the issue of French complicity in the Holocaust was avoided in public discourse. As a result, the new generation in France has been ignorant of the Holocaust; many former Vichy criminals evaded punishment; and France became a hotbed of Holocaust denial. Only in 1994 did Paul Touvier become the first Frenchman to be tried for crimes against humanity. In 1997 Maurice Papon was sentenced for similar crimes, but freed during the appeal process. What is at stake in Papon's trial is France's ability to address honestly its complicity in the Holocaust. *DR*

0544 Fraenkel, Jean-Jacques: *L'abus de confiance.* [Paris]: Biblieurope; [Kibbutz Lohamei Haghetaot]: Beit Lohamei Haghetaot, 1997. 175 pp. The author's name is given on the title page as Jjes. Fraenkel.

Memoirs of a Jew born in Paris in 1931. His father was a well-known dental surgeon; he was arrested in Paris in December 1941, and later died in Auschwitz. The extended family left Paris for Nice, where his mother worked for the Jewish resistance. In 1943 Fraenkel returned to Paris on "Aryan" papers. He lived with his parents' non-Jewish friends and later in a Catholic institution. His mother was deported to Auschwitz in November 1943 and killed; his sister Josette survived the war. *HV*

0545 Fralon, José-Alain: *Aristides de Sousa Mendes: Um herói português.* Trans.: Saul Barata. Lisboa: Editorial Presença, 1999. 125 pp. Originally published as "Aristides de Sousa Mendes: Le juste de Bordeaux" (Bordeaux: Mollat, 1998).

0546 Gervais-Marx, Danièle: *La ligne de démarcation.* Aigues-Vives, France: HB Editions, 1997. 317 pp.

Memoirs of a Jew born in Paris in 1932, relating her experiences during the Holocaust. Her parents owned a furniture shop which was Aryanized and handed over to their worker. The new owner sent money to the Marx family every month, which allowed them to survive. Danièle and her parents left Paris for Marseille in 1940. After the German occupation in November 1942, they moved to a small village, Hauterives, in Drôme, where they lived on false papers until the end of the war. Describes, as well, the fate of her grandparents, cousins, and other members of her large family. Some of them were deported to extermination camps and some of them survived the war helped by Christians. *HV*

0547 Guckes, Jochen: Le rôle des chemins de fer dans la déportation des Juifs de France. *Revue d'Histoire de la Shoah* 165 (Jan-Apr 1999) 29-110.

Presents a detailed analysis of the function of the transport system in the deportation of the Jews from France to Auschwitz. Describes the organization of the Reichsbann system and of the SNCF (Société Nationale des Chemins de fer Français), the administrative process, and the role of various institutions involved in placing the trains at the disposition of the Reich. Discusses Nazi control of the SNCF and the issue of financing the transports. It was only in France that the Jews themselves did not pay the costs of deportation. Emphasizes the responsibility of the French state and of the SNCF for the transports on French territory. *HV*

0548 Job, Françoise: *Racisme et répression sous Vichy: Le camp d'internement d'Ecrouves.* Paris: Messene; Centre de Documentation Juive Contemporaine, 1997. 249 pp.

Recounts the story of the camp of Ecrouves, which served as an internment camp in Lorraine between 1941-44. Originally founded by the Vichy government as a detention camp for French communists and members of the Résistance, it subsequently served as a transit camp for ca. 1,800 Jews, mostly from Alsace-Lorraine, during the German occupation. Describes daily life in the camp and outlines the history of the deportation of the Jews of the region. Includes lists of the Jews detained there (pp. 203-241). *TS*

0549 Korzec, Pawel; Burko, Jacques: *Le gouvernement polonais en exil et la persécution des Juifs en France en 1942: D'après des documents inédits.* Paris: Cerf, 1997. 234 pp.

Summarizes the relationship between France and Poland from the period prior to World War II until 1942 with respect to foreign Jewish residents subject to the laws of the Vichy government. Examines the policies of the French government-in-exile in London regarding the deportation of Polish

Jews living in the Free Zone. Argues that Polish diplomats, such as Stanisław Zabiello, utilized the limited means at their disposal to defend their Jewish compatriots during the period from July 1942 until the German occupation of the Free Zone in November 1942. Pp. 101-217 present documents, with annotations, in order to show that Polish diplomats did, in fact, try to help the Jews in this period. *TS*

0550 Laloum, Jean: La Caisse des dépôts et consignations et les avoirs des juifs spoliés pendant la guerre. *Archives Juives* 31, 2 (1998) 87-94.

Analyzes the protocols of meetings of the French authorities that directed the Caisse des dépôts et consignations (CDC), founded in 1941, regarding the whereabouts of Jewish property confiscated under Vichy Aryanization policy. Summarizes the circumstances which led to this policy. In reprisal for attacks against the German occupation forces, Jewish Poles were deported to the East; 100 Jewish communists and anarchists were executed, and a tax was placed on the French Jewish community of one billion francs, to be raised by the Union Générale des Israélites de France (UGIF). Discusses the solution of property confiscation proposed by the CDC to aid the UGIF, which was unable to amass such a sum. Assesses the ambiguous position of the CDC in liberated France; while lauded by some Jews for its efforts to assist the UGIF, others criticized the CDC for its exploitation of Jewish vulnerability. Estimates that the recently-established Mattéoli Commission, founded to study the circumstances surrounding the plunder of Jewish property under Vichy, has made progress in evaluating the role of the CDC. *TS*

0551 Lelaidier-Márton, Liliane: *A l'ombre de l'étoile: Réminiscences d'une enfant cachée.* Nice: Editions du Losange, 1997. 157 pp.

Memoirs of a Jew born in Paris in 1930. Her parents immigrated to France from Hungary in the 1920s. They lived in Blanc-Mesnil, close to Drancy. Her father was arrested and deported from Drancy to Auschwitz in March 1943, and her mother in August 1943. Liliane survived the war in hiding, helped by her mother's family (her uncle was a Protestant) and by Jewish members of the resistance, friends of her father. *HV*

0552 Lévi, Bernard: La condition des élèves juifs de l'Ecole polytechnique sous l'Occupation. *Archives Juives* 32, 1 (1999) 103-109.

Traces the deterioration of the situation of Jewish students at the Polytechnic University of Paris under the Vichy government. In 1940, the school moved to the Free Zone, and returned to Paris only in 1943. Outlines legislation aimed at the exclusion of Jews from positions of responsibility and relegating the students to inferior status, thereby denying them basic rights and tuition subsidies. A numerus clausus was declared in 1941, and by 1943 Jews could only study by correspondence. These measures were maintained throughout the war with the support of the student body, which had become increasingly pro-Pétain. Concludes with the return in 1945 of those who had escaped deportation, and their subsequent reintegration into the university with equal status. *TS*

0553 Lewendel, Isaac: *Not the Germans Alone: A Son's Search for the Truth of Vichy*. Evanston, IL: Northwestern University Press, 1999. xvi, 341 pp. Originally published as "Un hiver en Provence" (La Tour d'Aigues, France: Ed. de l'Aube, 1996).

0554 Mesnard, Philippe: Maurice Blanchot: *Le sujet de l'engagement*. Paris: L'Harmattan, 1997. 350 pp.

Analyzes the writings and political activities of Blanchot, the French writer and literary critic. Reflects on his extremist engagements. Pp. 15-66 discuss his involvement in the Action Française in the 1930s. At the time, Blanchot attacked the racial antisemitism of the Nazis, but he himself wrote antisemitic articles against the Jews as "strangers" in France and against the involvement of some of them in politics; one of his targets was Léon Blum. Pp. 225-243 discuss Blanchot's postwar involvement with the communist movement (in the years 1959-69) and, at the same time, his growing interest in Judaism, Jewish history, and the Holocaust. Blanchot wrote about Auschwitz as a symbol of evil and a product of essentially negative Western European culture. Relates, as well, to Blanchot's reflections on the expression of the Holocaust experience in literature. *HV*

0555 Poznanski, Renée: Life in the Jewish Camps in Southern France 1940-42. *Legacy* 1, 3 (Win 1998) 19-25. Appeared in Hebrew in "Bishvil ha-Zikaron" 18 (1996).

0556

פוזננסקי, רנה: להיות יהודי בצרפת, 1945-1939. תרגום: עדה פלדור. ירושלים: יד ושם, 1999. 752 עמ'.

[Poznanski, Renée: *To Be a Jew in France, 1939-1945*. Trans.: Ada Paldor. Jerusalem: Yad Vashem, 1999. 752 pp.] Originally published as "Etre Juif en France pendant la Seconde Guerre mondiale" (Paris: Hachette, 1994).

0557 Sanders, Paul: Anatomie d'une implantation SS: Helmut Knochen et la police nazie en France, 1940-1944. *Revue d'Histoire de la Shoah* 165 (Jan-Apr 1999) 111-145.

Discusses the activities of the SS in France and its relations with the Vichy French authorities in occupied France. Pp. 131-137 deal with its role in organizing the deportation of the Jews to the concentration camps. Describes the activities of Helmut Knochen, the SD commander in Paris, appointed in June 1940, who was responsible for rounding up the French Jews. Relates to his tactics of collaboration with the French police, and how he understood the specific situation in France. Knochen considered that it was difficult to develop among French people antisemitic feelings based on ideology. He decided not to proceed with a heavy hand, but to stimulate French initiatives to adopt antisemitic measures. Discusses the struggle between Knochen and Dannecker, who wanted to adopt more drastic methods, and the consecutive steps of the "Final Solution" in France as it was organized by Vichy and the SS, underlining divergences between the French and Nazi authorities. *HV*

0558 Schwartz, Laurent: *Un mathématicien aux prises avec le siècle*. Paris: Odile Jacob, 1997. 528 pp.

Memoirs of a Jew born in 1915 in Paris. In 1936 he joined the Trotskyists and remained active until 1947. Pp. 189-219 relate his experiences during the Nazi occupation of France. Conscious of his precarious situation as a Jew and a Trotskyist, he settled outside Grenoble in 1943 under a false identity. Schwartz was thus able to escape various searches for Jews conducted by the French police in the vicinity. He returned to Paris after the war; in 1950 he was awarded the prestigious Fields Medal for Mathematics. *TS*

0559 Schwed, Pierre: *De Riedwihr à Jérusalem*. [Jerusalem?: The Author?, 1997?]. 111, [8] pp.

Memoirs of a Jew from Alsace, born in 1932. Recounts his and his family's experiences during the Holocaust. Several members of the extended family were deported following the annexation of that region by the Germans. Schwed's father was arrested, but escaped and fled to the Free Zone. His wife and two sons were trapped in Alsace and then in Paris after an initial attempt to cross the demarcation line failed. The family was eventually reunited in Toulouse until that region, too, was occupied in 1942. They then separated; the boys were hidden by Italian farmers in the countryside until their Jewish identity was accidentally divulged. Then, the family fled together to a local priest, a member of the Résistance, who arranged a safe haven for them until their return to Alsace after the war. *TS*

0560 Stupp, François: *Réfugié au pays des "Justes": Araules 1942-1944*. Saint-Julien-Chapteuil: Editions du roure, 1997. 183 pp.

Memoirs of a French Jew (b. 1929) recounting his experiences during the Holocaust. In 1940 his family fled Belfort for Lyon, but in 1942 the requirement to wear the yellow star motivated Stupp's father to seek refuge for the children in the countryside. François and his sister Ida were placed with a Christian family, Tonia and Louis Ouillon, on a farm in the Massif Central. At the end of 1943 the local authorities were ordered to divulge the names of all Jews residing in the village, but they refused to do so. François and his sister remained in Araules until the end of the war, when they joined their family in Lyon. *TS*

0561 Taguieff, Pierre-André: D'un "prophète" l'autre: Drumont, Céline: L'antisémitisme revendiqué comme une tradition française (1940-1944). *Esprit* 250 (Feb 1999) 90-113.

Surveys anti-Jewish publications during the Vichy regime whose goal was to persuade readers that the racist and antisemitic doctrine was not imported but was of French origin. Drumont was the founder of a certain tradition, and even a "prophet" for antisemites of various orientations. One of the founders of the Ligue antimaçonnique de France, an antisemitic publicist since 1934, Count Armand de Puységur, praised Drumont as a writer who revealed the Jewish "conquest of France." Henri Labroue, professor of Jewish history at the Sorbonne (a chair created by Darquier de Pellepoix), in his book "Voltaire antijuif" (1942), argued that anti-Judaism is of French origin and was not imposed by the Nazis; to be anti-Jewish meant to be part of French political culture. States that such ideas strengthened French patriotic anti-German feelings. Another idea put forward in this period was the necessity to "purify" the French nation through sterilization of the Jews. Céline was a "prophet" of this ideology. *HV*

0562 Weinstein, Max: *Souvenirs, souvenirs*. Nice: Editions du Losange, 1997. 276 pp.

Memoirs of a Jew born in Nancy in 1927. His parents immigrated to France from Poland in the early 1920s. Describes his childhood in Nancy with his parents and two brothers. When the Nazis bombed Nancy in May 1940, the family fled to Roanne in the south of France. Weinstein's elder brother went to Lyon, and Max joined him there in 1943. They lived on "Aryan" papers and joined the Jewish communist movement, the Groupes de Combat Juifs of UJJ (Union de la Jeunesse Juive), part of MOI. Relates their resistance actions. The whole family survived the war. After the war Weinstein worked for the communist press. *HV*

0563 Weksler-Waszkinel, Romuald Jakub: Wokół testamentu Henri Bergsona [About the Will of Henri Bergson]. *Biuletyn Żydowskiego Instytutu Historycznego* 190 (June 1999) 33-42.

An abridged version of an article which appeared in the "Journal of Dharma" 22 (1997).

0564 Wildmann, Manfred; Wiehn, Erhard Roy, eds.: *Und flehentlich gesegnet: Briefe der Familie Wildmann aus Rivesaltes und Perpignan: Juedische Schicksale aus Philippsburg 1941-1943.* Konstanz: Hartung-Gorre, 1997. 204 pp.

Presents the correspondence of various members of the German Jewish family Wildmann during their internment in camps in southern France between November 1941-November 1943. The author (b. 1930), his parents, grandparents, two sisters, and a brother were deported from Philippsburg, Germany to Gurs in October 1940 and then to Rivesaltes in March 1941. The correspondence began when one of his sisters left Rivesaltes, in November 1941, to work in a home for children run by the Red Cross in Pringy, Haute Savoie; it ended when his father, Heinrich, was deported from the hospital in Perpignan to Auschwitz in 1943. Wildmann's mother and brother had already been deported in August 1942, via Drancy to Auschwitz. Only Manfred and his sisters survived (he was in the Red Cross home where his sister worked). In 1947 they emigrated to the USA. *ID*

0565 Zaretsky, Robert: In the Name of God and History: The Association Sully and Extreme Right-Wing Protestantism in France, 1933-1945. *Historical Reflections/Réflexions Historiques* 25, 1 (Spr 1999) 127-145.

The French Protestant pro-monarchist and anti-democratic Association Sully, established in 1925, somewhat paralleled its Catholic counterpart, Action Française. Pp. 133-135 discuss their shared antisemitism. Association Sully blamed Jews for both communism and democracy, and believed Jews had no place in France. Its Christian-based anti-Judaism began to emerge in 1936, and increased after the Nazi occupation and the establishment of the Vichy government. Although Sully's membership was not negligible, it was a marginal phenomenon within the basically liberal French Protestant tradition. *YC*

Germany

0566 *Am Wedding haben sie gelebt: Lebenswege juedischer Buergerinnen und Buerger.* Berlin: Metropol, 1998. 248 pp. On title page also: Berliner Geschichtswerkstatt (Hrsg.).

Essays on Jewish life in the neighborhood of Wedding, Berlin. Partial contents: Buehler, Annegret; Fuehre, Dorothea; Schroeter, Ursula: Juedisches Leben im Wedding [Including its liquidation by the Nazis.] (12-33);

Schroeter, Ursula: "Vor zwanzig Jahren haetten Sie forschen muessen" (34-36); Stange, Heike: Georg Benjamin: Arzt im Wedding (37-46); Buehler, Annegret: Uri Givon (Ulrich Greiner): "Meine Eltern wurden ermordet als begeisterte Anhaenger des Zionismus" (49-60); Fuehre, Dorothea: Marianne Kaufhold: "Man lebt ja nach vorne, nicht so mit der Vergangenheit, wenn man jung ist" – Oder: Die Muehe des Erinnerns (61-69); Schroeter, Ursula: Ruth Nathanson: "Dies ist ein Traditionshaus hier" (69-84); Id.: Schlomoh Pfennig: "Ich war der einzige Jude in meiner Klasse" [After arrest by the Gestapo, he fled to Belgium and spent the war in hiding and in the resistance, also participating in rescue of Jews from a transport.] (84-92); Id.: Helga Verleger: "Man kam sich auch selbst etwas schuldig vor, dass man hier geblieben ist" [At 16 she was deported to Estonia and passed through a series of labor camps; after the war she returned to Germany.] (92-133); Lefevre, Andrea: Die Enteignung der Familie Apt: Ein ganz legaler Beutezug im Spiegel der Akten des Oberfinanzpraesidenten (139-159); Stange, Heike: Die Baers: Eine Familien- und Firmengeschichte [On the Baer and Haber families.] (159-166); Buehler, Annegret: Abraham Buschke: 1906-1933 Arzt im Rudolf-Virchow-Krankenhaus (167-180); Id.: Martin Fleischmann: Mord auf der Pankebruecke [Fleischmann was a World War I veteran and married to an Aryan. In July 1944 two Nazis broke into his home and murdered him. Includes an interview with his son Michael (1992) and an account of the trials of his murderers in 1951 and 1953.] (181-198). Pp. 214-247 contain a list of the Jews deported from Wedding (giving the name, date and place of birth, and the date and destination of deportation). *RW*

0567 Ayalon, Moshe: *Das juedische Kinderheim der Flersheim-Sichel-Stiftung in Frankfurt am Main: Gruendung, Bluete, Bedrohung, Rettung, Untergang*. Frankfurt a.M.: Verein Jugendbegegnungsstaette Anne Frank; Historisches Museum Frankfurt am Main, 1996. 47 pp.

Surveys the fate of the Flersheim-Sichel Jewish children's home (founded in 1865) during the Nazi regime. In addition to orphans and half-orphans, children from small villages in the region also came to stay in the home; their number increased from 14 in 1927 to 32 in 1938. In March 1939, 28 of the boys left for England via the "Kindertransport" scheme. Efforts made by the directors in 1940-41 to send the 48 children then in the home failed, and they were deported together with the staff after the Aryanization of the home in 1942. None of them survived. *ID*

0568 Baeumer, Franz-Josef: "...weil Gesundheit, Rasse und Blut den 'Affekt des Herzens' der Zeit besitzen" (L. Bopp, 1937): Zur katholischen Pastoraltheologie im Nationalsozialismus. *Kirche und Israel* 11, 1 (1996) 62-73.

Based on a lecture held in Muenster, December 1995. Surveys Catholic periodicals and handbooks published during the Nazi period (to 1939). While some of these cautiously defend the Church and its doctrines against Nazi ideology, none show sympathy for the persecuted Jews. A standard Catholic dictionary repeats the usual antisemitic stereotypes: Jews dominate and corrupt the economy, the press, politics and culture. Many publications welcome Nazi suppression of Jewish influence, and the pastoral theologian Linus Bopp justifies it theologically. *RW*

0569 Bajohr, Frank: *"Arisierung" in Hamburg: Die Verdraengung der juedischen Unternehmer 1933-1945.* Hamburg: Christians, 1997. 415 pp. Reprinted in 1998. Based on the author's diss. – Universitaet Hamburg, 1996/1997.

Examines the process of Aryanization and liquidation of Jewish property and enterprises in Hamburg during the Nazi period. In 1938, 1,200 Jewish enterprises still existed in Hamburg; their Aryanization was accelerated as a result of the "Kristallnacht" pogrom. In its final stage, Aryanization in Hamburg resembled a race for enrichment, especially among Nazi functionaries. Between 1941-45, the property of 30,000 Jews from Germany and occupied Western Europe was sold at auction in Hamburg. Discusses the entire process which, from its beginning, was accompanied by violent attacks on Jews and Jewish businesses, organized boycotts, exclusion of Jews from professional associations, prohibition on Jewish physicians and lawyers to work, and dismissal of Jewish employees; in its final stages it led to the complete deprivation of the Jews and then to their annihilation. *ID*

0570 Barbian, Jan-Pieter; Brocke, Michael; Heid, Ludger, eds.: *Juden im Ruhrgebiet: Vom Zeitalter der Aufklaerung bis in die Gegenwart.* Essen: Klartext, 1999. 635 pp.

A collection of essays on Jewish life in the Ruhr region. The following deal with the Nazi period: Heid, Ludger: "Wahrhafter Seelsorger mit heiterem Gleichmut": Der Duisburger Rabbiner Manass Neumark [Neumark, b. 1875, served as rabbi in Duisburg for 37 years; he died in Theresienstadt in 1942.] (47-66); Aschoff, Diethard: Konzentrationslager in der Erinnerung juedischer Ueberlebender aus dem noerdlichen Ruhrgebiet (167-178); Moss, Christoph: Verfolgung und Vernichtung in juedischen Selbstzeugnissen: Jeanette Wolff und Marga Spiegel (179-194); Frankenthal, Hans: Der Weg eines Sauerlaender Juden von Schmallenberg nach Auschwitz und zurueck [Memoirs.] (195-209); Schmalhausen, Bernd: Dr. Rolf Bischofswerder: Ein Arzt aus Dortmund im "Dritten Reich" (211-223); Tietz, Manfred: "Shalom Solidarité": Juden aus dem Ruhrgebiet im Widerstand 1933-1945 (225-277); Wessel, Horst A.: Die Hahnschen Werke in Duisburg: Zur Geschichte eines "arisierten" Unternehmens (441-461); Zellner, Annette: "Cohen & Epstein": Das aelteste Kaufhaus Duisburgs als Opfer der "Arisierung" (463-466); Ellger-Ruettgardt, Sieglind: Das Israelitische Waisenhaus Dinslaken [Based on the memoirs of and an interview with survivor Yitzhak Sophoni Herz.] (503-522); Boeti, Pasquale: "Muskeljudentum": Der Turn- und Sportklub "Hakoah Essen" – ein juedischer Sportverein im Ruhrgebiet [Pp. 611-614 discuss the Nazi period.] (601-617). *ID*

0571 Bauer, Friedrich L.: Pringsheim, Liebmann, Hartogs – Schicksale juedischer Mathematiker in Muenchen. *Bayerische Akademie der Wissenschaften: Mathematisch-naturwissenschaftliche Klasse: Sitzungsberichte* (1997) 1-25.

Enumerates the mathematicians who were dismissed from universities in all of Germany, most for racial, some for political reasons, and describes their fate: emigration, or death in the Holocaust. Discusses the ascendancy of mediocre scholars who were Nazis, at the University of Munich and the Bavarian Academy of Sciences, and their agitation against Jewish colleagues. Traces the biographies, the scientific accomplishments, and the persecution of three Munich (Jewish) mathematicians: Pringsheim (1850-1941) emigrated to Switzerland in 1939; Liebmann (b. 1874) died a natural death in 1939; Hartogs (b. 1874) committed suicide in 1943 because he could no longer bear the constant humiliations. In November 1938 the Nazi-appointed president ousted all Jews from the Academy, despite the almost unanimous resistance of the other members. *RW*

0572 Baumer, Iso: War Antijudaismus/Antisemitismus unausweichlich? Ein Beispiel selbststaendigen Denkens und Redens 1925/27 und 1932. *Freiburger Rundbrief* 6, 3 (1999) 166-172.

Summarizes two articles and a lecture by Max, Herzog zu Sachsen, brother of the King of Saxony and controversial Catholic priest. He defended Zionism as the solution, promised in the Bible, for the persecuted Jewish people, and argued that it is perfectly compatible with loyalty to the country of residence. He expressed his faith, based on Romans 11, in the unabrogated election and future redemption of the Jewish people. And in a lecture in August 1932 he warned of the danger of current antisemitic movements. *RW*

0573 Beck, Gad: *An Underground Life: The Memoirs of a Gay Jew in Nazi Berlin.* Written with Frank Heibert. Trans.: Allison Brown. Madison: University of Wisconsin Press, 1999. 165 pp. Originally published as "Und Gad ging zu David: Die Erinnerungen des Gad Beck, 1923 bis 1945" (Berlin: Edition diá, 1995).

0574 Benz, Wolfgang: The Ousting of Jewish Lawyers from the German Nazi State. *Justice* 22 (Win 1999) 7-16. An abridged version of his German article which appeared in "Deutsche Juristen juedischer Herkunft" (Muenchen: C.H. Beck, 1993).

0575 Benz, Wolfgang: *Patriot und Paria: Das Leben des Erwin Goldmann zwischen Judentum und Nationalsozialismus: Eine Dokumentation.* Berlin: Metropol, 1997. 180 pp.

Chronicles Nazi measures to exclude "non-Aryans" from German society, and the conformity of both Churches. Traces the life of Goldmann (1891-1981), a fullblooded Jew, Protestant convert, and radical right-wing German nationalist. He was close to Naumann's Verband nationaldeutscher Juden. During the Nazi period, he lost his position as head of a public dental clinic in Stuttgart and, later, his license to practice. He was regional leader in the Paulus-Bund, an organization of nationalist non-Aryan Christians, until full Jews were expelled; then he worked in the Internal Mission with other non-Aryans. In 1940-43 he was an informer for the SD and the Gestapo. His marriage to a Christian saved him from deportation, but in November 1944 he was drafted for labor in a camp for "Mischlinge" in Wolfenbuettel, of which he was named headman. The postwar denazification tribunal imposed on him a severe sentence, but by 1950 he returned to normal life. Pp. 71-180 contain documents mainly from the proceedings of the denazification tribunal, especially Goldmann's own depositions. *RW*

0576 Bergerson, Andrew Stuart: Hildesheim in an Age of Pestilence: On the Political Anthropology of Birth, Death, and Resurrection of Normalcy. *Tel Aviver Jahrbuch fuer Deutsche Geschichte* 28 (1999) 303-340.

Mainly on the basis of interviews, compares nazification in Hildesheim with the plague in Camus' novel. Normal pre-Nazi life in Hildesheim was characterized by a certain neighborly civility in spite of religious and class differences. After the Nazis came to power, "normalcy" was radicalized; society became, almost without noticing it, a "Volksgemeinschaft" that fostered willing participation of "Aryan" Germans and excluded Jews. Gives examples of estrangement between Jews and former non-Jewish friends. Jews who thereby became aware of what was happening emigrated, thus also conforming to the Nazi plan. The persecution and deportation of those who remained were no longer noticed by "Aryan" Hildesheimers and inspired no feelings of guilt after the war. *RW*

0577 Blandford, Neil; Jones, Bruce: *Velichaishiye v mire zlodei* [The World's Most Evil Men]. Moskva: Kron-Press, 1997. 253 pp. Originally published as "The World's Most Evil Men" (London: Octopus Books, 1989).

Ch. 4 (pp. 133-167), "Natsisty" ("The Nazis"), deals mainly with the Holocaust. Dwells on Hitler's biography; briefly outlines the personalities of Himmler, Heydrich, and Eichmann – the main perpetrators of the genocide. Dwells, also, on the biography of Klaus Barbie, and describes briefly the deeds of Walter Rauff, Josef Mengele, and other Nazi criminals. *DR*

0578 Borries, Maria von: *Euer Name lebt: Zur Geschichte der Juden in der Region Bersenbrueck.* Bramsche: Rasch Verlag, 1997. 407 pp.

Surveys the history of Jews in Bersenbrueck (Lower Saxony) from the beginning of their settlement there in 1835 until the Holocaust. Pp. 201-353 deal with the Nazi period, describing the discrimination and persecution of the Jews, and finally their deportation. Mentions emigration of Jewish residents between 1933-39. During the "Kristallnacht" pogrom, Jewish men were arrested and deported to Buchenwald and Sachsenhausen. In December 1941, 14 residents were deported to Riga; in March 1942, three were deported via Piaski to Bełżec, where they were murdered. Of those remaining, ca. 30 were interned in April 1942 in a "Jewish house" in Alfhausen, from where they were deported in July 1942 to Theresienstadt; 11 were transported from there to Treblinka in September 1942 and 15 were transported to Auschwitz between January 1943-October 1944. *ID*

0579 Bradlow, Edna: "Kristallnacht": Broken Glass, Broken Lives. *Jewish Affairs* 54, 2 (Win 1999) 5-10.

The "Kristallnacht" pogrom in Germany was a carefully coordinated party and government action, rather than a spontaneous outburst of public rage. Many Germans participated in it. Public organizations, including the Churches, did not protest, due to fear of Nazi retaliation against those who dared to disagree. The riots were viewed with disgust by the rest of the world. The "Kristallnacht" pogrom marked the beginning of the annihilation of Jews in Europe. *DR*

0580 Bradlow, Edna; Hoff, Penelope: The Hanover Monument: A Unique Memorial to a Centuries-Old Community That Virtually Disappeared. *Australian Journal of Jewish Studies* 13 (1999) 117-130.

On 15 December 1991, exactly 50 years after the first deportation train left Hannover taking its Jews to ghettos and camps in the East, a group of Hannover citizens, mainly Christian, mooted the idea of erecting a memorial naming the victims. The memorial, designed by Michelangelo Pisteletto from Torino, was unveiled on 10 October 1994. Built of sandstone, its main feature is a broken square; around the base are inscribed the names of the 1,882 Hannover Jews who died in the death camps. Focuses on the history of the Hannover Jewish community during the Nazi regime, describing the rise of racial antisemitism in pre-Nazi Germany, the Nazi persecution of Jews, the "Kristallnacht" pogrom, and the deportations. *DR*

0581 Breitman, Richard David: "Mein Kampf" and the Himmler Family: Two Generations React to Hitler's Ideas. *Holocaust and Genocide Studies* 13, 1 (Spr 1999) 90-97.

Examines marginal comments found in Himmler's copy of "Mein Kampf" (volume 2), by Himmler and by his father Gebhard. Shows, inter alia, that Himmler was much more receptive to Hitler's racial and antisemitic ideas than his father, who was rather skeptical about them. Himmler took seriously Hitler's idea of "the possibility of de-miscegenation," which meant liquidation of "racially mixed" persons. Hitler and Himmler both had no doubt that Jews were the most dangerous source of racial infiltration and degeneration. *DR*

0582 Brinkmoeller, Karl, ed.: *Juedische Buerger in Bad Driburg, 1900 bis 1945*. [2nd ed.]. Bad Driburg: [The Editor], 1999. 156, 31 pp. First published in 1997. The last 31 pp. contain additions.

Discusses the Nazi period in Bad Driburg (Westphalia). Pp. 29-115 relate the fate of 16 Jewish families, former residents of the town, in the Holocaust. In 1936 the Jewish community consisted of 24 members; in 1938 only 13 remained. Three families emigrated; those remaining were deported to various camps. Eight families were annihilated in the Holocaust. *ID*

0583 Brix, Barbara: "*Land, mein Land, wie leb' ich tief aus dir": Dr. Walter Bacher – Jude, Sozialdemokrat, Lehrer an der Klosterschule*. Hamburg: Gymnasium Klosterschule; Doelling und Galitz, 1997. 67 pp.

A booklet written in memory of Walter Bacher and his wife Clara, who were murdered in Auschwitz. Bacher was born in 1893 in Halle to Jewish parents, but was baptized as a child. From 1925 he worked as a teacher in Berlin; he was dismissed in 1933. In 1935 he was employed in a Jewish school in Hamburg; in 1938 he became a member of the Jewish community and was employed in other Jewish schools. In the summer of 1942, Bacher and his wife were forced to move into a Jewish house; in July 1942 they were deported to Theresienstadt, and from there to Auschwitz in September-October 1944. *ID*

0584 Brocke, Michael, ed.: *Feuer an dein Heiligtum gelegt: Zerstoerte Synagogen 1938, Nordrhein-Westfalen*. Bochum: Kamp, 1999. xxvii, 674 pp. On title page also: Erarbeitet vom Salomon Ludwig Steinheim-

Institut fuer deutsch-juedische Geschichte. Opposite title page: Gedenkbuch der Synagogen Deutschland 1938.

A memorial book commemorating 349 synagogues which were destroyed in North-Rhine-Westphalia during the "Kristallnacht" pogrom in 1938. Presented in alphabetical order of the names of the towns, relates the history of each synagogue and includes illustrations and photographs of the buildings. The appendix (pp. 597-660) gives information for each town regarding population distribution in the 19th-20th centuries, location of the Jewish cemetery, bibliographic and photographic citations. *ID*

0585 Buettner, Ursula; Greschat, Martin: *Die verlassenen Kinder der Kirche: Der Umgang mit Christen juedischer Herkunft im "Dritten Reich."* Goettingen: Vandenhoeck & Ruprecht, 1998. 151 pp.

This book contains four essays. Contents: Buettner, Ursula: Von der Kirche verlassen: Die deutschen Protestanten und die Verfolgung der Juden und Christen juedischer Herkunft im "Dritten Reich" (15-69); Greschat, Martin: "Gegen den Gott der Deutschen": Marga Meusels Kampf fuer die Rettung der Juden (70-85); Greschat, Martin: Friedrich Weissler: Ein Jurist der Bekennenden Kirche im Widerstand gegen Hitler (86-122); Buettner, Ursula: "Wohl dem, der auf die Seite der Leidenden gehoert": Der Untergang des Dichters Jochen Klepper mit seinen juedisch-christlichen Angehoerigen [A revised version of her article in "...und ueber Barmen hinaus: Studien zur kirchlichen Zeitgeschichte" (Goettingen, 1995).] (123-149). *RW*

0586 Corbach, Dieter: *6.00 Uhr ab Messe Koeln-Deutz: Deportationen 1938-1945 = Departure: 6:00 a.m. Messe Koeln-Deutz: Deportations 1938-1945.* Koeln: Scriba, 1999. 804 pp.

A study of the deportation to concentration and extermination camps of 6,475 Jewish residents of Koeln during the Nazi period. Discusses the organization of the transports, which was preceded by financial statements on the property, resources, and housing conditions of the victims. P. 57 contains a list of the 23 deportations from Koeln-Deutz between October 1938-March 1945, to Zbąszyń, Łódź, Riga, Izbica, Minsk, Theresienstadt, and Auschwitz. Pp. 16-188 present the texts of 57 documents relating to the deportations, with connecting explanations. These include documents issued by the Nazis, by the Jewish community, and memoirs of deportees. Pp. 143-156 report on 14 deportations to Theresienstadt between June 1942-March 1945 and describe the conditions and daily life in the ghetto. Pp. 157-171 document the deportation to Minsk on 20 July 1942. Pp. 189-233 contain notes, a glossary, and indexes. Pp. 235-277 contain photographs of individuals who were deported. Pp. 279-578 include deportation lists with 6,335 names. Pp. 579-582 contain 133 additional names. Pp. 587-804 present the text of pp. 1-233 in English. *ID*

0587 Diehl, Katrin: *Die juedische Presse im Dritten Reich: Zwischen Selbstbehauptung und Fremdbestimmung*. Tuebingen: Max Niemeyer, 1997. xii, 362 pp.

During the Third Reich, the Nazis monopolized and dominated the press, Jewish and non-Jewish. Until 1938, 100 Jewish periodicals appeared in Germany. They were permitted to deal only with Jewish topics. The Jewish press was allowed to exist for economic reasons and also as a front to cover up the regime's anti-Judaism. Before the annihilation, the aim of the Nazis was isolation of the Jews, i.e. ghettoization, deprivation of rights, and discrimination. The renaissance of the Jewish press was doomed to failure from the start, since it was based on antisemitic policies. Jewish journalists responded to the Nazi oppression not only with a strong will to survive, but also with the aim of preserving their own culture. The Jewish press was part of the resistance; it encouraged people to persist. After the "Kristallnacht" pogrom the Jewish press ceased to exist. *ID*

0588 *Dimensions* 13, 2 (1999).

The section entitled "Race, Profit and Plunder: German Big Business and the Third Reich" contains essays on the culpability of German big business in the Holocaust, especially the exploitation of Jews for forced labor. Partial contents: Wiesen, S. Jonathan: German Industry and the Third Reich: Fifty Years of Forgetting and Remembering [On the firms of IG-Farben, Siemens, Krupp, and Bayer.] (3-8); Steinberg, Jonathan: Investigating Deutsche Bank (9-14); Reich, Simon: The Ford Motor Company and the Third Reich (15-17); Stallbaumer, Lisa M.: Friedrich Flick's Opportunism and Expediency (19-23); Gregor, Neil: Daimler-Benz, Forced Labor and Compensation (25-30). *RE*

0589 Dippel, John V.H.: *Bound upon a Wheel of Fire: Why So Many German Jews Made the Tragic Decision to Remain in Nazi Germany*. New York: Basic Books, 1996. xxx, 353 pp.

A study of six prominent Jews (Richard Willstaetter, Nobel laureate chemist; Max Warburg, banker; Hans-Joachim Schoeps, right-wing Jewish youth leader; Bella Fromm, Berlin society columnist; Robert Weltsch, Zionist editor; and Rabbi Leo Baeck) who remained in Germany during the early years of Hitler's regime. Notably, none of them were regarded by the Nazis as adversaries; Schoeps even sympathized with the Nazi goals. Their patriotism contributed to their decisions to stay in Germany, as well as a feeling that there was important Jewish communal work to be done, a hope that highly-placed connections would protect them, and an inability to understand that their world had turned upside down and the Nazi anti-Jewish policy would not be a temporary phenomenon. The "Kristallnacht" pogrom shook their hopes for Jewish-German coexistence in Hitler's Germany and caused some of them to emigrate. *REK*

0590 Dippel, John V.H.: *Die grosse Illusion: Warum deutsche Juden ihre Heimat nicht verlassen wollten.* Trans.: Angelika Beck. Introd.: Alfred Grosser. Weinheim: Beltz Quadriga, 1997. 538 pp. Originally published as "Bound upon a Wheel of Fire: Why So Many German Jews Made the Tragic Decision to Remain in Nazi Germany" (New York: Basic Books, 1996). Pp. 9-21 contain the introduction by Grosser.

0591 Dirschauer, Johannes: *Tagebuch gegen den Untergang: Zur Faszination Victor Klemperers.* Giessen: Psychosozial-Verlag, 1997. 229 pp.

An analysis of Klemperer's life and that part of his diary which covers the war years, 1933-45, published as "Ich will Zeugnis ablegen bis zum letzten" (1995). Klemperer was born in 1881, the son of a rabbi; he converted to Christianity in 1902, and married pianist Eva Schlemmer in 1906. The diary entries relate his dismissal from his position as a professor of romance languages in 1933, eviction from his home in 1940 and the move to a crowded "Jewish house," the wearing of the yellow star from September 1941, and forced labor in 1942. His marriage to an Aryan protected him from deportation until February 1945. Then Klemperer and his wife went into hiding and survived. Contends that Klemperer's diary notes during the Nazi period both endangered and protected him. There was permanent danger for himself and his wife, who hid the manuscripts, and for those mentioned in the diary. Chronicling the fascist terror required immense psychic strain, but was nevertheless an enterprise of rescue for the soul. *ID*

0592 Elias, Norbert: *The Germans: Power Struggles and the Development of Habitus in the Nineteenth and Twentieth Centuries.* New York: Columbia University Press, 1996. xvii, 494 pp.

Contains an account of German social development, focusing on the 20th century and showing that the rise of the Nazi regime grew out of Germany's past. Germany's unification took place through a series of wars, during which large sections of the middle classes abandoned humanistic values and adopted militaristic and authoritarian values. In pt. 4 (pp. 299-402), "The Breakdown of Civilization," Elias states that the extermination of the Jews is an example of the vulnerability of civilization and the dangers of growth processes, in which decay can predominate. Although every war is a regression to barbarism, the Holocaust was the deepest regression; even the minimum rules of civilized conduct were not observed. People in Germany and elsewhere thought that the Nazis, living in a "civilized" society, would not act as brutally as they talked; they could not believe that the Nazis would actually act in accordance with their barbaric ideology. *RE*

0593 Enders, Hermann; Hysky-Dambmann, Henny: *Geschichte der Juden in der Stadt Schotten*. Schotten: Vogelsberger Kultur- und Geschichtsverein Schotten, 1996. 14 pp. Reprinted from the book "Schotten und seine Stadtteile im Wandel der Zeiten" (Schotten: Stadt Schotten, 1996).

Chronicles the history of the Jews in Schotten, Hesse, from the Middle Ages, mentioning that in the 17th century they were not permitted to hold religious services. In later times they were well-accepted. Pp. 7-14 describe persecution in the Nazi period. In September 1935 the Schotten town council prohibited all friendly and business contact with Jews. In 1933 there were 88 Jews in the town. Most of them moved to larger cities (their subsequent fate is not mentioned); those who could, emigrated; 14 were deported from Schotten to concentration camps. *RW*

0594 Engelhard, Christoph, ed.: *Erinnerung stiftet Erloesung: Gedenkheft fuer die juedische Frauen, Maenner und Kinder aus Memmingen, die zwischen 1941 und 1945 verfolgt, verschleppt und ermordet wurden.* Memmingen: Stadtarchiv, 1999. 24 pp.

A memorial book commemorating 106 Jewish men, women, and children who lived in Memmingen until the Holocaust. Pp. 6-19 give information on those who were murdered between 1941-45, including date of birth, profession, date of deportation, and additional details on their fate. *ID*

0595 Erdmann, Ulrich: *Vom Naturalismus zum Nationalsozialismus? Zeitgeschichtlich-biographische Studien zu Max Halbe, Gerhart Hauptmann, Johannes Schlaf und Hermann Stehr.* Frankfurt a.M.: Peter Lang, 1997. 355 pp. Based on the author's diss. – Universitaet Stuttgart, 1996.

Traces the careers and especially the political development of four prominent German writers, all born in the 1860s and members of the naturalist school, including their attitude to Jews and antisemitism. All but Hauptmann were ultra-conservative, ultra-nationalist, and opposed to the Weimar Republic; all four were politically naïve with a tendency to "voelkisch" mysticism. Thus they were receptive to the ideology of Nazism and welcomed the Nazi rise to power. The four writers to some extent shared antisemitic stereotypes, especially those of Jews as materialistic and coldly intellectual, but opposed antisemitism. All were aided in their careers by Jewish publishers, theater directors, and patrons, and had Jewish friends. However, when in conflict with a Jew (e.g. a publisher who was a hard bargainer), they resorted to antisemitic epithets. During the Nazi period they remained loyal to Jewish friends and benefactors but tended to close their eyes to, or even justify, Nazi antisemitic policy. *RW*

0596 Ericksen, Robert P.; Heschel, Susannah, eds.: *Betrayal: German Churches and the Holocaust*. Minneapolis: Fortress Press, 1999. vi, 224, [10] pp.

A collection of essays. Contents: Ericksen, Robert P.; Heschel, Susannah: Introduction [Describing the attitudes of the German Churches toward Nazi policy and the Jews, shows that the Churches played a far more important role in Nazi atrocities than has hitherto been supposed; large segments of the Catholic and Protestant Churches supported Nazism with enthusiasm, and in fact blessed the Nazi racial policy and helped to create the moral atmosphere in Germany conducive to the Holocaust.] (1-21); Ericksen, Robert P.: Assessing the Heritage: German Protestant Theologians, Nazis, and the "Jewish Question" [Highlights the roles of P. Althaus, E. Hirsch, and G. Kittel in reshaping theology in accordance with Nazi goals.] (22-39); Bergen, Doris L.: Storm Troopers of Christ: The German Christian Movement and the Ecclesiastical Final Solution (40-67); Heschel, Susannah: When Jesus Was an Aryan: The Protestant Church and Antisemitic Propaganda [On the Institute for the Study and Eradication of Jewish Influence on German Religious Life, in Eisenach, and its director, Walther Grundmann.] (68-89); Baranowski, Shelley: The Confessing Church and Antisemitism: Protestant Identity, German Nationhood, and the Exclusion of the Jews (90-109); Barnes, Kenneth C.: Dietrich Bonhoeffer and Hitler's Persecution of the Jews (110-128); Lewy, Guenter: Pius XII, the Jews, and the German Catholic Church [First published in "Commentary" 37, 2 (1964).] (129-148); Lukens, Michael B.: Joseph Lortz and a Catholic Accommodation with National Socialism [Lortz (1887-1975) was an ordained priest, a church historian, a professor at Braunsberg and then Muenster during the Nazi period, and a member of the Nazi Party from 1933, and wrote propaganda on behalf of Catholic support for Nazism.] (149-168); Brumlik, Micha: Post-Holocaust Theology: German Theological Responses since 1945 (169-188). Pp. 189-219 contain the notes to the essays. *DR*

0597 Erler, Hans; Ehrlich, Ernst Ludwig; Heid, Ludger, eds.: *"Meinetwegen ist die Welt erschaffen": Das intellektuelle Vermaechtnis des deutschsprachigen Judentums: 58 Portraits*. Frankfurt: Campus, 1997. 555 pp.

A collection of biographical essays on the contributions to German culture and society of Jewish intellectuals who lived in Germany before and during the Nazi period. Some of them suffered extensively from Nazi persecution; two of them were murdered. In lieu of an epilogue, presents an essay by Dan Diner on three interpretative conceptions of the Holocaust. Partial contents: Brocke, Edna: Leo Baeck (1878-1956) (53-58); Herweg, Rachel Monika: Regina Jonas (1902-1944) (72-77); Mueller, Christoph: Hermann Heller (1891-1933) (345-361); Mueller, Ingo: Robert M.W. Kempner (1899-1993) (362-368); Sassenberg, Marina: Kaethe Leichter (1895-1942) (463-468); Diner, Dan: Zivilisationsbruch, Gegenrationalitaet, "gestaute Zeit": Drei interpretationsleitende Begriffe zum Thema Holocaust (513-520). *ID*

0598 Essner, Cornelia: Die Tagebuecher von Victor Klemperer (1933-1945). *Geschichte in Wissenschaft und Unterricht* 49, 7-8 (July-Aug 1998) 440-446.

Quoting extensively from Klemperer's diary, discusses his identity as a German and a Jew (although he converted to Protestantism in 1906) during the Nazi regime. A professor of romance languages, he was forced to retire in 1935. Because his wife was Aryan, he was exposed to serious antisemitic threats only from 1940, when the couple had to move to a "Jewish house." Childless, they were considered as a non-privileged mixed marriage. He had to wear a yellow star and lived in constant fear of arrest or deportation. Klemperer recorded all the events of his daily life, and also discussed works of Nazi racial theory, such as Rosenberg's "Mythos," Goebbels' "Das Reich," and Hitler's "Mein Kampf." *ID*

0599 Faber, Benedikt: "Alles, was ich fuer undeutsch gehalten habe, floriert hier": Die Auseinandersetzung Victor Klemperers mit seinem Deutschtum in den Tagebuechern 1918 bis 1945. *Sachor* 9 (1999) 49-59.

Klemperer saw himself entirely as a German; he had an aversion to many elements of Judaism and especially to Zionism. His attachment to Germany – the "real" Germany, which he distinguished from Nazism – was shaken by the anti-Jewish measures from which he suffered, and even more by the demeaning attitude of ordinary Germans – from academic colleagues to street urchins – who saw in him only the despised Jew. But his faith in the "real" Germany persisted, and after the war he saw it as his mission to participate in its reconstruction. *RW*

0600 Freund, Susanne: *Juedische Bildungsgeschichte zwischen Emanzipation und Ausgrenzung: Das Beispiel der Marks-Haindorf-Stiftung in Muenster (1825-1942)*. Paderborn: Ferdinand Schoeningh, 1997. xi, 403 pp. Based on the author's diss. – Westfaelische Wilhelms-Universitaet Muenster, 1995.

A history of the Marks-Haindorf-Stiftung, a Jewish institution established in 1825 in order to encourage Jews to acquire a profession in handicrafts. The institution also had an elementary school and a "Lehrerseminar." Pp.

291-325 discuss the fate of the school during the Nazi period. The school comprised, in 1935, eight grades, and accepted children from the nearby villages. In 1938 the number of pupils was 39; after the "Kristallnacht" pogrom and the ensuing wave of emigration, 29 remained. In March 1940 the school ceased to exist de jure; it was transformed into a so-called "Jewish house," and became de facto a place of refuge for Muenster's Jews. In July 1942 the 78 remaining Jews in Muenster were deported to Theresienstadt. *ID*

0601 Friedrich-Woessner, Sigrid: *Kultur und Lebensweise der Juden in Hof seit dem Ausgang des 19. Jahrhunderts*. Regensburg: Roderer, 1997. 177 pp. Based on the author's M.A. thesis – Universitaet Regensburg, 1988.

Surveys the history of the Jews in Hof, where they first settled in the medieval period; a Jewish community was reestablished in the second half of the 19th century. Pp. 147-153 deal with the Nazi period. In the late 1920s antisemitic agitation and propaganda increased; in 1928, attacks against Jewish persons or property took place in Hof. The Jewish community consisted of 25 families, which decreased to eight families by 1933. Young people emigrated; others sought refuge in larger towns. When the last Jew left in 1941, Hof was declared "judenfrei." *ID*

0602 Ganelin, Rafail Sh.: N.Ye. Markov 2-i o svoyom puti ot chernosotenstva k gitlerizmu [N.Ye. Markov the 2nd on His Path from Black-Hundredism to Hitlerism]. *Yevrei v Rossii: Istoriya i kultura: Sbornik nauchnykh trudov, Vol. 5* [Jews in Russia: History and Culture: Collected Articles, Vol. 5], ed. Dmitri Elyashevitch. Sankt-Peterburg: Peterburgskii Yevreiskii Universitet, 1998. Pp. 211-218.

N.Ye. Markov ("Markov the 2nd") was a leader of the Black Hundred movement before the Russian Revolution. In the interwar period, as an émigré from Russia, he became an ardent sympathizer with fascism and Nazism. In 1933 he went to Germany, eager to collaborate with the Nazi regime. As a monarchist, Markov had some disagreements with the Nazis, but his antisemitism helped bridge the ideological gap. It was also his antisemitism which helped Markov to mitigate his fear of Nazi anti-Slavism; he regarded the Germans as a lesser danger to the Russian people than the Jews. *DR*

0603 Gay, Peter: *Meine deutsche Frage: Jugend in Berlin 1933-1939*. Trans.: Ulrich Enderwitz, Monika Noll, Rolf Schubert. Muenchen: C.H. Beck, 1999. 230 pp. Originally published as "My German Question: Growing Up in Nazi Berlin" (New Haven: Yale University Press, 1998).

0604 *Geschichte juedischer Gemeinden in Sachsen-Anhalt: Versuch einer Erinnerung*. Magdeburg: Landesverband Juedischer Gemeinden Sachsen-

Anhalt, 1997. 310 pp. On back of title page: Autorengruppe: Baerbel Bugaiski, Ildiko Leubauer, Guenter Waesche.

Surveys the history of 55 Jewish communities in the region of Sachsen-Anhalt from the beginning of Jewish settlement in each town until World War II. Some of the reports include information on the Holocaust. *ID*

0605 Gohl, Beate: *Juedische Wohlfahrtspflege im Nationalsozialismus: Frankfurt am Main 1933-1943.* Frankfurt a.M.: Fachhochschulverlag, 1997. 99 pp.

Discusses Jewish welfare activities in Frankfurt in 1933-43. There were ca. 25,000 Jews in Frankfurt. Stresses the gradual impoverishment and aging of the Jewish population due to the emigration of young people. Even before the Nazi rise to power, a deterioration in the social situation of the Jews prompted the organization of public kitchens and other welfare activities. From 1938 the activities increased as a result of the discrimination against Jews, who then had to declare all their property; later, their bank accounts were suspended and their possessions were confiscated. In 1940 a special tax was imposed; from 1941, Jews fit for work had to do forced labor, and at the end of 1941 mass deportations began. Among the welfare activities carried out by the Jewish community were educational and professional advice, assistance to children, health care, maintenance of hospitals and homes for the elderly, aid in emigration, and cultural work. *ID*

0606 Goldhagen, Daniel Jonah: Willing Executioners: The Paradigm Challenged: Victim Testimony, Critical Evidence, and New Perspectives in the Study of the Holocaust. *Tikkun* 13, 3 (May-June 1998) 40-47.

Criticizes present historiography of the Holocaust as seriously distorting the perpetrators' attitudes toward the victims and the motives of their actions. There is a tendency to depict the perpetrators as having been compelled by external forces to act against their will, and personally having nothing against their victims. One of the reasons is the widespread dismissal of victims' accounts as a source of information regarding the perpetrators' motives and behavior. The Holocaust seems to be the only genocide whose historiography disregards the victims' testimonies. This leads to distortion of the truth and to the emergence of powerful myths on perpetrators and bystanders being forced to do what they did. Survivors show that the perpetrators were in no way reluctant killers, with no views of their own, and that their cruelty was anything but instrumental – they were motivated by antisemitism. Admits that this is not tantamount to condemnation of the entire German people or to the defamation of present-day Germany. See the response by Christopher Browning ["Tikkun" 14, 1 (Jan-Feb 1999) 56, 58], in which he argues that he has never followed the “old paradigm” of “irresistible external forces” which act on reluctant and puppet-like perpetrators, but Goldhagen's monocausal explanation of the murderers' motives (“genocidal culture” deeply imbedded in the German culture) is not sufficient. See also Goldhagen's reply (ibid., p. 57). *DR*

0607 Goldin, Milton: Financing the SS. *History Today* 48, 6 (June 1998) 28-34.

Explores Himmler's ambitions to establish the SS as a “state within a state,” and schemes the Nazis devised to finance the SS through industrial enterprise and plundered Jewish assets and property. Taking money from Jews for the benefit of the SS began already in the 1930s, when the inmates of concentration camps had to pay for their imprisonment. The mass murder of Jews in the course of “Operation Reinhard” was accompanied by the seizure of victims' personal properties. The most lucrative business was exploitation of slave labor in numerous camps operated by the SS. Highlights the financial aspect of the negotiations between Eichmann and Hungarian Jewish leaders in March-August 1944, and of an attempt made by Himmler in April 1945 to extort a ransom for the lives of the last Jews under Nazi control. *DR*

0608 Gruner, Wolf: The German Council of Municipalities (“Deutscher Gemeindetag”) and the Coordination of Anti-Jewish Local Politics in the Nazi State. *Holocaust and Genocide Studies* 13, 2 (Fall 1999) 171-199.

From the first days of the Nazi dictatorship, local governments in various parts of Germany initiated anti-Jewish measures which often preempted similar party and central government measures, and in some cases were illegal under the existing law. In May 1933 the German Council of Municipalities was founded, headed by Karl Fiehler. Municipal discriminatory measures against Jews were often coordinated by the Council, which also provided local governments with legal protection when their measures contradicted the law and could displease Berlin. Examines municipal government actions against Jewish property and businesses. Illustrates the activities of the “Gemeindetag” with its involvement in two local undertakings: the expulsion of Jewish merchants from city markets in 1933-38, and the organization of the project for municipal pawnbroking offices to “utilize” Jewish property after the “Kristallnacht” pogrom. *DR*

0609

גרונר, וולף: עוני ורדיפות: ההתאחדות הארצית של היהודים בגרמניה, היהודים והמדיניות האנטי-יהודית בגרמניה הנאצית בשנים 1939-1945. יד ושם 27 (1999) 17-46.

[Gruner, Wolf: Poverty and Persecution: The Reichsvereinigung, the Jewish Population, and Anti-Jewish Policy in the Nazi State, 1939-1945. *Yad Vashem* 27 (1999) 17-46.] Appeared in English in "Yad Vashem Studies" 27 (1999).

0610 Gruner, Wolf: Poverty and Persecution: The Reichsvereinigung, the Jewish Population, and Anti-Jewish Policy in the Nazi State, 1939-1945. *Yad Vashem Studies* 27 (1999) 23-60. Appeared in Hebrew in "Yad Vashem: Kovetz mehkarim" 27 (1999).

An expanded version of a paper delivered in German at an international conference on "Jews and Poverty," University of Leipzig, September 1997. Focuses on the welfare activities of the Reichsvereinigung, which was created in 1939. The main goal of the Nazis in setting up this organization was to oust Jews from the state welfare system and to establish a separate one for them. The Nazis needed such a system because growing pauperization of the German Jews hindered their emigration. The conditions in which the RV began its work were unfavorable – the communities were ruined by the "Kristallnacht" pogrom. Deals with the problems confronted by the RV, its ways of mobilizing money for relief, efforts to alleviate the plight of the needy by organizing work brigades for them, etc. The deportations which began in 1941-42 dealt a heavy blow to the RV's welfare activities. Relief work under the conditions of the Holocaust proved to be Sisyphean – most of Germany's Jews were deported and perished. *DR*

0611 Hahn, Joachim: *Juedisches Leben in Ludwigsburg: Geschichte, Quellen und Dokumentation.* Karlsruhe: Braun, 1998. 783 pp. On title page also: Herausgegeben von der Stadt Ludwigsburg – Stadtarchiv – und vom Historischen Verein fuer Stadt und Kreis Ludwigsburg e.V.

Traces the history of Jews in Ludwigsburg (Wuerttemberg) over the centuries, noting the good relations between the Jews and their Christian neighbors. Ch. 4 (pp. 207-278) deals with the Holocaust period. The Nazis imposed boycotts and Aryanization, spread antisemitic propaganda, and segregated the Jews. Describes the burning of the synagogue and the arrests of Jewish men during the "Kristallnacht" pogrom; many citizens were opposed to these actions, and some rescued Jews. Traces the emigration of individuals and families to whatever country would accept them, and the deportation of the remainder to Riga-Jungfernhof, Izbica, and Theresienstadt. Pp. 340-600 contain family histories and brief biographies. A preface by the mayor of Ludwigsburg, Christof Eichert, states that 146 Jews emigrated; 52 were deported, of whom only four survived. Pp. 593-600 contain lists of Jews who lived in Ludwigsburg at any time and who were deported, including many who were deported from other places; these apparently are not included in the mayor's statistics. *RW*

0612 Haller-Nevermann, Marie: *Jude und Judentum im Werk Anna Seghers': Untersuchungen zur Bedeutung juedischer Traditionen und zur Thematisierung des Antisemitismus in den Romanen und Erzaehlungen von Anna Seghers.* Frankfurt a.M.: Peter Lang, 1997. 279 pp. Based on the author's diss. – Freie Universitaet Berlin, 1996.

Seghers (née Netty Reiling) was born in 1900 in Mainz and died in 1983 in Berlin. Examines the influence of Seghers' rupture with Judaism and her turn to communism at the age of 26 on her writings, especially on her work in exile, e.g. "Das siebte Kreuz," "Transit," and "Der Ausflug der toten Maedchen." States that her writing is characterized by a general neglect and repression of Jewish themes, disregard of the antisemitism of the Weimar period, and avoidance of coping with the Holocaust. Claims that the rupture with her Jewish background served as a precondition for her decision to live and write as a communist, and involved denial and repression of anything Jewish. This rejection became a component of her self-concept, even to the extent of Jewish self-hatred. She left Germany in 1933 (for Paris), and in 1941 she fled to Mexico; the history and suffering of the Jews was part of her own biography. Her mother was deported and murdered in 1943. Thus, her work cannot be evaluated fully without consideration of this aspect of her life. *ID*

0613 Halmon, Raphael et al., eds.: *Gedenke – Remember – Yizkor.* Fuerth: Stadtarchiv; Tel-Aviv: Komitee zum Gedenken der Fuerther Shoah-Opfer, 1997. 481 pp.

Following introductions in German, English, and Hebrew, presents a list of 886 Jews who lived in Fuerth or were born there who died in the Holocaust as a result of Nazi persecution in Fuerth or after deportation. Each entry gives biographical data. Also contains lists of Nazi-designated "Jewish houses" and of the deportations. *RW*

0614 Hamerow, Theodore S.: *On the Road to the Wolf's Lair: German Resistance to Hitler.* Cambridge, MA: Belknap Press of Harvard University Press, 1997. xi, 442 pp.

The failed attempt on Hitler's life in July 1944 was the work of members of Germany's conservative elite, who sought to save their country from the disaster brought about by Hitler and the Nazis. Examines the ideals and motives of German resisters, many of whom were army officers, upper-level civil servants, or religious leaders. Chapter 4 (pp. 65-79), "That Perennial 'Jewish Question'," contends that most of the resisters supported racial antisemitism and believed that it was necessary to reduce Jewish influence in Germany. The most vocal antisemites among them were the Protestant clergy, and especially Otto Dibelius. The Catholic clergy were less hostile toward Jews, since they also suffered from discrimination, but they would not support the Jews for fear of endangering the Catholic community. Dietrich Bonhoeffer was the only resister who condemned Nazi antisemitism, but even he felt that there was a need to solve the "Jewish question." Most of the resisters were repelled by the Nazis' brutal treatment of the Jews, but their primary concern was for the future of Germany. *RE*

0615 Hassler, Marianne; Wertheimer, Juergen, eds.: *Der Exodus aus Nazideutschland und die Folgen: Juedische Wissenschaftler im Exil.* Tuebingen: Attempto, 1997. 347 pp.

Based on a series of lectures held at the University of Tuebingen in 1994-95. The following articles deal at greater length with the effects of antisemitism in the Weimar Republic and persecution under the Nazi regime on the careers of Jewish scholars, and with their reactions and their theories of antisemitism and totalitarianism: Hermann, Armin: Der Papst der Physik verlaesst die Alte Welt: Einsteins Emigration (19-31); Lerchenmueller, Joachim: Politische Verfolgung in der Weimarer Republik: Der Fall Georg Friedrich Nicolai (Lewinstein) an der Friedrich-Wilhelms-Universitaet Berlin 1919-1922 (91-123); Lueck, Helmut E.: Der Exodus und die Folgen fuer die Psychologie (157-176); Schmid Noerr, Gunzelin: Die Emigration der Frankfurter Schule und die Krise der Kritischen Theorie (228-245); Schoeps, Julius Hans: Der ungeliebte Aussenseiter: Zum Leben und Werk des Philosophen und Schriftstellers Theodor Lessing (286-303); Wertheimer, Juergen: Hannah Arendt (325-341). *RW*

0616 Healy, Marsha: The Holocaust, Modernity and the Enlightenment. *Res Publica* 3, 1 (1997) 35-59.

Disputes the popular thesis presented by Zygmunt Bauman, in his "Modernity and the Holocaust," that the Holocaust was made feasible by "the social norms and institutions of modernity" and by a mistaken stress on reason. Concurs with Gillian Rose (in her "Judaism and Modernity") in challenging the theory that modernism was a sufficient cause of Nazi bureaucratic mass murder, especially by "desk murderers." Criticizes the theory for its deterministic attitude based on the functionalist school of Holocaust historiography, and for confusing the "how?" and the "why?" of the Nazi genocide. Notes the theory's lack of attention to ideology, which is seen as having had a primary role in motivating mass murder, and its blurring of the distinction between potential and actual murderers. Considers the modernist theory to be nihilistic; it denies the possibility of moral choice and rejects the use of reason to learn what is taking place in society and to build political and social structures that protect human beings. Such a use of reason to protect humanity is presented as a response to Nazi dehumanization. *YC*

0617 Hecht, Ingeborg: *Invisible Walls: A German Family under the Nuremberg Laws, and To Remember Is to Heal: Encounters between Victims of the Nuremberg Laws.* Trans.: John Brownjohn, John A. Broadwin. Evanston, IL: Northwestern University Press, 1999. x, 259 pp.

A translation of two books: the author's memoirs, "Als unsichtbare Mauern wuchsen" (Hamburg: Hoffmann und Campe, 1984) [see the abstract for the French edition: "Les murailles invisibles" (La Tour d'Aigues: Ed. de l'Aube, 1993)] and "Von der Heilsamkeit des Erinnerns: Opfer der Nuernberger Gesetzes begegnen sich" (Hamburg: Hoffmann und Campe, 1991) which discusses reactions in West Germany to those memoirs. *SSC*

0618 Heer, Hannes, ed.: *Im Herzen der Finsternis: Victor Klemperer als Chronist der NS-Zeit.* Berlin: Aufbau-Verlag, 1997. 220 pp.

A collection of articles. Partial contents: Gerstenberger, Heide: "Meine Prinzipien ueber das Deutschtum und die verschiedenen Nationalitaeten sind ins Wackeln gekommen wie die Zaehne eines alten Mannes": Victor Klemperer in seinem Verhaeltnis zu Deutschland und zu den Deutschen (10-20); Nerlich, Michael: Victor Klemperer Romanist oder Warum soll nicht einmal ein Wunder geschehen? [Includes an account of the obstacles to Klemperer's university career in the Wilhelmine period and the Weimar Republic, partly because he was a Jew.] (35-48); Wildt, Michael: Angst, Hoffen, Warten, Verzweifeln: Victor Klemperer und die Verfolgung der deutschen Juden 1933 bis 1941 (49-72); Liebsch, Heike: "Ein Tier ist nicht rechtloser und gehetzter": Die Verfolgung und Vernichtung der juedischen Bevoelkerung Dresdens 1933-1937 (73-91); Goldenbogen, Nora: "Man wird keinen von ihnen wiedersehen": Die Vernichtung der Dresdener Juden 1938-1945 [An earlier version appeared in "Zwischen Integration und Vernichtung" (1996).] (92-109); Heer, Hannes: Vox populi: Zur Mentalitaet der Volksgemeinschaft (122-143); Kraushaar, Wolfgang: Karriere eines Boxers: Johannes Clemens: Vom Dresdener Gestapo-Schlaeger zum Doppelagenten des KGB im BND (152-169); Reemtsma, Jan Philipp: "Buchenwald wird von andern geschildert werden; ich will mich an meine Erlebnisse halten": Stenogramme aus der Vorhoelle [Includes a psychological analysis of the perpetrators.] (170-193). *RW*

0619 Herbst, Detlev: *Juedisches Leben im Solling: Der Synagogenverband Bodenfelde – Uslar – Lippoldsberg und die Synagogengemeinde Lauenfoerde*. Uslar: Schlieper-Druck und Verlag, 1997. 291 pp.

Surveys Jewish life in the region of Solling (Lower Saxony), from the beginning of Jewish settlement there in the 14th century until the Nazi period. Pp. 168-191, "Das Dritte Reich (1933-1945)," discuss the boycott of local Jewish enterprises, antisemitic agitation which led (among other things) to violence against Jewish children in school, Nazi legislation, Aryanization, the "Kristallnacht" pogrom, and the deportations. Pp. 221-276, "Aechtung – Vertreibung – Deportation," relate the fate of Jews in the Solling area: 70 emigrated, 39 were murdered in the Holocaust, and the fate of 16 is unknown. *ID*

0620 Hetzel, Marius: *Die Anfechtung der Rassenmischehe in den Jahren 1933-1939: Die Entwicklung der Rechtsprechung im Dritten Reich: Anpassung und Selbstbehauptung der Gerichte*. Tuebingen: Mohr Siebeck, 1997. xxii, 233 pp. Based on the author's diss., 1997.

Surveys court decisions on applications for the invalidation of mixed marriages for racial reasons. Wilhelmine law, which remained in force, stated that a marriage could be invalidated if one partner discovered in the other a personal defect (e.g. mental disease) which, had he known of it, would have prevented the marriage. Nazi judges held that Jewish blood was such a defect; but most of them ruled that it would have been known at the time of the marriage and thus was not a ground for invalidation. Some, however, argued that many Germans became aware only in 1933 of harmful Jewish racial characteristics. Hitler, for political reasons, opposed legislation facilitating the dissolution of mixed marriages. Thus, a new marriage law passed in 1938 did not expressly mention these. However, it provided for annulment of a marriage if one partner had been ignorant of the circumstances (rather than personal characteristics) of the other. Since the circumstances of the Jews were constantly changing for the worse, courts could now nullify almost any mixed marriage when desired by an Aryan partner. After 1941 this meant the almost certain death of the Jewish partner. Attributes the attitude of the judges, who enjoyed relative autonomy in the Nazi state (no judge was ever seriously penalized for his verdicts), to ideological solidarity and careerism. *RW*

0621 Heutger, Nicolaus: *Juedische Spuren in Niedersachsen*. Muenster: LIT, 1997. 160, [17] pp.

Surveys the history of the Jews in Lower Saxony from the beginning of their settlement there in 1074 until the Holocaust. Pp. 86-130 discuss the persecution of the Jews during the Third Reich. Deals with discrimination, the effects of the Nuremberg Laws, and the "Kristallnacht" pogrom, during which many Jewish men were deported to Buchenwald. Following the pogrom, most of the Jews attempted to emigrate. Mentions activities of clergymen who defended Jews and Judaism in their sermons and supplied food and aid, some of whom were imprisoned for their actions. In September 1941, 600 Jews of Hannover were crowded into 16 "Jewish houses"; the same occurred in all of the towns of the region. At the end of 1941 deportations began. Almost half of the 5,000 Jewish citizens of Hannover were murdered in the Holocaust, the same numerical proportion as everywhere in Lower Saxony. *ID*

0622 Hofstein, Irene S.: Irene: *Chronicle of a Survivor*. New York: Shengold, 1997. 177 pp.

A family chronicle, interspersed with personal reflections on Germany and its Jews. Hofstein was born in 1921 in Berlin to a Jewish family (Schlesinger); in July 1939 she and her mother emigrated to the USA. Her father survived the war in the Netherlands, to which he had emigrated in 1933, with his second (non-Jewish) wife. Hofstein's maternal grandmother, Jenny Pelz, died in Auschwitz in 1944. Pp. 42-108 present correspondence of Hofstein and her mother with close relatives left behind in Germany, especially her grandmother. Hofstein's recollections and her relatives' letters portray, inter alia, Jewish life in Nazi Berlin. *DR*

0623 Hohmann, Joachim S., ed.: *Chronik der juedischen Schule zu Tann (Rhoen): Mit einer kurzen Geschichte der israelitischen Gemeinde und zeitgenoessischen Lichtbildern.* Frankfurt a.M.: Peter Lang, 1997. 189 pp.

Presents the texts of documents from the Jewish school in Tann, Germany, between 1879-1934. Pp. 127-132 deal with the events in 1933-34, as well as pp. 185-188 of the "Kurze Geschichte..." (pp. 162-189). 84 Jews lived in Tann in 1934, 15 of whom were registered as pupils in the Jewish community school. During that year, 16 Jews emigrated; by 1939 the number of emigrants rose to 56. 18 Jews from Tann were deported to concentration camps in 1942-43, and murdered there. The fate of two persons is unknown. *ID*

0624 Housden, Martyn: Intentionalist versus Functionalist History: Where Do Historians Look for the Origins of the Holocaust? *Modern History Review* 10, 2 (Nov 1998) 9-12.

Discusses whether Nazi antisemitic policies were pre-planned or developed during the war. Intentionalist historian Hermann Graml describes four phases of Nazi policy – the removal of Jews' legal rights, their isolation through boycott and the Nuremberg Laws, expropriation of their property, and a "hate campaign which paved the way for genocide." He believes that these phases were generally preconceived and introduced gradually for tactical reasons. On the other hand, functionalist Hans Mommsen sees Nazism as a propaganda movement, devoid of ideology or political plans, in which party organizations and leaders introduced increasingly anti-Jewish policies in their competition for Hitler's favor. Examines intentionalist and functionalist views on the Nuremberg Laws and the "Kristallnacht" pogrom. Housden combines the two approaches, concluding that "a flexible form of intentionalism [was] able to take account of the chaos of the Nazi state." *RE*

0625 Housden, Martyn: *Resistance and Conformity in the Third Reich*. London: Routledge, 1997. xi, 199 pp.

Intended for use in teaching history in schools and universities, presents documents and excerpts from works by prominent historians of Nazism, accompanied by commentary. Ch. 6 (pp. 117-136), "Germany's Jewish Citizens: Like Lambs to the Slaughter?," deals with the reactions of German Jews, as well as Jews in Nazi-occupied Europe, to the Nazi persecution. The reactions varied from passivity and compliance to active resistance. Ch. 7 (pp. 137-160), "Exploring the Inexplicable: What Was the Relationship between Ordinary Germans and Racial Policy?," examines the attitudes of Germans toward racial antisemitism before, during, and after the Holocaust. *DR*

0626 Hutton, Christopher M.: *Linguistics and the Third Reich: Mother-Tongue Fascism, Race and the Science of Language*. London: Routledge, 1999. x, 416 pp.

Many of the ideas of pre-1945 mainstream linguistics had a political impact and, in particular, contributed to the formation of Nazi ideology. Deals with the ideology and cult of the mother-tongue, which in Nazi Germany was no less important than the cult of race. For the Nazis, language embodied the cultural and ethical values of the people and served as a boundary protecting them from assimilation. In the framework of this theory, Jews were a special case because they were regarded as lacking a sense of loyalty to their mother-tongue and having survived due to their uniquely strong racial instinct. Moreover, they seemed to constitute a threat to the German people because they lacked a healthy relationship with the German language and were bearers of universalizing ideologies, dangerous in the post-Versailles world. Dwells on German scholarship on the Yiddish language before 1945, especially on the views of two scholars – Lutz Mackensen and Peter-Heinz Seraphim – for whom Yiddish served as a demonstration of Jewish racial alienation and linguistic perversity. *DR*

0627 Ittner, Jutta: "Merkwuerdig unjuedisch?" Identitaet und Antisemitismus in Martin Gumperts Autobiographien. *Exil* 19, 1 (1999) 5-22.

Gumpert, a German-Jewish physician and writer for whom assimilation was almost an ideology, wrote a series of memoirs about his life before and after his emigration to New York in 1936. While he describes the Nazi takeover and Nazi brutality with great perceptiveness, he shows a strange emotional distance in his very brief references to antisemitic excesses and to measures that must have affected him personally. Attributes this partly to the deceptive inconsistency and ambiguity of early Nazi policy, but mainly to the need of assimilated Jews of his class to feel themselves wholly German, and to shut their eyes to the antisemitism about them – antisemitism that belonged to "uncultured" people and therefore was outside their "cultured" world. *RW*

0628 Jacob, Werner; Otto, Norbert: *Ich trage die Nummer 104953: Ein letztes Zeugnis: Werner Jacob, Lenhausen, im Gespraech mit Norbert Otto.* Olpe: Kreisarchiv, 1997. 208 pp. On title page also: Mit einer Einfuehrung zur Geschichte der Juden im Kreis Olpe von Christiane Mirgel und Dieter Troeps.

The introduction (pp. 9-35), surveys the history of the Jews in Olpe, Westphalia, from the beginning of their settlement there in the 13th century, with a focus on the Holocaust period (pp. 20-35). Pp. 53-186 contain the interview (held in 1989) with Werner Jacob, born in Lenhausen in 1920, describing his experiences during the Holocaust. Jacob and his father were arrested after the "Kristallnacht" pogrom and imprisoned for several months; his brother was deported to Buchenwald and shot on 9 November 1939. His parents and sister were deported in April 1942 to Zamość and murdered there. Jacob was taken for forced labor near Lenhausen; in 1943 he was deported to Auschwitz and put to work in the Buna plant. In 1945 he was taken to Buchenwald, but escaped. He returned to Lenhausen; he died in November 1992. *ID*

0629 Jersak, Tobias: Die Interaktion von Kriegsverlauf und Judenvernichtung: Ein Blick auf Hitlers Strategie im Spaetsommer 1941. *Historische Zeitschrift* 268, 2 (Apr 1999) 311-374.

Argues that Hitler's declaration in January 1939 that a world war would mean the extinction of the Jews exactly describes his choice of strategy. As long as he believed that he could attain world domination by defeating the European states and then the U.S., one by one, he planned that the Final Solution to the Jewish question (apparently by deportation, not extermination) should take place only after the war. At this stage, killing of Jews was relatively limited. On 15 August 1941, Hitler was alarmed by the Atlantic Charter and Roosevelt's and Churchill's message of support for the USSR. He imagined that his prophecy was coming true: a Jewish conspiracy was uniting the nations in a world war against him. World domination no longer seemed possible, and Hitler set more limited objectives for the war. Simultaneously, he ordered an immediate start to the extermination of the Jews. Thereafter, the order of precedence – pursuit of the war or extermination of the Jews – fluctuated with Hitler's perception of his chances for victory; but generally the greater emphasis remained on the extermination of the Jews. *RW*

0630 Jonca, Karol: Nazistowska polityka "ostatecznego rozwiązania kwestii żydowskiej" [The Nazi Policy for the "Final Solution of the Jewish Question"]. *Studia nad Faszyzmem i Zbrodniami Hitlerowskimi* 22 (1999) 229-284.

Discusses consecutive steps of the Nazi anti-Jewish policy between 1933-43. Various measures were adopted to separate the strongly assimilated German Jews from German society from 1933. The Nuremberg Laws caused the emigration of part of Germany's Jews. Deals with difficulties related to the emigration process, which was totally halted in October 1941. Describes eviction of Jews from their homes, and the deportation of German Jews to various places before and after the Wannsee Conference. Discusses the decisions taken at the Wannsee Conference on 20 January 1942. States that the German Jews were nearly totally annihilated; the number of survivors is not known. The exact date of the decision to exterminate the Jews is not known; there is much speculation about it in German historiography. Appendices (pp. 252-284) contain a list of the Jews from Wrocław deported to Kovno in November 1941, a list of those deported to the Lublin district in April 1942, and the protocol of the Wannsee Conference. *HV*

0631 *Justice* [Special Issue] (Fall 1999).

Contains lectures delivered at the conference "Remember Berlin" commemorating Jewish lawyers and jurists who perished in the Holocaust and their contribution to the law in their respective countries, convened by the International Association of Jewish Lawyers, the Berlin Bar Association, and the German-Israeli Jurists Association in Berlin, June 1999. Partial contents: Ben-Itto, Hadassa: Hope's Roots Lie Not in Ignoring the Past, but in Remembering It, in Studying the Lessons It Taught Us (2-4); Dombek, Bernhard: We Were Determined to Publish the List of Lawyers Who Suffered Anti-Semitic Persecution from 1933 [On the list of 1,785 names compiled and published as a book for this conference.] (5-7); Franssen, Everhardt: The Expulsion of the Jewish Lawyers Was at the Same Time an Extirpation of the Culture out of the Courtrooms (8-11); Ladwig-Winters, Simone: The Role of Jewish Lawyers in the Development of Law in Germany, and Their Subsequent Destiny until 1945 [Pp. 24-28 deal with the period 1933-45.] (21-29); Heinsohn, Gunnar: Why Was the Holocaust Different from All Other Genocides? [Argues that the rationale for the Holocaust was rooted in

Hitler's aim to delete from the German mind the Jewish code of ethics.] (30-32); Krach, Tillmann: The Expulsion of Jewish Lawyers from Their Profession (40-44); Koenigseder, Angelika: The Role of the Bar Association of Berlin in the Exclusion of Jewish Colleagues (45-52); Bugge, Steinar: Erich Meyer: Lawyer and Notary of Jewish Origin in Berlin 1933-1943 (59-63). *SSC*

0632 Kamber, Richard: Goldhagen and Sartre on Eliminationist Anti-Semitism: False Beliefs and Moral Culpability. *Holocaust and Genocide Studies* 13, 2 (Fall 1999) 252-271.

In "Hitler's Willing Executioners," Goldhagen advances the "cognitive" thesis that Germans who participated in the Holocaust were motivated primarily by their false belief that Jews, by their very nature, were harmful to Germany and thus deserved to be eliminated. Paradoxically, this thesis mitigates the moral culpability of the perpetrators – those who believed that they killed for the well-being of their country cannot be accused to the same degree as opportunistic killers. Besides, Goldhagen's thesis fails to explain how "ordinary" perpetrators could sustain their false beliefs in the face of horrific contrary evidence. Sartre, in whose conception antisemitism is a passion rather than an idea, and who considers knowing and not-knowing to be a matter of individual free choice, provides a better explanation of how the perpetrators could sustain their belief and action and preserve moral culpability. *DR*

0633

קראי, פליציה: "פגזים וחרוזים": הפעילות התרבותית של אסירות מחנה האסאג-לייפציג. ילקוט מורשת 64 (נוב' 1997) 67-78.

[Karay, Felicja: "Bombs and Beads": The Cultural Activity of the Women Imprisoned in the Camp HASAG-Leipzig. *Yalkut Moreshet* 64 (Nov 1997) 67-78.]

Describes cultural activities initiated by Jewish women from Poland who were taken for forced labor to the HASAG-Leipzig labor camp, a branch camp of Buchenwald, established in June 1944. Focuses on the activities of women transferred from the Skarżysko-Kamienna labor camp. Includes extracts in Polish and in Hebrew from songs and poems describing individual and collective camp experiences. *LFo*

0634 Keval, Susanna: *Die schwierige Erinnerung: Deutsche Widerstandskaempfer ueber die Verfolgung und Vernichtung der Juden*. Frankfurt a.M.: Campus, 1999. 291 pp. Based on the author's diss. – Universitaet Frankfurt, 1998.

A study based on a sequence of discussion groups held between January-March 1992 in Frankfurt with former German resistance fighters of different political and religious backgrounds, examining their attitude towards the persecution and annihilation of the Jews in the Holocaust at the time. States that although the topic was a focal point in Nazi politics, it occupied no central place in the resistance. Presents excerpts from the discussion groups, relating also the experiences of some of them with Jewish prisoners in concentration camps, latent antisemitism, failed rescue actions, experiences of fear, etc. Argues that, throughout the discussion, the notion of "resistance" remains vague vis-à-vis the Nazi persecution of the Jews. *ID*

0635 Kingreen, Monica, ed.: *"Nach der Kristallnacht": Juedisches Leben und antijuedische Politik in Frankfurt am Main, 1938-1945*. Frankfurt a.M.: Campus, 1999. 476 pp.

Contents: Stein, Harry: Das Sonderlager im Konzentrationslager Buchenwald nach den Pogromen 1938 (19-54); Kingreen, Monica: Von Frankfurt in das KZ Dachau: Die Namen der im November 1938 deportierten Maenner [Pp. 60-89 contain a list of 534 men, with data about them.] (55-89); Krohn, Helga: "Holt sie raus, bevor es zu spaet ist!" Hilfsaktionen zur Rettung juedischer Kinder zwischen 1938 und 1940 (91-118); Kingreen, Monica: Zuflucht in Frankfurt: Zuzug hessischer Landjuden und staedtische antijuedische Politik (119-155); Maierhof, Gudrun: Selbsthilfe nach dem Novemberpogrom: Die Juedische Gemeinde in Frankfurt am Main 1938 bis 1942 (157-186); Becker, Hanna: "...das Leben in die Tiefe kennengelernt...": Martha Wertheimer und ihr Wirken nach der "Kristallnacht" [On the Jewish writer and educator, b. 1890 in Frankfurt, deported to Sobibór in June 1942.] (187-210); Becht, Lutz: "Die Wohlfahrtseinrichtungen sind aufgeloest worden...": Vom "staedtischen Beauftragten bei der Juedischen Wohlfahrtspflege" zum "Beauftragten der Geheimen Staatspolizei..." 1938 bis 1943 (211-236); Eichler, Volker: Das "Judenreferat" der Frankfurter Gestapo (237-258); Gruner, Wolf: Der "Geschlossene Arbeitseinsatz" und die Juden in Frankfurt am Main von 1938 bis 1942 [On forced labor.] (259-288); Hebauf, Renate: Frankfurt am Main, Gausssstrasse 14: Ein Haus und seine juedischen Bewohner zwischen 1911 und 1945 (289-317); Daub, Ute: Die Stadt Frankfurt am Main macht sich "judenfrei": Zur Konzentrierung, Verbannung und Ghettoisierung der juedischen Bevoelkerung zwischen 1938 und 1943 (319-355); Kingreen, Monica: Gewaltsam verschleppt aus Frankfurt: Die Deportationen der Juden in den Jahren 1941-1945 (357-402); Opfermann, Charlotte Guthmann: "Im Hermesweg": Zur Taetigkeit in der Bezirksstelle der Reichsvereinigung in Frankfurt am Main von November 1942 bis Juni 1943: Ein Zeitzeugenbericht (403-413); Flesch-Thebesius, Marlies: "Wir sassen zwischen allen Stuehlen": Als Mischling zweiten Grades in

Frankfurt am Main (415-433); Kugelmann, Cilly: Befreiung – und dann? Zur Situation der Juden in Frankfurt am Main im Jahr 1945 (435-456); Heuberger, Georg; Drummer, Heike; Zwilling, Jutta: "Ensemble der Erinnerung": Gedenkstaette Neuer Boerneplatz, Namenfries am Alten Juedischen Friedhof, Datenbank im Museum Judengasse (457-469). *ID*

0636 Kirchheimer, Gloria DeVidas; Kirchheimer, Manfred: *We Were So Beloved: Autobiography of a German Jewish Community*. Pittsburgh, PA: University of Pittsburgh Press, 1997. xx, 367 pp.

Presents slightly edited texts of interviews conducted in 1981-82 among Jewish refugees from Nazi Germany who settled in the Washington Heights district of New York City. Part of the interviews were also included in Manfred Kirchheimer's documentary film "We Were So Beloved" (1986). Some of the interviewees left Germany before the "Kristallnacht" pogrom, some left immediately after, some failed to emigrate and were interned in ghettos and camps. They talk about their experiences, but at the same time insist that their German identities set them apart from others, e.g. Polish Jews. The book includes interviews conducted by Kirchheimer in the Saar region of Germany in 1991. Part of his interviewees, Saar Jews, emigrated in the Nazi period but returned after the war. Like them, Kirchheimer feels that he gradually regained his prewar German identity. Pp. 331-350 contain an afterword by Dan Bar-On, expressing doubts that pre-Hitler Germany was a loving homeland for the Jews and that the German identity is worth re-establishing. *DR*

0637 Klemperer, Victor: *To the Bitter End: The Diaries of Victor Klemperer, 1942-1945*. Trans.: Martin Chalmers. London: Weidenfeld & Nicolson, 1999. xv, 538 pp. The first volume (covering the years 1933-41) appeared in 1998. On title page also: Abridged and translated from the German edition. Originally published as "Ich will Zeugnis ablegen bis zum letzten: Tagebuecher" (Berlin: Aufbau, 1995).

0638 Klessmann, Eckart: M.M. *Warburg & Co.: Die Geschichte eines Bankhauses*. Hamburg: Doelling und Galitz, 1998. 206 pp.

Traces the history of this private bank (with headquarters in Hamburg) from its founding in 1798 to the present. Pp. 59-114 cover the Weimar and Nazi periods. Max M. Warburg, head of the bank during this period, served in numerous local and national commissions, including the German delegation to the Versailles peace talks; but as Jews, he and his colleague Carl Melchior were blamed for signing the reparations agreement (which in fact they had opposed) and received threats from antisemitic nationalists. After 1933 Warburg was expelled from all committees and it became dangerous even to greet him, but he felt himself a German and refused to close the bank and emigrate. However, he immersed himself in Jewish community work, especially funding emigration. He finally emigrated in 1938, after transferring the bank to men he could trust. In the following years, the Nazis forced the dismissal of Jewish employees and gradually confiscated all Jewish deposits and investments in the bank. *RW*

0639 Kohl, Christiane: *Der Jude und das Maedchen: Eine verbotene Freundschaft in Nazideutschland.* 3rd ed. Hamburg: Spiegel-Buchverlag, 1998. 384 pp. First published in 1997.

Relates an incident in 1935, in which the tenants of an apartment building in Nuremberg falsely suspected an Aryan girl of having an affair with a Jew, Leo Katzenberger. They denounced him, and he was tried and sentenced to death in March 1942. Surveys the events which occurred with the rise of the Nazi regime in Nuremberg: discriminations against Jews, boycott, the "Kristallnacht" pogrom, Aryanization of Jewish property, deportations. In the first transport on 25 November 1941, 512 Jews from Nuremberg were deported to Riga; 16 survived. From the second deportation to Izbica (near Lublin) in March 1942, which included Katzenberger's wife and brother, none of the 426 deportees survived. *ID*

0640 Kortlaender, Bernd: Le poète inconnu de la "Loreley": Le médiateur supprimé. *Romantisme* 101 (1998) 29-40.

Describes the reception in Germany of Heine's poem "Lorelei." It was written in 1823, but the text became very popular due to its musical adaptation by Friedrich Silcher in 1838. "Lorelei" became a part of German national culture. Traces the attitudes of the German bourgeoisie to Heine, which, although they accepted him, considered his views as too ironic toward tradition. Both antisemitic circles and the progressive intellectual elite held the same view, that Heine corrupted German poetic language. Quotes German literary critics, from the 19th century through the Nazi period, who saw Heine as someone who does not belong to the German community and who corrupts German people. Focusing on the Nazi period, relates that his writings disappeared from school and university textbooks. It was difficult to totally erase such a popular song as "Lorelei" from German culture, but a solution was found: "Lorelei" was published with a note that the author is anonymous. Mentions that in Nazi-occupied France, Heine's name did not appear in publications on German literature. *HV*

0641 Kosmala, Beate: Rettung von Juden im nationalsozialistischen Deutschland 1933-1945: Ein Forschungsprojekt am Zentrum fuer Antisemitismusforschung der Technischen Universitaet Berlin. *Jahrbuch der Historischen Forschung in der Bundesrepublik Deutschland 1998* (1999) 36-44.

Surveys previous research on assistance to Jews in Nazi Germany and occupied Europe, noting its fragmentary character. Discusses issues of definition (help due to humane motives or as political resistance vs. help in exchange for work or money); sources (interviews, official Nazi and other archives, memoirs and other published works; the research team attempts to document each case from more than one source); periodization (contact with Jews became a punishable offense in October 1941); regional differences; and characteristics of helpers, their political and religious affiliation, and whether they acted alone or as part of a group. The project will cover all of Germany (in its pre-1937 borders) and will be available in a data bank. *RW*

0642 Kreuter, Maria-Luise: Rettung von Juden im nationalsozialistischen Deutschland 1933-1945: Ein Dokumentationsprojekt mit Datenbank am Zentrum fuer Antisemitismusforschung der Technischen Universitaet Berlin. *Zeitschrift fuer Geschichtswissenschaft* 46, 5 (1998) 445-449.

Reports on a project undertaken by the Zentrum fuer Antisemitismusforschung in 1997. The purpose of the project is to cover systematically all known cases of rescue and to make accessible heretofore unknown cases. The plan calls for a table of contents, indexes of places, names, and additional data on the rescuers and the rescued, and evaluation of the data. *ID*

0643 Lange, Thomas, ed.: *"L'chajim": Die Geschichte der Juden im Landkreis Darmstadt-Dieburg.* Darmstadt: Landkreis Darmstadt-Dieburg, 1997. 370 pp.

The following articles deal with events in the region before and during the Nazi period: Reinhold-Postina, Eva: Verbrannt, verwuestet, vergessen und verdraengt: Auf der Suche nach alten Synagogen im Landkreis Darmstadt-Dieburg (61-105); Lange, Thomas: Judenfeindschaft und Antisemitismus zwischen 1890 und 1933 – auch eine Spurensuche (139-168); Steffens, Gerd: Ausgrenzung, Verfolgung, Enteignung, Deportation, Vernichtung: Die Leidensgeschichte der juedischen Bevoelkerung im Gebiet des heutigen Landkreises Darmstadt-Dieburg 1933 bis 1945, mit einem Rueckblick auf die zwanziger Jahre (169-270). *ID*

0644 *Leben und Schicksale juedischer Buerger aus Coesfeld: Texte und Zeugnisse der Ausstellung, 29.10-13.11.1998, ehemalige Synagoge.* Coesfeld: Das Stadtarchiv, 1999. [ca. 280] pp.

Presents photographs and facsimiles of documents shown in an exhibition of 30 posters describing the history of the Jews in Coesfeld from the Middle Ages to post-World War II. Posters 3, 4, and 9 depict antisemitism in the medieval, early modern, and modern periods. Posters 10-27 deal with the Holocaust, describing anti-Jewish measures between 1933-44, the systematic exclusion of Jewish residents from public life, the "Kristallnacht" pogrom, and Nazi legislation. 26 Jews emigrated before 1939. Relates the murder of mentally-ill Jews in September 1940, among them four from Coesfeld, and discusses the deportation of 19 Jews in December 1941 to Riga. Altogether, 53 Jews from Coesfeld were murdered in the Holocaust. *ID*

0645 Lichtenstein, Heiner: Betrifft: "Aktion 3": Wie juedische Nachbarn "verwertet" wurden. *Tribuene* 149 (1999) 31-36.

Describes the auctions of the property of Jews deported from Germany, which permitted many individual Germans and institutions, as well as the authorities who conducted the auctions, to enrich themselves. Argues that this contradicts the claim that most Germans did not know what was happening to the Jews. The auction records were found recently by Prof. Wolfgang Dressen and shown in an exhibition. *RW*

0646 Lindemann, Gerhard: *"Typisch juedisch": Die Stellung der Ev.-luth. Landeskirche Hannovers zu Antijudaismus, Judenfeindschaft und Antisemitismus 1919-1949.* Berlin: Duncker & Humblot, 1998. 1037 pp. Based on the author's diss. – Universitaet Heidelberg, 1997.

Even before 1933, the heads of the Hannover Lutheran Church were conservative and antisemitic. A minority, influenced by Barth, fought antisemitism. In the early 1920s, the Church tolerated the antisemitic activities of Pastor Muenchmeyer on the resort island of Borkum. Under the Nazis, the Hannover Church remained affiliated with the Confessing Church; but its head, Bishop August Marahrens, sympathized with Nazi ideology and held to the Lutheran "two-realms" doctrine. He did not protest against antisemitic measures or the "Kristallnacht" pogrom, and gave no effective backing to the four ministers of part-Jewish descent who were forced to retire. (The Hannover Church did not adopt an "Aryan paragraph" but rather a regulation enabling it to remove clergy whose positions had become "untenable.") An "Aryan" pastor who aroused the wrath of local Nazis by his pro-Jewish stance was transferred to a remote rural parish. The majority of parishioners stood up for their pastors but were ignored by the Church. In December 1941 the Church excluded "non-Aryans" from worship with "Aryans." After the war, only one of the retired ministers was reinstated, and only after a struggle. Suggests that the Church would have seen any restitution for its "non-Aryan" members as a confession that it had done wrong. But Marahrens was forced to retire in 1947. *RW*

0647 Loew, Andrea; Robusch, Kerstin: "Es lebe das Leben!" Die Freimarks: Eine deutsch-juedische Familie. *Sachor* 9 (1999) 37-48.

Traces the lives of the Freimark family from Bochum. Before the Nazi period they were German patriots; Simon was a World War I invalid. He was manager of a textile firm; when the firm was Aryanized, the new owner kept him on. In 1938 the parents sent the two teen-age children to an uncle in America. They themselves were reluctant to emigrate; when they finally tried, it was too late. In 1942 they were deported to Theresienstadt. Describes their life and survival in the ghetto and later in a DP camp, as they told it, with the utmost reserve, to their children, whom they were able to join in 1946. *RW*

0648 Marssolek, Inge; Davids, Wiebke, eds.: *"Man haengt immer zwischen Himmel und Erde...": Juedische Emigrantinnen und Emigranten (1933-45) aus Bremen berichten.* Bremen: Selbstverlag des Staatsarchivs Bremen, 1997. 235 pp.

In the framework of a project spurred by a seminar given by Marssolek dealing with the emigration of non-prominent victims of the Nazi regime between 1933-39, a questionnaire was sent in 1996 to several such persons. Presents the responses (seven of them in English), edited and annotated, of 15 Jews, former residents of Bremen, who now live in Great Britain, Israel, Peru, and the USA. They relate, also, their experiences in Germany before their emigration. The respondents are: Elise Aber Aftergut, Rachel Aber Schlesinger, Ingrid Esslinger Altman, Walter Baer, Martin Bialystock, Lisa Markus Cohen Grubel, Peter and Walter Jacobsohn, Rita Posnansky Kahn, Charlotte Abraham Levy, Joachim Markreich, Rosel Lubleski Pinthus, Werner Renberg, Margaret Fink Rosenbaum, David Wiener. *ID*

0649 Matthaeus, Juergen: Kampf ohne Verbuendete: Der Centralverein deutscher Staatsbuerger juedischen Glaubens 1933-1938. *Jahrbuch fuer Antisemitismusforschung* 8 (1999) 248-277.

Using the recently found files of the Centralverein (CV) in the Moscow archives, describes changes in the organization's ideology, aims, and activities between 1933-35. At first the CV hoped it would be possible to come to an arrangement with the new regime. It issued declarations of loyalty and condemnations of the "Greuelpropaganda" spread by Jews abroad. Even the boycott of April 1st did not change this position. However, the CV began to cooperate more closely with other Jewish organizations. While others helped Jews to emigrate, the CV tried to aid those who stayed behind, using its contacts in the bureaucracy and appealing to the courts on behalf of Jews dismissed from their jobs, refused service in shops, and otherwise discriminated against beyond what the government decreed. Lack of clarity about the government's priorities and the competencies of the various agencies made this work difficult. The Nuremberg Laws convinced some of the leadership that the CV must remove from its name and its platform the identification of the Jews as Germans; but this was not done until 1936. *RW*

0650 Mehnert, Gottfried: Wilhelm Freiherr von Pechmann: Sein Eintreten fuer die Juden 1933-1945. *Freiburger Rundbrief* 6, 4 (1999) 241-249.

Pechmann (1859-1948), a layman who had filled leading roles in governing bodies of the German Evangelical Church, tried again and again between March 1933-November 1941 to induce it, preferably jointly with the Catholic Church, to issue a clear statement against the persecution of the Jews. His efforts remained unsuccessful. In April 1934 he gave up his membership in the Church; in an open letter he blamed the violation of the Church's independence and its silence in the face of "all the grief and suffering brought into countless 'non-Aryan' hearts and homes, Christian and Jewish." *RW*

0651 Meyer, Beate: *"Juedische Mischlinge": Rassenpolitik und Verfolgungserfahrung, 1933-1945*. Hamburg: Doelling und Galitz, 1999. 494 pp. Based on the author's diss. – Universitaet Hamburg, 1998.

Examines the fate of the "Mischlinge" (part-Jews) in the Nazi period. Analyzes the prohibition of mixed marriages under Nazism, the development of possibilities to avoid the stigma and persecution (recognition of a "special status" for some persons), and the policy as carried out in practice – actions and reactions. Pp. 266-371 describe interviews conducted with 61 former "Mischlinge," including excerpts from the interviews. Notes that the number of "Mischlinge" who chose to emigrate, go into hiding, escape, or resist, was small. Enumerates reasons for this fact, including their belonging to the middle class; their belief that they could maintain their place in society through education, competence, and efficiency; their being extremely law-abiding and adaptable to changing rules; their ability to obtain various exceptions to the regulations; and the hope of males for re-integration in 1939 by being drafted into the army. *ID*

0652 Michalka, Wolfgang, ed.: *Vtoraya mirovaya voina: Diskussii, osnovnye tendentsii, rezultaty issledovanii* [The Second World War: Discussions, Main Tendencies, Results of Research]. Moskva: Ves Mir, 1996. 700 pp. A translation of "Der Zweite Weltkrieg: Analysen, Grundzuege, Forschungsbilanz" (Muenchen: Piper, 1989).

Partial contents: Pingel, Falk: Uzniki kontsentratsionnykh lagerei: Unichtozheniye i prinuditelnyi trud [Concentration Camp Inmates: Annihilation and Productive Labor] (619-630); Moghareh-Abed, Hamid: Rasovaya gigiyena, ili yevgenika: Ideologicheskaya predraspolozhennost i prakticheskaya motivatsiya k genotsidu [Racial Hygiene, or Eugenics:

Ideological Propensity or Practical Motivation for the Genocide] (631-644); Petter, Wolfgang: Natsional-sotsialisticheskaya evtanaziya: Istoki i predely [National-Socialist Euthanasia: Origins and Limits] (645-655); Kampe, Norbert: "Okonchatelnoye resheniye" cherez ekspatriatsiyu? Protivorechivye tselevye ustanovki antisemitskoi politiki do 1941 g. ["Final Solution" through Expatriation? The Contradictory Aims in Antisemitic Policy before 1941] (656-669); Madajczyk, Czesław: Sushchestvuyet li sinkhronnost mezhdu "Generalnym planom Ost" i "Okonchatelnym resheniyem yevreiskovo voprosa"? [Is There Synchronism between the "Generalplan Ost" and the "Final Solution of the Jewish Question"?] (670-680). *DR*

0653 Minczeles, Henri: Quelques réflexions sur la responsabilité collective des allemands. *Revue d'Histoire de la Shoah* 167 (Sept-Dec 1999) 102-131.

Reflects on German responsibility for the Shoah, beginning with the dichotomy of individual vs. collective guilt. Discusses the theory of collective guilt by comparing those actively involved in the genocide to the silent majority. Divides the notion of responsibility into four categories: legal, political, moral, and metaphysical. Poses the question whether there is a German cultural predisposition to antisemitism. Analyzes the role of antisemitism as a fundamental aspect of Nazi ideology. Discusses the "historians' debate" of the 1980s. Concludes that the Nazi era could be described as the epitome of a modern process of developing technology while simultaneously demonstrating moral regression. Believes that the concept of "lest we forget" means neither revenge nor pardon but rather memory as synonymous with the constant need for the preservation of human dignity. *TS*

0654 Monteath, Peter: The Fuehrer's Decision. *History Today* 48, 9 (Sept 1998) 4-6.

Expounds the viewpoint of Christian Gerlach ["WerkstattGeschichte" 18 (1997)], according to which Hitler indeed gave a verbal order for the murder of all European Jews. Gerlach dates the announcement of this decision to the entry of the USA into the war as a result of the attack on Pearl Harbor. In Hitler's view, it was the Jews who turned the European conflict into a world war, and they had to be punished. Gerlach states that the initial aim of the Wannsee Conference (planned for 9 December 1941) was to solve the question of the fate of German Jews. Gerlach's viewpoint was criticized by Hans Mommsen. *DR*

0655 Mueller, Ingo: Juristen und die nationalsozialistische Judenverfolgung. *Trumah* 8 (1999) 63-72.

States that after Jewish jurists were removed from office and their books from the libraries, Nazi law went its own way. National and local authorities issued 1,973 anti-Jewish laws, ordinances, guidelines, and decrees, which were no longer subject to judicial review. Moreover, in the absence of a relevant law, courts often based anti-Jewish decisions on conformity to Nazi racial ideology. In 1936 Jews were ruled to be without rights, i.e. legally dead. The term "sexual relations" in the context of the Nuremberg Laws was interpreted so broadly that even exchange of letters might violate the law; and the Jewish culprit was automatically sent to a concentration camp after his release from prison. In 1943 all matters concerning Jews were transferred from the courts to the police. *RW*

0656 Mueller-Wesemann, Barbara: *Theater als geistiger Widerstand: Der Juedische Kulturbund in Hamburg 1934-1941*. Stuttgart: M & P – Verlag fuer Wissenschaft und Forschung, 1997. 579 pp. Based on the author's diss. – Universitaet Hamburg, 1995.

Surveys the activities of the Juedischer Kulturbund in Hamburg, founded in 1934 due to the dismissal of Jewish artists from cultural institutions. Ca. 5.3% of German Jews were involved in theatrical professions, such as writers, editors, actors, stage-managers, and producers. In 1933 there were 16,973 Jews in Hamburg. During its existence, the Kulturbund employed ca. 400 artists, clerks, and technical staff. In 1935-36, 57 artistic events took place in Hamburg: 28 theater productions, 16 concerts, 11 soirées, a Hanukkah event, and an event for youth. The Kulturbund functioned until the end of 1938, when it was liquidated by the German authorities. All Jewish artists had to register with the national federation of Juedische Kulturverbaende (in Berlin), the activity of which was controlled; in May 1939 a branch was also opened in Hamburg. At the same time all Jewish bookshops, publishing houses, and the press were closed down. With the increase in emigration, German authorities lost their interest in controlling Jewish cultural life, which had almost disappeared. On 11 September 1941 the federation was disbanded. *ID*

0657 Naarmann, Margit: *"Von ihren Leuten wohnt hier keiner mehr": Juedische Familien in Paderborn in der Zeit des Nationalsozialismus*. Koeln: SH-Verlag, [1997?]. 607 pp.

Pp. 25-102 present a history of the Jewish community of Paderborn during the Nazi period. In 1933 the Jewish community consisted of 310 members. Describes the Nazi anti-Jewish measures, the wave of emigration after the "Kristallnacht" pogrom, persecution and deportation. Pp. 103-506 contain the history of 127 Jewish families who lived in Paderborn and their fate in the Nazi period. Pp. 522-602 contain lists of names. *ID*

0658
נידרלנד, דורון: יהודי גרמניה – מהגרים או פליטים? עיון בדפוסי ההגירה בין שתי מלחמות העולם. ירושלים: הוצאת מאגנס, האוניברסיטה העברית, 1996. 257 עמ'.

[Niederland, Doron: *German Jews – Emigrants or Refugees? Emigration Patterns between the Two World Wars.* Jerusalem: Magnes Press, Hebrew University, 1996. 257 pp.] Based on the author's diss. – Hebrew University of Jerusalem, 1988.

During the interwar period in Germany, many Jews emigrated for economic reasons, especially in 1922-25 and 1930-32. Under the Weimar Republic, Jews emigrated because of inflation and unemployment; in the Nazi period, economic difficulties were caused by antisemitic legislation, e.g. limitations on employment, boycott, and Aryanization. Nazi policy, especially in regard to Jews buying foreign currency and taking it out of the country, made emigration difficult. Before the "Kristallnacht" pogrom, German Jews continued to hope that their situation would change; therefore, their main considerations in emigration were economic. Those who emigrated before 1938 were not refugees; their emigration was only partly forced. The desire of Jews to emigrate became widespread only after the "Kristallnacht" pogrom, when Nazi policy turned them into refugees. *RE*

0659 Omland, Sabine: *Zur Geschichte der Juden in Drensteinfurt, 1811-1941.* Warendorf: Kreisgeschichtsverein Beckum-Warendorf, 1997. 368 pp.

A history of the Jews in Drensteinfurt from the early 19th century to 1941. Pp. 152-216 deal with the persecution of the Jews during the Nazi period: the gradual stages of discrimination, the implementation of the Nuremberg Laws, the "Kristallnacht" pogrom, the Aryanization of Jewish property, emigration, and deportation. In 1928 the Jewish community consisted of 21 members; there is no data for 1933. Between 1933-39, 16 persons, including most of the Terhoch family, managed to emigrate to Uruguay. A deportation list of 1941 includes the names of 10 Jews from Drensteinfurt. Includes a report on the fate of Herta Herschcowitsch, who was deported to the Riga ghetto, and later to Stutthof, but survived; the fate of the other deportees is not known. *ID*

0660 Ott, Hugo: Martin Heidegger und sein ambivalentes Verhaeltnis zum Judentum. *Trumah* 8 (1999) 29-41.

Based on a lecture held at the Hochschule fuer Juedische Studien in Heidelberg, January 1998. Discusses the evidence for Heidegger's identification with Nazi antisemitism, especially in the negative opinions he submitted on Jewish candidates for university positions. As early as 1929, he added to his recommendation of a certain (non-Jewish) candidate the warning that German intellectual life must be preserved from "Verjudung." His phraseology, even when he did not expressly mention a candidate's Jewish origin, betrayed his antisemitism. With a few exceptions, he had no sympathy for the Jewish scholars dismissed, under Nazi law, during his rectorship at the University of Freiburg, and distanced himself from his teacher Husserl. Both Husserl and Jaspers judged him antisemitic. After the war he refused to admit guilt. *RW*

0661 Papen, Patricia von: *"Scholarly" Antisemitism during the Third Reich: The Reichsinstitut's Research on the "Jewish Question," 1935-1945.* Ann Arbor, MI: UMI Dissertation Services, 1999. 391 pp. Reproduction of the author's diss. – Columbia University, 1999.

Analyzes the activities of two quasi-academic institutions in Nazi Germany – the Reichsinstitut fuer Geschichte des Neuen Deutschland, founded in 1935 (and especially its Munich branch, the Forschungsabteilung Judenfrage, opened in 1936), and the Institut zur Erforschung der Judenfrage, established by Alfred Rosenberg in Frankfurt am Main in 1941. Their field was antisemitic historical research, and in 1935-44 they were engaged in legitimizing the end of Jewish emancipation and the Final Solution. Their experts were drawn from a wide political and religious spectrum; thus conservative, religious, and social anti-Jewish stereotypes may be traced in the publications of these institutions. Contends that the institutions bear responsibility for the implementation of the Final Solution because they underpinned it scholastically and in some cases influenced the decisions of the Nazi leadership (e.g. they characterized non-Ashkenazi Jews in occupied Europe as "racial Jews"). Dwells on the careers of such experts as K.A. von Mueller, W. Frank, W. Grau, and W. Euler. The Frankfurt Institute collaborated with anti-Jewish institutions abroad, and it participated in the confiscation of Judaica collections from many European libraries. *DR*

0662 Paucker, Arnold: *Deutsche Juden im Widerstand 1933-1945: Tatsachen und Probleme.* Berlin: Gedenkstaette Deutscher Widerstand, 1999. 47 pp. An expanded and updated edition of the booklet originally published as "Juedischer Widerstand in Deutschland" (Berlin: Gedenkstaette Deutscher Widerstand, 1989).

0663 Paul, Gerhard; Gillis-Carlebach, Miriam, eds.: *Menora und Hakenkreuz: Zur Geschichte der Juden in und aus Schleswig-Holstein, Luebeck und Altona (1918-1998).* Neumuenster: Wachholtz Verlag, 1998. 943 pp. On title page also: Eine gemeinsame Publikation des Forschungsprojektes "Zur Sozialgeschichte des Terrors" am Institut fuer schleswig-holsteinische Zeit- und Regionalgeschichte an der Bildungswissenschaftlichen Hochschule Flensburg – Universitaet (Schleswig) und des Joseph-Carlebach-Instituts an der Bar Ilan-Universitaet (Ramat Gan) Israel.

Partial contents: Goldberg, Bettina: "...und vieles bleibt ungesagt": Die Israelitische Gemeinde zu Kiel vor und nach 1933: Versuch einer Annaeherung (49-66); Gillis-Carlebach, Miriam: "...damit die Kinder in die Schule gehen" (Jer. Talmud): Die letzte Phase der Israelitischen Gemeindeschule in Altona (1928-1938) (121-133); Paul, Gerhard: "Die Juden sind unser Unglueck!" Zum Antisemitismus in einer fruehen Hochburg der NSDAP (183-204); Jakob, Volker: Wilhelm Spiegel: Jude – Anwalt – Sozialist: Das erste Mordopfer der antisemitischen Gewalt (205-213); Uhlig, Ralph: "...da die erforderlichen Messungen s.Z. nicht vorgenommen wurden": Die Verdraengung von juedischen Gelehrten an der Kieler Christian-Albrechts-Universitaet nach 1933 (215-235); Wieben, Matthias: "Aus Deutschland sind sie ausgespien fuer alle Zeiten": Die Vertreibung juedischer Studenten der Christian-Albrechts-Universtitaet zu Kiel (237-243); Goldberg, Bettina: "Am preiswertesten und am besten bei Gerstel": Die Geschichte einer Muetzenfabrik und der Vertreibung ihrer Besitzer [On the Gerstel cap factory in Kiel.] (245-263); Schreiber, Albrecht: "Ich seh schon, ich muss mir noch allein begroben": Das Ende der juedischen Gemeinde Luebeck (265-282); Kirschninck, Harald: "Wer beim Juden kauft, ist ein Volksverraeter!" Der Untergang der juedischen Gemeinde Elmshorn (283-295); Pingel, Fiete; Steensen, Thomas: "Es gab einmal Juden in Nordfriesland": Juedisches Leben und Antisemitismus in Friedrichstadt und im uebrigen Nordfriesland (297-315); Dettmer, Frauke: Hinaus aus der Festung! Der Niedergang der juedischen Kleingemeinden am Beispiel Rendsburg (317-330); Mussdorf, Torsten: "...ich beabsichtige, mich auf dem Hausboden zu erhaengen": Die Verdraengung juedischen Lebens in Bad Segeberg (331-343); Obst, Carsten: "...muss wegen seiner juedischen Abstammung die Gilde verlassen": Judenverfolgung in Neumuenster (345-352); Stokes, Lawrence Duncan: "...die Tretmuehle der Nazi-Buerokratie wurde in Schwung gehalten": Judenverfolgung in Eutin 1933-1945 (353-362); Lopau, Christian: "Die Stadt Ratzeburg hat ein erhebliches Interesse an dem Erwerb dieses Grundstuecks": Judenverfolgung im Kreis Herzogtum Lauenburg (363-368); Koch, Erich: Was nuetzt einem die Assimilation, wenn man Horwitz oder Weinberg heisst: Schicksale juedischer Familien in Schleswig (369-386); Warkus-Thomsen, Dirk: "Juedische Kinder gehoeren in juedische Heime": Von einem juedischen Kinderheim und einer "Judenaustreibung" in Wyk auf Foehr (387-396); Pritzlaff, Christiane: "Zu guten Juden und zu guten Deutschen": Die juedische Volksschule in Luebeck (1934-1940) (397-410); Philipsen, Bernd: "Atempause auf der Flucht in ein Leben mit Zukunft": Der Kibbuz auf Gutshof 'Jaegerslust' bei Flensburg (1934-1938) (411-424); Paul, Gerhard: "Nur Shanghai war noch offen": Der juedische Exodus aus Schleswig-Holstein (1933-1941) (437-458); Philipsen, Bernd: "...voellig ueberfluessige Versammlungshaeuser": Die Pogromnacht vom 9./10. November 1938 in Schleswig-Holstein (469-480); Paul, Gerhard: Klassenfoto 1939: Das Schicksal der Schueler und Lehrer der juedischen

Volksschule in Kiel (481-490); Paul, Gerhard: "Betr.: Evakuierung von Juden": Die Gestapo als regionale Zentralinstitution der Judenverfolgung (491-520); Yashek, Richard J.: "Ich habe immer noch vor Augen, wie der Schnee sich rot faerbte": Kindheitserinne rungen an Bad Schwartau, Luebeck und Riga [Memoirs of a German Jew, b. Juergen Jaschek in 1929 in Luebeck.] (521-529); Paul, Gerhard: "Es ist direkt zum Verzweifeln mit unserer Auswanderung": Aus den Briefen von zwei Luebecker Schwestern via Sibiria nach Shanghai (1939-1941) [On the sisters Berta and Dora Lexandrowitz, who wrote letters to Shanghai in an attempt to flee Germany; they failed, and both died in the Holocaust.] (531-540); Guttkuhn, Peter: "Sie sieht nicht sehr juedisch aus": Leben, Leiden und Sterben der Luebecker Lehrerin Emma Gruenfeldt (541-547); Gillis-Carlebach, Miriam: "Licht in der Finsternis": Juedische Lebensgestaltung im Konzentrationslager Jungfernhof [On Jews from Hamburg-Altona deported to the concentration camp near Riga, based on interviews.] (549-563); Bajohr, Frank: "...protzten ploetzlich mit einer neuen Wohnungseinrichtung": Schleswig-Holsteiner als materielle Nutzniesser der Shoah (565-571); Paul, Gerhard: "Nein, den kenn' ich, der ist Italiener!" Wie Menschen aus Schleswig-Holstein verfolgten Juden beistanden (573-589); Friess, Kaethe: "Es war ein Zug des Jammers": Erinnerungen an die Evakuierung juedischer KZ-Haeftlinge im April 1945 nach Kiel (591-603); Klingner, Klaus: Ignoranz statt Gerechtigkeit: Die schleswig-holsteinische Nachkriegsjustiz und die Judenverfolgung [On war criminals and war crimes trials.] (723-728); Bussenius, Sieghard: "Es muesste doch moeglich sein, diesen Friedhof zu schuetzen!" Schaendungen juedischer Friedhoefe (729-741); Kahl, Guenter: Brandnester: Die Anschlaege auf die Luebecker Synagoge [In March 1994 and May 1995.] (763-770); Kugler-Weiemann, Heidemarie: "Kennen Sie jemanden, der im KZ war?" Zur Behandlung des Holocaust in der Geschwister-Prenski-Schule Luebeck (771-783). Pt. 3 (pp. 605-719) deals with Jews who emigrated and their lives abroad (1930-1998). Pt. 4 (pp. 721-811) deals with the postwar period in Germany. Pp. 813-943 contain notes, a brief bibliography and indexes. *ID*

0664 Pine, Lisa: *Nazi Family Policy, 1933-1945*. Oxford: Berg, 1997. xii, 208 pp.

Discusses the relation of eugenics to Nazi policy in terms of the ideal "Kinderreich family" and of "asocials," such as Gypsies, prostitutes, homosexuals, criminals, etc. Ch. 5 (pp. 147-183), "The Jewish Family," deals with the impact of Nazi ideology, legislation, and persecution (including the "Kristallnacht" pogrom) on the Jewish family, e.g. changes in the roles of men and women, the breakup of the family with the emigration of some members (especially children), suicides, pauperization, etc. Concludes that no clear correlation is seen between the effects of the persecution of Jewish families and their responses before

1939, but that later, in the camps, parents were unable to protect their children and the latter lost respect for their parents. *YC*

0665 Pollmeier, Heiko: Inhaftierung und Lagererfahrung deutscher Juden im November 1938. *Jahrbuch fuer Antisemitismusforschung* 8 (1999) 107-130.

Based on ca. 200 memoirs, most of them unpublished (chiefly from the Wiener Library, London), describes the internment of German Jews in Sachsenhausen, Buchenwald, and Dachau in the wake of the "Kristallnacht" pogrom. Most of the memoirs, written immediately after emigration from Germany, reflect the difficulty of coping with these memories; often the experiences are described as though by an observer. Even more than from the extreme physical hardships, the Jews suffered from the brutality and humiliation at their reception, which was calculated to obliterate their identity as human beings and as individuals; the camp life, where they were treated "like cattle"; the reversal of all values; the unpredictability of punishment and death or, on the other hand, of release. States that this experience brought home to the victims that they must leave Germany, but did not change their previous attitude to their Jewishness. *RW*

0666 Recker, Klemens-August: *"Wem wollt ihr glauben?" Bischof Berning im Dritten Reich.* 2nd, rev. ed. Paderborn: F. Schoeningh, 1998. 528, [16] pp. First published in 1998.

Bishop Wilhelm Berning of Osnabrueck was known as the "Nazi bishop"; he was instrumental in carrying out Cardinal Bertram's policy of close relations with the Nazi regime and opposition to Nazi measures through confidential petitions rather than public protest. Argues that this image can be corrected through study of Berning's sermons, in which he spoke out against racism. After 1934 he progressively distanced himself from the regime and from Bertram. Pp. 307-353 analyze his attitude to the persecution of the Jews. He accepted traditional anti-Jewish doctrines and current theories of race. But he insisted that all human beings were created by God and that natural law demands respect for their rights and their lives. He attempted to intervene for non-Aryan Catholics and, as head of the St. Raphael Verein, to help them emigrate. After learning of the massacres of Jews and conditions in the concentration camps, he was among the initiators of a series of attempts to draft pastoral letters and petitions to the Fuehrer condemning the infringement of Church rights and of human rights. Bertram either blocked these initiatives or struck out the reference to Jews. Criticizes Berning's stand after the war: he denounced Nazi crimes but did not acknowledge the Church's and his own collaboration. *RW*

0667 Reller, Gisela; Grosinski, Klaus; Roder, Bernt, eds.: *Leben mit der Erinnerung: Juedische Geschichte in Prenzlauer Berg.* Berlin: Hentrich, 1997. 447 pp.

A collection of articles published to accompany an exhibition held in the Prenzlauer-Berg-Museum (Berlin) in 1997, tracing Jewish life in this district. The articles, based on oral or written testimonies, relate the fate of some of the survivors from amongst the ca. 18,000 Jews who lived in Prenzlauer Berg after 1933. The following survey Jewish life in the Nazi period, and deal with discrimination, worsening living conditions, Aryanization of Jewish property, emigration, forced labor, and deportation: Scheer, Regina: Evas Haus: Alice Loewenthal und ihre Toechter (16-24); Jung, Ulla: "Ich werde mich wehren": Werner Jacobowitz, ein Ueberlebender des Auerbach'schen Waisenhauses (49-56); Scheer, Regina: Der zerbrochene Stein: Vera und Hella Frankenberg (56-62); Heinrich, Christa: Berlin – Lissabon – New York: Ruth Carvalho (71-84); Scheer, Regina: Die Untergetauchte: Martha Ullendorff und ihre Helferin Else Rosenfeld (84-88); Scheer, Regina: Brueder: Familie Danelius (89-96); Scheer, Regina: Eine Treppe ins Nichts: Josepha Gutmann (97-103); Roder, Bernt: "Die interessierte nur, ob ich gut spielen konnte": Der Musiker Horst Prentki (103-111); Jung, Ulla: "In unserer Strasse wusste jeder, was mit den Juden geschieht": Hedwig und Erika Hildemann (111-117); Roder, Bernt: Nachrichten von Onkel Gershon: Simon Mandel (127-147); Scheer, Regina: Totentanz: Martin Friedlaender (147-168); Id.: "Wo ist mein Leben geblieben?" Das Verstummen der Gerda Victoria Wolff (169-177); Roder, Bernt: Illegal nach Holland: Bernard Wassermann (177-188); Pachnicke, Jeanne; Hinze, Sibylle: Der Schoss ist fruchtbar noch...: Peter B. (188-192); Hinze, Sibylle: Der Anfang vom Ende: Zerstoerung der Lebensgrundlagen juedischer Buerger des Prenzlauer Bergs in den Jahren 1933-1936 (286-296); Zunzer, Daniela: "Die deutsche Geschichte kommt als Gespenst zurueck": Die "Arisierung" juedischen Hausbesitzes am Helmholtzplatz (316-332); Id.: Die "Arisierung" von juedischem Gewerbe: Das Kaufhaus Lamm in der Danziger Strasse 98 (332-335); Jung, Ulla: Ein Stern aus Blech, gelb angemalt: Zwangsarbeit in der Ruestungsfirma Ehrich & Graetz (335-339). *ID*

0668 Ribbe, Wolfgang, ed.: *Die Lageberichte der Geheimen Staatspolizei ueber die Provinz Brandenburg und die Reichshauptstadt Berlin 1933 bis 1936. Vol. 1: Der Regierungsbezirk Potsdam.* Koeln: Boehlau, 1998. xx, 467 pp. On title page also: Mit ergaenzenden Materialien bearbeitet, eingeleitet und erlaeutert von Sibylle Hinze.

A collection of the original Gestapo reports. Many of them contain brief mention of the expanded activities of Jewish organizations, especially the formation of youth groups; an agricultural training school preparing young Jews for emigration to Palestine; the unwillingness of Jews in general to emigrate; Germans who continue to shop in Jewish stores, and big concerns – even the military – that continue to give large orders to Jewish firms; but also growing antisemitism among Germans, expressed in signs "Jews not wanted here" as well as in unlawful vandalism and violence that needs to be stopped. In the introduction (pp. 1-20), the editor explains that in this mainly suburban and rural area surrounding Berlin, Jews formed only 2.8% of the population in 1933 and steadily less through the 1930s, as they moved into Berlin for greater safety in anonymity. Accordingly, references to them in these reports are mostly brief. *RW*

0669 Rink, Thomas: "In erster Reihe Deutscher": Fritz Rathenau im Spannungsfeld zwischen deutscher und juedischer Identitaet. *Sachor* 9 (1999) 27-36.

Fritz Rathenau (1875-1949), a cousin of Walther Rathenau, held high-ranking positions in the Prussian civil service. He saw himself as completely German and did not think his loyalty to Germany was inconsistent with his loyalty to Judaism, which he considered a religion, not a nationality. He resented Zionists because they seemed to put this dual loyalty in doubt and thus supplied fuel to the antisemites. In 1935 he was dismissed from his post. In 1939 he emigrated to the Netherlands, from where he was deported in 1942 to Theresienstadt. He survived and returned to the Netherlands, refusing ever to return to Germany. After his emigration he no longer believed in the possibility of dual loyalty, and supported the idea of a Jewish state. *RW*

0670 Roseman, Mark: Surviving Memory: Truth and Inaccuracy in Holocaust Testimony. *Journal of Holocaust Education* 8, 1 (Sum 1999) 1-20.

Discusses the veracity of details of Holocaust testimony, based on a comparison between Nazi records and the testimony of Marianne Ellenbogen, who survived the Holocaust by hiding in various places in Germany. Most of Ellenbogen's testimony is corroborated by Nazi documents, including Gestapo and municipal records, although Gestapo files were sometimes meant to mislead the reader. However, survivors' memories are occasionally inaccurate and can change over time. For example, in Ellenbogen's memory, circumstances surrounding traumatic events took on slightly larger dimensions, while other events were diminished or underplayed; on the other hand, many unlikely things mentioned by her were corroborated by Nazi records. The discrepancies resulted from Ellenbogen's attempts to impose some control on unbearable memories and from her inability to confront the past as it was. *RE*

0671

רוזנבאום, רון: היטלר: מסע אל שורשי הרשע. תרגום: ברוך קורות. תל-אביב: מטר, 1999. 408 עמ'.

[Rosenbaum, Ron: *Hitler: A Journey to the Roots of Evil.* Trans.: Baruch Korot. Tel-Aviv: Matar, 1999. 408 pp.] Originally published as "Explaining Hitler: The Search for the Origins of His Evil" (New York: Random House, 1998).

0672 Rosenbaum, Ron: *Explaining Hitler: The Search for the Origins of His Evil.* New York: Random House, 1998. xlvi, 444 pp.

Discusses various explanations of Hitler's personality (e.g. that he lacked self-confidence, had sexual problems, was a battered child, or was mentally ill) and of his antisemitism, as explicated by scholars and thinkers. Examines their views on Hitler's personal responsibility for the Holocaust, including those of Hugh Trevor-Roper, Alan Bullock, Claude Lanzmann, Emil Fackenheim, Yehuda Bauer, and Lucy Dawidowicz. They disagree as to whether Hitler knew he was evil or actually thought that he was acting for the good of mankind, but they agree that the Holocaust would not have taken place without him. Lanzmann strongly objects to explanation of Hitler, since the arguments tend to absolve him of guilt and even turn him into a victim. Discusses, also, the views of Fritz Michael Gerlich, editor of the anti-Nazi German periodical "Der gerade Weg," who was killed in Dachau in 1934; David Irving, the revisionist who absolves Hitler of guilt; Louis J. Micheels, a Jewish survivor who was a doctor at Auschwitz and feels that he has a right to search for an explanation of the Holocaust; George Steiner, who blames the Jews for bringing hatred upon themselves; Hyam Maccoby, who sees Christian antisemitism as the root of the Holocaust; and Daniel J. Goldhagen, who blames traditional German antisemitism. *RE*

0673 Rosenbaum, Ron: *Die Hitler-Debatte: Auf der Suche nach dem Ursprung des Boesen.* Trans.: Suzanne Gangloff, Holger Fliessbach. Muenchen: Europa Verlag, 1999. 672 pp. Originally published as "Explaining Hitler: The Search for the Origins of His Evil" (New York: Random House, 1998)

0674 Ruetten, Thomas: Hitler with – or without – Hippocrates? The Hippocratic Oath during the Third Reich. *Korot* 12 [1996-1997] (1998) 91-106.

Although the Nazis tried to abolish the Hippocratic Oath, eventually they realized that, since physicians had internalized it, it was necessary to reinterpret the Oath in accordance with Nazi ideology. They therefore claimed that the annihilation of the Jewish race was necessary in order to "heal" the German people, applying the Oath to the nation rather than to individuals. An SS doctor at Auschwitz, Fritz Klein, told a Jewish prisoner that due to his respect for human life he needed to remove the Jews, who were "a gangrenous appendix in the body of mankind." *REK*

0675 Rupprath, Gisela, ed.: *Die juedische Gemeinde Hennef-Geistingen*. Hennef: Stadt Hennef, 1999. 137 pp.

A collection of essays on Jewish life in Hennef (Rhineland) inspired by an exhibition held in 1998 commemorating 60 years since the "Kristallnacht" pogrom. Partial contents: Fischer, Helmut: Die Juden in der heutigen Stadt Hennef [A survey of the history of Jews in the town from the Middle Ages to the present, focusing on the Holocaust period.] (1-40); Jakoby, Hannelore: Unbeschwerte Kindheit [Memoirs. Jakoby's Jewish father was arrested during the "Kristallnacht" pogrom and interned in Dachau; several members of her extended family perished in the Holocaust.] (41-46); Fischer, Helmut: Die Zerstoerung der Synagoge in (Hennef-)Geistingen [The synagogue was destroyed in the "Kristallnacht" pogrom. Presents an interview with an eyewitness, carried out in 1980.] (47-51); Id.: Der letzte Brief – die erste Karte [On the Wolff family and their fate.] (53-58). Pp. 61-135 contain photographs and facsimiles of documents. Pp. 124-127 present a list of 54 residents who were murdered in the Holocaust, including birth dates, and date and destination of deportation. *ID*

0676 Scarre, Geoffrey: Understanding the Moral Phenomenology of the Third Reich. *Ethical Theory and Moral Practice* 1, 4 (1998) 423-445.

Argues that if one accepts Daniel Goldhagen's thesis that inherited negative stereotypes of Jews were prime causal factors contributing to the genocide, the Germans of the Third Reich can not be considered responsible for the undeniable evil of the Holocaust. Stressing Thomas Nagel's idea of the "paradox of moral luck," concludes that the Germans were dupes of an "hallucinatory ideology," and that, in similar circumstances, other people would have been as susceptible to error as were Hitler's Germans. *YC*

0677 Schmalhausen, Bernd: Geachtet und geehrt – geaechtet und verfolgt: Max und Martha Liebermann im Dritten Reich. *Max Liebermann und die franzoesischen Impressionisten*, eds. G. Tobias Natter, Julius H. Schoeps. Koeln: DuMont, 1997. Pp. 50-54.

Soon after coming to power, the Nazis began to remove works by Jewish artists, including Liebermann, from the museums. Many were sold in order to obtain foreign currency for German rearmament. Liebermann no longer saw a future for Jews in Germany, although he himself was too old to emigrate. He was increasingly isolated; when he died in 1935 at the age of 87, only a few brave souls attended the funeral. His widow was successively deprived of the Liebermanns' country villa, their house in the center of Berlin, and their savings, and lived in great want. At the last moment she attempted, with the help of friends and relatives, to escape abroad; but the Nazis kept raising the exorbitant emigration tax they demanded of her. When she was ordered to prepare for deportation, she committed suicide. *RW*

0678 Schoeps, Julius Hans: "Ick kann jar nich so viel fressen, wie ick kotzen moechte": Max Liebermann, die Nazis und das Scheitern der deutsch-juedischen Symbiose. *Max Liebermann und die franzoesischen Impressionisten,* eds. G. Tobias Natter, Julius H. Schoeps. Koeln: DuMont, 1997. Pp. 43-49.

Liebermann, who saw himself as equally a German and a Jew, felt his world collapse with the rise of Nazism. 85 years old in 1933, he himself, who had been head of the Prussian Academy of the Arts and the leading influence in the German art world, was denigrated as a "representative of Semitic art" and the adversary of "German" art. After the Nazi takeover of the Academy and the expulsion or resignation of political opponents, and before the expected expulsion of the Jewish members, Liebermann resigned. During the last two years of his life he was almost completely isolated. He wrote to a friend that German Jews must give up their dream of assimilation, and those young enough should emigrate to Palestine. *RW*

0679 Schuchalter, Jerry: Representing the Unrepresentable: Victor Klemperer's Holocaust Diaries. *Nordisk Judaistik* 19, 1-2 (1998) 7-32.

Klemperer's diary (1995) has been seen as illustrating a counter-argument to Goldhagen's thesis of German "eliminationist antisemitism," showing that antisemitism in Germany was a mere aberration that found little resonance among ordinary Germans. A close reading of the diary shows that Klemperer's account is that of an assimilated Jewish intellectual, alienated both from Jewry and from the majority of Germans. Discusses Klemperer's representation of Nazism and antisemitism, showing where it diverges from reality. Contends that, as a source illuminating the Holocaust, the diary must be used with caution. Klemperer's creation of a dichotomy between Nazis and Germans is due to wishful thinking on the part of German Jews and reinforces revisionist historiography that wishes to exonerate Germans of their collective guilt. *DR*

0680 Schueler-Springorum, Stefanie: Fear and Misery in the Third Reich: From the Files of the Collective Guardianship Office of the Berlin Jewish Community. *Yad Vashem Studies* 27 (1999) 61-103. Translated from the German: "Zeitschrift fuer Geschichtswissenschaft" 45 (1997). Appeared in Hebrew in "Yad Vashem: Kovetz mehkarim" 27 (1999).

0681

שולר-שפרינגורום, סטפני: מצוקה ופחד ברייך השלישי: מתיקי האפוטרופסות הקהילתית של הקהילה היהודית בברלין. יד ושם 27 (1999) 47-80.

[Schueler-Springorum, Stefanie: Fear and Misery in the Third Reich: From the Files of the Collective Guardianship Office of the Berlin Jewish Community. *Yad Vashem* 27 (1999) 47-80.] Translated from the German: "Zeitschrift fuer Geschichtswissenschaft" 45 (1997). Appeared in English in "Yad Vashem Studies" 27 (1999).

0682 Seeger, Andreas: *"Gestapo-Mueller": Die Karriere eines Schreibtischtaeters*. Berlin: Metropol, 1996. 231 pp. Based on the author's diss. – Universitaet Hamburg.

A biography of Heinrich Mueller, the chief of the Gestapo. Ch. 7 (pp. 74-172), "Verantwortungsbereiche," shows, inter alia, that Mueller was a key figure in the Nazi "Final solution of the Jewish question." It was on his initiative that the yellow badge for the Jews in Germany was introduced, and he shares responsibility for the "Kristallnacht" pogrom with Goebbels and other top Nazi leaders. As Eichmann's superior, Mueller was also responsible for the deportations and the genocide of the Jews. He organized deportations of Jews as early as 1940 (e.g. from Stettin), and later deportations from France, the Balkans, and Denmark; he was enraged by the failure of the latter operation. Mueller disappeared without a trace in April 1945. *DR*

0683 Seeger, Andreas: *"Gestapo-Myuller": Karyera kabinetnovo prestupnika* ["Gestapo-Mueller": The Career of a Desk Criminal]. Rostov-na-Donu: Feniks, 1997. 381 pp. Originally published as "'Gestapo-Mueller': Die Karriere eines Schreibtischtaeters" (Berlin: Metropol, 1996).

0684 Seemueller, Ulrich: *Das juedische Altersheim Herrlingen und die Schicksale seiner Bewohner*. Blaustein: Gemeinde Blaustein, [1997?]. 195 pp.

Surveys the history of the Jewish home for the elderly in Herrlingen (near Ulm) and traces the fate of its residents in the Holocaust. The institution was established in April 1939 and functioned until summer 1942. Many elderly Jews were left behind when their families emigrated after the "Kristallnacht" pogrom. With the gradual deterioriation of the situation of the Jews during 1940, older people also attempted to emigrate; among the residents of the home, only one person succeeded. Reports on residents who were victims of the euthanasia program in 1940-41. Relates the deportation to Riga in December 1941, in which nine members of the staff were taken and probably killed. In summer 1942 the home was closed and its residents were moved to Oberstotzingen, from where they were transported to Theresienstadt. Many were later deported and murdered in Treblinka. Only four out of 96 residents of the home survived the war. Pp. 127-184 present brief biographies of residents and members of the staff. *ID*

0685 Sharfman, Glenn Richard: The Dilemma of German-Jewish Youths in the Third Reich: The Case of the Bund deutsch-juedische Jugend 1933-1935. *Shofar* 16, 3 (Spr 1998) 28-41.

The Bund deutsch-juedische Jugend, formed in 1933 and forced to disband in 1936, was the largest non-Zionist Jewish youth movement at that time. It was based on the idea that Jews had a future in Germany; it opposed assimilation and sought to strengthen Jewish self-knowledge and self-consciousness as a means of resistance to the pressures of the Nazi environment. However, its educational emphasis was on German culture and only to a lesser degree on Judaism. Its occupational training stressed "non-Jewish" occupations such as farming, crafts, and machine repair, as a means of undercutting the sources of antisemitism and achieving a more healthy and natural way of life. Local Nazi authorities at first did not interfere with the movement's activities, but in 1935 they declared that groups which did not encourage emigration were unwelcome. In 1936 the Nazis cracked down on "assimilationist" groups, and both the adult and youth movements of the Bund were dissolved. *REK*

0686

שילוני, ישראל: אפשרי ובלתי אפשרי: זיכרונות. עורכת: תלמה שילוני. תפן: המוזיאון הפתוח, 1998. 337, [7] עמ'.

[Shiloni, Yisrael: *Das Moegliche und das Unmoegliche: Erinnerungen.* Ed.: Talma Shiloni. Tefen [Israel]: Das Offene Museum, 1998. 337, [7] pp.] In German and Hebrew.

After an introductory chapter describing the situation of the Jews in Germany before Emancipation, and the many restrictions under which they suffered, the author (born as Hans Herbert Hammerstein in 1901 in Berlin) tells the story of his life. He was active in the Zionist youth group Blauweiss and a teacher. Describes his teaching in Jewish schools in Bonn and Stettin during the Nazi period. During the "Kristallnacht" pogrom he was arrested and interned in Sachsenhausen. In 1939 he emigrated to England, hoping to bring his wife and children after him; but the war broke out and they were deported to Lublin and to their deaths. Shiloni was interned as an "enemy alien" in England, and sent to Australia on the ship "Dunera." From there he was eventually able to reach Palestine. He founded and organized a museum on the history of German Jewry, first in Nahariya and now in Tefen. *RW*

0687 Smith, Robert B.: Anti-Semitism and Nazism: Reconciling Fromm and Goldhagen. *American Behavioral Scientist* 41, 9 (June-July 1998) 1324-1362.

Erich Fromm's "Arbeiter und Angestellte am Vorabend des Dritten Reiches" (first published in German in 1980 and in English in 1984) was the initial survey in the research program that culminated in "The Authoritarian Personality" by Adorno et al. Based on replies to a questionnaire, the study indicated that many German workers in 1929-30 were not antisemitic. Goldhagen's "Hitler's Willing Executioners" finds a high level of antisemitism among ordinary Germans, including the working class. To resolve the apparent contradiction in these works, Smith undertook to study the process of nazification of the German public by estimating the changing relative levels of antisemitism prior to Hitler's gaining power, describing how Hitler attained power, and explicating how antisemitism and social influence facilitated the perpetrators' killing of Jews. By 1941 many Germans had internalized the Nazi worldview, of which eliminationist antisemitism was an intrinsic component. Shows that, in addition to any hypothetical, historical prejudices, the Nazis' threat of coercion, the public's perceptions of the regime's economic and international achievements, and antisemitic propaganda all influenced the Germans' manifest levels of antisemitism and Nazism. *REK*

0688 Spelthahn, Gabriele, ed.: *An der Synagoge – Juelich und der Holocaust: Die Aufarbeitung: Festakt zur Umbenennung der Gruenstrasse: Zur Erinnerung an die Einwohner juedischen Glaubens aus Aldenhoven, Inden, Juelich, Linnich und Titz.* Juelich: Verlag des Juelicher Geschichtsvereins 1923 e.V., 1997. 160 pp.

A book commemorating the Jewish community of Juelich (Rhineland), published on the occasion of the re-naming of a street as "An der Synagoge" in 1996. Relates that in 1933 ca. 120 Jews lived in Juelich. The Nazi discriminatory measures against the Jews led to a wave of emigration which increased after the "Kristallnacht" pogrom. Pp. 28-31 list the names of 26 Jews who were arrested on 9 November 1938 and interned in Sachsenhausen; most of them were released after a few weeks. Pp. 14-18 contain the names of 32 members of the community who were deported in March 1941, first to an assembly point at Villa Buth in nearby Kirchberg and then to concentration camps in the East. Other lists (pp. 69-122) contain the names of 232 victims of the Holocaust from the area, including 71 from Juelich. *ID*

0689 Stallbaumer, Lisa M.: Big Business and the Persecution of the Jews: The Flick Concern and the "Aryanization" of Jewish Property before the War. *Holocaust and Genocide Studies* 13, 1 (Spr 1999) 1-27.

Based on the records of the Friedrich Flick Concern and archival material from the German Ministries of Finance, Economics, and Foreign Affairs, examines Flick's involvement with the Aryanization of three industrial enterprises: Simson Waffenfabrik in 1934, Hochofenwerk Luebeck in 1933-38, and the Petschek properties in Germany, Austria, and Czechoslovakia in 1938-39. The owners of the Flick Concern were motivated by business considerations, not by racism per se. However, the cases examined show that the more successful cases of Aryanization cannot be attributed so much to friendly relations between industry and state as to their common goals. Therefore Friedrich Flick and his associates were not so much "Aryan predators" as opportunists willing to involve themselves in the persecution of Jews if it was advantageous to business. *DR*

0690 Standhartinger, Ernst: Juden in Biebesheim – 1933 bis 1945. *Die Geschichte der Juden in Biebesheim.* Riedstadt: Forum Verlag, 1997. Pp. 38-56.

Discusses the Jewish community in Biebesheim (in the region of Gross-Gerau) between 1933-45, derived mostly from letters and oral testimonies. In 1933 there were five families in Biebesheim – 24 Jews in all. Deals with the gradual deterioration of the Jews' situation – exclusion from social life, discrimination, boycott, Aryanization, and the "Kristallnacht" pogrom – leading to their emigration between 1935-38. Six Jewish persons who were either born in Biebesheim, lived there temporarily, or were descendants of Jews from Biebesheim perished in the Holocaust. *ID*

0691 Stein, Harry, ed.: *Konzentrationslager Buchenwald, 1937-1945: Begleitband zur staendigen historischen Ausstellung.* Goettingen: Wallstein, 1999. 320 pp. On title page also: Herausgegeben von der Gedenkstaette Buchenwald.

Accompanies the permanent exhibition at the Buchenwald concentration camp. Describes the installation and administration of the camp, and daily life between 1937-43, specifying the categories and nationalities of the prisoners, among them over 9,000 Jews in November 1938. Discusses, also, the special camp for Jews which was established at the outbreak of the war for 1,035 Polish Jewish prisoners, and the deportation and murder of Jews in February 1942. In October 1942, the 405 remaining Jews were deported to Auschwitz. Between 1942/3-1945 the number of prisoners in Buchenwald rose from ca. 8,400 to 84,505, from all over Europe. In mid-1944, ca. 13,000 Jews were brought from Auschwitz, 4,500 from Dachau, and thousands from other places, in order to exterminate them through hard labor. At the end of 1944, with the approach of the Soviet army, prisoners in camps in Poland and eastern Germany were taken on death marches westward. Among 16,000 Jews who left Auschwitz in January 1945 only ca. 5,000 reached Buchenwald. In total, ca. 56,000 persons perished in Buchenwald. Includes many facsimilies and photographs. *ID*

0692 Steiner, John M.; Cornberg, Jobst von: Willkuer in der Willkuer: Befreiungen von den antisemitischen Nuernberger Gesetzen. *Vierteljahrshefte fuer Zeitgeschichte* 46, 2 (Apr 1998) 143-187.

The Nuremberg Laws and their derivative regulations gave Hitler the right to confer the status of "German blood" on individual Jews and half-Jews. Applications were processed in the Ministry of the Interior and the Reich Secretariat before being submitted to Hitler. Hitler apparently endorsed no more than ca. 1,300 such applications, which were of two kinds: for full citizenship or for the right to serve and be promoted in the army. In the first category were high-ranking officials, prominent artists and scientists, and especially deserving veterans of the Nazi movement; in the second were soldiers who had distinguished themselves. Half-Jews often alleged illegitimate birth from an Aryan father. Bernhard Loesener of the Interior Ministry, though a convinced Nazi, advised and helped them. In 1943 the responsibility passed to the RSHA, and no more exceptions were made. *RW*

0693 Steppe, Hilde: *"...den Kranken zum Troste und dem Judenthum zur Ehre...": Zur Geschichte der juedischen Krankenpflege in Deutschland.* Frankfurt a.M.: Mabuse-Verlag, 1997. 396 pp.

Examines the history of Jewish nursing services in Germany, especially in Frankfurt, between 1880-1940. Pp. 71-75 discuss the development of antisemitism in Imperial Germany and its intensification in the Weimar Republic. Pp. 76-79 and 157-178 examine various aspects of Nazi persecution in the sphere of Jewish nursing. Discusses the role of German nurses who largely obeyed the Nazi rules; in accordance with their ideal of following orders at all times, many became assistants to murder, while only a minority resisted. In 1934 six Jewish hospitals were still functioning. Surveys the gradual exclusion of Jewish nurses from 1933; from 1938 they were allowed to work only in Jewish institutions. Pp. 176-178 relate the fate of five Jewish nurses. Pp. 236-245 survey events in the Jewish hospital in Frankfurt, where 400 patients were sheltered in 1942, until its closure in October of that year. Most of the patients were deported to Theresienstadt. *ID*

0694 Strachotta, Gerd: *Juden in der Wesermarsch 1933-1945.* Oldenburg: Isensee Verlag, 1997. 165 pp. On title page also: Herausgegeben vom Landkreis Wesermarsch.

A history of the persecution of the Jews in the region of Wesermarsch, Germany, during World War II. Ch. 1 (pp. 8-28) surveys Jewish history in the Wesermarsch from the Middle Ages to the Nazi period, including antisemitism. The rest of the book examines persecution under the Nazi regime: discrimination, exclusion, boycott, Aryanization of Jewish property, the Nuremberg Laws, the "Kristallnacht" pogrom (which claimed three victims from the area). Many of the men who were arrested and sent to concentration camps during the pogrom did not return home after their release, preferring to move with their families to larger towns. At the outbreak of the war, most Jewish residents had left the Wesermarsch area for other countries in Europe or overseas. Pp. 128-141 relate the fate of individuals. 19 Jews from the region were murdered in concentration camps. *ID*

0695 Strauss, Herbert Arthur: *In the Eye of the Storm: Growing Up Jewish in Germany, 1918-1943: A Memoir.* New York: Fordham University Press, 1999. xii, 262 pp.

Strauss was born in Wuerzburg; his mother was a convert from Catholicism. In 1936 he went to Berlin to study at the Lehranstalt fuer die Wissenschaft des Judentums (first called Hochschule). He studied there until the Hochschule was closed in July 1942. Describes the courses he took, the personalities of the teachers, and their fate, including that of the head of the school, Leo Baeck. Between 1936-38 he also worked for the Jewish Jugendverband in Berlin. In January 1942, Strauss was drafted for forced labor (street-sweeping) in Berlin. In October he and his future wife, Lotte Schloss, went into hiding in Berlin, aided by many people. In May 1943 Lotte escaped to Switzerland; Strauss followed six weeks later. After he completed his Ph.D. in Switzerland they emigrated to the USA, where Strauss became a well-known historian. Strauss' father was deported to Warsaw in April 1942, and did not survive. His mother survived the war in Wuerzburg, and joined them in New York. His sister emigrated to Palestine in 1938. Lotte's parents were deported to Riga in 1942; they were murdered, along with the other deportees, while in transport. *DR*

0696 Strauss, Herbert Arthur: *Ueber dem Abgrund: Eine juedische Jugend in Deutschland, 1918-1943.* Trans.: Bettina Abarbanell. Berlin: Ullstein, 1999. 325 pp. Originally published as "In the Eye of the Storm: Growing Up Jewish in Germany, 1918-1943: A Memoir" (New York: Fordham University Press, 1999).

0697 Strauss, Lotte: *Over the Green Hill: A German Jewish Memoir, 1913-1943.* New York: Fordham University Press, 1999. x, 180 pp. The German translation was published first: "Ueber den gruenen Huegel: Erinnerungen an Deutschland " (Berlin: Metropol, 1997).

0698 Teicher, Wilfried: Der Anteil der juedischen Aerzte an der Spezialisierung im ersten Drittel dieses Jahrhunderts in Preussen. *Medizinische Wissenschaften und Judentum,* eds. Nora Goldenbogen et al. Dresden: Verein fuer regionale Politik und Geschichte Dresden, 1996. Pp. 14-29.

Suggests that many Jewish physicians became specialists, preferably in newly developing fields, in order to overcome the disadvantages imposed by antisemitism. This is shown by directories published in the Nazi period in which Jews are marked as such. An additional indication is the declining ratio of specialists to general practitioners during this period, as Jews were dismissed or emigrated. *RW*

0699 Trapp, Frithjof et al., eds.: *Handbuch des deutschsprachigen Exiltheaters, 1933-1945.* Vol. 1-2/2. Muenchen: K.G. Saur, 1999. x, 525; xxx, 548-1092 pp.

Vol. 1, "Verfolgung und Exil deutschsprachiger Theaterkuenstler," contains essays discussing the changes which occurred in German theatrical life after 1933, the Nazi polemics against "Jewish-Marxist" influences, the gradual exclusion and persecution of Jewish artists, and their annihilation. Vol. 2, pts. 1-2 contain a lexicon (divided alphabetically: A-K, L-Z) of German (including Jewish) theatrical personalities who were persecuted by the Nazi regime; some of them established a new career in exile. The following essays relate to persecution of Jews: Vol. 1: Mittenzwei, Werner: Verfolgung und Vertreibung deutscher Buehnenkuenstler durch den Nationalsozialismus (7-79); Freeden, Herbert: "Auf Abruf" – das Theater des juedischen Kulturbundes im "Dritten Reich" 1933-1941 [An earlier version appeared in "Exil" 13 (1993).] (81-95). Vol. 2/1: Trapp, Frithjof: Verfolgung und Exil deutschsprachiger Theaterkuenstler zwischen 1933 und 1945 (vii-xxx). *ID*

0700 Vogt, Annette: Issai Schur – als Wissenschaftler vertrieben. *Menora* 10 (1999) 217-235.

Traces the life of the well-known mathematician who contributed to the worldwide reputation of the Berlin University, where he was a professor for thirty years. Schur was born in 1875 in Mogilev, studied mathematics in Berlin from 1882, and graduated in 1901. In 1903 he began to teach at the university; he became a full professor only in 1921. Presumably, the reason for the delay was latent antisemitism. In 1935 Schur was dismissed from his position at the university. In 1939 he left for Palestine, via Switzerland, where his daughter had emigrated several years before. Schur never got over his dismissal. He died in 1941 of a heart attack. *ID*

0701 Walter, Dirk: *Antisemitische Kriminalitaet und Gewalt: Judenfeindschaft in der Weimarer Republik.* Bonn: J.H.W. Dietz, 1999. 349 pp.

A study of the development of antisemitic violence during the Weimar Republic, especially in Bavaria. States that between 1918-33 antisemitism was very common in the form of discrimination and disdain toward Jews. Discusses social distance from Jews, their exclusion from schools, universities, and professions, the campaign against Ostjuden in 1920, Jews held as hostages during Hitler's "putsch" in Munich, synagogue and cemetery desecrations, ritual murder propaganda, reports on violence of the SA and other right-wing groups at the end of the 1920s, the development of "voelkisch" antisemitism, and the orientations of the NSDAP. Although, traditionally, antisemitism was one of the major ideological principles of the Right before 1933, varying between passive affirmation of a "Jewish problem" and approval of violence, both public and political antisemitism became more radical. Antisemitism was a basic principle in Nazism before the war, and developed its murderous destructiveness during and after the "Kristallnacht" pogrom. *ID*

0702 Wecht, Martin Johannes: Jochen Klepper: *Ein christlicher Schriftsteller im juedischen Schicksal.* Duesseldorf: Archiv der Evangelischen Kirche im Rheinland, 1998. xiv, 658 pp.

A biography of Klepper (1903-1942), a poet and a person of deep religious faith. In 1931, in Breslau, he married Johanna Stein-Gerstel, a Jewish woman and mother of two girls, Brigitte and Renate. Johanna converted to Christianity in 1938. Excerpts from Klepper's diary (published in 1956) reflect the influence of the Nazis' rise to power on his professional and private life. He was dismissed from work at the Berlin radio in 1933 and at the Ullstein broadcasting company in 1935. Discusses Klepper's views on the events in Nazi Germany and the role of the Church, especially the Confessing Church. The efforts of Klepper and his wife to arrange for the emigration of her daughters were successful for Brigitte, who left for England, but failed for Renate. Between 1940-41 Klepper served in the army, but in September 1941 he was dismissed. In December 1942 Renate received a summons for deportation. On 10 December 1942 Klepper, his wife, and Renate committed suicide. Pp. 395-581 contain correspondence of Klepper with theologians, critics, and friends. *ID*

0703 Wehler, Hans-Ulrich: The Goldhagen Controversy: Agonizing Problems, Scholarly Failure and the Political Dimension. *German History* 15, 1 (1997) 80-91. An expanded version of an article published in German in "Die Zeit" (24 May 1996).

Contends that a book such as Goldhagen's "Hitler's Willing Executioners" could cause such a stir, in part, because scholars in Germany and other countries have been silent on many crucial aspects of the Holocaust. For this reason his book must not be ignored; rather, its defects must be discussed. Goldhagen's approach is simplistic and mono-causal – his one and comprehensive explanation of the Holocaust is the demonic antisemitism of the German people. Presenting the Holocaust as a unique event, he avoids comparisons and does not mention antisemitism in countries other than Germany, other genocides in this century, or even the fact that Jews were not the only victims of the Nazis. He also ignores a great number of scholarly works on this issue. Concludes that Goldhagen has done great damage to the study of the Holocaust; the negative reaction to his book in Germany makes serious discussion of many unresolved problems of German history which his book aroused impossible. See the response by Michael Pinto-Duschinsky [in "German History" 16 (1998) 397-411], in which he states that Goldhagen rejects German collective guilt and does discuss antisemitism in other countries and Nazi persecution of other groups; Wehler's criticism ignores what is actually in Goldhagen's book. *DR*

0704 Welch, Steven R.: "Mischling" Deserters from the "Wehrmacht." *Leo Baeck Institute Year Book* 44 (1999) 273-324.

Describes Nazi Germany's policy concerning the military service of male offspring of mixed marriages ("Mischlinge" of the first degree), as well as of "pure Aryans" married to Jewish women. This policy was erratic, ambivalent, and contradictory; during the war, the Wehrmacht underwent several campaigns of purging its ranks from "Mischlinge" or of drafting them. Many "Mischlinge" were willing to serve on the front line, hoping that they would be re-classified as "full Germans" after the war. Examines eight cases of desertion by "Mischlinge" from the Wehrmacht in wartime, brought before German military courts. In at least four cases, the soldiers' "Mischling" status was a motivating factor in their desertion."Mischlinge" were clearly disadvantaged in the military courts, and their racial background had an impact on their punishment. The Wehrmacht's policy toward "Mischlinge" and the harsh punishment of "Mischlinge" deserters show how deeply the German army was involved in the implementation of Nazi racial policy. Pp. 316-326 contain facsimiles of documents. *DR*

0705 *Widerstand und Verfolgung im Muenchner Westend, 1933-1945: Ein Stadtteilfuehrer*. Muenchen: A-B Verlag, 1997. 104 pp.

Relates experiences of residents of the Westend district in Munich, both Jews and non-Jews, who were persecuted by the Nazi regime in 1933-45. The events are listed in alphabetical order of the names of the streets where the persecuted were living. 26 non-Jews offered various forms of resistance against the regime or were politically persecuted; five of them were murdered. Relates experiences of 15 Jews; some of them emigrated and eight perished in the Holocaust. *ID*

0706 Wielepp, Christoph; Luehr, Hans-Peter, eds.: *Leben in zwei Diktaturen: Victor Klemperers Leben in der NS-Zeit und in der DDR.* Dresden: Friedrich-Ebert-Stiftung, Buero Dresden, [1997]. 81 pp. On title page also: Eine Tagung der Friedrich-Ebert-Stiftung in Zusammenarbeit mit dem Verein "Gegen Vergessen – fuer Demokratie," am 19. und 20. September 1997 in Dresden.

Partial contents: Nemitz, Kurt: Viktor Klemperer und die juedische Alltagsexistenz im NS-Staat, 1933 bis 1941 (28-38); Goldenbogen, Nora: Zur Chronologie des Alltags der Tyrannei: Die Verfolgung und Vernichtung der juedischen Bevoelkerung Dresdens, 1933 bis 1945 (39-51); Dirschauer, Johannes: Zur Faszination Victor Klemperers (70-76). *RW*

Great Britain

0707 Cesarani, David: Le "Jewish Chronicle" de Londres et la Shoah. *Revue d'Histoire de la Shoah* 163 (May-Aug 1998) 183-202.

Examines the items in the "Jewish Chronicle" during the Second World War relating to the Holocaust. States that the size and placement of the articles were uneven, and editorial commentaries in 1941-42 were minimal. Concludes that the reports of the "Jewish Chronicle" on the Final Solution facilitated occultation of the fate of the Jews. Official and non-official censorship also limited the scope of information given. Directives of the Ministry of Information from 1939 asked not to describe the sufferings of the Jews in Europe so as not to disturb British policy in the Middle East, nor the Arabs. The tragedy of the "Jewish Chronicle"'s journalists was that they were part of a system – press, culture, and society – for which the Holocaust was happening too far away; for a long time it was not considered an important subject. The journalists, as representatives of British Jewry, were highly assimilated; because of their loyalty to Great Britain, they did not want to focus on Jewish suffering when all Britons were in danger. *HV*

0708 Kochavi, Arieh Joseph: British Officialdoms and European Jewry during the 1940s. *Immigrants & Minorities* 17, 3 (Nov 1998) 34-54.

In 1939-48 the main issue which dominated the British authorities' treatment of the Jewish problem in Europe was the dispute over Palestine. Britain refused to open the gates of Palestine, or her own gates, for Jewish refugees from Nazi-controlled Europe. The British government took no note of the tragic plight of the Jews in Europe; it regarded the refugees as pawns in the hands of the Zionists, who could antagonize the Arabs against London. Both during and after the war, when the scope of the Nazi genocide became known, the British disregarded the uniqueness of the Jewish tragedy under the Nazis, downplayed postwar antisemitism in Eastern Europe (and even placed responsibility for it on the Jews themselves), and impeded Jewish immigration to Palestine. George Rendel, of the Foreign Office, determined Britain's policy toward Jewish DPs. The British wartime policy vis-à-vis the refugee problem and immigration to Palestine caused a deterioration in relations between the Zionists and Britain. *DR*

0709 Palmer, Andrew: Orwell and Antisemitism. *The Jewish Quarterly* 172 (Win 1998-1999) 41-45.

Discusses George Orwell's non-fiction writings before the mid-1930s as reflecting the "polite" antisemitic prejudice of his class, presenting Jews via common stereotypes. Once he realized the nature of Nazism, ca. 1934, his attitude became ambivalent: he sympathized with Jewish victims of fascism but retained an unacknowledged distaste for "the Jews." In the early war years he continued to stereotype Jews even while criticizing public disinterest in their fate. To maintain this duality, he distinguished between British and Nazi strains of antisemitism, the former being essentially non-consequential. Palmer suggests that English (including Orwell's) thought on antisemitism was double-think, like that depicted in his novel "1984." During the war years, Orwell began struggling with his own antisemitic stereotyping, as is reflected in "1984." *YC*

0710 Prokopov, Aleksandr Yu.: Antisemitskaya kampaniya Soyuza fashistov Osvalda Mosli [The Antisemitic Campaign of Oswald Mosley's Union of Fascists]. *Vestnik Yevreiskovo Universiteta* 20 (1999) 129-155.

Antisemitism was not prominent in the ideology of the British Union of Fascists, but came to the fore in 1933, in the wake of the Nazi takeover in Germany. Among the reasons for this shift, besides the desire for relations with Germany's Nazis, were the fact that London Jews had attacked BUF members and, most of all, Mosley's desire to construct the image of a formidable enemy, which might be helpful in mobilizing the masses for the BUF. In 1934-37 the East End of London, with its large Jewish population, became the center of BUF activities; there the BUF tried to draft new members into the party, held mass rallies, etc. In the provinces, the organization's level of activities, as well as the popularity of its anti-Jewish policies, was much lower. Jewish organizations in the East End were at the forefront of the resistance to the advance of fascism in the city. In October 1936 Jews in the East End, together with non-Jewish leftists, thwarted a fascist march. Despite its antisemitic electoral campaign, the BUF lost the 1937 election to the London City Council. After this defeat, the BUF began to soft-pedal antisemitism in its ideology. *DR*

Hungary

0711 Braham, Randolph L.: *A népirtás politikája: A Holocaust Magyarországon.* Vol. 1-2. 2nd, rev. ed. Trans.: Tamás Zala et al. Budapest: Belvárosi Könyvkiadó, 1997. 1126 pp. Originally published as "The Politics of Genocide: The Holocaust in Hungary" (New York: Columbia University Press, 1981). A revised English edition appeared in 1994. First appeared in Hungarian as "A Magyar Holocaust" (Budapest: Gondolat; Wilmington, DE: Blackburn International, 1988). This is a revision of that edition, based on the English revised edition.

A comprehensive history of the Holocaust in Hungary, relating also to the history of Hungarian Jewry from its emancipation in 1849 to the present. The first anti-Jewish laws in the 1920s and the increase of antisemitism in the 1930s were followed by persecution, forced military occupation by the Nazis, ghettoization, and deportation. Discusses the events of the Holocaust according to six territorial zones: the Carpathians and the northeastern parts of Hungary; northern Transylvania; northern Hungary; southeastern Hungary; western Hungary; Budapest and environs. Describes the fate of the Jews in many locations. Relates, inter alia, to the military labor service units, the Jewish Council, Jewish rescue operations, the Nazis' actions and the cooperation of the Hungarian authorities, and the actions of the Arrow Cross, as well as international interventions and rescue actions of the Christian Churches. *MA*

0712 Cole, Tim: Constructing the 'Jew,' Writing the Holocaust: Hungary 1920-45. *Patterns of Prejudice* 33, 3 (July 1999) 19-27.

Examining Hungarian anti-Jewish legislation from 1920 (the Numerus Clausus Law) to 1944 (the law on wearing the yellow star), shows how the category "Jew" was constructed in Hungary. This category has always (including in interwar and wartime Hungary) been a social rather than a biological, religious, etc., construct.. Due to the failure of scholars to recognize the artificiality of the category "Jew," they posit the existence of the Jews as a group and of a "Jewish problem." Therefore, Randolph Braham and Hannah Arendt were wrong when they condemned the behavior of those Hungarian Jews who exploited the opportunity to gain a privileged status (exempting them from wearing the yellow star) and thus did not share the fate of other "Jews." These people were free to choose their own identity. *DR*

0713 Don, Yehuda: Reports to the Nazi Authorities as a Quantitative Source in History. *Studia Judaica* 7 (1998) 71-88.

In April 1944 the German occupying forces in Hungary distributed a questionnaire among the Jewish communities, demanding an assessment of their properties, funds, foundations, and the aggregate property taxes paid by the community members in the fiscal year 1943-44. 740 communities responded, among them 460 which are in the present-day borders of Hungary. 106 of the latter submitted data about the property tax and its base. These data are used in order to calculate the real value of unclaimed Jewish real estate in Hungary; states that it amounts to 222,888,000 pengő. *DR*

0714

ארז, צבי: "שינדלר הונגרי" – האמנם? ילקוט מורשת 68 (אוקט' 1999) 85-109.

[Erez, Zvi: "A Hungarian Schindler" – Is That So? *Yalkut Moreshet* 68 (Oct 1999) 85-109.]

Describes, in detail, the fate of the Jews of Budapest between October 1944-February 1945, during the rule of the Arrow Cross party and the Soviet siege of Budapest, when many of the Jews in Budapest were murdered by Arrow Cross gangs. Traces the involvement of Pál Szalai – an Arrow Cross party member who was nominated in November 1944 as the coordinator between the party and the Hungarian police in Budapest – in some rescue acts related to Budapest and its ghettos, some of which were initiated by Raoul Wallenberg. Contends that although Szalai is connected to some rescue acts, he was just one factor in a long and complicated line of circumstancial actions which were not initiated by him. Therefore, he could hardly be a candidate for the award of "Righteous among the Nations." *LFo*

0715 Guttmann, David: *Schwierige Heimkehr: Leben und Leiden in Ungarn, dann auf der 'Exodus' und zurueck ueber Bergen-Belsen nach Tel-Aviv: Juedische Schicksale 1944-1948.* Konstanz: Hartung-Gorre, 1997. 131 pp. Translated from the English ("Homecoming"). Published in Hungarian in Budapest: Kodex, 1995.

Memoirs of a Jew born in 1932 in a small town and raised in Budapest. Focuses on events in 1944, following the Nazi invasion of Hungary, when he was interned, together with his mother and two sisters, first in a crowded "Jewish house" and later in the ghetto. Guttmann and his younger sister were moved to a protected house run by the Swiss Embassy, but fled back to their family in the ghetto. Later they heard that the protected house was overtaken by the Nazis and none of the children survived the Holocaust. After the war, Guttmann and his sisters left

Hungary and passed through DP camps in Austria and Germany. The older sister left for Palestine alone. In 1947, David, his younger sister, and his mother (who had joined them) sailed to Palestine on the ill-fated "Exodus." Relates their trials and tribulations until they reached Palestine in April 1948. Guttmann later became a professor of social work at the University of Haifa. *ID*

0716

גוטמן, דוד: השיבה הביתה. תל-אביב: צ'ריקובר, 1999. 146 עמ'.

[Guttmann, David: *Homecoming*. Tel-Aviv: Tcherikover, 1999. 146 pp.] Appeared in German as "Schwierige Heimkehr: Leben und Leiden in Ungarn, dann auf der 'Exodus' und zurueck ueber Bergen-Belsen nach Tel-Aviv: Juedische Schicksale 1944-1948" (Konstanz: Hartung-Gorre, 1997).

0717 Horák, Magda, ed.: *"Ősi hittel, becsülettel a hazáért!" OMIKE – Országos Magyar Izraelita Közművelődési Egyesület, 1909-1944* ["In Traditional Belief and Honor for Our Country!" OMIKE – Hungarian Jewish Cultural Society, 1909-1944]. Budapest: Háttér Kiadó, 1998. 601, [23] pp.

A documentary history of OMIKE, founded in 1909 by Chief Rabbi Simon Hevesi in Budapest in order to help Jewish writers and artists. Comprises reprints of articles from the Jewish press (in chronological order), mainly from the weekly "Egyenlőség." In the period of the numerus clausus (1920s-30s), OMIKE helped Jewish students to continue their studies outside of Hungary, and in the period of the persecutions its Artists' Action enabled Jewish actors to continue giving performances. Pp. 258-570 deal with the period 1933-45. *MA*

0718 Komoróczy, Géza, ed.: *Jewish Budapest: Monuments, Rites, History*. Trans.: Vera Szabó. Budapest: Central European University Press, 1999. xii, 597 pp. Originally published as "A zsidó Budapest: Emlékek, szertartások, történelem" (Budapest: Városháza; MTA Judaisztikai Kutatócsoport, 1995).

0719

ליברמן, יצחק: ואל מכאוב חלליך יספרו (תהלים ס"ט). פתח-תקוה: המחבר, 1999. 141 עמ'.

[Liebermann, Jichak: *And They Tell of the Pain of Those Whom Thou Hast Wounded (Psalms 69)*. Petach-Tikva: The Author, 1999. 141 pp.]

Memoirs of an Orthodox Jew, born in 1924 in Békéscsaba, Hungary. Pp. 32-56 contain a brief history of the Holocaust describing the Nazi invasion of Hungary in March 1944 and the subsequent deportations of Jews. Pp. 57-78 relate the fate of the community and of Liebermann's family, as well as his own experiences in the Holocaust. Most of the Jews of Békéscsaba (ca. 2,500 in 1941) were deported to Auschwitz in June 1944; very few survived. Liebermann's parents and two of his brothers perished in Auschwitz. One brother was inducted into a military labor service unit and perished; Liebermann and two other brothers, who survived, were also inducted, but Liebermann fled from his unit just before the liberation. He was then imprisoned by the Russians and freed through bribery by his brother; he emigrated to Eretz Israel in 1947. *RE*

0720 Lowinger, Maurice: *Miracle in the Ashes*. Ed.: Yaakov Lavon. Southfield, MI: Targum Press, 1999. 191 pp.

Memoirs of a Jew born in 1920 in Mezőtúr, eastern Hungary, to an Orthodox Jewish family. His father died in the mid-1930s and the family (eight children) moved to Győr, where Lowinger became the main breadwinner. In 1940 he was dismissed from his job due to the Hungarian anti-Jewish laws and moved to Budapest. He attemped to rescue the families of two sisters and a brother who had settled in Slovakia by smuggling them back into Hungary, but he was arrested and tortured. In October 1942 he was drafted into a military labor service unit and sent to the Püspökladány camp, where he suffered many hardships. In 1943 he returned to Budapest, where he experienced the German occupation in 1944 and the Nyilas terror. He and his new wife were interned in the ghetto of Budapest, where he tried to help his fellow inmates, using bribes, personal connections, etc. Lowinger and his wife survived the ghetto; his mother and seven siblings (along with the spouses and children of those who were married) perished in the Holocaust. *DR*

0721 Milland, Gabriel: The BBC Hungarian Service and the Final Solution in Hungary. *Historical Journal of Film, Radio and Television* 18, 3 (Aug 1998) 353-373.

For a long time, during World War II, the BBC Hungarian Service avoided broadcasting on Jewish issues lest it be identified with Jews. After the Nazi occupation of Hungary in 1944, it changed its attitude and devoted much time to coverage of the situation of Hungarian Jewry. Its broadcasts contained reports on events, warnings addressed to Hungary's rulers, and appeals to the Hungarian people to sabotage the genocide, help Jews, and thereby recover Hungarian national honor. However, the BBC Hungarian Service failed to influence events in Hungary or to save any lives. Following the directives of the Political Warfare Service, it avoided stressing Jewish issues, as well as broadcasting news that had no definite confirmation (like gassings of deportees). The appeals for Hungarian-

Jewish brotherhood came too late, and many other broadcasts were ill-timed. *DR*

0722 Pierik, Perry: *Hungary 1944-1945, the Forgotten Tragedy: The Last German Offensives of the Second World War: The Loss of the Last Jewish Community in Europe.* Trans.: D.E. Butterman-Dorey. Nieuwegein, Netherlands: Aspekt, 1996. 304 pp. Originally published as "Hongarije 1944-1945, de vergeten tragedie: De laatste Duitse offensieven van de Tweede Wereldoorlog: De ondergang van de laatste joodse gemeenschap in Europa" (Nieuwegein: Aspekt, 1995).

Deals with events in and around Hungary from 18 March 1944 (the Nazi invasion) until the war's end in 1945. Pp. 73-105, "The Annihilation of the Hungarian Jews," dwell on the deportation and murder of the majority of the country's Jews. Eichmann arrived in Budapest some days after the occupation, claiming that the "Hungarian operation" would be the most successful in his career. The anti-Jewish operation was carried out quickly, with Hungarian acquiescence and collaboration. States that the Vatican tried to protest against the deportation of Jews from Hungary. Mentions rescue efforts by Kasztner and his aides, as well as by Wallenberg, Lutz, and Perlasca. *DR*

0723 Pierrejean, Claudine; Pierrejean, Daniel: *Les secrets de l'affaire Raoul Wallenberg: Du juste de Budapest au premier martyr de la guerre froide.* Paris: L'Harmattan, 1998. 302 pp.

Chronicles the life and activities of Wallenberg. Pp. 23-162 describe the events in Hungary from the Nazi occupation in March 1944, and Wallenberg's pivotal role in the effort to rescue Jews in Budapest. The rest of the book deals with his arrest by the Russians in January 1945, the possible reasons for this act, and the efforts since then to discover what happened to him. *TS*

0724 Raab, Ernest: *Violin of Stone.* Ed.: Corinne Copnick. Trans.: Jeremy Fox. Toronto: Lugus, 1999. 277, [32] pp. Translated from the French manuscript.

Memoirs of a Holocaust survivor, now a well-known Canadian sculptor. Raab (whose Hebrew name is Moshe Yehudah) was born in 1926 in Komárom (then Czechoslovakia). In October 1938, following the Munich agreement, his town was accorded to Hungary, which marked the beginning of the sufferings of Komárom's Jews. In 1944, when Hungary was occupied by the Germans, Raab was sent to the Komárom labor camp, from which he escaped. He and his brother went to Budapest, where he managed to provide both of them with "Aryan" papers and tried to help the Jews of the Budapest ghetto. They were liberated in January

1945 by the Soviet army. Raab's parents perished in the Holocaust, but his two sisters survived Auschwitz. *DR*

0725

ראובני, שרי: חסידי אומות העולם והמחתרת החלוצית בהונגריה. ילקוט מורשת 64 (נוב' 1997) 47-55.

[Reuveni, Sári: The Righteous among the Nations and the Underground of the Zionist Youth Movements in Hungary. *Yalkut Moreshet* 64 (Nov 1997) 47-55.]

An expanded version of a lecture delivered in December 1996 at Yad Vashem. Dwells on the factors that affected rescue activities in Hungary after the Nazi occupation in March 1944. Also discusses misapprehensions regarding rescue activities initiated by the Zionist youth movements and the acts of Righteous Gentiles in Hungary. States that rescue operations were impeded mainly by the swift and systematic process of deportation rather than by local antisemitism and the danger involved in rescue. The rescue operations of the Zionist youth movements were overshadowed after the war by the focus on ghetto uprisings and on partisan warfare. *LFo*

0726

ראובני, שרי: הנסיבית שקידמו ומנעו הצלה בקרב יהודי הונגריה. דפים לחקר תקופת השואה 14 (1997) 313-325.

[Reuveni, Sári: The Circumstances Which Facilitated and Which Impeded Rescue of Jews in Hungary. *Dapim le-Heker Tekufat ha-Shoah* 14 (1997) 313-325.]

Rescue in Hungary was affected by the situation of the Jews in various parts of the country. For example, the Jews in the provinces were speedily segregated and deported; they did not have an opportunity to plan or implement rescue actions. The Jews of Budapest were interned for a longer time and in familiar surroundings, where they could utilize their communal organizations as well as their relations with friendly non-Jews. The families deported to Austria for forced labor had not been segregated from the rural population beforehand. Their specific circumstances made it possible to receive aid from friendly non-Jews. *LFo*

0727

רון, אלכסנדר: התנועה הציונית בהונגריה בין שתי מלחמות העולם. דיס' — האוניברסיטה העברית בירושלים, 1996. [8], 365, 19, [4] עמ'.

[Ron, Alexander: *The Zionist Movement in Hungary between the Two World Wars.* Diss. – Hebrew University of Jerusalem, 1996. 8, 365, 19, 4 pp.] With an English summary (19 pp.).

Ch. 3 (pp. 55-98) discusses the Zionist movement in Hungary against the background of growing antisemitism after World War I. Hungarians identified Jews with communists because many Jews had supported the short-lived communist revolution of Béla Kun, which ended in August 1919. Hungary was the first country to pass an anti-Jewish law in 1920, a numerus clausus quota for university students. Ch. 6 (pp. 167-213) relates the difficulties faced by Zionist youth movements due to anti-Jewish legislation in 1938-40. In September 1938 the formation of new Zionist groups was forbidden. In July 1940 Zionist activity was banned altogether. Activities for illegal immigration continued underground, but with little success. *RE*

0728 Szita, Szabolcs: Az 1944-1945. évi deportáció, a Nyugat-Magyarországi erődítő kényszermunka áldozatai [The 1944-1945 Deportation: Victims of the Fortification Forced Laborers in Western Hungary]. *Holocaust Füzetek* 6 (1996) 106 pp.

Pp. 7-12 describe the circumstances in which the Hungarian Jewish men and women named in the lists presented here lost their lives between November 1944-March 1945 during forced labor on fortification lines built by the German army during its retreat from the western border of Hungary and eastern Austria. The workers were former Hungarian Jewish military forced laborers who were turned over by the Hungarian army to the Todt Organization, the SS, the Gestapo, and the Wehrmacht, and Hungarian Jews who were deported, by foot, over the Austrian border and turned over to various Nazi military bodies. The lists are held by the Hungarian Military Archives in Budapest. They are interspersed with correspondence of Hungarian authorities, and Hungarian Jewish organizations and individuals, regarding the search for persons mentioned in the lists and preparation of the lists. The names are arranged according to the names of the work camps, as well as mass graves found near camps and work sites. *LFo*

0729

צור, ראובן: בריחה מהגטו: פרקי זכרונות. תל-אביב: הקיבוץ המאוחד. 106 עמ'.

[Tsur, Reuven: *Escape from the Ghetto: Memoirs.* Tel-Aviv: Hakibbutz Hameuchad, 1996. 528 pp.]

Memoirs of a Jew born in 1933 in Nagyvárad (now Oradea Mare, Romania). Soon after the German occupation of Hungary in March 1944, he and his family (parents and sister) were forced to move into the ghetto.

When the deportations began, the director of the ghetto hospital claimed that there was a typhus epidemic there; all those in the hospital, including Tsur's family, were allowed to remain behind. Some months later, when the family heard that those in the hospital were to be deported, they fled from the ghetto and over the Romanian border. After the war, they returned to Nagyvárad; they emigrated to Israel in 1948. *RE*

Italy

0730 Barozzi, Federica: I percorsi della sopravvivenza: Salvatori e salvati durante l'occupazione nazista di Roma (8 settembre 1943-4 giugno 1944). *La Rassegna Mensile di Israel* 64, 1 (Jan-Apr 1998) 95-144.

Discusses, on the basis of testimonies of Jewish survivors of the Holocaust, the fate of the Jews of Rome during the Nazi occupation. Describes Jewish individual strategies for survival, as well as the support of the local Italians who helped them by passive resistance or active involvement. Recalls the roundup in the ghetto of Rome on 16 October 1943, resulting in the arrest of some 20% of the Jews. States that Pope Pius XII and the Vatican did not protest publicly but only took some diplomatic steps (without result) to stop the arrests, as seen in the letters of Germany's ambassador to the Vatican, Ernst von Weizsacker. *AA*

0731 Biagini, Furio: *Mussolini e il sionismo*. Milano: M&B, 1998. 172 pp.

Asserts that until 1938 Mussolini's policy toward the Zionist movement was ambiguous. Internationally, he would sympathize with Zionist claims, in view of Italy's interests in the Middle East. Internally, he opposed Zionism since it stood in conflict with the nationalistic, integrative program enunciated in fascist ideology. In 1938 Italy's relations with Nazi Germany and the enactment of the racial laws made explicit the change in Mussolini's attitude and fascist ideology from the concept of nation to the concept of race, initiating persecution of the Jews and of the Zionist movement. *MZ*

0732

קארפי, דניאל: בניטו מוסוליני – על יהודים, ציונים וציונות (1917-1922). אלפיים 17 (1999) 208-221.

[Carpi, Daniel: Benito Mussolini – on Jews, Zionists and Zionism (1917-1922). *Alpayim* 17 (1999) 208-221.]

Examines Mussolini's attitude as reflected in the articles he published in the periodical "Il Popolo d'Italia" when he was already an ardent nationalist but still not in a powerful position. His reactions to events in

Russia, the Soviet Union, Germany, Palestine and the Middle East manifest an ambivalent and contradictory attitude toward Jews, including suspiciousness regarding their motives, fear of their powerful influence on international and national politics, intolerance of their uniqueness, along with reluctance to confront them openly. *LFo*

0733 *Cinéma et enseignement de l'italien: "Il Giardino dei Finzi-Contini."* Trans.: Giancarlo Gerlini, Elisabeth Kertesz-Vial. Paris: CIRRMI, Université de la Sorbonne Nouvelle-Paris III, 1998. 401 pp.

A collection of articles produced to accompany the film "The Garden of the Finzi Continis" for use by high school teachers to promote a greater understanding of Italian culture and politics during World War II. Partial contents: Kertesz-Vial, Elisabeth: Historique de l'antisémitisme [In Italy during the Nazi period.] (39-46); Groppo, Bruno: Notes sur la politique antisémite du fascisme italien (1938-1945) [A lecture delivered at a conference at the University of Caen.] (67-75); Groppo, Bruno: Primo Levi: Il dovere di memoria e i limiti della testimonianza (373-381). Pp. 47-66 contain documents, including the original text of the Italian racial laws of 1938. *TS*

0734 Cohen, Kate: *The Neppi Modona Diaries: Reading Jewish Survival through My Italian Family.* Hanover, NH: University Press of New England, [for] Dartmouth College, 1997. x, 274 pp.

A biography based on extensive quotes from the diaries of Aldo Neppi Modona (1895-1985), a Jewish professor from Florence, and his son Leo (1932-1986), and accounts of interviews with Aldo's wife, Rachel, and his daughter, Lionella. Aldo fought in World War I and was a member of the Fascist Party, which was not initially antisemitic. The family moved to Rome in 1934. Describes the onset of racial antisemitism in Italy in summer 1938, followed by anti-Jewish legislation which directly affected the family, since Aldo lost his teaching position. The family then returned to Florence, where Aldo taught in a Jewish school which was newly-organized for Jewish children who were not permitted to attend Italian schools. After the German occupation in 1943, the family fled to the countryside, posing as non-Jews; they returned to Florence after the liberation. *RE*

0735 Marchione, Margherita: *Yours Is a Precious Witness: Memoirs of Jews and Catholics in Wartime Italy.* New York: Paulist Press, 1997. ix, 259 pp.

Numerous Jews, both Italian and non-Italian, who happened to be in Italy due to the vicissitudes of the war or were in areas occupied by the Italian army, were rescued by Italians from all social strata, including the military. The Catholic Church (e.g. convents belonging to various monastic orders, the Holy See) was especially active in rescuing Jews and assisting them. Pope Pius XII condemned the German anti-Jewish policy, as well as its Italian and Vichy French imitators. He refrained from taking a public stance against the Nazis because he did not want to endanger the lives of other people. Gives many examples of aid rendered to Jews by Italians. Dwells on the wartime activities of Marchione's own monastic order, the Religious Teachers Filippini. *DR*

0736 Pinkus, Karen: "Black" and "Jew": Race and the Resistance to Psychoanalysis in Italy. *Annali d'Italianistica* 16 (1998) 145-167.

Discusses how the 20th-century racist Italian society tended to reject pyschoanalysis as a "Jewish" and "Black" science. The Italian self-image, especially under fascism, is viewed as rejecting the "pessimism" of Black-Jewish analysis of the individual personality for a more "healthy" collective therapy. However, a case study discussed shows this approach to have been based on self-deception. *YC*

0737 Pugliese, Stanislao G.: Bloodless Torture: The Books of the Roman Ghetto under the Nazi Occupation. *Libraries & Culture* 34, 3 (Sum 1999) 241-253.

A paper first presented at an international conference, "The Holocaust and the Book," held at Drew University (Madison, NJ), November 1996. A burning of Jewish books in Italy took place even before the German occupation – in Turin in 1938. With the German occupation of Rome, in September 1943, representatives of the Einsatzstab Reichsleiter Rosenberg appeared at the synagogue looking for Jewish works of art and books. The great book and manuscript collection, some of the items dating from before the Common Era, was plundered and consigned to Frankfurt, to the Institut zur Erforschung der Judenfrage. In December 1943, the book collection of the Collegio Rabbinico in Rome was also plundered. Many of the books were returned after the war, but some were lost forever. *DR*

0738 Sarfatti, Michele: Żydzi w faszystowskich Włoszech [The Jews in Fascist Italy]. *Biuletyn Żydowskiego Instytutu Historycznego* 185-186 (Jan-June 1998) 75-84.

Surveys the political situation in Italy before the fascist regime came to power. Underlines that the unification of Italy occurred against the will of the clergy; one of its consequences was the promotion of Jews in Italian political life. All of the parties accepted Jews as members. Antisemitic propaganda began to appear in 1922. Some Jews became members of the Fascist Party, but many of them were active in anti-fascist organizations.

The first restrictions of citizens' rights were introduced in 1937. Anti-Jewish decrees were signed in 1938. Describes consecutive steps in the antisemitic policy of the fascist regime up to the fall of Mussolini in July 1943, and later, between 1943-45, in the new political situation (the north occupied by the Nazis, the south – a new fascist state). Describes the deportations of Italian Jews. There were ca. 50,000 Jews in Italy before the war. 35,000 survived the war due to various forms of aid given them by non-Jews. *HV*

0739 Sellin, Volker: Die Juden Italiens zwischen Emanzipation und Holocaust (1796-1945). *Trumah* 8 (1999) 73-95.

An expanded version of a lecture delivered at the Hochschule fuer Juedische Studien in Heidelberg, January 1998. Focuses on the period 1938-45. States that Mussolini's anti-Jewish laws, internment of Jews, and apparent cooperation in their deportation seem to be in contradiction with the total integration of Jews in Italian society until 1938. This integration was due to the democratic and anti-clerical principles of the Risorgimento as well as to demographic factors. Originally, antisemitism played no part in fascism. But in 1938 Jews were an obstacle to Italian solidarity with Nazi Germany. Mussolini's measures were aimed at forcing them to emigrate. Instead, the Jewish population was constantly swelled by refugees. Italy interned many of them, but refused to deliver them to the Germans. The Italians also protected the large Jewish populations in territories occupied by them; after Italy's surrender, these territories fell into German hands. Argues that Mussolini's puppet Salo Republic rounded up and interned Jews in camps in the vain hope of forestalling their deportation, but this actually made it easier for the Germans to seize them. *RW*

Latvia

0740 Ezergailis, Andrew: *Holokausts vācu okupētajā Latvijā, 1941-1944* [The Holocaust in German-Occupied Latvia, 1941-1944]. Trans.: Dina Plandere, Inese Skrīvere, Elga Zālīte. Rīga: Latvijas Vēstures Institūta Apgāds, 1999. 591 pp. Originally published as "The Holocaust in Latvia, 1941-1944: The Missing Center" (Riga: Historical Institute of Latvia; Washington, DC: United States Holocaust Memorial Museum, 1996).

0741 Gordon, Frank: Mezhdu molotom i nakovalnei [Between Hammer and Anvil]. *Rizhskii Almanakh* 5 (1997) 5-34.

An excerpt from his book "Latvians and Jews between Germany and Russia" (1990). This section focuses on the Holocaust and Latvian rescuers of Jews. *DR*

0742 Schneider, Hubert; Vestermanis, Margers: Die Ermordung der lettischen Juden: Ein aktuelles Thema? Margers Vestermanis im Gespraech mit Hubert Schneider. *Sachor* 7 (1997) 100-105.

Discusses the role of various groups, particularly the Wehrmacht and the Latvian militia, in the extermination of the Jews of Latvia. Some local Wehrmacht commanders did not wait for orders from the central authorities but massacred Jews on their own initiative; others did not, and the attitude of ordinary soldiers to Jews who worked for them was often decent. The Latvian militia did not murder Jews before the German occupation; but when encouraged by the SD, they murdered with a will. After the war, the Latvian population was too immersed in its own suffering from Soviet terror to deal with its guilt toward the Jews, especially since the trials of war criminals were not publicized. Today, the Russian population of Latvia credits the USSR with saving the Jews; this causes the Latvian population to justify its alliance with the Nazis as a fight for the freedom of Latvia, while downplaying or denying the role of Latvians in Nazi crimes. *RW*

0743 Tsin, Basya: *Vyzhit, chtoby vernutsya: Dokumentalnaya povest* [To Survive in Order to Return: A Documentary Novel]. Tel Aviv: Biblioteka Matveya Chornovo, 1997. 259 pp.

Memoirs of a Jew (née Drizine), born in 1912 in Kraslava, southeastern Latvia. During World War II she, together with the rest of Kraslava's Jews, was interned in the ghetto of Daugavpils. She survived several mass executions, but her daughter died. In October 1943 Tsin was transferred to Kaiserwald, a camp in Riga; then she was sent to the labor camp Dundaga, and later to Stutthof. In January 1945, during the death march from Stutthof, the overseer of her column allowed the prisoners to flee. After the war Tsin settled in Latvia. In 1966 she testified at the war crimes trial held in Riga. The defendants were Latvian collaborators who took part in the transfer of Kraslava's Jews to Daugavpils (during which many Jews were killed) and other atrocities. In the 1990s, Tsin emigrated to Israel. *DR*

0744 Tsin, Basya: *Wie ein grauenhafter Traum: Vier Jahre zwischen Leben und Tod: Juedische Schicksale aus Lettland 1941-1945.* Trans.: Kurt Foss. Konstanz: Hartung-Gorre, 1998. 48 pp. First published in Russian in "Vestnik Yevreiskoi Kultury" (Spr 1989). See the abstract for the expanded Russian version: "Vyzhit, chtoby vernutsya" (Tel Aviv: Biblioteka Matveya Chornovo, 1997).

Lithuania

0745 Abramowicz, Dina: Guardians of a Tragic Heritage: Reminiscences and Observations of an Eyewitness. *Proceedings of the 33rd Annual Convention of the Association of Jewish Libraries, Philadelphia, PA, June 21-24, 1998,* eds. Barbara Y. Leff, Laurel S. Wolfson. New York: Association of Jewish Libraries, 1999. Pp. 11-19. Appeared also as a separate booklet, published in New York: National Foundation of Jewish Culture, [1998].

Discusses the efforts of Hermann Kruk to preserve the Jewish library, known as the Hevrah Mefitse Haskalah Library, in Vilna after the Nazi occupation. Abramowicz was one of Kruk's assistants. By chance, the library happened to be in the territory of the ghetto. By 1942, the ghetto library had circulated 100,000 books, an indication of the importance of "book therapy" in the face of the Holocaust. When German cultural looting was organized by the Einsatzstab Reichsleiter Rosenberg in July 1941, some Jews resisted. Kruk's team continued trying to save Jewish cultural treasures. He, his assistant Zelig Kalmanovitch, and most of their comrades were killed by the Nazis. After the war, Abraham Sutzkever and Schmerke Kaczerginski succeeded in unearthing some materials that had been buried for safety. Part of the famous YIVO collection from Vilna was recovered from Germany; the part that remained in Vilna had a complex fate not described here. *YC*

0746 Ceitlinas, Jevsejus, ed.: *Vilniaus getas: Kalinių sąrašai = Vilnyusskoye getto: Spiski uznikov = Vilnius Ghetto: Lists of Prisoners.* Vol. 1-2. Vilnius: Lietuvos valstybinis žydų muziejus, 1996-1998. 432; 336 pp.

Presents lists of names, accompanied by essays in Lithuanian, Russian, and English. Contents: Vol. 1: Guzenberg, Irina: Vilniaus getas ir 1942 metų gyventojų surašymas = Vilnyusskoye getto i perepis naseleniya 1942 goda = The Vilnius Ghetto and the Population Census of 1942 (8-23; 28-46; 51-68). Pp. 71-386 contain lists of the Vilnius ghetto inmates from the population census conducted in the "Generalkommissariat" Lithuania in May 1942; the lists are given street by street. Kalasauskienė, Leonora Ona: Sąrašų publikavimo tvarka (archeografinis komentaras) = Osobennosti publikatsii spiskov (arkheograficheskii kommentarii) = Order of the Publication of the Lists (Archaeographical Commentary) (387-405). Vol. 2: Guzenberg, Irina: Vilniaus geto darbo stovyklos ir 1942 m. gyventojų surašymas = Rabochiye lagerya Vilnyusskovo getto i perepis naseleniya 1942 goda = The 1942 General Population Census in Lithuania: The Labor Camps of Vilnius Ghetto (5-64). Pp. 67-171 contain lists of Jewish prisoners in the labor camps at the "Kailis" factory, in Baltoji Vokė, Bezdonys, Kena, Riešė, "Giesler" in Pabradė, Petešа, Užtrakis and at Vokė farmstead. Pp. 180-329 contain an alphabetical index of the names of the ghetto residents mentioned in these two volumes. *DR*

0747 Eliach, Yaffa: *There Once Was a World: A Nine-Hundred Year Chronicle of the Shtetl of Eishyshok.* Boston: Little, Brown, 1998. xiv, 818 pp.

A study of Jewish life in the Polish-Lithuanian town of Ejszyszki (Eišiškes), based largely on interviews with former residents of the town. Pt. 6 (pp. 557-697), "The Bitter End: 1933-41," deals with the Holocaust period. The story of the destruction of the Jewish community is interlaced with the story of Eliach's family, the Sonensons, and their experiences during the war. Many Jews were killed by the Nazis and local collaborators on 25-26 September 1941. The Sonensons were among the few who managed to escape; they passed through some small ghettos and were then hidden by a Polish farmer, Kazimierz Korkuc. They returned to Eišiškes after the liberation, but Eliach's mother and baby brother were murdered in their home by rampaging Polish AK partisans in October 1944. Contends that, despite the relatively large number of rescuers, the majority of Poles were antisemitic and wished for a "Poland without Jews." The Nazis would not have been able to carry out the genocide of the Jews without widespread Polish and Lithuanian collaboration. *DR*

0748 Elkes, Joel: *Values, Belief and Survival: Dr Elkhanan Elkes and the Kovno Ghetto: A Memoir.* London: Vale, 1997. xii, 100 pp. On title page also: The inaugural Elkhanan and Miriam Elkes memorial lecture, delivered at the Stanley Burton Centre for Holocaust Studies, the University of Leicester, October 31st, 1991.

A biography of Elkes' father (b. 1879), a physician, who was the head of the Judenrat in Kovno following the Nazi occupation of Lithuania. Describes the establishment of the ghetto in July 1941 and of the Judenrat, the massacre of ca. 9,000 Jews in October 1941, the ghetto's transformation into a concentration camp in November 1943. After a roundup in March 1944, in which ca. 2,000 people were killed, the members of the Judenrat were arrested. In July the ghetto was liquidated. Elkes was sent to the Landsberg labor camp, where he died. His wife was sent to Stutthof and survived. Although hardly a man of resistance, Elkes was respected by the Jews of Kovno and tried to alleviate their plight. *DR*

0749

פייטלסון, אלכס (אלתר): מעשי הרצח של משטרת העזר הליטאית. ילקוט מורשת 64 (נוב' 1997) 150-145.

[Faitelson, Aleks (Alter): The Acts of Murder of the Lithuanian Auxiliary Police. *Yalkut Moreshet* 64 (Nov 1997) 145-150.]

Presents a section from Faitelson's memoirs not included in the Hebrew edition ("Be-sufah u-ve-ma`avak," 1994), but which did appear in the German (1998) and French (1999) translations. Relates the involvement of the Lithuanian Auxiliary Police in the murder of Jews and partisans in Lithuania and Belarus. *LFo*

0750 Faitelson, Aleks (Alter): *Courage dans la tourmente en Lituanie 1941-1945: Mémoires du ghetto de Kovno.* Trans.: Eve Line Blum-Cherchevsky. Paris: L'Harmattan, 1999. 431 pp. Translated from the English: "Heroism and Bravery in Lithuania 1941-1945" (Jerusalem: Gefen, 1996). Originally published in Yiddish as "In shturm un gerangel" (Vilnius: Lituanus, 1993).

0751 Jofė, Chaimas: Žydų gyvenimas ir žūtis [Life and Death of Jews]. *Jurbarkas: Istorijos puslapiai* [Jurbarkas: Pages of History], ed. Arnoldas Piročkinas. Vilnius: Pradai, 1996. Pp. 174-188.

Pp. 183-188 discuss the period of World War II and the Holocaust in Jurbarkas. Ca. 2,300 Jews lived in the town in 1940, and only 76 in 1945. All but those Jews who were deported by the Soviets prior to the German invasion in June 1941 were murdered by the Nazis and their local helpers in summer-fall 1941. *DR*

0752

לוין, הדסה: בת-עמי: נתיב היסורים והאמונה של בת יהדות ליטא בימי השואה. עורך: יוסף אריה אליעז (לוין). בני-ברק: הוצאת בת-עמי, 1996. כג, 424 עמ'.

[Levin, Hadassah: *A Daughter of My People: The Path of Suffering and Faith of a Lithuanian Jewish Woman in the Time of the Holocaust.* Ed.: Yosef Arieh Eliaz (Levin). Bnei-Brak: Bat-Ami, 1996. 23, 424 pp.]

Fictionalized memoirs of an Orthodox Jewish woman from Panevežys, Lithuania, born in 1912, who was a writer and teacher in Telšiai and Kovno. Levin married in 1938 and gave birth to a son in December 1939; her husband was killed by the Germans in 1941. Describes the German occupation of Telšiai and the massacres which followed, as well as life in the ghetto and Levin's escape before its liquidation. She hid her son, Yosef Arieh, with a priest, and after hiding in many villages, arrived at the Šiauliai ghetto. Describes conditions in the ghetto – forced labor, sickness, murders – and her escape before its destruction. She hid in forests and on farms until the liberation, and then reclaimed her son. The story ends with emigration to Palestine; however, Levin actually died of tuberculosis in a Jewish hospital in Munich in 1946. She wrote this book before she died, and asked in her will that it be published. *RE*

0753

לוין, הלל: רשימת סוגיהרה — בחיפוש אחר הטוב: סיפורו של הדיפלומט היפני שסיכן את חייו כדי להציל אלפי יהודים מן השואה. תרגום: גידי אביבי. תל-אביב: ידיעות אחרונות — ספרי חמד, 1999. 324 עמ'.

[Levine, Hillel: *Sugihara's List – in Search of the Good: The Story of the Japanese Diplomat Who Risked His Life to Rescue Thousands of Jews from the Holocaust.* Trans.: Gidi Avivi. Tel-Aviv: Yediot Ahronot – Hemed Books, 1999. 324 pp.] Originally published as "In Search of Sugihara: The Elusive Japanese Diplomat Who Risked His Life to Rescue 10,000 Jews from the Holocaust" (New York: Free Press, 1996).

0754

סג"ל, י: בעוד לילה: הרבנית בת שבע שורץ ע"ה, בת הרב יהושע ברקמן הי"ד מטלז. בני ברק: מרכז פראגר — מכון למורשת ישראל, גנזך קידוש השם, 1996. 256 עמ'.

[Segal, Y.: *When It Was Still Night: The Rabbanit Bat Sheva Schwartz, Daughter of Rabbi Yehoshua Brackman of Telz.* Bnei Brak: Merkaz Prager, 1996. 256 pp.]

A biography of Schwartz, a Jewish woman from Telz (Telšiai), Lithuania. Pp. 37-109 relate her experiences in the Holocaust. The German occupation of the town in summer 1941 was followed by a massacre of the Jewish men, older women, and children. The young women, including Schwartz, were herded into the ghetto, where they suffered from hunger and disease. Schwartz and one of her sisters escaped from the ghetto in the winter, just before its liquidation, and hid with various non-Jews in the area until their liberation in November 1944. After the war they emigrated to Palestine. *RE*

0755

שנר-נשמית, שרה: רוצחים וחסידי אומות העולם ליטאים. דפים לחקר תקופת השואה 14 (1997) 273-312.

[Shner-Neshamit, Sara: Lithuanian Murderers and Righteous among the Nations. *Dapim le-Heker Tekufat ha-Shoah* 14 (1997) 273-312.]

A study based on data found in the 179 case-files relating to Lithuania (up to 1995) opened by the public committee of Yad Vashem for the recognition of "Righteous among the Nations," and on data compiled from material deposited at Beit Lohamei Haghetaot and in published testimonies. Characterizes Lithuanian Righteous Gentiles and their activities, e.g. their motives, their origins, their background. Describes acts of individuals and groups, such as the acts of four families in the village Sarnele. Includes a description of the anti-Jewish propaganda and actions of the Lithuanian Activist Front, the extreme nationalists who collaborated with the Nazis and blamed the Sovietization of Lithuania between 1940-41 on the Jews. Attributes some rescue activities of Lithuanians to their disappointment with the Nazis who refused to grant Lithuania independence. *LFo*

0756 Šuras, Grigorijus: *De joden van Wilno: Een kroniek 1941-1944*. Trans.: Jacqueline Godfried. Amsterdam: Jan Mets, 1997. 222 pp. Unseen. Translated from the Russian. The author's name is given as Grigori Szur. See the abstract for the German edition: "Die Juden von Wilna: Die Aufzeichnungen des Grigorij Schur, 1941-1944" (Muenchen: Deutscher Tas chenbuch Verlag, 1999).

0757 Šuras, Grigorijus: *Die Juden von Wilna: Die Aufzeichnungen des Grigorij Schur, 1941-1944*. Ed.: Wladimir Porudominskij. Trans.: Jochen Hellbeck. Muenchen: Deutscher Taschenbuch Verlag, 1999. 220 pp. On the title page, the author's name is given as Grigorij Schur. Translated from the Russian: "Yevrei v Vilno: Khronika 1941-1944 gg." (Sankt-Peterburg: "Obrazovaniye – Kultura," 2000).

A chronicle of the Vilna ghetto written during the war by an inmate. The editor's introduction (pp. 9-28) describes the life of Zvi Hirsh Shur (1898-1945) and the conditions under which he wrote (he worked in a factory repairing uniforms), and discusses the moral dilemmas facing the ghetto population. In summer 1941 the Jews moved into the ghetto; periodically, groups were rounded up and shot in Ponary. Eventually, the selection was taken over by the Jewish Council, from July 1942 by Jacob Gens, and the arrests carried out by the Jewish police. Gens believed that by sacrificing some Jews he could save others. Jews working for the Germans were temporarily spared, along with their families. Oskar Glick, a Jew posing as an "Aryan," socialized with Nazi officials and secured a large increase in the number of workers and their transfer to special blocks where they were somewhat protected from the Gestapo. He was denounced and killed in January 1942. In September 1943 the ghetto was liquidated, under the command of Bruno Kittel. Gens was shot. Only the Jews in the workers' blocks remained; they were shot shortly before the liberation. Shur was interned in Stutthof and, when that camp was liquidated, taken with hundreds of other prisoners out to sea and drowned. *RW*

Netherlands

0758 Assan, Aryeh: *The Name of Solomon: Memoirs 1940-1945*. Trans.: Sheila Assan. Jerusalem: [The Author], 1997. 75 pp. Translated from the Dutch.

Memoirs of a Jew from Arnhem, Netherlands, born in 1923 as Louis Jacob van Essen. Relates his father's activities in helping Jewish refugees from Germany in 1935-36. Describes how he and his family (parents and three sisters) coped with the Nazi occupation from May 1940. When he was ordered to report for forced labor in September 1942, the family went into hiding in various locations near their home, helped by friends. They were supported by revenue from their shop which was transferred to the ownership of their maid. While in hiding, Assan organized a group which gathered intelligence and passed on information to the Allies. During the battle of Arnhem, they were forced to leave the city. The family found refuge with the Methorst family, near Groenendaal. Assan converted an old schoolhouse into an Evacuation Post for Dutch people fleeing the front lines, and that remained the cover for the family's existence until the liberation. He emigrated to Israel in 1949. *REK*

0759

בנימין, יגאל: פעולות ההצלה של המחתרת החלוצית בהולנד. דפים לחקר תקופת השואה 14 (1997) 77-107.

[Benjamin, Yigal: The Rescue Activities of the Pioneering Youth Movements' Underground in the Netherlands. *Dapim le-Heker Tekufat ha-Shoah* 14 (1997) 77-107.]

In response to the intensification of anti-Jewish measures in the summer of 1942, the vocational training groups of the Zionist youth movements in the Netherlands, comprised mainly of refugees from Central Europe who were preparing for emigration to Palestine, were forced to change their focus and help their members to live in hiding or to escape from the Netherlands. Describes the activities of the various groups, including the help provided by Jewish communal organizations, the local population, and resistance groups, as well as the aid given by Jewish and non-Jewish underground groups in France (to where many escaped en route to Spain or Switzerland). *LFo*

0760 Bing, Dov: The Nazi Persecution of the Jewish Community of Holland: Collaboration and Heroism. *Letters and Texts of Jewish History*, ed. Norman Simms. Hamilton, New Zealand: Outrigger, for the Waikato Jewish Studies Seminar, 1998. Pp. 1-8.

Disputes the myth of widespread Dutch resistance to the deportation of the Jews. Only about 22,000 of 140,000 Dutch Jews survived the Holocaust; 16,000 were rescued by non-Jews. The myth was initiated by Abel Herzberg's "critical" academic study of 1950. Jacob Presser's "The Destruction" (1965) challenged the myth but also blamed the Jews, especially the Jewish Council, for their fate. Louis de Jong, in pt. 8 of "The Kingdom of the Netherlands in the Second World War" (1969-92), was also critical of the Jewish Council but noted the ambivalent role of the government and the Dutch underground, and the general lack of concern about Jews. Nanda van der Zee, in "To Prevent Worse" (1997), openly criticized the Dutch elites for their moral failures: this included Queen Wilhelmina and the government for fleeing and leaving a political vacuum that was filled by the Nazi-installed governor, whose orders regarding the Jews were faithfully followed by Dutch ministers, the Supreme Court, and the police. Concludes that the storm of protest evoked by van der Zee's work indicates that the Dutch are still incapable of coming to terms with their past. *YC*

0761 Groenleer, Gert: *In veilige haven: Joodse onderduikers in Colijnsplaat, 1942-1944* [In Safe Haven: Jewish Persons in Hiding in Colijnsplaat, 1942-1944]. Vlissingen: ADZ, 1997. 64 pp.

Relates the story of the survival of the Polak family (Mozes and Rosa, and their son and daughter, Lion and Ada) during the war. Originally from Middelburg, a small town in the province of Zeeland, they were evacuated to Amsterdam in January 1942, with other Jews of the area. Their eldest daughter went to Zutphen to be with her fiancé. In August 1942 they avoided a roundup, and with the help of the Groenleer family (the author's grandfather and other family members) they were sent into hiding in the village of Colijnsplaat in Zeeland, where they remained until their liberation in November 1944. The eldest daughter survived as well. *MF*

0762 Grüter, Regina: *Een fantast schrijft geschiedenis: De affaires rond Friedrich Weinreb* [A Beguiler Invents History: The Friedrich Weinreb Affairs]. Amsterdam: Balans, 1997. 445 pp. With an English summary (pp. 439-444).

Weinreb (1910-1988), a Jew, was born in Lvov; his family fled to the Netherlands during World War I. He became an economist and senior staff member of the Nederlands Economisch Instituut. At the time of the Nazi occupation, he had a wife and two daughters. In November 1941, Weinreb lost his job due to the anti-Jewish measures. He then started an emigration swindle, telling Jews in distress that he had permission from the Nazis to organize Jewish emigration to unoccupied France and Portugal; he charged 100 guilders per adult. The Nazis arrested him in September 1942; after three days he was freed. He then became an informer for the Nazis. After the liberation he was accused of betrayal and arrested; he was sentenced to six years in prison, but was released after two months. In the 1960s a furor arose in the Netherlands as to whether Weinreb was a wartime hero (as he described himself in his memoirs) or a collaborator. In 1976, after a thorough investigation, the RIOD published a report concluding that he was a liar and a swindler, and that his collaboration with the Germans had resulted in 70 deaths. *SSC*

0763 Haan, Ido de: The Netherlands and the Novemberpogrom. *Jahrbuch fuer Antisemitismusforschung* 8 (1999) 155-176.

A lecture delivered at a conference at the Zentrum fuer Antisemitismusforschung, Berlin, November 1998. While many groups in the Netherlands sharply condemned Nazism and the persecution of the Jews, others, especially of the Christian Churches, although rejecting the Nazis' brutal methods, professed understanding of their wish to get rid of a destructive, exploitative foreign element, or even argued that the Jews were receiving their just punishment. The government, dominated by a right-wing Protestant party, warned the press and the public to restrain their criticism in order not to alienate Germany. With more and more refugees seeking asylum, the government strictly limited the number of those admitted. Even when the "Kristallnacht" pogrom aroused a popular wave of protest against the quotas, they were raised only slightly, and that on condition that the Jewish community pay all the expenses of the internment camps. The situation led to isolation of the Jewish community, whose interests seemed to diverge from those of the general population. In the Jewish leaders it bred a habit of compliance with the authorities. All this pointed toward the behavior of Dutch Jews and non-Jews during the Holocaust. *RW*

0764 Heymans, Albert: *Jood zonder ster* [Jew without a Star]. Westervoort: Van Gruting, 1999. 126 pp. Originally published as "Al saf ha-avadon be-olam ha-tselalim: Badad ba-ma`arakhah, 1940-1945 be-Holland" (Nahariya: The Author, 1995).

0765 Lipschits, Isaac: *Tsedaka: Een halve eeuw joods maatschappelijk werk in Nederland* [Tsedakah: Half a Century of Jewish Social Work in the Netherlands]. Zutphen: Walburg Pers, 1997. 462 pp.

Ch. 2 (pp. 45-63), "Sjoa en Tsedaka," relates that at the time of the Nazi invasion there were 1,015 Jewish organizations engaged in social and financial aid. The Germans united most of them; in May 1941 there were 239 organizations. From August 1941, all Jewish households had to put their savings in one bank, Lippmann, Rosenthal & Co., and each was allowed only 250 Hfl. a month. From January 1943, all the accounts were joined together, and only the Jewish Council could take out money. In July 1942 the Council decided to help and support, as much as possible, those Jews slated for deportation and those left behind. Jews were asked to give, collect, and distribute clothing for the needy. Between September 1942-February 1943 ca. 7,000 families received some help from the Council. In November 1942 the Council was instructed by the Nazis to clear all hospitals, mental institutions, and rest homes of their inhabitants. In September 1943 the last members of the Council were deported to Westerbork; organized aid was no longer possible. *MF*

0766 Metselaar, Menno; Rol, Ruud van der: *Een geschiedenis voor vandaag: Anne Frank* [A History for Today: Anne Frank]. Amsterdam: Anne Frank Stichting, 1996. 94 pp.

A booklet of photographs relating to Anne Frank's story and the Nazi persecution of the Jews of Europe, accompanied by explanations of the photographs, quotations from Anne Frank's diary, and brief testimonies by people who knew her or had similar experiences during the war. *RE*

0767

מצלר, מנו; רול, רוד ואן דר: היסטוריה להיום – אנה פראנק. תרגום: רונה גרינברג.
אמשטרדם: בית אנה פרנק, 1996. 94, [2] עמ'.

[Metselaar, Menno; Rol, Ruud van der: *A History for Today – Anne Frank*. Trans.: Rona Grinberg. Amsterdam: Anne Frank Stichting, 1996. 94, 2 pp.] Originally published as "Een geschiedenis voor vandaag: Anne Frank" (Amsterdam: Anne Frank Stichting, 1996).

0768 Michman, Dan: Preparing for Occupation? A Nazi "Sicherheitsdienst" Document of Spring 1939 on the Jews of Holland. *Studia Rosenthaliana* 32, 2 (1998) 173-189.

Historians are generally of the opinion that the Nazis formulated plans for dealing with Dutch Jews only after the occupation in May 1940, but this would have been contrary to what is known about Third Reich policies in general, and in regard to Jewish issues in particular. In fact, the Nazis were already planning how to deal with Netherland's Jews by early 1939, as evidenced by a detailed report on the Dutch Jewish community found in the Bundesarchiv in Koblenz, which is accompanied by a cover letter from the SD-Fuehrer des SS-Oberabschnittes Nord-West in Hamburg, dated 28 March 1939. The report contains demographic, historical, and political information on the Jewish community and its organizations, as well as information on Jewish leaders, all of which would facilitate the implementation of anti-Jewish measures. Discusses, also, power struggles within the Nazi regime. There was no Jewish Department of the SD in the Netherlands because Seyss-Inquart insisted on controlling the various aspects of the German occupation himself. Nevertheless, he used the favored SD organizational tool, the Judenrat, to carry out his plans. He may have also used the 1939 report, a facsimile of which appears on pp. 185-189. *REK*

0769 Orlow, Dietrich: A Difficult Relationship of Unequal Relatives: The Dutch NSB and Nazi Germany, 1933-1940. *European History Quarterly* 29, 3 (July 1999) 349-380.

Originally, the ideology of the Dutch Nationaal Socialistische Beweging in the interwar period did not include antisemitism. The party had, however, a racist nationalist wing, prone to antisemitic ideas. With growing pressure from the Nazi party in Germany in the 1930s, and following the NSB's failure in the election of 1937, the racist wing gained more support in the party; its leader, Mussert, and his "old guard" accepted Nazi-style antisemitism. This ideological shift in the late 1930s did much to destroy the party's credibility as an indigenous and not German-operated fascist movement, and led to the political isolation of the NSB in the Netherlands before the Nazi occupation. *DR*

0770 Richardson, Annette: Expurgating Jewish Faculty during the Netherlands' Nazi Occupation: 1940-1943. *Political Crossroads* 7, 1-2 (1999) 111-128.

Nazi anti-Jewish policy in occupied Netherlands was met in many cases by public protests. In particular, the overt discrimination against and subsequent dismissal of Jewish lecturers and officials from the universities of Leiden, Amsterdam, Delft, and other schools of higher learning were followed by protests on the part of professors and sometimes of students. Some protesters paid for this resistance with their lives. However, they were unable to prevent the genocide of Netherland's Jews. *DR*

Poland

0771 Abramson, Henry Maurice: The Prince in Captivity: Reading Hasidic Discourses from the Warsaw Ghetto as Social and Intellectual History. *Journal of Genocide Research* 1, 2 (June 1999) 213-225.

Presents a reading of "Esh kodesh" by Rabbi Kalonimus Kalmish Shapira, a collection of Sabbath and holiday discourses written in Nazi-occupied Warsaw, as historical evidence and as a theological treatise attempting to explain the Holocaust and its meaning. Remarkably, Shapira did not resort to traditional Jewish theodicy in his discourses. "Esh kodesh" sheds much light on the social and psychological strategies employed by a faith community in coping with Nazi oppression. *DR*

0772

אדלר, יוסף; רותם, יהודית: חיים פשוטים. [בצרה: יוסף אדלר], 1997. 50 עמ'.

[Adler, Yosef; Rotem, Yehudit: *A Simple Life*. Batzra [Israel]: Yosef Adler, 1997. 50 pp.]

Pp. 13-26 describe the Holocaust experiences of Adler's wife, Bella, who owned a hotel in Warsaw that was confiscated by the Germans in 1940.

Her first husband was deported and did not return; she and her young son, Yossi, were forced to move into the ghetto. They escaped to the "Aryan side" in spring 1943 and hid separately in various places. Bella was arrested as a non-Jew and sent for forced labor to Germany, where she worked on a farm. She was liberated by American forces in spring 1945 and rejoined her son, who had been hidden by a Polish friend. Yosef, in Palestine since 1935, joined the British army in 1941. He met Bella in Italy after the war, where they underwent a fictitious marriage in order to allow Bella and her son to immigrate to Palestine. Later, they met again and remained together. *RE*

0773 Alekseyev, Valentin M.: *Varshavskovo getto bolshe ne sushchestvuyet* [The Warsaw Ghetto Does Not Exist Any More]. Moskva: Zvenya, 1998. 160 pp.

A history of the Warsaw ghetto, from its establishment until the end of the ghetto uprising in 1943. Briefly describes the events that preceded the transfer of the Jews to the ghetto. Dwells on the early period, when there was still a semblance of normal existence, and even a social gap between rich and poor. Later, the situation worsened and the ghetto became a realm of starvation and disease. Focuses on the issue of collaboration with the Nazis, both Jewish and Polish. Describes the establishment of the Jewish resistance organizations in the ghetto – the Antifascist Bloc, Jewish Fighting Organization, and Jewish Military Union; the deportation of Jews to Treblinka; and the uprising. Discusses the attitudes of Poles toward the events in the ghetto. Many antisemites supported the Nazi murder of Jews, whereas many others revised their antisemitism; a significant number of Poles tried to help Jews. Pp. 3-12 contain a preface by Aleksandr Kh. Gorfunkel, relating the story of Alekseyev's book: it was written in 1964, but was rejected for publication. *DR*

0774 Appignanesi, Lisa: *Losing the Dead*. London: Chatto & Windus, 1999. 232 pp.

A biography of Hena and Aron Borenstejn, written by their daughter (née Elizabeta) following her trip to Poland in 1997. Her maternal grandparents, Sara and David Lipszyc, lived near Warsaw, with their children Adolf (Adek, b. 1911) and Hena (b. 1915). The family lived together until Hena's husband Aron was taken for forced labor in 1940. Hena and her son Staszek (Stanley, b. 1939) lived as Aryans with Adek, who managed to spirit his parents out of the ghetto to Adamów. Following David Lipszyc's death from typhus while in hiding, Sara joined the others. In 1941 Aron escaped from the Warsaw ghetto and joined the family as well. In 1944, Aron was arrested by the Germans but was freed and settled in Łódź until 1945, when he was arrested by the Russians. He was imprisoned in Radom until 1946, when he was reunited with his family. Following Lisa's birth in 1946 they emigrated to Paris, and then to Montreal in 1951. Lisa also describes antisemitism in Quebec after the war. *TS*

0775 Bacall-Zwirn, Alina; Stark, Jared: *No Common Place: The Holocaust Testimony of Alina Bacall-Zwirn.* Lincoln, NE: University of Nebraska Press, 1999. xix, 124 pp.

A memoir composed of excerpts from interviews by Stark and the Fortunoff Video Archive with Alina Bacall-Zwirn (born Alinka Handszer, 1922, Warsaw) and selections from the unpublished manuscript of her husband, Leo Bacall (born Leshek Bakalczuk in Davidgrodek, Poland, 1915). They were married in the Warsaw ghetto in September 1941, and remained together for much of the war. They were put on a train to Treblinka, but jumped off and returned to Warsaw. After the ghetto uprising in 1943 they were sent to Majdanek, and months later to Auschwitz, where Leo worked in the Buna factory. Alina survived a three-day march and was then transported to Ravensbrueck and Neustadt-Glewe. Both ended the war in Bergen-Belsen. Alina's greatest suffering was giving birth to a child in Auschwitz and having it taken away from her. She partly owed her survival to help from non-Jews and to her camp sister, Ruth Fontak Kugelman, who was also interviewed. *YC*

0776 Bachmann, Klaus: Tykocin im Herbst: Randnotizen aus einem Schtetl ohne Juden. *Tribuene* 152 (1999) 176-185.

Relates the history of several Polish families in Tykocin who sheltered some twelve Jewish fugitives from nearby Kolno in 1941. The Jews were discovered and deported to the Białystok ghetto; all of the Sadowski family, who had hidden six of them, were executed. Two of the Jews, Pejsach Zdrojewicz and his wife, escaped from Białystok and returned to Tykocin; they were taken in by the Chwiesinski family. Although all the neighbors knew about the hidden Jews, they did not betray them. But at war's end, after the Soviet takeover, when the Polish population identified Jews with the hated communists, life in Tykocin became dangerous for both Pejsach and his rescuers, and they had to leave town. *RW*

0777 Bałabuch, Henryk: Antyżydowskie zaburzenia w Lublinie 24-25 kwietnia 1919 roku [Anti-Jewish Disturbances in Lublin on 24-25 April 1919]. *Żydzi w Lublinie: Materiały do dziejów społeczności żydowskiej Lublina, Vol. II*, ed. Tadeusz Radzik. Lublin: Wyd. Uniwersytetu Marii Curie-Skłodowskiej, 1998. Pp. 269-288.

Describes the disturbances in Lublin in which a number of Jews were killed and many Jewish shops and houses were destroyed. Underlines that during this conflict there were also victims among the Poles. The research is based mainly on the Lublin press of the period because police archival material is lacking. States that the press gave various interpretations of the pogrom, each according to its own ideology. The Endecja press presented this pogrom, and other pogroms in Poland in this period, as a "Jewish provocation." It argued that the Jews were not interested in an independent Poland, and that they wanted to fight against it, even if a number of Jews would be killed. The left-wing press saw in the pogroms a provocation of "right-wing elements" that wished, through antisemitic ideology, to distort real Polish problems, e.g. the class struggle. Concludes that tensions between Poles and Jews were perceived by the press only as provocations; the press accounts served the ideological struggle, and no one tried to discover the real causes. *HV*

0778 Berendt, Grzegorz: Rezultaty hitlerowskiej akcji usuwania Żydów z byłego Wolnego Miasta Gdańska [Results of the Nazi Policy of Removing the Jews from What Was Once the Free City of Gdańsk]. *Studia nad Faszyzmem i Zbrodniami Hitlerowskimi* 22 (1999) 285-327.

Analyzes demographic changes in the Jewish population of Gdańsk between 1929-45. The number of Jews constantly decreased because of the policies of Albert Forster, NSDAP leader in Gdańsk. Various extant sources indicate different statistics for the number of Jews in the city. Discusses the causes of the differing data, and attempts to determine the actual numbers by comparing Jewish and non-Jewish sources. States that in 1929 there were ca. 10,500 Jews in Gdańsk and its environs. According to the Jewish community list, in August 1939 there were about 1,660 Jews. Most of the Jews left Gdańsk because of the persecutions which began in 1933. Traces the fate of those who remained in Gdańsk during the war. Pp. 298-311 deal with the fate of mixed families and their children ("Mischlinge"). Two lists are appended: Gdańsk Jews deported to Theresienstadt, and Jews who were in Gdańsk on 1 September 1939 and were found on the list of Jews in Gdańsk in 1945-46. *HV*

0779 Berendt, Grzegorz: *Żydzi na terenie Wolnego Miasta Gdańska w latach 1920-1945 (działalność kulturalna, polityczna i socjalna)* [Jews in the Territories of the Free City of Gdańsk between 1920-1945 (Cultural, Political and Social Activities)]. Gdańsk: Gdańskie Towarzystwo Naukowe, 1997. 327 pp.

Examines various aspects of Jewish life in Gdańsk and its surroundings against the background of political events in this period. The first antisemitic parties were founded at the beginning of the 1920s, but an important change in policy toward the Jewish population began in 1930 when Albert Forster became head of the local NSDAP, which then began to infiltrate all spheres of public life. Many Jews lost their jobs, especially in the fields of administration and culture. In 1937 the local NSDAP decided at its congress to open an anti-Jewish front. The first pogroms took place in October 1937. Describes the Aryanization of businesses, the expropriation of Jewish assets, the liquidation of Jewish institutions, the demolition of the Great Synagogue, etc. Discusses the Jewish reactions to these events: the founding of self-defense organizations and appeals addressed to the Senate and the League of Nations. In November 1939 there were about 1,500 Jews in Gdańsk as opposed to the more than 7,000 living there in 1923; most of the Jews had left the city before the war because of the persecutions. The last chapter (pp. 258-289) deals with the fate of the Jews during the Holocaust. *HV*

0780 Bergmann, Olaf: *Narodowa Demokracja wobec problematyki żydowskiej w latach 1918-1929* [National Democracy and the Jewish Problem in the Years 1918-1929]. Poznań: Wyd. Poznańskie, 1998. 400 pp.

Examines the place of the "Jewish question" in National Democracy (Endecja) ideology, based on the Endecja press and books and pamphlets written by party ideologues. Surveys, inter alia, the views of Endecja regarding the participation of Jews in Polish political life, which it considered as negative, especially the role of the Jews during the Peace Treaty negotiations. In economic life, the role of the Jews was perceived as disastrous because of the large number of Jews in trade and crafts. Endeks demanded "dejudaization" of the Polish economy. Polish culture was considered to be spoiled by the Jews. The Endeks connected socialist and communist ideas, and the Bolshevik Revolution, with Jewish influence in tsarist Russia. Describes various proposals by Endecja to curtail Jewish influence in Poland, such as segregation between Jews and Poles in all spheres, and mass Jewish emigration from Poland. Endecja ideologues believed that all Jews were anti-Polish, and made no distinction even between assimilated Jews and Hasidim. Discusses Endecja propaganda, and the stereotype of the Jew in this propaganda. *HV*

0781 Bogner, Nahum: The Convent Children: The Rescue of Jewish Children in Polish Convents during the Holocaust. *Yad Vashem Studies* 27 (1999) 235-285. Appeared in Hebrew in "Yad Vashem: Kovetz mehkarim" 27 (1999).

Among Catholic institutions, it was mainly the convents that engaged in education and welfare that participated in rescuing Jews in Poland. Examining testimonies and memoirs by both rescuers and rescued, discusses the problems involved in the rescue of Jewish children by the convents. The Jewish public awakened to this possibility only when the genocide was under way; even then, many Jews were against placing Jewish children in these bastions of Christianity. The path to the convents for Jews was difficult; the conditions of life in them were hard. Besides, the convents in Poland were not protected against Gestapo raids. The nuns involved in rescue were in a difficult situation; they did not want to pressure the children to convert, but the convents were missionary institutions and the participation of a non-baptized Jew in Catholic rites could be a sacrilege in their eyes. Many hidden children converted; examines the complicated factors that made them do so. It is difficult to estimate the number of Jewish children rescued by the convents in Poland; most probably, more than 1,000. *DR*

0782

בוגנר, נחום: ילדי המנזרים: הצלת ילדים יהודים במנזרים בפולין בתקופת השואה. יד ושם 27 (1999) 189-226.

[Bogner, Nahum: The Convent Children: The Rescue of Jewish Children in Polish Convents during the Holocaust. *Yad Vashem* 27 (1999) 189-226.] Appeared in English in "Yad Vashem Studies" 27 (1999).

0783 Caumanns, Ute; Niendorf, Mathias: Święty Maksymilian Maria Kolbe a sprawy żydowskie na łamach prasy niepokalanowskiej (1918-1939) [The Saint Maksymilian Maria Kolbe and Jewish Issues in the Niepokalanów Press (1918-1939)]. *Biuletyn Żydowskiego Instytutu Historycznego* 188 (Dec 1998) 20-33. An abridged and revised version of an article which appeared in German in "Der Fremde im Dorf" (1998).

Analyzes the Jewish themes in two Catholic journals published by the Niepokalanów monastery near Warsaw. Kolbe (1894-1941) was the head of the monastery and as such he was responsible for its publications. The monthly journal "Rycerz Niepokalanej" ("Knight of the Immaculate") was founded by Kolbe in 1922, and he was its editor-in-chief. The journal considered liberals, freemasons, socialists, communists, and Jews as the main enemies of Catholicism. The Jews were accused of manipulating the other groups for their own aim – a conspiracy to conquer the world. An ideology of materialism was perceived as common to these groups which threatened Christian faith. "Mały Dziennik" ("Small Daily"), the cheapest Polish Catholic daily, was published in the second half of the 1930s; Kolbe was its supervisor. Jews were presented there in a very stereotypical way, as strangers in Poland and as the symbol of evil. The paper spread belief in the Jewish world conspiracy. It advocated the emigration of Jews from Poland, but not to Palestine which was a holy place for Christians. *HV*

0784 Caumanns, Ute; Niendorf, Mathias: Von Kolbe bis Kielce: Ein Heiliger, seine Presse und die Geschichte eines Pogroms. *'Der Fremde im Dorf': Ueberlegungen zum Eigenen und zum Fremden in der Geschichte: Rex Rexheuser zum 65. Geburtstag,* eds. Hans-Juergen Boemelburg, Beate Eschment. Lueneburg: Institut Nordostdeutsches Kulturwerk, 1998. Pp.

169-194. Unseen. See the abstract for the abridged Polish version: "Biuletyn Żydowskiego Instytutu Historycznego" 188 (1998).

0785

צ'רניאבסקי, אירית: חסידי אומות העולם הפולנים — היבטים סטטיסטיים. דפים לחקר תקופת השואה 14 (1997) 339-365.

[Cherniavsky, Irit: The Polish Righteous among the Nations – Statistical Aspects. *Dapim le-Heker Tekufat ha-Shoah* 14 (1997) 339-365.]

A study based on data found in the minutes of meetings of the public committee for the recognition of "Righteous among the Nations" at Yad Vashem and in files of the committee. The 673 files surveyed amount to about one-quarter of the case-files collected up to 1996. Analyzes the data according to the number of cases of rescue in various regions, the percentage of rescue cases compared to the percentage of Jews in those regions, the type of location (e.g. rural, urban), the sex and age of the rescued, the socioeconomic position of the rescuers, the type of aid provided, etc. Points out that the results do not accurately reflect the rescue situation in Poland, since the records of the committee contain mostly data on applicants who were more educated, or who were in need and hoped to get help from the committee. *LFo*

0786 Cymet, David: Polish State Antisemitism as a Major Factor Leading to the Holocaust. *Journal of Genocide Research* 1, 2 (June 1999) 169-212.

Contends that the Polish official policy of elimination of the Jews both before and during World War II prepared the way for Hitler's extermination of the Jews, carried out by Germany. Prior to the war, Nazi German and Polish aims concerning the Jews coincided to a large extent. The main reason why the Nazis chose Poland as the killing ground of European Jewry was that the Nazi program would not encounter opposition on the part of the non-Jewish population. Examines Polish policy toward the Jews before and during the war, and shows that this policy was not a sporadic outburst of popular antisemitic prejudice but was a well-directed national project to eliminate the Jews. Neither the Home Army nor the government-in-exile did anything to help the Jews in Poland. All of the interwar Polish governments were united in their goal to force the Jews out of Poland. The postwar restoration of Polish independence was accompanied by bloody pogroms perpetrated by the army. Communist and post-communist Poland preserved the antisemitic stance of interwar Poland. *DR*

0787 Davies, Norman: *Auschwitz and the Second World War in Poland = Auschwitz i druga wojna światowa w Polsce: A Lecture Given at the "Representations of Auschwitz" International Conference, at the*

Jagiellonian University, July 1995. Trans. into Polish: Ewa R. Nowakowska. Kraków: Universitas, 1997. 75 pp. In English and Polish.

Analyzes misunderstandings related to the memory of wartime events in Poland due to lack of proper information about their complexity. Few people outside of Poland are aware that Auschwitz was composed of three camps, and that in Auschwitz I thousands of Polish prisoners died. Remarks that it is unfortunate that Auschwitz has been adopted as the international symbol of the Holocaust. Treblinka would be more proper as its symbol. States that those who accuse the non-Jewish population of not helping the condemned Jews are not aware of the terror which affected all the inhabitants of occupied Poland. Discusses, as well, racial segregation, collaboration, and resistance in Poland. Few people know that, after the war, Nazis and Polish non-communist fighters (e.g. members of the Home Army) were imprisoned together by the new communist regime. Soviet misinformation about recent history also led to false understanding of the tragedies which occurred on Polish soil. *HV*

0788 Draenger, Gusta Davidson: *Das Tagebuch der Partisanin Justyna: Juedischer Widerstand in Krakau.* Trans.: Victor Mika. Berlin: Elefanten Press, 1999. 317 pp. Translated from the Polish. Originally published as "Pamiętnik Justyny" (Kraków: Jewish Historical Commission of the Central Committee of Polish Jews, 1946).

The German edition includes the diary (pp. 13-121), along with the additional fragments found in Beit Lohamei Haghetaot – see the English edition: "Justyna's Narrative" (1996). Pp. 123-269 contain a lengthy essay, "Juedischer Widerstand in Krakau," by Jochen Kast, Bernd Siegler, and Peter Zinke. It describes Jewish life in Kraków during the interwar period, the activities of the Zionist youth movements in the city, the Nazi occupation, the Jewish resistance, information about the fighters – those who died and those who survived, Polish-Jewish relations during the war, and Polish postwar historiography. Pp. 271-311 present a translation of excerpts from the underground newsletter "Hehalutz Halohem" printed between August-October 1943 by the resistance fighters (including Draenger and her husband) who escaped from the ghetto into the forest. *SSC*

0789 Eckhaus, Wiktor: *Witus en de jaren van angst: Een reconstructie* [Witus and the Years of Anguish: A Reconstruction]. Lubberhuizen: Uitg. Bas, 1997. 182 pp.

Memoirs of a Jew born in Warsaw in 1930. He, his mother, and his sister survived the war in hiding and moving from place to place. His father was deported, but survived the war. In 1947 the family emigrated to the Netherlands. *MF*

0790 Esch, Michael G.: Die Politik der polnischen Vertretungen im Deutschen Reich 1935 bis 1939 und der Novemberpogrom 1938. *Jahrbuch fuer Antisemitismusforschung* 8 (1999) 131-154.

Because of the economic crisis, the Polish government encouraged emigration, and especially of Jews, who were considered particularly undesirable; it was said that many were paupers, others politically subversive, and all formed a harmful foreign element in the Polish nation. Certainly the Poles opposed the return of those who had already emigrated. Soon after the Anschluss, when the return of Polish Jews, especially from Austria, became a flood, the Foreign Ministry instructed the consulates to prepare the cancellation of Polish citizenship of large categories of Jews living in Germany and Austria; on 31 March the Sejm passed a law to this effect. Although on 28-29 October, when Germany arrested and expelled Polish Jews, some Polish consuls protected them, and although the consuls were appalled at the "Kristallnacht" pogrom, they continued the process of expatriation. *RW*

0791 Furth, Hans G.: One Million Polish Rescuers of Hunted Jews? *Journal of Genocide Research* 1, 2 (June 1999) 227-232.

Contends that the number of Poles who rescued Jews during World War II is greatly underestimated. Lists eight factors leading to this underestimation: not all the rescuers survived, not all the rescued survived, many rescued and rescuers never spoke about it after the war, etc. Believes that the number of Polish wartime rescuers may be estimated at 720,000; and, together with short-term helpers, at 1,200,000. Criticizes the Jewish writings which emphasize the abandonment of the Jews. *DR*

0792 Geldman, Adam: *Smertnyi prigovor odinnadtsati millionam* [A Death Sentence to Eleven Million]. San Francisco: [The Author], 1997. 218 pp.

A collection of essays dealing with the Holocaust. Geldman was born in Lublin in 1922; in September 1939 he fled to Soviet-occupied territory and survived the Holocaust, but the rest of his family remained in Lublin and perished. The essays deal with the Wannsee Conference; victims and survivors of the Holocaust, such as Dawid Rubinowicz, Janusz Korczak, Hanna Szenes, Mordechaj Gebirtig, Itzhak Katzenelson, Hirsch Glik, and Shlomo Perel; the Majdanek concentration camp; the annihilation of Lublin's Jewish community; and the uprising in the Warsaw ghetto. The texts are interspersed with Geldman's personal recollections and photographs of his family, as well as those made during Geldman's trip to Poland in 1994. *DR*

0793 Gołota, Janusz: Losy Żydów ostrołęckich w czasie II wojny światowej [The Fate of the Jews in Ostrołęka during World War II]. *Biuletyn Żydowskiego Instytutu Historycznego* 187 (Sept 1998) 23-35.

In the 1930s there were about 7,000 Jews in Ostrołęka (north of Warsaw) and its surroundings, about one third of the whole population. Describes events after the occupation of the town by the Nazis. In October 1939 this region was annexed to Germany. The Jews were driven out of the town. Some of them were taken to labor camps and some escaped to Vilna, Łomża, Białystok and other localities occupied by the Soviets. Some remained in these localities and some were deported by the Soviets to Siberia. In June 1941 many of these towns were occupied by the Nazis. Ghettos were founded and thousands of Jews were killed; those who survived the "actions" were deported to Treblinka when the ghettos were liquidated in 1943. Traces the fate of several Jewish families from Ostrołęka whose members survived the war, mainly due to deportation to Siberia. Gołota interviewed them in Israel in the mid-1990s. Less than 8% of Ostrołęka's Jews survived the war. *HV*

0794 Gross, Jan Tomasz: *Upiorna dekada: Trzy eseje o stereotypach na temat Żydów, Polaków, Niemców i komunistów, 1939-1948* [Ghastly Decade: Three Essays about Stereotypes on Jews, Poles, Germans and Communists, 1939-1948]. Kraków: Universitas, 1998. 124 pp.

Contents: "Ten jest z Ojczyzny mojej...", ale go nie lubię ["He Is My Compatriot...", but I Don't Like Him. An expanded version of an article which appeared in "Aneks" 41-42 (1986), and in English in "Archives Européennes de Sociologie" 27 (1986) and in "Dissent" 34 (1987).] (25-60); "Ja za takie oswobodzenie im dziękuję i proszę ich żeby to był ostatni raz" [For Such Liberation I Thank Them and I Ask That It Not Be Repeated. Describes the situation of the Jews in eastern Poland following the Soviet occupation in September 1939 and reflects on Polish stereotypes related to the behavior of the Jews.] (61-92); Cena strachu [The Price of Fear. On antisemitism in Poland in the years 1945-1946.] (93-113). *HV*

0795

הורביץ, אריאל: מצעד המוות של יהודי חלם והרובייישוב לעבר הנהר בוג בדצמבר 1939.
ילקוט מורשת 68 (אוקט' 1999) 51-68.

[Hurwitz, Ariel: The Death March of the Jews of Chełm and of Hrubieszów to the River Bug in December 1939. *Yalkut Moreshet* 68 (Oct 1999) 51-68.]

Ca. 4,000 Jewish men aged 15-60, from Chełm and Hrubieszów, were forced to march during 29 November-5 December 1939 toward the Bug, to the border between Nazi-occupied and Soviet-occupied Polish lands, probably as part of the Nazi effort to drive Jews into Soviet-held territories. The SS-commander of the operation took this opportunity to implement his self-initiated murderous actions. Ca. two-thirds of the Jews were murdered during the forced march. Those remaining were driven over the border and back, again and again, until the Nazi formation in charge of the march left the area. Testimonies by survivors of the march and Holocaust survivors from that area relate to the march, but it is not mentioned in German wartime records and in postwar war crimes trials. *LFo*

0796

אימבר, שולמית ואחרים: חיי היום-יום בגטו ורשה 1941: יחידת לימוד לחטיבת הביניים ולחטיבה העליונה: מדריך למורה וחוברת לתלמיד. ירושלים: יד ושם, המדור לפיתוח תוכניות – מחלקת החינוך, 1996. 80; 55 עמ'.

[Imber, Shulamit et al.: *Day-to-Day Life in the Warsaw Ghetto 1941: A Study Unit for Junior High and High Schools: Teacher's Guide and Student's Manual.* Jerusalem: Yad Vashem, Education Department, 1996. 80; 55 pp.]

In the teacher's guide, pp. 5-34 describe the content of slides to be shown to students depicting life in the ghetto. Pp. 36-80, "Ha-ghetto" by Yisrael Gutman, is a shorter version of his essay "Yehudei Varsha bi-yemei ha-Shoah," which appeared in "Toledot yehudei Varsha" (Jerusalem: Keter, 1991). Gutman describes life in the ghetto, including the smuggling of food and other goods, the role of the Judenrat and the Jewish police, deprivation, and death, as well as educational, religious, and cultural activities which took place despite the ghetto conditions. The student's manual contains photographs of life in the ghetto, appropriate excerpts from books on the subject, and questions for discussion. *RE*

0797

כהן, אברהם: עסייען. מעלבורן: פארלאג "אונדזער געדאנק", 1996. 130, 56 עמ'.

[Kahan, Arcadius: *Essays*. Melbourne: "Undzer Gedank," 1996. 130, 56 pp.]

A collection of articles by Kahan (Vilnius, 1920 – Chicago, 1982), in English and in Yiddish, and articles in memory of Kahan, in Yiddish. Kahan was active in the Bund. Includes articles about economic and social issues of the Jewish people, the Jewish labor movement in various countries (including the Bund), YIVO, the Jewish community of Vilnius, and the Holocaust period. The essay "A Day in the Ghetto" (pp. 18-35) relates to the ghettos of Warsaw and Łódź during the Nazi occupation. It deals with the German administration and organization of the two ghettos, the daily routine and conditions in the ghettos, and differences between the two ghettos.

0798 Konieczny, Alfred: Żydowski obóz pracy przymusowej w Jeleniowie z lat 1943-1944 (ZAL Gellenau) [The Jewish Labor Camp in Jeleniów in the Years 1943-1944 (ZAL Gellenau)]. *Studia nad Faszyzmem i Zbrodniami Hitlerowskimi* 22 (1999) 411-426.

States that according to the information given by the Red Cross Office in Arolsen in 1979, the labor camp in Jeleniów (Lower Silesia) functioned between 1941-44. Based on analysis of documents, concludes that the camp was probably organized in March 1943 (and not 1941), and that Jewish women were brought there the same month through the Organisation Schmelt. They worked for the Dierig textile factory. Describes the organization of the work there, and deals with the financial aspects of the forced labor; quotes sums earned by Organisation Schmelt due to the forced labor of Jewish women. In March 1944 the women were moved to the labor camp in Bielawa (Langenbielau), a sub-camp of Gross-Rosen. The Jeleniów camp was closed in summer 1944 because of the liquidation of the Organisation Schmelt. A list of 192 Jewish women who worked in Jeleniów is appended (pp. 421-426). States that it is a unique list, because other lists from the Organisation Schmelt have not been found up to now. *HV*

0799 Korboński, Stefan: *Polacy, Żydzi i Holocaust* [Poles, Jews and Holocaust]. Warszawa: Fundacja Pomocy Antyk; "Wydawnictwo Antyk – Marcin Dybowski," 1999. 176 pp. Originally published as "The Jews and the Poles in World War II" (New York: Hippocrene Books, 1989).

0800 Landau-Czajka, Anna: The Jewish Question in Poland: Views Expressed in the Catholic Press between the Two World Wars. *Polin* 11 (1998) 263-278. This article is the second part of the essay "The Image of the Jew in the Catholic Press during the Second Republic," which appeared in "Polin" 8 (1994).

Surveys Catholic periodicals of the interwar period and analyzes their attitude to the Jewish minority. Presents demands made by the Catholic press concerning Jews, like boycott of the Jews, separation of Poles from Jews by introducing a numerus clausus and numerus nullus, and even introduction of closed ghettos for Jews. The complete exclusion of the Jews from Polish society was to be a step toward the final goal of emigration of the Jews from Poland. Catholic writers expressed strong opposition to Jewish settlement in Palestine and to the Zionist movement, considering these a threat to Christian rights in the Holy Land. Discussing extensively the need for "de-Jewing" Poland, the writers reflected on whether Christian teachings and the doctrine of loving one's neighbor are compatible with the anti-Jewish campaign. Shows how various journalists solved this problem. Readers of the Catholic press were persuaded that the struggle against the Jews was a legitimate act of self-defense. The number of antisemitic articles in the Catholic press grew substantially in the second half of the 1930s. *HV*

0801 Lasman, Noach: *Die Strasse: Erinnerungen eines juedischen Zwangsarbeiters an eine "ganz normale Firma."* Trans.: Danuta Dombrowska. Muenster: Waxmann, 1999. 175 pp. Originally published as "Ha-kvish" (Tel-Aviv: Ministry of Defence, 1994).

0802 List-Pakin, Jeanne: La campagne autour de Kielce: Le Journal de David Rubinowicz. *Revue d'Histoire de la Shoah* 166 (May-Aug 1999) 166-194.

Quotes and comments on large fragments from the diary by Dawid Rubinowicz, a Jewish boy from Krajno, a village near Kielce. He began to write his diary in March 1940 when he was 12 years old; the last pages were written in June 1942. In September 1942 the Jewish population of the Kielce region was killed in the forests. Part of the population was deported to Treblinka. The diary was found in Poland in 1957 and published there in 1960. *HV*

0803 List-Pakin, Jeanne: Le ghetto d'une ville de province – Kielce: Son instauration et sa liquidation. *Revue d'Histoire de la Shoah* 166 (May-Aug 1999) 144-165.

Recapitulates and comments on the chapter dealing with the fate of the Jews in Kielce during World War II from Krzysztof Urbański's "Kieleccy Żydzi" (1993). Adds information on life in the Kielce ghetto from the letters of Gertrude Zeisler ("I Did Not Survive: Letters from the Kielce Ghetto," Jerusalem, 1981) and from various testimonies from ŻIH (the Jewish Historical Institute in Warsaw). *HV*

0804 Markowski, Mieczysław Bolesław; Gapys, Jerzy: Konflikty polsko-żydowskie w województwie kieleckim w latach 1935-1936 [Polish-Jewish Conflicts in the Kielce Region in the Years 1935-1936]. *Biuletyn Żydowskiego Instytutu Historycznego* 192 (Dec 1999) 41-93.

States that there were two periods of sharp conflict between Poles and Jews in interwar Poland: the first, in the immediate postwar years (1918-1921), the second – in the period of growing nationalism (1935-1939). Describes the situation in the Kielce region in the years of crisis and growing influence of the Endecja party, which provoked violent anti-Jewish demonstrations. Pp. 46-93 contain documents from the Kielce State Archives, including police reports and files of the Regional Office in Kielce concerning information on disturbances in Odrzywół in November

1935, in Kłobuck and Truskolasy in January 1936, and in Kielce in April 1936. *HV*

0805 Mueller-Madej, Stella: *Le livre de Stella: Jeune fille de la liste de Schindler.* Trans.: Marcelo Gandaras. Paris: Editions du Félin, 1997. 300 pp. Unseen. Originally published as "Oczami dziecka" (Kraków, 1991). See the abstract for the German translation: "Das Maedchen von der Schindler-Liste" (Augsburg: Oelbaum Verlag, 1994).

0806 Nelken, Halina: *And Yet, I Am Here!* Trans.: Halina Nelken, Alicia Nitecki. Amherst: University of Massachusetts Press, 1999. xii, 277 pp. Originally published as "Pamiętnik z getta w Krakowie" (Toronto: Polski Fundusz Wydawniczy w Kanadzie, 1987). See the abstract for the German translation: "Freiheit will ich noch erleben: Krakauer Tagebuch" (Gerlingen: Bleicher, 1996).

0807 Ney, Arthur: Pamiętnik [A Memoir]. *Biuletyn Żydowskiego Instytutu Historycznego* 189 (Mar 1999) 49-70.

Memoirs of a Jew born in Warsaw in 1930. The memoirs were written immediately after the war, and given to the Jewish Historical Institute in Warsaw by the author in 1998. Describes the life of his family (parents and an older sister) before the war in Warsaw and later in the Warsaw ghetto. Ney would leave the ghetto periodically in search of food on the "Aryan side." The last time he left the ghetto was the night of 18 April 1943, the night before the uprising. He does not know if his parents and sister died during the uprising or were deported afterwards. He survived in the villages outside of Warsaw, working for peasants. In 1944 he returned to Warsaw and took part in the insurrection of August 1944. After the war he left Poland, went to France, and later settled in Canada. *HV*

0808 Oehrl, Christina: Ueberleben in den Waeldern: Facetten der juedisch-polnischen Widerstandsbewegung waehrend der deutschen Okkupation von 1939 bis 1945. *Sachor* 7 (1997) 67-79.

Explains that the Polish government-in-exile, as well as the PKP and the Armia Krajowa in Poland, opposed partisan activity in general (as a dangerous provocation of German reprisals), and communist and Jewish partisan activity in particular (as a threat to their political position and to public order, and due to antisemitism). Jewish and communist partisans were equated to criminal gangs. This attitude encouraged Poles to denounce Jewish partisans to the Germans. Only the communist partisan groups were willing to accept Jews as members, provided they were fit to fight, and to cooperate with Jewish groups. Notes the impossibility of surviving alone after escaping from a ghetto or camp, and the odds against locating a partisan group and being accepted. Describes partisan activity in the Zamość region after the Germans decided to expel the Poles, some of whom fled to the forest and cooperated with Jewish and communist groups already active there. Another form of partisan group, especially in Belarus, was the family camp, e.g. that of the Bielski brothers, which accepted women, old and frail people, and children. *RW*

0809

פפירבלט, מרדכי: מבריח הקרפיונים: 900 יום במחנות אושוויץ: יומן אישי. [תל-אביב]: משרד הבטחון, 1996. 500 עמ'.

[Papirblat, Mordechai: *The Carp Smuggler: 900 Days in the Auschwitz Camps: A Personal Diary.* Tel-Aviv: Ministry of Defence, 1996. 500 pp.]

Memoirs of a Jew who was born in 1925 in Radom and grew up in Warsaw. Describes the Nazi invasion in 1939 and life in the Warsaw ghetto until his escape in April 1941. He wandered through several towns, working as a non-Jew, until he was arrested in July 1942 and sent to Auschwitz. Papirblat was supervisor of a group that prepared fish ponds for carp, among other tasks; he and his workers sometimes ate carp secretly or smuggled them into the camp. He was eventually caught smuggling, punished with a beating, and sent for harder labor with another group, and later to the Jaworzno labor camp. In January 1945 the prisoners were taken on a death march to the Blechhammer camp, from where Papirblat escaped with a few others. They hid in forests and villages until the arrival of the Russian army. After the war Papirblat emigrated to Palestine. *RE*

0810 Pasztor, Maria: Położenie mniejszości żydowskiej w Polsce w świetle francuskiej korespondencji dyplomatycznej z lat 1921-1939 [The Situation of the Jewish Minority in Poland in the Light of French Diplomatic Correspondence in the Years 1921-1939]. *Biuletyn Żydowskiego Instytutu Historycznego* 188 (Dec 1998) 3-19.

Surveys reports written by French diplomats in Poland concerning the Jewish minority and their treatment by the Polish state. Hector de Panafieu, J. Laroche and Léon Noël were consecutive French ambassadors in Poland. Most of the reports show a negative attitude on the part of the diplomats toward the Jewish minority and their role in Poland. Jewish politics was judged as particularistic, balancing between left and right for their own profit. The Jews in general were perceived by ambassadors as unfriendly to Poland. Discusses, among others, the descriptions of antisemitic incidents in Poland, the attempts to introduce a numerus clausus at the universities, and the growth of antisemitism after Piłsudski's death, including its causes. States that only Léon Noël, who served in Poland between 1935-39, observed the real problems of the Jews. Concludes that the opinions of the ambassadors were evidently influenced by their own stereotypical perception of the Jews and by the opinions of Endecja politicians who were among their friends. *HV*

0811 Polen, Nehemia (Nathan): Die Kinder und das Antlitz Gottes: Das Zeugnis des Rabbi Kalonymos Kalmish Shapira (1889-1943). *Kultur allein ist nicht genug: Das Werk von Elie Wiesel – Herausforderung fuer Religion und Gesellschaft,* eds. Reinhold Boschki, Dagmar Mensink. Muenster: LIT, 1998. Pp. 190-197.

Summarizes Shapira's last work (published after the war as "Esh kodesh"), a mystical interpretation of the Holocaust, written in the Warsaw ghetto. He states that the Nazis' war against the Jews is a war against God, whose yoke they bear; their suffering ennobles them. God himself suffers, but in secret; were his suffering revealed, it would destroy the world. Notes that Shapira was particularly concerned for the children in the ghetto, who were even more defenseless than the adults (and often had to support the adults by smuggling food). Discusses and quotes from his sermon of 27 June 1942; it, too, is an affirmation of God as the source and end of all, but also an outcry against the slaughter of the children. *RW*

0812

רגב, מנחם, עורך: לגעת באדם: דמותו והגותו של יאנוש קורצ'אק כפי שהן משתקפות בכתביו ובכתבי אחרים. ירושלים: אקדמון, 1996. 336 עמ'.

[Regev, Menachem, ed.: *To Touch the Man: The Figure and Philosophy of Janusz Korczak as Reflected in His Writings and in the Writings of Others.* Jerusalem: Akademon, 1996. 336 pp.]

A collection of excerpts from writings, by Korczak himself and by others, on his life and his educational philosophy. On pp. 146-149, Korczak describes, inter alia, antisemitic experiences in his childhood. Pp. 179-209 contain excerpts on the Holocaust period, by Korczak and others, relating his activities on behalf of the children in the Warsaw ghetto and particularly those in his orphanage. Korczak attempted, through his writings for children and his plays which were performed at the orphanage, to teach the children the importance of leading a moral life and facing death with dignity. Describes Korczak's efforts to maintain his own human dignity, as well as a semblance of normal life in the orphanage. Korczak and the children were deported to Treblinka, where they were murdered. *RE*

0813 Reinke, Andreas: *Judentum und Wohlfahrtspflege in Deutschland: Das juedische Krankenhaus in Breslau, 1726-1944.* Hannover: Verlag Hahnsche Buchhandlung, 1999. xii, 351 pp.

A study of the Jewish welfare system in Breslau between 1726-1944. Pp. 240-300 discuss the role and functioning of the Jewish Hospital during the Nazi period. Describes the consequences for the hospital of the Nazi boycott and persecution of the Jews, the "Kristallnacht" pogrom in Breslau, and the confiscation of the hospital at the outbreak of World War II. The hospital suffered from a declining number of physicians and nurses; patients were eventually transferred to a small hospital for incurables, and at the end of 1939 to a building of the Jewish community. In 1941 ca. 9,000 Jews lived in Breslau; after the deportations in October 1942, 3,800 Jews remained, under the care of only 17 medical staff members. Eventually, physicians and nurses were requested by the Gestapo to control the persons assigned for deportation, and to prepare them for it. In June 1943 the hospital was closed; the remaining patients were deported to Theresienstadt. *ID*

0814 Rokicki, Jarosław: Ks. dr Stanisław Trzeciak (1873-1944): Szkic biograficzny [The Priest Dr. Stanisław Trzeciak (1873-1944): A Biographical Essay]. *Biuletyn Żydowskiego Instytutu Historycznego* 190 (June 1999) 43-54.

Presents the figure of Stanisław Trzeciak, a Polish priest who was well-known in the interwar period for his extremely antisemitic opinions. He published numerous articles and pamphlets, and delivered public lectures, on the "Jewish question." He promoted isolation of the Jews from Polish society through boycotts and the numerus clausus. He believed that the persecuted Jews would emigrate from Poland en masse. In the mid-1930s, Trzeciak participated in the meetings of Weltdienst (Institut zur Erforschung der Judenfrage) in Erfurt. He upheld the authenticity of the "Protocols of the Elders of Zion," and fought against Jewish "shechitah" (ritual slaughter). He praised Hitler for his war against the Jews, but he was opposed to physical violence. During the war, Trzeciak took part in a conspiracy against the Nazis and gave moral support to the nuns who hid Jewish children. He was shot and killed during the Warsaw uprising in 1944. *HV*

0815 Sakowska, Ruta, ed.: *Archiwum Ringelbluma: Konspiracyjne Archiwum Getta Warszawy. Vol. 1: Listy o Zagładzie* [Ringelblum Archives: Clandestine Archives of the Warsaw Ghetto. Vol. 1: Letters about the Holocaust]. Warszawa: Wydawn. Naukowe PWN, 1997. xxix, 381 pp. Polish, German, and Yiddish.

Presents 149 letters from the Ringelblum Archives. They were sent from various places, mostly ghettos and labor camps outside of Warsaw, to family and friends in the Warsaw ghetto. All the letters were written in 1942 and give information, sometimes cryptically, about the mass murder of Jews or the fate of many Jews. Part of them are addressed to the

Judenrat and contain requests for information about the address or fate of somebody, mainly a member of the family. Six letters were thrown out from the train during the transport of Płońsk Jews to Auschwitz. Most of the letters were written in Polish or Yiddish, some of them in German. Includes facsimiles. Pp. xi-xxix contain the editor's preface, dealing with the figure of Emanuel Ringelblum and the story of the "Oneg Shabbat" organization and its archives. *HV*

0816

שחר, דוד: הטבח ביהודי קרסנושילץ והקמת אתר ההנצחה לזכרם. ילקוט מורשת 68 (אוקט' 1999) 35-49.

[Shachar, David: The Massacre of the Jews of Krasnosielc and the Foundation of the Memorial Site to Commemorate Them. *Yalkut Moreshet* 68 (Oct 1999) 35-49.]

Reconstructs the chain of events related to the mass murder perpetrated in Krasnosielc, Poland on 5 September 1939, during the Nazi invasion of the country, when ca. 50 Jews were murdered by SS and Wehrmacht units. The murderers were charged with unjustified murder; they were court-martialed by the German military authorities and sentenced to imprisonment. Introduces German records related to the case, collected by the author, a son of one the Jewish victims; he was nine years old when he witnessed the events. States that the massacre was probably the first case of mass murder of Jews in Poland during the Holocaust, and the only one for which the perpetrators were charged with unjustified murder by the authorities. Also relates his successful activities during the 1980s-90s to commemorate the victims. *LFo*

0817

שחם, ישראל: עתונות-המחתרת היהודית בווארשה. כרך ששי: מארס 1942-יולי 1944. ירושלים: יד ושם, 1997. 632 עמ'.

[Shaham, Israel: *The Jewish Underground Press in Warsaw. Vol. 6: March 1942-July 1944*. Jerusalem: Yad Vashem, 1997. 632 pp.] Vols. 1-5 were published in 1979-1993.

Presents 51 issues of various Warsaw ghetto periodicals found in the Ringelblum Archives. All but two of the issues appeared between March 1942-January 1943; most of them appeared before the first deportation in July 1942. They contain ghetto news, as well as reports on the murder of Jews in concentration camps, especially in Chełmno, and the murder of Jews in the East by the Einsatzgruppen. Ca. 60% of the issues presented are youth movement periodicals. These were less youth-oriented and directed more at the ghetto inhabitants in general in late 1942-early 1943; they include calls for rebellion and armed resistance. *RE*

0818 Shay, Arnold: *From Bendzin to Auschwitz: A Journey to Hell.* Hanover, MA: Christopher Pub. House, 1996. 320 pp.

Memoirs of a Jew from Będzin, Poland, born in 1922 as Abram Szyjowicz. Describes the German occupation, the Judenrat (headed by Moniek Merin), the establishment of the ghetto in the Kamionka suburb on 1 January 1942, murders of Jews in the town, and deportations. He condemns the Jewish collaborators (the Judenrat and Jewish police) who facilitated the Nazis' murder of Jews. Shay and his father obtained work at a German tailor shop. The ghetto was liquidated in August 1943. Shay was sent to Auschwitz; he survived, but his family perished. Mentions some events in Sosnowiec and other towns in the vicinity (the establishment of Jewish Councils, ghettos, deportations, etc.). After the war he settled in the USA. Includes illustrations and facsimiles of documents, as well as poetry and other writings by Shay. *DR*

0819

שרגר, רבקה: הבריחה אל החיים (זכרונות ילדות). [חולון: המחברת], 1996. 277 עמ'.

[Shrager, Rivkah: *The Flight to Life (Memoirs of Childhood)*. Holon: The Author, 1996. 277 pp.]

Memoirs of a Jew born in 1933 in Brzozów, Poland. Describes the Nazi occupation in 1939. Her father fled to Soviet-occupied territory; in 1941 he was killed by a German bomb. Most of the Jews in Brzozów were killed by the Nazis in August 1942. Shrager fled with her mother and younger brother to the countryside. Mentions that Polish acquaintances who had been paid in advance refused to help them; one family that agreed to hide them stole their money after a few months and threw them out. They fled to the Przemyśl ghetto, where they suffered from hunger and disease, and where they hid in a bunker with some other Jews for eleven months. Shrager's mother died there; the others were liberated in April 1945. Shrager and her brother emigrated to Israel in 1948. *RE*

0820

שטרנברג, שמואל: זכרונות מעולם שאבד וקורות חיי. חלק א: דוקלא. פתח תקוה: [המחבר], 1996. 160 עמ'.

[Shternberg, Shmuel: *Memoirs from a Lost World and the Story of My Life. Part 1: Dukla.* Petach Tikvah: The Author, 1996. 160 pp.]

Describes Jewish life in the town of Dukla (southeastern Poland), where ca. 1,200 Jews lived in 1939. Pp. 110-160 relate events from the German occupation in September 1939; they include memoirs of Shternberg (b. 1924), an Orthodox Jew, and testimonies of other residents. Despite their persecution and humiliation, they were determined to maintain religious practices. Most of the Jews were expelled from the town shortly after the occupation, including Shternberg and his father, who fled to Soviet-controlled territory. His mother and two brothers did not manage to get out in time and remained in Dukla. Describes the suffering of the Jews in Dukla and their deportation in summer 1942 to various camps, where most of them were killed, including Shternberg's mother and brothers. *RE*

0821 Stark, Jared: The Task of Testimony: On "No Common Place: The Holocaust Testimony of Alina Bacall-Zwirn." *History & Memory* 11, 2 (Fall-Win 1999) 37-61.

Reflections of a former worker at Yale University's Fortunoff Video Archive for Holocaust Testimonies who, in 1993-97, worked with Holocaust survivor Alina Bacall-Zwirn and her family on her memoirs, "No Common Place: The Holocaust Testimony of Alina Bacall-Zwirn" (1999). Stark decided to base the book on verbatim transcripts of Bacall-Zwirn's testimony, to avoid the distorting mediation of any less neutral mode of recording her tale. Reflects on the relation between the survivor's traumatic past and the attempt to retell the story. A gap between memory and narration remains even in a taped account, and Stark's experience with Bacall-Zwirn shows how the witness tries to overcome this gap. Illustrates this with an excerpt from Bacall-Zwirn's text, relating her escape from a transport train on its way to Treblinka; her text is juxtaposed with an excerpt on the same episode taken from Lawrence Murphy's "One Small Candle," based on Bacall-Zwirn's life story. Concludes that, with all its shortcomings, oral testimony surpasses all other ways of preserving and commemorating the personal Holocaust experience. *DR*

0822 Stiffel, Frank: *Aquele anel trouxe de volta meus dedos: Memórias de um sobreviviente do inferno.* Trans.: Dirce de Assis Cavalcanti. Rio de Janeiro: Exodus, 1997. xv, 328 pp. Originally published as "The Tale of the Ring – a Kaddish: A Personal Memoir of the Holocaust" (Wainscott, NY: Pushcart Press, 1984).

Memoirs of a Jew (b. 1916) describing his experiences in the Holocaust – his deportation from Lvov to Treblinka, his escape from the camp, and his recapture and internment in Auschwitz until the end of the war. *SSC*

0823 Tych, Feliks: Witnesses of Shoah: The Extermination of Jews in Polish Diaries, Memoirs and Reminiscences. *Biuletyn Żydowskiego Instytutu Historycznego* 192 (Dec 1999) 3-18.

Analyzes the results of research on more than 400 published and unpublished diaries and memoirs of the Second World War written by Poles. Surveys those parts of the texts which refer to the fate of the Jews.

Many authors, from all the political orientations, including left-wing well-known personalities, fail to mention the Holocaust. Some authors insist that Jews and Poles shared a common fate under the Nazi occupation. Discusses the common view that there are two truths relating to Polish-Jewish relations during the Nazi occupation: a Polish truth and a Jewish truth. Asserts that there are two Polish truths: that of those who saved Jews and that of those who had little idea concerning the fate of the Jews and the rescue activities. Notes a variety of attitudes towards the Jews and the Holocaust, and the moral devastation of Polish society caused by the Holocaust. Stresses psychological aspects of the witnessing of Nazi crimes, and the detrimental effect on the population. States that before the war, before the Holocaust, a pogrom such as occurred in Kielce in 1946 would have been impossible. *HV*

0824 Urbański, Krzysztof: *Powstanie, rozwój i zagłada gminy żydowskiej w Modrzejowie* [The Foundation, Development and Annihilation of the Jewish Community in Modrzejów]. Kielce: Agencja "JP," 1998. 83, [6] pp. Includes an English summary (pp. 67-74).

A history of the Jewish community in Modrzejów, a small Silesian town, today a neighborhood of Sosnowiec. The first Jews settled there at the beginning of the 18th century. In 1939 there were 5,882 inhabitants, 1,670 of them Jews. Pp. 51-58 deal with the Holocaust. The Nazis occupied Silesia in September 1939 and annexed it to the Third Reich. The ghetto was founded in Modrzejów in November 1939. Transports to Auschwitz took place between May 1942-May 1943; most of the Jews of Modrzejów were murdered there. Some young Jews were taken to labor camps in Germany; about 60-80 of them survived the war. Pp. 75-80 contain lists of the Jewish inhabitants of Modrzejów in the 1930s. *HV*

0825 *"Les vrais riches" – Notizen am Rand: Ein Tagebuch aus dem Ghetto Łódź (Mai bis August 1944).* Eds.: Hanno Loewy, Andrzej Bodek. Leipzig: Reclam, 1997. 165 pp.

This diary of an anonymous young man was written on the blank pages and in the margins of the novel "Les vrais riches," alternately in English, Polish, Hebrew, and Yiddish. (A facsimile is appended on pp. 101-159.) He describes the torments of hunger and expresses the despair that makes him wish for his own death and the destruction of the whole world. News of the alleged assassination of Hitler and the progress of the Soviet army inspire momentary hope, and he plans his life after the war. The diary ends with the speech of the Nazi commander of the Łódź ghetto, Biebow, calling on all the remaining Jews to report for deportation. The editors' preface (pp. 5-34) traces the history of the Łódź ghetto and discusses the many Holocaust diaries that have been discovered, especially those from Łódź. *RW*

0826 Weinstock, Nathan, ed.: La grande déportation de Varsovie (juillet-septembre 1942). *Revue d'Histoire de la Shoah* 162 (Jan-Apr 1998) 164-181.

Presents three texts from the Ringelblum Archives written by Yisroel Lichtensztajn, who worked with Ringelblum in the Warsaw ghetto. Lichtensztajn participated in the action to save the archives. The first two texts describe the deportation from the ghetto in summer 1942, the conditions in the ghetto, and the people he knew. The third text is his own last testament. Lichtensztajn's wife and daughter were murdered in Treblinka in 1943; the date of his death is unknown. Weinstock translated the texts from Yiddish, and describes the figure of Lichtensztajn (pp. 164-171). *HV*

0827

ויסברג, דב (ברוניסלב): התעוררתי משריקת הקטר. תל-אביב: פרוינד, 1998. 173 עמ'.

[Weissberg, Dov (Bronisław): *I Awoke from the Engine's Whistle*. Tel-Aviv: Freund, 1998. 173 pp.]

Memoirs of a Jew born in Lvov in 1929. In August 1942 his mother was deported with ca. 50,000 other Jews from Lvov. In December his father sent him to Warsaw to live with two uncles who had escaped from a labor camp and were living on the "Aryan side" with false documents. Weissberg, who also had Aryan papers, stayed with various families in Warsaw. Describes the Polish uprising in summer 1944, after which he was evacuated from the city with the other inhabitants. He was sent to a refugee camp near Częstochowa, and then lived with a Polish family in the area. After the war he returned to Warsaw and lived with relatives until 1950, when he emigrated to Israel. His parents and brother perished in the Holocaust. *RE*

0828 Weissberg, Dov (Bronisław): *I Remember....* London: Freund, 1998. viii, 184 pp. Originally published as "Hitorarti mi-shrikat ha-katar" (Tel-Aviv: Freund, 1998).

0829 Weissberg, Dov (Bronisław): *Zbudzil mnie gwizd parowozu* [I Awoke from the Engine's Whistle]. Londyn: Freund, 1999. 190 pp. Originally published as "Hitorarti mi-shrikat ha-katar" (Tel-Aviv: Freund, 1998).

Portugal

0830 Janeiro, Helena Pinto: *Salazar e Pétain: Relações luso-francesas durante a Segunda Guerra Mundial (1940-44)* [Salazar and Pétain: Luso-French Relations during the Second World War (1940-1944)]. Lisboa: Cosmos,

1998. xviii, 272 pp. Based on the author's M.A. thesis – Universidade Nova de Lisboa, 1995.

Discusses similarities and mutual influences between the regimes of Salazar in Portugal and Pétain in Vichy France. States that Pétain looked to the corporative Estado Novo in Portugal as a model. Pp. 91-102, "O limiar do antisemitismo," deal with the attitude of Salazar's regime toward the Jews. As early as 1939 Salazar ordered that entry visas be denied to Jews seeking refuge; he severely punished Mendes, the consul in Bordeaux, for disregarding this order. When in October 1942 the property of Portuguese Jews residing in France was about to be Aryanized, Salazar and his ambassador to Vichy, José Caeiro da Mata, appointed Portuguese administrators in order to keep this property in Portuguese hands after the Jews were deported. On the other hand, thousands of Jewish refugees found asylum in Portugal as a transit point for emigration out of Europe. *AA*

0831 Milgram, Avraham: Portugal, the Consuls, and the Jewish Refugees, 1938-1941. *Yad Vashem Studies* 27 (1999) 123-155. Appeared in Hebrew in "Yad Vashem: Kovetz mehkarim" 27 (1999).

Portugal was not regarded by Jewish refugees as a possible haven, but rather as an important transit point on their way to the Americas. Portugal was reluctant to admit foreigners into the country, although the motives of Salazar and his officials were more isolationist than antisemitic. Nonetheless, some Portuguese ambassadors and consuls disregarded their instructions and allowed many Jewish refugees into Portugal – e.g. Alberto da Veiga Simões, ambassador to Germany, and Aristides de Sousa Mendes, consul in Bordeaux. In stark contrast with other neutral countries, such as Brazil, the Portuguese consular corps proved to be free of antisemitic sentiments and sensitive to the fate of Jews. *DR*

0832

מילגרם, אברהם: פורטוגל, הקונסולים מטעמה והפליטים היהודים, 1938-1941. יד ושם 27 (1999) 95-122.

[Milgram, Avraham: Portugal, the Consuls, and the Jewish Refugees, 1938-1941. *Yad Vashem* 27 (1999) 95-122.] Appeared in English in "Yad Vashem Studies" 27 (1999).

Romania

0833 Carmilly-Weinberger, Moshe: *Út a szabadság felé! Zsidó menekültek megsegitésének története a Holocaust idején: Erdély, 1936-1944* [The Road to Freedom! The Rescue Operation of Jewish Refugees during the

Holocaust: Transylvania, 1936-1944]. Cluj-Napoca: Sincron, 1999. 233 pp. Originally published as "The Road to Life: The Rescue Operation of Jewish Refugees on the Hungarian-Romanian Border in Transylvania, 1936-1944" (New York: Shengold, 1994).

0834

דוד, שלמה, עורך: דורוהוי — דורות של יהדות וציונות: סאוון, מיכאילן, דאראבאן, הרצה, רדאוץ-פרוט. כרך ג. קרית ביאליק: ארגון יוצאי דורוהוי והסביבה בישראל, 1996. 376, 384 עמ'.

[David, Shlomo, ed.: *Dorohoi – generaţii de iudaism şi sionism: Dorohoi, Săveni, Mihăileni, Darabani, Herţa, Rădăuţi-Prut (România).* Vol. 3 [Dorohoi – Generations of Judaism and Zionism]. Kiryath Bialik: Organizaţia Israelienilor Originari din Oraşul şi Judeţul Dorohoi, 1996. 376, 384 pp.] Vols. 1-2 were published in 1992-93.

Includes memoirs of Jews (in Hebrew or Romanian), former residents of Dorohoi and the surrounding area, many of whom relate their experiences in the Holocaust. Ca. 10,000 Jews from the area were transported to Transnistria in November 1941; many of them perished there. Ca. 6,000 returned in December 1943; they were persecuted by local authorities until the liberation in April 1944. *LV*

0835

דוד, שלמה, עורך: דורוהוי — דורות של יהדות וציונות: סאוון, מיכאילן, דאראבאן, הרצה, רדאוץ-פרוט. כרך ד. קרית ביאליק: ארגון יוצאי דורוהוי והסביבה בישראל, 1998. 292, 468 עמ'.

[David, Shlomo, ed.: *Dorohoi – generaţii de iudaism şi sionism: Dorohoi, Săveni, Mihăileni, Darabani, Herţa, Rădăuţi-Prut (România).* Vol. 4 [Dorohoi – Generations of Judaism and Zionism]. Kiryath Bialik: Organizaţia Israelienilor Originari din Oraşul şi Judeţul Dorohoi, 1998. 292, 468 pp.] Vols. 1-2 were published in 1992-93; vol. 3 in 1996.

Includes memoirs and life stories (in Hebrew or Romanian) of Jewish former residents of Dorohoi and the surrounding area. Many of them relate experiences in the Holocaust, including the effects of anti-Jewish laws enacted after the Germans entered Romania and the hardships of life in the camps of Transnistria. *RE*

0836 Finkelstein, Albert: *Etre ou ne pas naître: Chronique de l'Holocauste en Roumanie.* Paris: Etoile de la Pensée, 1997. 295 pp.

Memoirs of a Jew from Bucharest (b. 1920), relating his experiences during the Holocaust period. Describes events during the pogrom in Bucharest in 1941, particularly the attack on the chief rabbi, Zvi Gutman, and the murder of his two sons (pp. 171-183). Recounts the history of the numerous efforts to smuggle Jews out of Romania to Palestine. Relates his own escape on the boat Dor de Val in November 1941, along with his aunt, uncle, and cousin. They were shipwrecked in Turkey, where they were held until the British interned them in Cyprus. There, they awaited visas for Palestine, granted in March 1944. That summer, his sister Sophie, who had remained in Bucharest, joined a group of refugees bound for Palestine aboard the Mefkure; she was among the 195 who perished when the boat was attacked and sank. Chapters of the book also deal with the pogrom in Iaşi, the deportation of the Jews of Bessarabia and Bukovina to Transnistria, and the sinking of the Struma and the Mefkure. *TS*

0837

גלר, יעקב: הצלת פליטים יהודים מפולין ומהונגריה בתקופת השואה על-ידי יהודי רומניה.
ילקוט מורשת 64 (נוב' 1997) 29-45.

[Geller, Iaacov: The Rescue of Jewish Refugees from Poland and from Hungary by the Jews of Romania during the Holocaust. *Yalkut Moreshet* 64 (Nov 1997) 29-45.]

Describes rescue and relief organizations initiated by Jewish communal institutions. Dwells on the number of refugees who entered Romania, their origins, financial aspects of the relief and rescue activities, and acts of various individuals. Also discusses the stance of the Romanian authorities toward the refugees and toward the involvement of the Jewish communal organizations in their rescue. Describes negotiations which resulted in officials' disregard of deportation, imprisonment, and expulsion orders, thereby enabling the Jewish organizations to rescue the refugees who fled to Romania during the war. *LFo*

0838 Haimovitch-Hirsh, Mali: *Kindheit und Jugend im Schatten der Schoáh: Von Radautz durch Transnistrien nach Israel und zurueck: Juedische Schicksale aus der Bukowina 1928-1990.* Trans.: Rosemarie Wiedmann, Zwi Smoliakov. Konstanz: Hartung-Gorre, 1999. 61 pp. Originally published as "Le-Transylvania u-ve-hazarah" (Tel-Aviv: The Author, 1988). Translated from the Yiddish (published in 1993).

0839 Hausleitner, Mariana: War Antonescu ein Kriegsverbrecher? Zur aktuellen Diskussion ueber den Nationalismus in Rumaenien. *Jahrbuch fuer Antisemitismusforschung* 8 (1999) 312-328.

After tracing Antonescu's career, describes the crimes against Jews committed by the Romanian army and security forces, sometimes in cooperation with German Einsatzgruppen, during the Second World War: the pogrom in Iaşi, the massacres in Bessarabia and Bukovina, the deportations to Transnistria, and the massacre in Odessa. Even where some of these atrocities were initiated by the army, as in Iaşi and Odessa, it was Antonescu who gave orders to continue and intensify them. After the war, he was tried and executed for a wide range of war crimes, including the Odessa massacre. After the fall of Ceauşescu, Romanian revisionist historians, as well as prominent officers in the army and security services, promoting the rehabilitation of Antonescu as part of a campaign against national minorities, revived the allegations of Jewish collaboration with the enemy with which Antonescu justified his actions against them. *RW*

0840 Iancu, Carol: La Shoah en Roumanie. *Les Temps Modernes* 599 (May-June 1998) 43-65.

Surveys the rise of antisemitic movements in Romania in the beginning of the 1920s and their main leaders. Underlines that important Romanian intellectuals (e.g. Mircea Eliade, Nae Ionescu, Emil Cioran) joined the fascist movement and contributed to the development of its ideology. Adds that Romanian Orthodox clergymen also supported the extreme right. Discusses consecutive governments in Romania in the 1920s-30s and their support for or indifference to the fascist and antisemitic movements. Relates to the antisemitic measures during Antonescu's dictatorship, including the pogroms in 1941 in Bucharest and Iaşi and the deportations of the Jews of Bessarabia, Bukovina, and part of Moldova to camps in Transnistria. More than 60% of them died or were killed there. Ca. 40% of Romania's Jews survived the war. *HV*

0841 Manea, Norman: Un témoin de la “felix culpa”: Mihail Sebastian. *Les Temps Modernes* 599 (May-June 1998) 25-42.

Discusses the situation in Romania after 1989: a lack of real discussion among intellectuals on Romania's past, on antisemitism in the interwar period, and on the Holocaust period. The publication of the “Journal” of the Romanian-Jewish writer Mihail Sebastian in 1996 provoked a vast amount of discussion about this book, and proved once more the incapacity of most Romanian intellectuals to cope with the past. Relates to the figure of Sebastian, an assimilated Jew, and to his “Journal.” Discusses problems regarding Sebastian's friendship with Mircea Eliade, who was a fascist ideologue of the Iron Guard in the late 1930s in Romania. *HV*

0842

מנדלוביץ', אביגדור: עדות של נער מגדות נהר דימבוביצה — בוקרסט. תל-אביב: ירון גולן, 1996. 288 עמ'.

[Mendelovitch, Avigdor: *Testimony of a Youth from the Banks of the Dâmbovița River – Bucharest.* Tel-Aviv: Yaron Golan, 1996. 288 pp.]

Memoirs of a Jew, born in Bucharest in 1925. Pp. 109-173 recount his own wartime experiences, as well as those of his family and of the Jewish community. Mendelovitch's father was expelled from the Romanian Army, along with all Jewish soldiers, in 1940. Describes the pogrom carried out by the Iron Guard in Bucharest in January 1941, during which many Jeews were killed and synagogues and houses were destroyed. Discusses anti-Jewish measures, including the sending of all Jewish men to forced labor in the city or in nearby labor camps from late 1941. Mendelovitch worked at forced labor in the area until spring 1944, when he was to be sent to a labor camp. Instead, he went into hiding and joined the Gordonia Zionist youth movement. He emigrated to Eretz Israel in 1945. *LFo*

0843 Sebastian, Mihail: Journal. *Les Temps Modernes* 599 (May-June 1998) 1-24.

Excerpts from the "Journal," covering the period March 1935-October 1944. The full text of the journal was first published in the original Romanian in București: Humanitas, 1996. The French translation appeared in Paris: Stock, 1998. The excerpts printed here were selected by Claude Lanzmann and are accompanied by a brief introduction by him (p. 1). *HV*

Scandinavia

(see also Denmark)

0844 Kirchhoff, Hans: "Doing All That Can Be Done" – the Swedish Foreign Ministry and the Persecution of Jews in Denmark in October 1943: A Study in Humanitarian Aid and Realpolitik. *Scandinavian Journal of History* 24, 1 (1999) 1-43. An abridged and partly revised version of his article published in Danish in "Historisk Tidsskrift" 97 (1997).

It was the Foreign Ministry (UD) which defined Sweden's policy in regard to refugees before and during World War II. The Swedish policy toward Jewish refugees in the 1930s-1942 was no better than that of other democratic countries; the restrictions were enforced due to considerations of economy and security, as well as some measure of xenophobia and antisemitism. The change in Sweden's policy in 1943 was motivated not only by humanitarian considerations, but also by the turn of the war and the wish to whitewash the tarnished image of Sweden as Germany's passive ally in the eyes of the United Nations and, especially, of other Scandinavian countries. The goodwill motifs underlying the UD's decision to accept the Danish Jews in October 1943 must not be dismissed; Swedish policy here was a continuation of the rescue action for Norwegian Jews and the struggle to bring Swedish Jews home from Europe. Besides, public opinion in Sweden was pro-refugee. The Swedish decision was applauded worldwide, by Western governments and in the Jewish and non-Jewish press. *DR*

Slovakia

0845 Breznitz, Shlomo: *Polia spomienok: Svedectvo vojnového dieťaťa* [Memory Fields: A Testimony of Wartime Events]. Trans.: Margita Ančicová. Bratislava: SNM – Múzeum Židovskej Kultúry, 1999. 107 pp. Translated from the English. Appeared in Hebrew as "Sdot ha-zikaron" (Tel-Aviv: Am Oved, 1993).

0846 Jelinek, Yeshayahu Andrej: *Židia na Slovensku v 19. a 20. storočí: Zborník statí, I. časť* [Jews in Slovakia in the 19th and 20th Centuries: A Collection of Articles, Part I]. Bratislava: Judaica Slovaca, 1999. 156 pp.

A collection of six articles. The following deal with antisemitism and the Holocaust: Hladanie identity: Slovenské Židovstvo a nacionalizmus (1918-1938) [In Search of Identity: Slovakian Jewry and Nationalism (1918-1938). Appeared in English in "A Social and Economic History of Central European Jewry" (1990).] (75-92); Komunistická strana Slovenska a Židia: Desať rokov (1938-1948) [The Slovak Communist Party and Jews: Ten Years (1938-1948)] (93-113); "Konečné riešenie" – Slovenská verzia ["Final Solution" – Slovak Version] (115-126); Židia na Slovensku od jesene 1942 do jesene 1944 [The Jews of Slovakia between Fall 1942-Fall 1944] (127-155). *SSC*

0847 Trančík, Martin: *Zwischen Alt- und Neuland: Die Geschichte der Buchhaendlerfamilie Steiner in Pressburg: Ein mikrohistorischer Versuch.* Bratislava: Vydavatel'stvo PT, 1996. 272 pp.

A study of four generations of the Steiner family, bookdealers in Pressburg (Bratislava) from 1849, between 1821-1948. Pp. 127-241 discuss the Nazi period. Only a few personal documents of the nine brothers and sisters of the third generation are extant; therefore, the chapter on 1918-38 is based on information from interviews with their descendants carried out in 1994. Pp. 208-241 survey the period of the war in independent Slovakia, characterized by increasing persecution of the Jews. The Steiner bookstore was Aryanized. At first, the family members remained as employees, but in June 1942 the proprietor transmitted to the authorities a list of 22 names of Jews working in the bookstore, and indicated that he no longer required their services. Some of them went into hiding and the others were deported. In August 1944 Slovakia was occupied by the Germans. By March 1945, 13 transports had left Slovakia, with 13,500 deportees. Only one family member of the third generation of Steiners survived; from the fourth generation, eleven survived and five were murdered. *ID*

Switzerland

0848 Altermatt, Urs: *Katholizismus und Antisemitismus: Mentalitaeten, Kontinuitaeten, Ambivalenzen: Zur Kulturgeschichte der Schweiz, 1918-1945.* Frauenfeld: Huber, 1999. 414 pp.

Analyzes the attitudes of Swiss Catholics during the critical period when Switzerland refused to serve as a refuge for persecuted Jews. Describes traditional anti-Judaism based on the deicide accusation, which was instilled by the Holy Week liturgy and customs and by passion plays. Argues that modern antisemitism formed a continuity with this anti-Judaism. Describes a "dual antisemitism": the Church feared Jews as agents of modernism and socialism, but opposed racism as inconsistent with doctrine. It condemned the persecution in Germany, but espoused "self-defense" against supposed Jewish dominance in culture and the economy. In the 1930s-40s, with the growing influx of refugees, the majority of Catholics, like other Swiss, supported the government's decision to close the borders in order to prevent "Ueberfremdung" – flooding with strangers (a code word for Jews). Summarizes the publications of influential theologians, and editorials and columns in leading Catholic newspapers and family journals, which inculcated traditional anti-Judaism and often modern antisemitism as well. Declarations of regret for the part played by Catholics and Catholic theology in making the Holocaust possible appeared in Switzerland only in the 1990s. *RW*

0849 Ferrero, Shaul: Switzerland and the Refugees Fleeing Nazism: Documents on the German Jews Turned Back at the Basel Border in 1938-1939. *Yad Vashem Studies* 27 (1999) 203-234. Appeared in Hebrew in "Yad Vashem: Kovetz mehkarim" 27 (1999).

Switzerland was traditionally considered a haven for the persecuted, but ceased to be such for Jewish refugees from the Reich in 1938-39. In October 1938, Germany introduced a "J" stamped in all of the passports of Jews leaving Germany, a measure initiated by the Swiss police. Relates the history of restrictive Swiss policies regarding admission of Jewish refugees from 1919 to 1938. In 1939 Switzerland closed its border to Jewish refugees. It is true that some cantonal authorities implemented federal policy toward the refugees with a certain laxity. Documents received recently at the Yad Vashem Archives from the Municipal

Archives of Basel show the reasons given by the border police of the canton when it turned back Jewish refugees from Germany. Pp. 225-234 present a sample of the documents from Basel, translated into English. *DR*

0850

פררו, שאול: שווייץ ופליטי הנאציזם: מסמכים הנוגעים ליהודים הגרמנים שגורשו לגבול בזל בשנים 1938-1939. יד ושם 27 (1999) 159-187.

[Ferrero, Shaul: Switzerland and the Refugees Fleeing Nazism: Documents on the German Jews Turned Back at the Basel Border in 1938-1939. *Yad Vashem* 27 (1999) 159-187.] Appeared in English in "Yad Vashem Studies" 27 (1999).

0851 Hoerschelmann, Claudia: *Exilland Schweiz: Lebensbedingungen und Schicksale oesterreichischer Fluechtlinge 1938 bis 1945.* Innsbruck: Studien Verlag, 1997. 478 pp.

Discusses the fate of Austrian refugees in Switzerland during World War II. The Swiss immigration restrictions had fatal consequences for persecuted Austrian Jews who escaped after the "Anschluss." Between 1933-45 only 29,000 Jewish refugees, overall, were admitted into Switzerland, while almost as many were rejected. Mentions reports on 3,655 refugees from Austria admitted between 1933-57, but according to the latest statistics only 870 Austrian refugees (799 Jews) were registered between September 1939-May 1945. The exact number cannot be established, partly because of illegal entries. Discusses the pressure of the illegal immigration on the cantons in the northeast, emphasizing the especially hostile attitude of the canton of Thurgau, which in 1938 did not accept even one refugee. On the other hand, relates acts of rescue in which Swiss citizens opposed the official policy, e.g. police captain Paul Grueninger, who facilitated the entry of more than 2,000 Austrian Jews into Switzerland. Pp. 238-402 contain 250 biographies of Jews who managed to cross the border; some of them relate their stories of escape. *ID*

0852 Koch, Peter Ferdinand: *Geheim-Depot Schweiz: Wie Banken am Holocaust verdienen.* Muenchen: List, 1997. 320 pp.

Examines the cooperation of Swiss banks with Nazi Germany, emphasizing Switzerland's role as financier of the Third Reich. Although pretending to be neutral, Switzerland actually helped and supported the Nazi regime, and accepted its dictates. Switzerland profited from Germany's war policy, maintaining relative independence and avoiding occupation. Criticizes the criminal behavior of Swiss bankers after the war; they pretended not to have known about Jewish accounts because they were inactive, although the accounts were balanced every year. No

Swiss bank tried, on its own initiative, to find the heirs of dormant accounts. Includes a chapter by Richard Chaim Schneider, "Die Schweiz will vom Dritten Reich nichts wissen" (pp. 227-265), on the discovery in 1995 of the Swiss bank scandal regarding the dormant accounts. *ID*

0853 Luz, Ulrich: Die Schweiz, die Juden und die Kirchen. *Evangelische Theologie* 58, 1 (1998) 83-88.

Deals with Switzerland's behavior toward the Jews during World War II, defining it as the most important topic of discussion in Switzerland in 1997. Among the many criticisms were the traffic with Nazi Germany, the reception of gold which the Nazis had taken by force from the victims, the turning back of thousands of refugees who tried to cross the Swiss border, passive collaboration with the Nazis by allowing transport of prisoners through Swiss territories, holding stolen Jewish property at Swiss banks, and the Swiss economy's profiting from the war. Among the reactions to these revelations, mentions the effort to cope with the past made by the parliament, establishment of historians' commissions, and creation of funds for aid by banks on the one hand and the rise of crude antisemitism on the other. Examines the role of the Protestant Church in Switzerland during the Nazi period, tracing the errors made, and stresses the need for reparations and a change in theological orientation. *ID*

0854 Roschewski, Heinz: *Rothmund und die Juden: Eine historische Fallstudie des Antisemitismus in der schweizerischen Fluechtlingspolitik, 1933-1957.* Basel: Helbing & Lichtenhahn, 1997. 86 pp. On title page also: Herausgeber: Schweizerischer Israelitischer Gemeindebund.

Examines the influence of Heinrich Rothmund on the restrictive Swiss immigration policies during World War II. Rothmund served as head of the coordinating office for foreign policy, and later as head of the Justice and Police Department until 1954. States that documents and personal records clearly show Rothmund's antisemitic inclinations, which had fatal consequences for Jewish refugees who tried to escape Nazi Germany by crossing into Switzerland. At least 30,000 Jews were sent back to Germany and exposed to deportation to concentration camps. Argues that even "discreet" anti-Judaism can have disastrous results when circumstances allow, and when an average antisemite is placed in a key position. Pp. 48-65 deal with the period 1945-57, discussing the continuing antisemitism and xenophobia evinced by Rothmund toward Jews and toward the new State of Israel. *ID*

0855 Schuett, Julian: "Vermoechte man ueber Hofmannsthals halbjuedische Abkunft hinwegzukommen...": Die Schweizer Hochschulgermanistik und die nationalsozialistischen Judenverfolgungen. *Trumah* 8 (1999) 7-27.

Swiss scholars of German literature strove to maintain a strictly unpolitical attitude to Nazi Germany in order not to endanger their chances to publish in that country. The main advocate of this approach was Emil Staiger of the University of Zurich, who had been a member of a pro-Nazi front in the 1930s, and was the most influential literary critic of his time. Using a text-immanent analysis which supposedly ignored all extraneous factors, he "happened" to favor the same writers favored by the Nazis, and denounced Weimar, especially Jewish, writers as degenerate. Swiss universities had a "black list" of writers not to be taught. Most Swiss scholars were hostile to refugees, due to antisemitism and fear of competition; they supported the government's policy to refuse them asylum and employment. Describes the isolated position of the few Jewish scholars of German who had been at Swiss universities since well before 1933. One, Jonas Fraenkel, was dismissed as editor of the works of Gottfried Keller under pressure from Germany and from Swiss nationalists. *RW*

0856 Wagner, Meir: *Die Gerechten der Schweiz: Eine Dokumentation der Menschlichkeit.* Tel Aviv: Eli Meir, 1999. 154 pp. On title page also: Aufgezeichnet von Mosche Meisels.

Relates the wartime rescue activities of 35 Swiss citizens recognized by Yad Vashem as Righteous of the Nations. They saved the lives of Jews in Switzerland, France, Belgium, Germany, and Hungary. Politically neutral, Switzerland represented the diplomatic interests of 43 nations during World War II, which enabled the rescue work of Swiss diplomats (e.g. Carl Lutz in Budapest) and facilitated the humanitarian activities of the Swiss Red Cross, several of whose delegates are profiled in this book. But most of the stories are of common, courageous people who risked their lives to help fellow human beings in need. *ID*

Ukraine

0857

אגמון, פנחס, עורך: עדות מאוחרת: ניצולי שואה מאוקראינה מספרים. תרגום [מאוקראינית]: עמנואל ביחובסקי. תל-אביב: משרד הבטחון; קיבוץ לוחמי הגיטאות: בית לוחמי הגיטאות, 1997. 239 עמ'.

[Agmon, Pinchas, ed.: *Late Testimony: Survivors of the Holocaust from the Ukraine Relate.* Trans. [of Ukrainian material]: Emanuel Bykhovsky. Tel-Aviv: Ministry of Defence; Kibbutz Lohamei Haghetaot: Beit Lohamei Haghetaot, 1997. 239 pp.]

A collection of interviews with 19 Holocaust survivors, conducted between 1992-95 in Ukraine and in Israel. They relate their experiences in various ghettos, labor camps, and partisan units. *RE*

0858 Berenshtein, Leonid Ye.; Yelisavetskii, Ster Ya.: *Yevrei – geroi soprotivleniya v podpolnoi i partizanskoi borbe protiv natsistskikh okkupantov na Ukraine (1941-1945)* [Jews – Resistance Heroes in the Underground and Partisan Struggle against the Nazi Occupiers in the Ukraine (1941-1945)]. Tel-Aviv: Krugozor, 1998. 336 pp.

Deals with the participation of Jews in the Soviet urban underground and partisan warfare, and resistance in ghettos and camps (mainly in Volhynia and in the area of Kiev). Although there were manifestations of antisemitism in partisan units in Ukraine, this accusation cannot be raised against the whole Ukrainian partisan movement. Dwells on partisan commanders – e.g. Sidor Kovpak, Aleksei Fyodorov, Aleksandr Saburov, Nikolai Konishchuk – in whose formations there was a particularly large number of Jewish fighters and who made efforts to prevent antisemitism among the rank-and-file and the subaltern officers. Describes the actions of Jewish heroes of the resistance, such as Leonid Berenshtein (one of the authors), Moshe Gildenman, and others. Pp. 246-332 contain a list of Jewish underground fighters and partisans in Ukraine. *DR*

0859 Bilinkys, Lazar: Hromadzyanska viina na Ukrayini ta yevreyi: Fragmenty [The Civil War in the Ukraine and the Jews: Excerpts]. *Khronika-2000* 21-22 (1998) 234-251.

Excerpts from a manuscript held in the Vernadsky Central Library in Kiev. The date of its compilation is unclear. Presents an eyewitness account of the pogrom in the town of Talne, near Uman. Contends that the pogroms in Talne and its vicinity broke out on the background of the deeply-rooted contempt of the population toward the Jews, but were provoked by the influx of Jews to the hated Soviet administration and the Jews' abuses of power. In May 1919 a revolt broke out in Talne which was accompanied by murder of Jews. In particular, the rebels demanded that the Jews give up weapons allegedly left behind by the retreating German army in 1918 for the use of Jews. Having failed to receive the weapons, the rebels collected the Jews in the central square; a number of them were killed. The rebels were supported by Petlyura's commander, Tyutyunnik. A similar pogrom occurred at the same time in nearby Ivanki. The arrival of the Red Army and later of Denikin's White Army stopped the murder of Jews, but not the theft of their property. *DR*

0860 Finkelshtein, Iosif M.: *Kniga muzhestva i skorbi (yevrei Vinnitsy v gody Velikoi Otechestvennoi voiny)* [The Book of Courage and Mourning (the Jews of Vinnitsa in the Years of the Great Patriotic War)]. Vinnitsa:

Vinnitskii Regionalnyi Yevreiskii Obshchinnyi Tsentr "Mishpakha," 1999. 107 pp. On title page also: Vypusk 1-yi.

Pp. 4-9, "Skorb (kak byl 'reshyon' yevreiskii vopros v Vinnitse)" ["Mourning (How the Jewish Question in Vinnitsa was 'Solved')"], briefly describe the Holocaust in Vinnitsa in 1941-42. Pp. 10-18, "Muzhestvo (Pravedniki Mira. Yevrei-uchastniki dvizhenii soprotivleniya. Na frontakh Velikoi Otechestvennoi voiny)" ["Courage (Righteous Gentiles. Jewish Participants in the Resistance Movements. At the Fronts of the Great Patriotic War)"], discuss, inter alia, the rescue of Jews by non-Jews in Vinnitsa during the Nazi occupation and Jews who took part in the Soviet resistance in the area. Pp. 19-62 contain a partial list of Jews murdered by the Nazis in Vinnitsa. According to the Ch.G.K. estimates, 15,000 Jews perished in the city during the Nazi occupation. *DR*

0861 Gon, M.M.: Yevrei Vostochnoi Galitsii v usloviyakh ekonomicheskovo krizisa 1929-1933 gg.: Tendentsii ekonomicheskoi i obshchestvenno-politicheskoi zhizni [The Jews of Eastern Galicia in the Conditions of the Economic Depression of 1929-1933: Tendencies of Economic and Social-Political Life]. *Berastseiski Khranohraf* 2 (1999) 34-44.

Violent antisemitism was not characteristic of Ukrainians in Galicia before 1929. The Depression created interethnic strife in the economic sphere. The Ukrainian cooperative movement, and the campaign to close taverns, struck hard at the Jewish middle class; these measures were perceived by Jewish leaders as motivated partly by antisemitism. 1929 also saw a tendency to oust Jews from the countryside, acts of arson and other forms of violence against Jewish farmers, and calls to boycott Jewish trade. The Jewish-Ukrainian conflict was aggravated by a pro-Ukrainian declaration by Hitler and by an increase of pro-German sentiments among the Polish Ukrainians. The Depression broke the alliance between the Jews and other ethnic minorities in the Second Polish Republic. Jews discriminated against also by Polish law, found themselves pressed on two sides. *DR*

0862

גוברין, יוסף: החיים בצל האבדון: זכרונות על טרנסניסטריה. ילקוט מורשת 64 (נוב' 1997) 79-105.

[Govrin, Yosef: Life in the Shadow of Death: Memoirs about Transnistria. *Yalkut Moreshet* 64 (Nov 1997) 79-105.]

Memoirs of a Jew born in 1930 in Czernowitz, who lived in Edineţi, Moldavia, until the Nazi invasion of the Soviet Union in June 1941. Describes life in Czernowitz and in the ghetto until his deportation to Transnistria in the autumn of 1941. Relates his experiences in the ghetto of Mogilev until the liberation in March 1944. He emigrated to Palestine in 1947. *LFo*

0863 Grinberg, Efraim: *Yevrei Bukoviny: Istoricheskii ocherk* [The Jews of Bukovina: A Historical Essay]. Ierusalim: Ierusalimskii Izdatelskii Tsentr, 1997. 30 pp.

Covers the period from the 14th century to the 1990s. Inter alia, describes the antisemitic policy of the Romanian government in the late 1930s, and anti-Jewish atrocities perpetrated by the Romanian army during its retreat from northern Bukovina in June 1940. The Soviet policy in this area in 1940-41 also had anti-Jewish overtones. The Jews of Bukovina experienced their blackest period in 1941-44, when thousands were deported to Transnistria. At the same time, the Jews of Bucharest and even some Romanian officials tried to alleviate the plight of Bukovina's Jews. After the war, the Jews of northern Bukovina experienced the Soviet anti-Jewish campaigns of 1945-53 and of the Brezhnev period. *DR*

0864 Guensberg, Salomon: Mémoires du docteur Salomon Guensberg de Stanisławów. *Tsafon* 35-36 (Fall-Win 1998-1999) 31-50.

A text presented (on pp. 31-33) and translated by Paweł Śmieciuch, from the manuscript found in the ŻIH archives. Memoirs by a Jewish physician from Stanisławów, who survived the Holocaust. Describes events in Stanisławów after the Nazi occupation in July 1941. The Judenrat was founded and the registration of the Jews began. Jewish professionals were registered separately and murdered in the nearby forest in August 1941, but some physicians were not taken (including Guensberg). In October 1941 the Germans murdered about 12,000 Jews in Stanisławów. The ghetto was founded in November. Describes ghetto life, and the "actions" of murder of the Jews of Stanisławów and those brought to the ghetto from nearby villages. The ghetto was liquidated in February 1943. Most of the Jews were killed on the spot; some of them were sent to Bełżec. *HV*

0865

הרצוג, פסח: בצל הנשר השחור: עיירתי טארנופול מתבוססת בדמה: עדות אישית. תל-אביב: ירון גולן, 1996. 414 עמ'.

[Herzog, Pesach: *In the Shadow of the Black Eagle: My Town Tarnopol Is Drenched in Blood: A Personal Testimony*. Tel-Aviv: Yaron Golan, 1996. 414 pp.]

Memoirs of a Jew from Tarnopol (b. 1922), relating his fate and that of his family and community in the Holocaust. Describes the pogrom of July 1941, shortly after the German occupation, in which 5,000 Jews were killed by Germans and Ukrainians. Relates the difficulties of life in the ghetto, and states that the Judenrat and the Jewish militia cooperated with the Nazis more than was necessary. Herzog's mother and sister were deported and killed in summer 1942, and his father soon after. Herzog escaped from the ghetto and, posing as an Aryan, worked on a local farm until the liberation in March 1944. He joined the Polish army in order to fight the Nazis; in 1946 he emigrated to Australia, and in 1954 to Israel. *RE*

0866 Heymann, Florence: Au-delà du Dniester, jusqu'aux plaines du Nil. *Les Cahiers du Judaïsme* 4 (Spr 1999) 68-77.

Gives a brief account of the fate of the Jews of northern Bukovina and Bessarabia during the Holocaust, based on archival material and interviews with survivors of the Transnistria camps and ghettos. From July 1941 these territories were under Romanian occupation. Describes the deportation of the Jews from Czernowitz and other towns and villages. The Jews were sent to Mogilev and other places. In June 1942 the ghettos were founded and Judenraete were organized. States that the Judenraete contributed to better life conditions for the inhabitants of the Transnistria camps. However, more than half of the Jews deported died of hunger and illness, or were killed in the camps. *HV*

0867 Khonigsman, Yakov S.: *600 let i dva goda (istoriya yevreyev Drogobycha i Borislava)* [600 Years and Two Years (History of the Jews in Drohobych and Boryslav)]. Lvov: Bnei-Brit "Leopolis"; Lvovskoye obshchestvo yevreiskoi kultury im. Sholom Aleikhema, 1997. 66 pp.

A historical essay on the Jews of two neighboring Galician towns. Pp. 46-59 deal with the Holocaust in Drohobych and Boryslav. Dwells on the establishment of Jewish Councils and ghettos in the vicinity and on the deportations of Jews to Bełżec. There were Ukrainian families who hid Jews escaping from the ghettos. Describes an attempt to establish an underground organization in the ghetto of Boryslav. In July 1943 both of the towns were declared "free of Jews." 27,000 Jews were murdered; only 126 survived. *DR*

0868 Kruglov, Aleksandr I.: *Unichtozheniye yevreiskovo naseleniya Ukrainy v 1941-1944 gg.: Khronika sobytii* [The Annihilation of the Jewish Population in Ukraine in 1941-1944: A Chronicle of the Events]. Mogilyov-Podolskii: [The Author], 1997. 98 pp.

Presents a month-by-month chronicle of the mass murder of Jews perpetrated by the German and Romanian occupiers and their local collaborators. The chronicle covers the period from June 1941 to 1944 in the territory of present-day Ukraine, including Crimea. *DR*

0869 Kruglov, Aleksandr I.: *Unichtozheniye yevreiskovo naseleniya v Vinnitskoi oblasti v 1941-1944 gg.* [The Annihilation of the Jewish Population in the Vinnitsa Region in 1941-1944]. Mogilyov-Podolskii: [The Author], 1997. 87 pp.

The Vinnitsa region was occupied by the German and Romanian armies; the Romanian zone was annexed to Transnistria. Discusses the extermination of Jews in the German occupation zone, and lists the mass executions that took place there. Describes the situation in the ghettos of Transnistria. Pp. 24-89 contain statistical data on the Jews of the Vinnitsa region before the war, documents pertaining to the genocide of the Jews in this area, and statistics on and lists of the Jews murdered in Mogilyov-Podolsk, Yaryshev, and other places. *DR*

0870 Levitas, Ilya M., ed.: *Kniga pamyati* [Memorial Book]. Kiev: Yevreiskii Sovet Ukrainy, 1999. 300 pp.

Pp. 7-14 contain a historical essay on the mass murder of the Jews in Babii Yar. Pp. 15-45 contain photographs, taken mainly at several mass rallies commemorating the mass murder (in 1989, 1991, and others) and from visits of various statesmen to the Babii Yar memorial complex in the 1990s. Pp. 47-76 are dedicated to the Righteous Gentiles who rescued Jews in Kiev, containing a list of names, photographs, and facsimiles. Pp. 77-288 contain lists and photographs of Jews who perished in Babii Yar in September 1941. Pp. 289-295 contain a list and photographs of those who survived the massacre at Babii Yar. *DR*

0871 Margulis, Iosif: Vtoroye dykhaniye [Second Breath]. *Vtoroye dykhaniye: Literaturnyi almanakh* [Second Breath: Literary Almanach], eds. Aleksandr Nakhshin, Mikhail Ronkin. Petach-Tikva: Krugozor, 1997. Pp. 39-62.

Memoirs of a Jew born in 1924 in Vydybor, near Zhitomir. He tried to evacuate in the summer of 1941 but failed and returned to the village. He was the only Jew who survived the massacre of the Jews of Vydybor. He wandered throughout northern Ukraine passing as a straggling Red Army soldier. Later he joined a partisan unit where he also concealed his Jewishness. *DR*

0872

מילך, ברוך: ואולי השמים ריקים. תרגום ועריכה: שוש מילך-אביגל, אפרים סטן. ירושלים: יד ושם; תל-אביב: ידיעות אחרונות, ספרי חמד, 1999. 184 עמ'.

[Milch, Baruch: *And the Heavens May Be Void*. Trans. & eds.: Shosh Milch-Avigal, Ephraim F. Sten. Jerusalem: Yad Vashem; Tel-Aviv: Yediot Ahronot, Hemed Books, 1999. 184 pp.]

This memoir is a combination of Milch's writings during the Holocaust, which were left in Poland and retrieved after his death, with those written in the 1980s. Born in 1907 in Podhajce, Poland, Milch worked as a doctor in his hometown. After the Soviet occupation in 1939 he was forced to move to the town of Tłuste (now Tovste in Ukraine) with his wife and young son. Describes the murder of Jews there by Ukrainians during the German invasion in 1941; the Nazi anti-Jewish measures; the corruption of the Judenrat, of which he was a member for a short time; and the suffering and murder of Jews in the Jewish quarter. Milch's three-year-old son was killed during a massacre in May 1943. Milch went into hiding on a farm with his wife, but she was discovered and murdered; he then hid in another village, where he wrote his first memoirs and where he remained until the liberation in March 1944. His parents and four siblings perished in the Holocaust. Milch returned to Tłuste after the war, then fled to Germany. In 1948 he emigrated to Israel with his second wife and daughter. *RE*

0873 Pavlyuchenkov, Sergei A.: *Voyennyi kommunizm v Rossii: Vlast i massy* [Military Communism in Russia: Power and the Masses]. Moskva: RKT-Istoriya, 1997. 270 pp.

On pp. 251-263, "Vmesto zaklyucheniya: Yevreiskii vopros v revolyutsii, ili O prichinakh porazheniya bolshevikov na Ukraine v 1919 godu" ("Instead of a Conclusion: The Jewish Question in the Revolution or On the Causes of the Bolsheviks' Defeat in Ukraine in 1919"), maintains that the over-representation of Jews in the Bolshevik government in Ukraine from 1919 on precipitated the disappointment of the Ukrainians with the Soviet regime. Ukrainian villages were already infected with antisemitism. The fact that the Jews were the only group in the country who were ready to support the Bolsheviks set the peasants against them even more. The Bolsheviks had no choice in 1919 except to put Jews in key positions; thus, the struggle against the Soviet regime took the shape of fighting the Jews. The brutality of the pogroms in 1919 exceeded the average degree of brutality in the Civil War. Moscow took measures to exchange Jewish dignitaries in Ukraine for Russians. Attempts to draft masses of Jews into the Red Army failed – the Red Army was infected with antisemitism no less than the Ukrainian peasantry. *DR*

0874 Podolskii, Anatolii Ye.: Yevreyi Ukrayini v roky natsistskoyi okkupatsiyi: Sproba istoriohrafichnoho narysu [The Jews of Ukraine in the Years of the Nazi Occupation: An Attempt at a Historiographical Overview]. *Khronika-2000* 21-22 (1998) 392-399.

Briefly surveys Holocaust historiography in Ukraine, Soviet historiography of the Holocaust, Ukrainian post-1991 and émigré historiography, and Israeli and Western writings. Relates, also, to Holocaust denial. *DR*

0875 Privler, Maks: *Rasstrelyannyi trizhdy: Avtobiograficheskaya trilogiya* [Thrice Execution by Shooting: An Autobiographical Trilogy]. Tel-Aviv: [The Author], 1998. 240 pp.

Memoirs of a Jew, born Meyer Privler, in 1930 in the village Mikulichin, near Yaremcha, Western Ukraine. During the Nazi occupation he narrowly escaped being shot three times: in 1941, when he was taken to be executed together with his father; in 1942, in the ghetto of Stanisławów, during the second round-up; and in fall 1942, when he escaped from the ghetto and was arrested near Tarnopol. After his escape, Privler lived in a Ukrainian village in the Carpathian mountains, passing as a Ukrainian. In winter 1942 he joined Soviet partisans, and later served as a young reconnoiterer in a Red Army unit. After the war, Privler settled in Eastern Ukraine; he was confronted with antisemitism many times, immediately after the war, in 1953, and in 1970. In 1990 he emigrated to Israel. *DR*

0876 Redlich, Shimon: Jewish-Ukrainian Relations in Inter-War Poland as Reflected in Some Ukrainian Publications. *Polin* 11 (1998) 232-246.

Discusses the political orientations of the Ukrainian minority in Poland and their press. Surveys Jewish themes in some important journals and weeklies. These themes are also related to Poles and the Polish state. Polish-Jewish cooperation was perceived by the Ukrainians as Jewish collaboration with Poles against Ukrainian interests. An important issue was the assassination of Simon Petlyura who became, after his death, a national hero and martyr. Petlyura's murder provoked an outburst of antisemitic items in the press. Another popular issue was an accusation of Jewish economic supremacy on Ukrainian ethnic territory. Moreover, increasing anti-Soviet and pro-Nazi sentiments had an important impact on Ukrainian attitudes towards the Jews who were accused of influencing various states and societies by plotting against humanity in a Judeo-Bolshevik conspiracy. Concludes that although there were various opinions on the Jews in the Ukrainian press, the antisemitic voices were prevalent and contributed to Ukrainian receptivity to Nazi antisemitic ideology and propaganda. *HV*

0877 Shissel, Aleksandr M.: *...i na bereg vytashchen, k schastyu...: Povest-svidetelstvo* [...And Was Pulled to the Shore, Fortunately...: Novel-Testimony]. Yaroslavl: A. Rutman, 1999. 173 pp. Part I (pp. 5-68) was published previously as "Rodnye kamni mostovoi" (Yaroslavl: A. Rutman, 1997).

Fictionalized memoirs on the Romanian occupation of Odessa in 1941-44. Shissel (b. 1930), his mother, and his brother were arrested in October 1941 together with thousands of other Jews and sent to prison in reprisal for an explosion on Romanian army premises in Odessa. Shissel and his brother managed to flee from the prison and were hidden by non-Jews. Pt. II of this book begins in January 1942 when the two were interned in the newly-established ghetto, from which they escaped and were hidden again by non-Jews. Because of informers, they were forced to change their hiding place all the time, and Shissel's brother was killed. In the same year, Shissel managed to obtain forged papers and was taken to a Ukrainian orphanage. For the rest of the occupation period, Shissel survived leading a semi-criminal way of life, sometimes being arrested and fleeing from prisons. He witnessed both the warm-heartedness and meanness of people with whom he lived through the war years. Shissel's mother was killed in a massacre at Dalnik, in October 1941. *DR*

0878 Skorokhod, Yevgenii: *Doroga na tot svet (Babii Yar) = The Road to Death (Babiy Yar).* New York: "Kiev," 1997. 135 pp. In Russian and English.

Pp. 9-43 contain the poem "Doroga na tot svet" ("The Road to the Other World") describing the day of the Babii Yar massacre in Kiev (29 September 1941), and how Skorokhod's mother, Anna Vigdorovna Skorokhod, rescued him and herself by escaping when they were halfway from Kiev to Babii Yar. Pp. 44-68 contain an essay, "Posleslovie k tragedii" ("An Afterword to the Tragedy"), on the mass murder in Babii Yar and on postwar Soviet efforts to maintain silence about this tragedy. Pp. 89-129, "The Road to Death," describes the day of the massacre in Kiev and the Skorokhods' rescue. *DR*

0879 Sushon, Leonid P.: *Transnistriya: Yevrei v adu* [Transnistria: Jews in Hell]. Odessa: RIO AO Kinokompaniya "Yug," 1998. 431 pp.

Memoirs of a Jew born in 1931 in Odessa, interspersed with excerpts from recollections by other survivors of Transnistria from documents and historical works. During the Romanian occupation of the city, Sushon and his family were confined in a ghetto in the Slobodka-Romanovka neighborhood. In June 1942 they were transferred to the Akhmechetka camp (following brief internments in other camps), and in February 1942 to the labor camp of Pechora. Before his liberation by the Soviet army in March 1944, Sushon passed through several labor camps in northern Transnistria (Karlovka, Viktorovka). Dwells on cases of rescue of Jews by righteous non-Jews and on the fate of children in Transnistria. Sushon's family also survived the camps. Pp. 355-426 contain lists of Jews who survived Transnistria. *DR*

0880 Trubakov, Zakhar: *Taina Babyevo Yara: Dokumentalnaya povest-khronika* [The Secret of Babii Yar: A Documentary Chronical Narrative]. Tel Aviv: Krugozor, 1997. 184 pp.

Memoirs of a Jew from Kiev, born Zalman Trubakov. In September 1941, when the German authorities in the city ordered a "resettlement" of the Jews, Trubakov did not report to the assembly point. He continued to live in Kiev with his Ukrainian wife, provided with "Aryan" papers in the name of Zakhar Trubakov. In 1942 he was arrested for having contacts with the anti-Nazi underground, and was confined in the Syrets camp on Kiev's outskirts. In 1943 all the Jews and communists of Syrets were taken to the execution site at Babii Yar to work on exhuming and burning all the bodies buried there. Trubakov was among the more than 300 prisoners who, chained together, succeeded to escape from these "works." He was hidden in Kiev by his Ukrainian relatives until the liberation. *DR*

0881 Umanskii, Semen S.: *Juedisches Glueck: Bericht aus der Ukraine, 1933-1944.* Trans. & ed.: Ingrid Damerow. Frankfurt a.M.: Fischer Taschenbuch Verlag, 1998. 183 pp. Translated from the Russian manuscript.

Memoirs of a Jew born in 1929 in Tulchyn. Mentions antisemitism among the Ukrainian, Russian, and Polish inhabitants. Describes anti-Jewish measures after the Nazi occupation. In November 1941 the Jews were rounded up; the old and weak were massacred and the others were deported to Pechora. Umanskii's family fled and hid in the forests until overcome by cold and hunger; then they too entered Pechora. This was a lightly-guarded camp in which Jews were left without food or hygienic facilities. Many died of starvation or typhus, including Umanskii's mother. He, his father, and his sister fled to Romanian-controlled Transnistria and entered the Bershad ghetto. The ghetto was open and life somewhat normal; Umanskii earned a living by trading and odd jobs. In 1943 he was taken, with some hundred others, for forced labor near Odessa. After five months, with the German defeat imminent, they were released. He returned to Bershad and was liberated there by the Red Army. *RW*

0882 Urban, Marek: *Polska... Polska...* [Poland... Poland...]. Warszawa: Żydowski Instytut Historyczny, 1998. 180 pp.

Memoirs of a Jew born in Bychawa (Lublin area) in 1912. Describes his family and his childhood in Lubartów and later in Lublin. Between 1935-39 he studied agriculture in Warsaw. Mentions antisemitic incidents which he witnessed or of which he was a victim during the interwar period. At the beginning of the war, he went with his girlfriend and her father to Lvov. Describes his life and work under Soviet occupation and later under Nazi occupation. His girlfriend and her father were taken by the Soviets to Siberia. Urban survived the war on "Aryan papers," working in various places in the area of Lvov. After the war he returned to Lublin. His parents and three sisters were murdered by the Nazis, probably in Majdanek. His older brother survived the war in the Polish Army in the Soviet Union. Describes, as well, his postwar life in Poland. *HV*

0883 Vynnychenko, Volodymyr: Yevreiske pytannya na Ukrayini [The Jewish Question in the Ukraine]. *Khronika-2000* 21-22 (1998) 153-165. Reprint of an article published in "Nova Ukrayina" (Prague – Berlin, 1923).

Holds that the "Jewish question" is a social and economic question rather than religious or national. The main bearers of antisemitism have always been economic elites who regarded the Jews as competitors. In Ukraine, it is the middle class and bourgeois intelligentsia, and never rural and urban workers, who are anti-Jewish. Antisemitism in Ukraine arose in the harsh period of the Ukrainian revolution in 1919, when bourgeois and old elite elements took over the leadership of Petlyura's army. Petlyura, seeking popularity among the new commanders, condoned the pogroms; they were always perpetrated by the military and never supported by the populace. Since the victory of Bolshevism in Ukraine, antisemitism is increasingly penetrating the toiling masses. Contends that the cause lies in the influx of Jews into the ranks of the Bolsheviks and in the politics of Moscow, which laid the implementation of its agrarian policy and struggle against religion on Jews. Believes that restoration of democracy in Ukraine and Russia will be beneficial for the Jews. *DR*

0884 Yaker, Yevgenii: Shest syuzhetov iz moyei zhizni [Six Plots from My Life]. *Yehupets* 5 (1999) 268-283.

Presents fragments of his memoirs. The first four of the "six plots" deal with Yaker's life in Transnistria under Romanian occupation during World War II. Describes the deportation of his family from Bessarabia to Transnistria in July 1941; on the way, the family was saved by a Romanian officer. After a stay in the camp Ozarinetskaya Gora, Yaker was transferred to the ghetto of Shargorod; he escaped in summer 1942. *DR*

0885 Yones, Eliyahu: *Die Strasse nach Lemberg: Zwangsarbeit und Widerstand in Ostgalizien, 1941-1944.* Frankfurt a.M.: Fischer Taschenbuch Verlag, 1999. 239 pp. Originally published as "Al pi habor" (Jerusalem: Yad Vashem, 1960). On title page also: Bearbeitet von Susanne Heim. Aus dem Hebraeischen uebersetzt im Auftrag der Zentralen Stelle der Landesjustizverwaltungen zur Verfolgung der NS-Verbrechen, Ludwigsburg.

Memoirs of a Jew born in 1915 in Vilna. Yones was active in the Bund. With the Soviet occupation of Vilna, he fled in order to escape arrest. (After the Nazi occupation in 1941, his father and two brothers were murdered by the Nazis.) Yones managed to reach Lvov, where he experienced a pogrom incited by the Germans in July 1941, during which several thousand Jews were murdered. Yones was one of only 80 Jews who were arrested and survived. In November 1941 he was deported to the Kurowice labor camp, where he worked on the construction of a road between Lvov and Tarnopol. In August 1943, he managed to escape shortly before the liquidation of the camp and the murder of all the prisoners. He joined Jewish partisans in the forests who were fighting against the Germans and also against Ukrainian informers who hunted the victims for money. In 1944 he volunteered for the Soviet Army, was almost caught twice by the Soviet secret service, and finally returned to Vilna in spring 1946. He worked in DP camps until his emigration to Israel in 1950. Pp. 229-237 contain an epilogue by Susanne Heim, summarizing Yones' life and activities. *ID*

0886 Yones, Eliyahu: *Yevrei Lvova v gody vtoroi mirovoi voiny i katastrofy yevropeiskovo yevreistva 1941-1944* [The Jews of Lvov in the Years of the Second World War and of the Holocaust of European Jewry 1941-1944]. Trans.: Svetlana Shenbrunn. Moskva: Rossiiskaya Biblioteka Kholokosta, 1999. 488 pp. A translation of the author's diss. – "Yehudei Lvov bi-tekufat ha-Shoah 1941-1944" (Hebrew University of Jerusalem, 1993).

0887 Yones, Eliyahu: *Żydzi Lwowa w okresie okupacji, 1939-1945*. Trans.: Wiesława Promińska. Łódź: Oficyna Bibliofilów, 1999. 293 pp. Unseen. Translated from the Russian: "Yevrei Lvova v gody vtoroi mirovoi voiny i katastrofy yevropeiskovo yevreistva 1941-1944" (Moskva: Rossiiskaya Biblioteka Kholokosta, 1999).

0888 Zenner, Hermann: *Wie Hermann Zenner aus Kolomea den Nazi-Krieg in 1941-44 ueberlebte: Hundert mal dem Tod entgangen*. Ed.: Tuviah Friedman. Haifa: Institute of Documentation in Israel for the Investigation of Nazi War Crimes, 1999. 222 pp.

Memoirs based on a diary written between 1941-44 by Zenner, a Jew born in 1906 in Kolomyya, Ukraine. He served in 1926 in Vilna in the legionary regiment of the Polish army, together with Max Schaerfer, then a non-commissioned officer, who later was appointed commander of the SS in Poland. In 1941 Schaerfer was in charge of a shooting action in Horodenka, near Kolomyya. He recognized his former comrade Zenner, who was hiding his son under his coat. Schaerfer shot the child on the spot, but allowed Zenner to live. Zenner subsequently escaped many "actions." During a transport to a labor camp he jumped from a train near

Bolshovtsy, where he entered the ghetto, was taken for forced labor, and spent part of the time in hiding, but survived. His wife and daughter were killed in Tluste in June 1943. Friedman met Zenner in Vienna in 1947, where he was testifying at a war crimes trial. Later, he emigrated to the U.S. *ID*

USSR

(see also Belarus; Ukraine)

0889 Asher, Harvey: The "Black Book" and the Holocaust. *Journal of Genocide Research* 1, 3 (Nov 1999) 401-416.

A concise history of the "Black Book." The idea to publish a collection of eyewitness accounts on the Holocaust in the USSR emerged in the USA in 1942; it was to be a joint Soviet-American edition. The Soviet version of the book was prepared in late 1945-early 1946, and some of its materials were sent abroad, but the publication was suppressed by the authorities. Only in 1980 was a version of the "Black Book" published, for the first time, in Israel. In the 1990s, the restored "Black Book" was published in Lithuania, Ukraine, France, and Germany. Reflects on the question why the Soviet authorities suppressed the publication of the "Black Book" in the 1940s. Among the reasons were the emphasis in the manuscript on the anti-Jewish aspects of the Nazi horrors, the stress on the heroism of Jews, and frequent mention of non-German participants in the genocide. *DR*

0890 Beizer, Michael: Antisemitism in Petrograd/Leningrad, 1917-1930. *East European Jewish Affairs* 29, 1-2 (Sum-Win 1999) 5-28.

Traces the ups and downs of popular antisemitism in Leningrad from 1917 until the establishment of Soviet totalitarianism. Notes the gradual growth of antisemitism after the February (democratic) revolution granted equal rights to Jews. The October Bolshevik coup led to anti-Jewish incitement, as Jewish and communist opposition to antisemitism were perceived as further evidence of Jewish influence. The NEP period saw resentment against Jews in the wake of a large Jewish migration to the city and the view that Jews were responsible for anti-religious policies. Jews were stereotyped as shady businessmen and viewed as overrepresented in party, government, and economic life. Factory workers scapegoated Jews, as did many students. Even a murderer of a Jew received much popular sympathy. The government campaign against antisemitism peaked in 1929-30 and ended in 1932-33. Its propaganda had little effect. Antisemitism finally declined not because of propaganda but with the suppression of civil liberties and dissent when Stalin consolidated his power. *YC*

0891
ביכלר, יהושע רוברט: יחידות שיטור מקומיות – Schutzmannschaften – בשטחים
הכבושים של ברית המועצות. ילקוט מורשת 68 (אוקט' 1999) 69-84.

[Buechler, Yehoshua Robert: Local Police Units – Schutzmannschaften – in the Occupied Territories of the Soviet Union. *Yalkut Moreshet* 68 (Oct 1999) 69-84.]

Describes the SCHUMA – the Schutzmannschaften – which were auxiliary police units of collaborators constituted by the Nazis as part of the German police responsible for public order in the civil administration in the Nazi-occupied territories of the USSR. Many of the recruits were ardent antisemites and anti-Soviets. They were employed by the Nazis in the majority of the mass murders of Jews and of communists. Deals with the Nazi administrative acts related to the constitution of the SCHUMA, the conditions of service of the auxiliary police forces, their structure, and their relation to the German police forces, as well as the acts in which they were involved. *LFo*

0892 Gershman, Mikhail et al., eds.: *Skvoz ogon Katastrofy, 1939-1945: Letopis maloletnikh uznikov kontslagerei i getto* [Through the Fire of the Holocaust, 1939-1945: A Chronicle of Juvenile Inmates of Concentration Camps and Ghettos]. Ashkelon: The Editors; Ashkelon Municipality, 1997. [60] pp.

Short biographical sketches of more than 60 persons, most of them Jews, most of them born in the USSR, who at a young age survived ghettos, concentration camps, and forced labor camps in Germany, and now reside in Ashkelon, Israel. *DR*

0893 Herbeck, Ulrich: *Antisemitismus in der fruehen Sowjetunion.* Berlin: Institut fuer Internationale Politik und Regionalstudien des Fachbereichs Politische Wissenschaft der Freien Universitaet Berlin, 1998. 95 pp. (Arbeitspapiere des Instituts ..., 22).

Describes the role of antisemitism in tsarist Russia and chronicles its suppression by the Bolsheviks. Summarizing and frequently differing from previous (mostly American) research, argues that during the 1920s-early '30s the regime did not act from antisemitic motives. The quick assimilation and success of the emancipated Jews actually reduced antisemitism in the population, except among opponents of the regime. In a government and party organized territorially, Jews were at a disadvantage, but the yevsektsias represented them effectively. The campaign against the Jewish religion, the Hebrew language, and bourgeois Jewish community structures were not discrimination against Jews but part of the general anti-religious and anti-bourgeois campaign.

Jewish institutions and Yiddish culture were encouraged. In the late 1920s, the regime launched a vigorous campaign against antisemitism through propaganda, education, and disciplinary measures. Questions the extent of antisemitic motivation in Stalin's purge of Trotsky, in the choice of unattractive Birobidzhan as a Jewish territory, and in the show trials of the late 1930s. But the trials created an infrastructure of suspicion and conspiracy beliefs on which antisemitism could later build. *RW*

0894 Kozerod, Oleg V.: *Perelomnye gody: Yevreiskaya obshchina Ukrainy v pervoye poslevoyennoye desyatiletiye (1919-1929 gg.)* [Years of Rupture: The Jewish Community of Ukraine in the First Postwar Decade (1919-1929)]. Kharkov: Yevreiskii Mir, 1998. 152 pp.

The Ukrainian Civil War (1918-20) and the ensuing Sovietization of the country changed Jewish life there drastically. At the end of 1918 pogroms were perpetrated by Petlyura's Ukrainian nationalist army, and from summer 1919 by Denikin's White Army. Forces led by Makhno, Grigoriev, Struk, Bulak-Balakhovich, etc., were also pogromist. The main motive for their zeal was the identification of Jews with Bolsheviks. The Soviet regime did not perpetrate pogroms, but classified the majority of Ukrainian Jews as a “bourgeois” element and subjected them to economic and legal restrictions. 1921-29 were years of relentless struggle against the Jewish religion and the Zionist movement that undermined the foundation of Jewish life. Many Soviet reforms only enhanced popular antisemitism. Pp. 118-133 contain excerpts from the book by G. Ledat, “Antisemitizm i antisemity” (1929), examining the social and professional structures of Soviet Jewry and debunking widespread Jewish stereotypes connected with that structure. *DR*

0895

רדליך, שמעון: החוג של אנשי הוועד היהודי האנטי-פאשיסטי בברית המועצות ויחסם לגרמניה, לנאציזם ולשואה. ילקוט מורשת 68 (אוקט' 1999) 111-119.

[Redlich, Shimon: The Members of the Jewish Anti-Fascist Committee in the Soviet Union and Their Attitude toward Germany, Nazism and the Holocaust. *Yalkut Moreshet* 68 (Oct 1999) 111-119.]

Examines attitudes expressed by members of the Committee and by Jewish associates in public meetings and in closed conferences, as well as in publications. Many did not distinguish between Germans and Nazis in their statements against Germany's hostile acts. They pointed out the uniqueness of the anti-Jewish acts compared to acts against other people. For most of them, the Holocaust period provided opportunities to manifest their Jewish identity and to identify with the Jewish people – issues for which they were persecuted later on. *LFo*

0896 Romanovsky, Daniel: The Holocaust in the Eyes of "Homo Sovieticus": A Survey Based on Northeastern Belorussia and Northwestern Russia. *Holocaust and Genocide Studies* 13, 3 (Win 1999) 355-382.

Based on interviews conducted in 1985-87 with Jewish survivors and non-Jewish (Belorussian, Russian) eyewitnesses of the Holocaust in the areas which belonged to the USSR before 1939 (not those annexed in 1939-40). The accounts of the non-Jewish witnesses reflect many negative stereotypes of Jews widespread in the area both before the war and in the mid-1980s, the most characteristic being the cowardly and passive Jew. These stereotypes influenced, also, the accounts of some Jewish survivors. For the non-Jews, the stereotypes help not only to explain the "baffling" behavior of Jews in extreme situations, but also to justify their own behavior. Many Jewish survivors are no less prejudiced against the non-Jews. The narratives show that despite twenty years of prewar Soviet indoctrination and the postwar mythos of the Great Patriotic War, the witnesses, both Jewish and non-Jewish, are less influenced by the Soviet mythos than would have been expected. *DR*

0897 Rozenblat, Grigorii: *Yevrei v partizanskom soyedinenii S.A. Kovpaka* [Jews in S.A. Kovpak's Partisan Formation]. Moskva: Izdatelstvo Fonda "Kholokost," 1998. 108 pp.

A collection of brief essays (some of them published previously) dealing with the participation of Jews in Soviet partisan warfare in Ukraine and Belarus, especially Kovpak's partisan division in Ukraine. Jewish partisan warfare was an aspect of Holocaust resistance. Pp. 87-106 present a list of the Jewish partisans in Kovpak's division. *DR*

0898 Stivelman, Michael: *A marcha* [The March]. Rio de Janeiro: Nova Fronteira, 1998. 231, [20] pp.

Memoirs of a Jew born in 1928 in Secureni, Bessarabia. He was 13 when his hometown, then in the USSR, was invaded by the German and Romanian armies in July 1941. The Jewish community in Secureni experienced a violent pogrom, after which the remaining Jews were taken on a march – to Briceni, Vertujeni, back to Secureni, then to Otaci, Mogilev-Podolsk, Bershad, and Obodovka in Transnistria. Many of the deportees died on the way. Stivelman and his family (parents and brothers) were interned in the Obodovka ghetto, and then the Verkhovka ghetto, where he was almost adopted by a local Ukrainian family. During the summer of 1943, he spent a few months with the partisans. Verkhovka was liberated in 1944, and Stivelman with his mother (his father and brothers perished) went back to Secureni. From there they moved to Romania, and in 1948 to Brazil. *AA*

Yugoslavia

0899

הלפרט-זמיר, לילי; נידרלנד, דורון: בין היסטוריה למבדה: שואת יהודי יוגוסלביה.
ירושלים: מכון מופ"ת; יד ושם, 1999. 103 עמ'.

[Halpert-Zamir, Lily; Niederland, Doron: *Between History and Fiction: The Holocaust of the Jews of Yugoslavia.* Jerusalem: Makhon Mofet; Yad Vashem, 1999. 103 pp.]

A guidebook for educators who wish to teach this subject in high schools and teachers' seminars. The book combines historical information and literary sources. Pp. 19-59 discuss the fictional works of Danilo Kiš. Kiš's father and other relatives were murdered in the Holocaust; his works focus on the fate of Yugoslav Jewry. Pp. 60-80 relate the fate of the Jews in the region of Vojvodina, which was occupied by Hungary in 1941. Describes anti-Jewish measures carried out by the Hungarian authorities: a massacre of Jews in Novi Sad in January 1942; the induction of ca. 50,000 Jewish men into Hungarian military labor service units, where many perished; the deportation of the Jews in spring 1944; and Jewish participation in partisan groups. Discusses, also, the fate of Croatian Jews. Pp. 87-92 contain suggestions for activities. *RE*

0900

לבל, ג'ני: פתאום שונה, פתאום אחרת: סיפורה של נערה יהודייה ביוגוסלוויה הכבושה.
ירושלים: יד ושם, 1997. 248 עמ'.

[Lebl, Ženi: *Suddenly Different, Suddenly Other: The Story of a Jewish Girl in Occupied Yugoslavia.* Jerusalem: Yad Vashem, 1997. 248 pp.]

Memoirs of a Jewish woman from Belgrade, born in 1927. Describes the German invasion in April 1941 and the hardships which followed, including anti-Jewish laws, forced labor, and food shortages. Lebl's father, who was drafted into the Yugoslav army, was taken prisoner, her brother joined a partisan group, and her mother was deported and killed. Lebl escaped to Niš, where she had lived for a while, and hid with a former teacher who was an active member of the underground. The two were later betrayed by an informer. Lebl was arrested, tortured, and deported to a labor camp in Germany. She escaped, but was caught and taken to another camp, and later sent to a prison in Berlin, where she was eventually liberated. She returned to Belgrade, as did her father and brother; in 1954 she emigrated to Israel. *RE*

0901 Mošić, Aleksandar: *Die Juden in Belgrad: Eine kurzgefasste Uebersicht ihrer Geschichte.* Berlin: Neue Wege, 1997. 28 pp.

A revised version of a lecture delivered in Berlin and in Dresden in April 1997. Pp. 19-23 deal with the Holocaust period. With the Nazi occupation of Yugoslavia in April 1941, the activities of the Jewish community came to an abrupt end. Describes the increasing discrimination, the order from April 1941 that Jews must be distinguished by a "J" or a yellow star, forced labor, and deportation. Mentions three concentration camps – Topovske Šupe, Banjica, and Sajmište – which were situated close to Belgrade, unlike in other occupied countries where camps were established far from the large cities. 10,000 Jews were murdered in these three camps by June 1942, when Serbia was declared "judenfrei." *ID*

1919-1945: Latin America

Argentina

0902 Feierstein, Daniel; Galante, Miguel: Argentina and the Holocaust: The Conceptions and Policies of Argentine Diplomacy, 1933-1945. *Yad Vashem Studies* 27 (1999) 157-201. Appeared in Hebrew in "Yad Vashem: Kovetz mehkarim" 27 (1999).

Examination of messages sent by Argentine ambassadors and other diplomatic officials from Germany and countries occupied by or allied with Germany (e.g. Hungary, Italy, Bulgaria, France) between 1933-45 shows that the Interior Ministry, and thus the government of Argentina, had extensive information about events in Europe. Some diplomats (such as Eduardo Labougle, ambassador to Germany in 1933-39), who tried to justify the Nazi measures against the Jews, used German official sources uncritically. Some others, however, deplored the Nazi measures. The full knowledge of what was going on in Europe did not cause the Argentine government to revise its "closed-door" policy toward refugees. Moreover, Argentine officials in Europe often refused to grant diplomatic protection to naturalized Jewish citizens of Argentina, condoning their deportation. Contends that sympathy for German antisemitic policies played a role in the determination of Argentina's immigration policy, for fear of the same problems arising concerning the Jews. *DR*

0903

פיירשטיין, דניאל; גלנטה, מיגל: ארגנטינה והשואה: התפיסות שבבסיס הדיפלומטיה הארגנטינית, 1945-1933. יד ושם 27 (1999) 157-123.

[Feierstein, Daniel; Galante, Miguel: Argentina and the Holocaust: The Conceptions and Policies of Argentine Diplomacy, 1933-1945. *Yad Vashem* 27 (1999) 123-157.] Appeared in English in "Yad Vashem Studies" 27 (1999).

Mexico

0904
אבני, חיים: מקסיקו, קרדנס והפליטים היהודים בשנים 1940-1938. מקסיקו לאחר המהפכה: חברה ופוליטיקה, 1952-1910, עורכים צבי מדין, רענן ריין. תל אביב: מפעלים אוניברסיטאיים, 1999. עמ' 177-155.

[Avni, Haim: Mexico, Cárdenas, and the Jewish Refugees in 1938-1940. *Mexico le-ahar ha-mahapekhah: Hevrah u-politikah, 1910-1952* [Mexico after the Revolution: Society and Politics, 1910-1952], eds. Tzvi Medin, Raanan Rein. Tel Aviv: University Publishing Projects, 1999. Pp. 155-177.] Appeared in Spanish in "Estudios Interdisciplinarios de América Latina y el Caribe" 3 (1992).

Netherlands Antilles

0905 Lansen, Oscar E.: Victims of Circumstance: Jewish Enemy Nationals in the Dutch West Indies 1938-1947. *Holocaust and Genocide Studies* 13, 3 (Win 1999) 437-458.

In May 1940 the Dutch Colonial Government in Curaçao arrested and interned enemy nationals, of whom 77 were Jews. In a camp set up on the island of Bonaire, Jews were interned together with German and Dutch Nazis and were treated similarly. This policy is contrasted with that in Surinam and the Dutch East Indies. In Curaçao, the women and children were freed in July 1940, but Jewish males were liberated only in August 1941. Even after the war, Jewish refugees were subjected to various restrictions. The harsh treatment of the Jewish refugees was motivated by a desire to stop their influx into the colony and by economic considerations, as well as by antisemitism. The conflict between the Jewish leadership in Curaçao and the authorities over the refugees only fueled the antisemitism. The wartime policy toward Jewish refugees broke the longstanding cordial relations between Protestants and Jews in the colony. *DR*

1919-1945: North America

USA

0906 Abzug, Robert H., ed.: *America Views the Holocaust, 1933-1945: A Brief Documentary History.* Boston: Bedford/St. Martin's, 1999. xv, 236 pp.

A selection of documents illustrating the varied texture of Americans' reactions to the Holocaust. Presents press accounts, editorials, personal letters, government reports, and other records reflecting these reactions at three stages: the rise of Hitler (1932-35), the crescendo of persecution of the Jews before the U.S. entered the war (1935-41), and the wartime grappling with the knowledge of the mass murder of the Jews. The epilogue (pp. 207-213) shows a change in perception of the Holocaust by Americans after the war, which found its expression in the writings of historians and in mass culture. *DR*

0907

אחי שמשון רפאל (פסויד'): די הצלה-ארבעט פון אגודת-ישראל אין אמעריקע. דאס אידישע ווארט 337 (ניסן תשנז) 56-60.

[Achi Shimshon Raphael (pseud.): The Rescue Activities of Agudath Israel in America. *Dos Yiddishe Vort* 337 (Nissan 5757 = Apr 1997) 56-60.]

Surveys the involvement of various Agudath Israel personalities and divisions in the USA (including the Agudath Israel Youth Council and the World Union of Agudath Israel) in rescue activities. Discusses joint activities with other organizations, such as the War Refugee Board and rescue committees in the USA and around the world. *LFo*

0908

באומל, יהודית תידור: בין אידאולוגיה לתעמולה: משלחת האצ"ל בארצות הברית ושורשי הימין היהודי המיליטנטי באמריקה. ירושלים: הוצאת מאגנס, האוניברסיטה העברית, 1999. 240 עמ'.

[Baumel, Judith Tydor: *Between Ideology and Propaganda: The "Irgun" Delegation and the Origins of American-Jewish Right-Wing Militancy.* Jerusalem: Magnes Press, Hebrew University, 1999. 240 pp.]

Discusses the activities of a delegation sent by the Irgun (military wing of the Revisionist Zionists) in Eretz-Israel to the U.S., between 1939-48, which included attempts to establish a Jewish army and homeland. Ch. 4 (pp. 109-154) discusses the delegation's rescue activities in 1943-44, including the establishment by Hillel Kook of the Vaad ha-Herum le-Hatzalat Yehudei Europa (Emergency Committee to Save the Jews of Europe). The Committee attempted to influence U.S. government policy through lobbying, press advertisements, etc. A conference organized by Kook in July 1943 and a march on Washington in October 1943 received wide press coverage, but failed to impress President Roosevelt. In 1944 the Committee increased its lobbying activities and received support from the Congress and Henry Morgenthau; however, their efforts were opposed by Roosevelt and the State Department, as well as by many American

Jewish and Zionist organizations and leaders, especially Stephen Wise. The Committee worked in conjunction with the Orthodox movement Agudath Israel; they saved ca. 200,000 Jews. *RE*

0909 Jenkins, Philip: *Hoods and Shirts: The Extreme Right in Pennsylvania, 1925-1950.* Chapel Hill, NC: University of North Carolina Press, 1997. 343 pp.

Surveys antisemitic attitudes common among extreme-right fringe groups in the 1930s in Pennsylvania, especially in Philadelphia, described as a "national center for far-Right political activism." The antisemitism included both home-grown elements and those imported from Nazi Germany. Discusses activities of various groups, including the German American Bund, the Christian Front, the Ku Klux Klan, and the Silver Shirts. Although these groups had few members, they influenced many people and organized rallies that were attended by thousands. Ch. 5 (pp. 113-135), "Silver Shirts among the Brown: The Rise of Political Anti-Semitism," discusses the history of antisemitism in Pennsylvania, where Jews were excluded from various clubs and hotels for many years; the antisemitic views of Pennsylvania Congressman Louis McFadden; William Dudley Pelley's Silver Shirts organization, which held Jews responsible for all of the world's evils; Bessie Burchett's Anti-Communism Society (ACS), which contained elements of antisemitism, since Jews were thought to control communism; and anti-Jewish propaganda, including publications on conspiracy theories and leaflets urging non-Jews to boycott Jewish businesses. *RE*

0910 Melnick, Ralph: *The Life and Work of Ludwig Lewisohn.* Vol. I-II. Detroit: Wayne State University Press, 1998. 754; 596 pp.

A biography of Lewisohn (1882-1955), an American Jewish writer, editor, and critic. In vol. I, pp. 137-141 discuss Lewisohn's rejection for a university teaching post due to antisemitism. Pp. 278-282 relate to WASP critiques of this Jew for presuming to think he could understand American culture. Pp. 612-633 deal largely with his public criticism, from 1933, of the Nazi regime in Germany, including its genocidal attitude toward Jews. He hoped that the West would be moved to fight Hitler and provide refuge for Jews. In vol. II, ch. 40 (pp. 297-324), "Holocaust Revealed," highlights Lewisohn's Zionism as a reaction to Jewish assimilation and to the Western, Christian world's indifference to the fate of the Jews in the Holocaust. He criticized the West's failure to oppose the rise of Nazism and to provide safe havens, as Roosevelt had promised, to Jews (e.g. in Hungary) whom Hitler had not yet murdered. Lewisohn's novel "Breathe upon These" (1944) blamed the British for closing the gates to Palestine in the faces of Jews who might have found refuge there. *YC*

1945-1990: General

0911 Bell, Vikki: Mimesis as Cultural Survival: Judith Butler and Anti-Semitism. *Theory, Culture & Society* 16, 2 (Apr 1999) 133-161.

In discussing the feminist theory of Judith Butler, explores the concept of gender, noting connections between her ideas and the psychological and sociological theories of antisemitism of Sartre, on the one hand, and Adorno and Horkheimer, on the other. Explores mimicry of "hegemonic" values in terms of fascist societies and culturally determined gender identities. Points out parallels and differences between the ideas of Butler and those of Jonathan and Daniel Boyarin about ethnicity (their focus is on diaspora Jewishness) and cultural survival. *YC*

0912 Birnbaum, Pierre: Réflexions peinées: Sur "Réflexions sur la question juive." *Les Cahiers du Judaïsme* 3 (Fall 1998) 93-106.

Evokes the memory of the lecture given by Sartre in June 1947, organized by the Alliance Israélite Universelle, during which he summarized his recently-published work, "Réflexions sur la question juive." Rejects Sartre's definition of Jewish identity: that authentic Jews are those who accept the identity imposed upon them and inauthentic Jews are those who deny it. Compares the text of the book with the lecture (published in this issue of the journal), where the term "Israélite" replaces the term "Juif" utilized in the book. Exposes Sartre's stereotypes of the relationship between Jews and the economy, the physical traits of Jews, and their identity as a culture vs. race. Criticizes Sartre's ignorance of Jewish history and his description of antisemitism in France. Notes his silence in the book regarding Vichy collaboration with the Nazis. Concludes that while most French Jews, whether "authentic" or "inauthentic," were abandoned by their compatriots during the occupation, a minority of French citizens defied antisemitic policy and came to their aid. Regrets that Sartre ignored both of these facts in his work. *TS*

0913 Judaken, Jonathan: The Queer Jew: Gender, Sexuality and Jean-Paul Sartre's Anti-Antisemitism. *Patterns of Prejudice* 33, 3 (July 1999) 45-63.

Paradoxically, Sartre resorted, in his "Réflexions sur la question juive," to some covert anti-Jewish clichés and stereotypes (e.g. he admitted the existence of a Jewish race). To explain why this work produced an "antisemitic effect," turns to Sartre's conception of the relations between Self and Other as elaborated in his "L'Etre et le Néant." Sartre compares the relations between "the inauthentic Jew" and the "antisemite" to those between a masochist and a sadist; both kinds of relations are sexually tinged. Hatred is the most extreme manifestation of sadism. Sartre's view is masculinist, and he construes the relations between Self and Other in terms of domination and submission. Suggests that immanent in Sartre's position is another way of conceiving of the relation between Self and Other, which embraces difference, singularity, and particularity. *DR*

0914 Lyosov, Sergei V.: *Popytka ponimaniya: Izbrannye raboty* [An Attempt at Understanding: Selected Works]. Moskva: Universitetskaya Kniga, 1999. 575 pp.

A collection of articles, most of them published previously. Partial contents: Natsionalnaya ideya i khristianstvo [The National Idea and Christianity. On the influence of the Holocaust on postwar Christianity. Appeared in "Oktyabr" 10 (1990).] (231-254); Novyi Zavet i yudophobiya [The New Testament and Judeophobia. Reprint of an article by Hyam Maccoby which appeared in Russian translation in "Strana i Mir" 43 (1988), with a brief introduction by Lyosov.] (255-276); Glazami ubitykh detei: Mir Eli Vizelya [In the Eyes of Murdered Children: The World of Elie Wiesel. Appeared as an afterword to the Russian edition of Wiesel's "Night. Dawn. Day" (Moscow, 1993).] (277-283); "Yevreiskii vopros" v russkoi intellektualnoi zhizni, 1985-1995 [The "Jewish Question" in Russian Intellectual Life, 1985-1995. Appeared in "Znamya" 9 (1996).] (292-300); Gitlerovskii genotsid i yevreiskaya identichnost [Hitler's Genocide and Jewish Identity] (301-304). *DR*

0915 Martin, Thomas: The Role of Emotion in Sartre's Portrait of Anti-Semitism. *Australasian Journal of Philosophy* 76, 2 (June 1998) 141-151.

Examines Sartre's portrait of the antisemite as presented in his work "Anti-Semite and Jew" in light of his theory of emotion. According to Sartre, antisemitism is a kind of Manichaean worldview, with the Jew as the enemy, based on passion (emotion). Sartre's view maintains that emotional conduct is an active response to a situation which involves a magical transformation of an all-too-difficult world (thwarting us in our aims) into a more congenial world, through the removal of the normal constraints of instrumentality and determinism. The antisemite's Manichaeism is such a magical transformation of the world. Antisemitism is also driven by fear of freedom. The role played by actual Jews in Sartre's account of antisemitism is marginal, almost incidental. "Anti-Semite and Jew" shows little appreciation of the social and political dimensions of antisemitism. Nevertheless, his account does contain some valuable insights for an understanding of contemporary racism. *DR*

0916 *October* 87 (Win 1999).

This issue is entitled "Jean-Paul Sartre's 'Anti-Semite and Jew': A Special Issue," ed. by Denis Hollier, and is composed mainly of papers delivered at a conference at New York University, April 1998, on the 50th anniversary of the American publication of Sartre's book. Most of the papers were translated from the French. Partial contents: Hollier, Denis: Introduction (3-6); Vidal-Naquet, Pierre: Remembrances of a 1946 Reader (7-23); Levinas, Emmanuel: Existentialism and Anti-Semitism [Translation of Levinas' introduction to fragments of Sartre's lecture (see below), published in the June 1947 issue of "Les Cahiers de l'Alliance."] (27-31); Sartre, Jean-Paul: Reflections on the Jewish Question: A Lecture [Translation of the transcript of the lecture delivered, on the invitation of the Alliance Israélite Universelle, in Paris on 3 June 1947.] (33-46); Contat, Michel: The Intellectual as Jew: Sartre against McCarthyism – an Unfinished Play [Discusses the text of a manuscript of an unfinished play by Sartre, found in the Bibliothèque Nationale, dealing with the case of Abraham Feller, an American Jew and an intellectual, a general adviser to the UN, who committed suicide in New York in 1952.] (47-62); Misrahi, Robert: Sartre and the Jews: A Felicitous Misunderstanding (63-72); Traverso, Enzo: The Blindness of the Intellectuals: Historicizing Sartre's "Anti-Semite and Jew" [A paper delivered at a colloquium at New York University, April 1998.] (73-88); Birnbaum, Pierre: Sorry Afterthoughts on "Anti-Semite and Jew" (89-106); Schor, Naomi: Anti-Semitism, Jews, and the Universal (107-116); Petrey, Sandy: Reflections on the "Goyishe" Question [The response of a non-Jew.] (117-127); Suleiman, Susan Rubin: Rereading Rereading: Further Reflections on Sartre's "Réflexions" (129-138); Hollier, Denis: Mosaic: Terminable and Interminable (139-160); Rybalka, Michel: Publication and Reception of "Anti-Semite and Jew" (161-182). *SSC*

Responses to the Holocaust

0917 Baccarini, Emilio; Thorson, Lucy, eds.: *Il bene e il male dopo Auschwitz: Implicazioni etico-teologiche per l'oggi: Atti del Simposio Internazionale (Roma, 22-25 settembre 1997)*. Milano: Paoline, 1998. 451, [7] pp.

Contents: Baccarini, Emilio: Introduzione (11-25); Fackenheim, Emil Ludwig: L'aggressione all'alleanza di Abramo: La necessità di una teologia in seguito all'Olocausto: Ebraica, cristiana e anche musulmana (29-51); Metz, Johann Baptist: Tra la memoria e l'oblio: La Shoah nell'epoca dell'amnesia culturale [Appeared in German in "Das Juedische Echo" 46 (1997).] (52-61); Bemporad, Jack: Cosa possiamo dire di Dio noi ebrei dopo l'olocausto? (62-80); Carucci Viterbi, Benedetto: "Chi non esperimenta il nascondersi del volto non fa parte di loro": Il nascondersi di Dio, il bene e il male (81-92); Pawlikowski, John T.: Dio: La questione etica fondamentale dopo l'Olocausto [Appeared in "SIDIC" 31 (1998).] (93-111); Fritz, Maureena: Amore saldo e verità dopo Auschwitz (112-123); Ansorge, Dirk: Dio tra misericordia e giustizia: La provocazione di

Auschwitz e la speranza di riconciliazione universale (124-141); Baccarini, Emilio: Tra volontà di potenza e derelizione: Dire l'uomo dopo la Shoah [Appeared in "Amicizia Ebraico-Cristiana" 32 (1997).] (145-165); Haas, Peter Jerome: Fare etica in un'età di scienza (166-178); Pollefeyt, Didier: La moralità di Auschwitz? Confronto critico con l'interpretazione etica dell'Olocausto di Peter J. Haas (179-204); Lepicard, Etienne: Auschwitz nella prospettiva di Norimberga: Implicazioni mediche ed etiche (205-223); Rigobello, Armando: Principi morali in situazione limite: Auschwitz e la verità sulla condizione umana (227-234); Dalmasso, Gianfranco: Il bene e il male dopo Auschwitz: Il giudizio (235-244); Levi Della Torre, Stefano: Il limite e l'illimitato (245-256); Dupuy, Bernard: Il problema del bene e del male negli scritti di Hans Jonas e Hannah Arendt (257-272); Kajon, Irene: Auschwitz come esperimento cruciale: Il "servo sofferente di Dio" (Isaia 49-57) nell'interpretazione di André Neher ed Emmanuel Levinas (273-285); Fleischner, Eva: Ricordo e responsabilità: Soccorritori di ebrei durante la Shoah (289-303); McGarry, Michael B.: Il coraggio dopo la Shoah: Esame esplorativo di una virtù cristiana (304-328); Giuliani, Massimo: "Aberrazione della libertà ed empio eroismo": Osservazioni su coscienza e sospensione della valutazione etica nel caso-Auschwitz (329-339); Meghnagi, David: L'elaborazione del lutto tra memoria e storia: Alcune considerazioni su un'intervista di Marek Edelman, vicecomandante dell'insurrezione del ghetto di Varsavia (340-360); Bernauer, James William: Lo spirito e la carne: Verso un'etica encarnazionista per il tempo dopo la Shoah e per l'epoca postmoderna (361-376); Hoeckman, Rémi: Il bene e il male dopo Auschwitz nell'insegnamento del Papa (379-394); Blumenthal, David Reuben: La banalità del bene e del male: Comportamento antisociale, comportamento pro-sociale e insegnamenti religiosi ebraici (395-415); Halpérin, Jean: Dopo Auschwitz l'etica rappresenta una responsabilità prioritaria (419-427); Sievers, Joseph: Riflessioni di vittime del nazismo sul futuro: Testi scelti e introdotti [Excerpts from the diaries of Moshe Flinker, Etty Hillesum, and Dietrich Bonhoeffer.] (428-440). *MZ*

0918 Barnéa, Naomy: Après la Shoah: Deuil et mémoire, un "travail" sans précédent. *Revue d'Histoire de la Shoah* 167 (Sept-Dec 1999) 132-160.

Discusses the efforts made by survivors and the general Jewish population to deal with the horrors of the Shoah as a paradigm of trauma in apocalyptic dimensions. Compares the calculated cruelty of the perpetrators with the incredible will to live of the victims. Describes the complex process by which the trauma and its after-effects have been transformed into living memory as an appeal to the consciences of humanity. Concentrates on the testimonies of 50 French-speaking families in Belgium, France, and Israel - including first, second, and third generations of survivors - collected between 1988-92. Background information was assembled from three sources: historians' documentation found after the fall of the Reich, evidence provided during the trials of war criminals, and studies of places of memory such as museums, memorials, symposia, and ceremonies held on Yom Hashoah. Four major issues are treated: Jewish identity, the State of Israel, universalistic themes, and the search for meaning after the Shoah. Finally, the role of the Jewish people among the nations is discussed. *TS*

0919 Barnouw, David: *Anne Frank – vom Maedchen zum Mythos*. Trans.: Erik & Mary Alder-Sijmons. Muenchen: Econ & List Taschenbuch Verlag, 1999. 128 pp. Originally published as "Anne Frank voor beginners en gevorderden" (Den Haag: Sdu, 1998).

0920 Bernstein, Richard J.: Responsibility, Judging, and Evil. *Revue Internationale de Philosophie* 53, 2 (June 1999) 155-172.

Stresses that Hannah Arendt's "Eichmann in Jerusalem" opened public discussion of important questions. She is credited for new thinking about unprecedented phenomena, such as the crimes of the Nazis, and the criminal-bureaucrat. She was misread as exonerating Eichmann and blaming the victims for their own deaths, but was in fact attempting to discuss varying degrees of responsibility, including that of the "bystanders." Her criticism of the Judenrats is clarified: she espoused judging their arguments, not the people themselves. Until her death in 1975, Arendt continued to explore the concept of true thinking ("judging"), in contrast to the shallow, banal thinking of Eichmann that failed to question the murderous immorality of his society. *YC*

0921 Boschki, Reinhold; Mensink, Dagmar, eds.: *Kultur allein ist nicht genug: Das Werk von Elie Wiesel – Herausforderung fuer Religion und Gesellschaft*. Muenster: LIT, 1998. 429 pp. On title page also: Unter Mitarbeit von Britta Frede.

Based on papers delivered at a conference held in Stuttgart, May 1995. Partial contents: Mensink, Dagmar: Zur Einfuehrung: Elie Wiesel lesen (13-21); Boschki, Reinhold: Zum Autor: Elie Wiesel im Spiegel seiner Autobiographien (22-37); Wiesel, Elie: Zum Geleit: Kultur allein ist nicht genug [On his relationship to German culture and on religious implications of the Holocaust.] (38-42); Feinberg, Anat: Das Unbeschreibliche beschreiben: Literarische Annaeherungen an den Holocaust (47-57); Lenzen, Verena: "Reden ist schwer, schweigen unmoeglich" (Elie Wiesel): Sprache und Schweigen im Schatten der Schoah [An abridged version of a chapter in her book "Juedisches Leben und Sterben im Namen Gottes" (1995).] (58-71); Bergida, Ruth: Die Sprache des Schweigens (72-80); Abrahamson, Irving: "Das Testament": Fakt, Fiktion und Wahrheit (81-90); Langer, Lawrence L.: Leichname im

Spiegel: Eltern und Kinder in Wiesels "Nacht" und in anderen Holocaust-Zeugnissen (91-102); Thomas, Jean-François: Elie Wiesel und die Traurigkeit (103-118); McCord, Janet Schenk: Seelenqual, Selbstmord und die heilende Kraft der Freundschaft im literarischen Werk Elie Wiesels (119-131); Rosen, Alan: Der Anfang der Zeugenschaft und das Zeugnis des Anfangs (132-138); Berger, Alan L.: Elie Wiesel und der Zeuge der Zweiten Generation: Die Fackel der Erinnerung weitergeben [Appeared in English in "Telling the Tale: A Tribute to Elie Wiesel" (1993).] (139-154); Soelle, Dorothee: Nach der Shoah: Wie unsere theologischen Entwuerfe sich aendern muessen (159-167); Rendtorff, Rolf: Ist in Auschwitz das Christentum gestorben? (168-180); Friedlander, Albert Hoschander: Biblische und talmudische Dimensionen bei Elie Wiesel (198-209); Berg, Gundula van den: Elia: Elie Wiesels Rezeption des Propheten (210-233); Cargas, Harry James: Positive Ambiguity – die Gegensaetze aushalten: Elie Wiesels religioeses Denken [First published in English in "Spiritual Life" (Summer 1990).] (234-244); Eckardt, Alice Lyons: Leiden: Herausforderung des Glaubens – Herausforderung Gottes [In particular, the suffering of the Jews in the view of Elie Wiesel and of Christian theologians.] (245-261); Kuschel, Karl-Josef: Verweigerung der Theodizee – Warten auf Theodizee: Zu Elie Wiesels Drama "Der Prozess von Schamgorod" [First published in "Das Gegenteil von Gleichgueltigkeit ist Erinnerung" (1995).] (273-288); Metz, Johann Baptist: Der juedische Geist und die Zukunft Europas [Europe must integrate in its culture of ideas the Jewish culture of memory, which the Nazis sought to wipe out. Only through memory, first and foremost of the Holocaust, can it restore a universal morality based on sensitivity to the suffering of the Other, and overcome the contemporary God-crisis.] (293-305); Muenz, Christoph: Erinnerung und Gedaechtnis als Grenzscheide zwischen Judentum und Christentum: Thesen zu Geschichtsinsensibilitaet und Gedaechtnisarmut des Christentums und der christlich gepraegten Kulturen (306-319); Roth, John King: Der Schatten von Birkenau: In der Oeffentlichkeit von Religion sprechen [On Wiesel's lecturing and writing on the Holocaust.] (320-332); Kanofsky, Joseph A.: Elie Wiesel als Rebbe des zeitgenoessischen amerikanischen Judentums (333-346); Schneider, Werner: Abwehrstrategien in der Auseinandersetzung mit dem Holocaust: Eine Problemanzeige (361-367); Knigge, Volkhard: Erinnerungsarbeit in der Gedenkstaette Buchenwald (368-387). Brumlik, Micha: Elie Wiesel – Ein Bote des Unheimlichen (391-397); Boschki, Reinhold; Rosen, Alan: Elie Wiesel – Ein Bote der Toten, ein Bote fuer die Lebenden: Eine Entgegnung auf Micha Brumlik (398-401); Literaturverzeichnis [Works by and about Elie Wiesel.] (402-425). *RW*

0922 Braiterman, Zachary Joseph: *(God) after Auschwitz: Tradition and Change in Post-Holocaust Jewish Thought*. Princeton, NJ: Princeton University Press, 1998. 208 pp. Based on his diss. – "Theodicy and

Antitheodicy: Tradition and Change in Post-Holocaust Jewish Theology" - Stanford University, 1995.

0923 *Le Débat* 98 (Jan-Feb 1998).

A section of this issue, entitled "Unicité de la Shoah?," contains three articles discussing the book by Jean-Michel Chaumont, "La concurrence des victimes: Génocide, identité, reconnaissance" (1997), and a reply by Chaumont. Contents: Bouretz, Pierre: Cette fumée-ci, pourtant, ils ne savent pas... (156-166); Marienstras, Richard: La stupeur du monde (167-176); Wieviorka, Michel: La transformation des Juifs (177-180); Chaumont, Jean-Michel: Le prisme de l'extrême (181-186). *HV*

0924 Drury, Shadia B.: *Leo Strauss and the American Right*. Basingstoke, Hants: Macmillan, 1997. xiii, 239 pp.

Examines the influence of the political thought of Strauss (1899-1973) on neoconservatism in America. The German-Jewish philosopher fled Germany when the Nazis came to power. He settled in the U.S. and taught political science at the University of Chicago. Ch. 2 (pp. 31-64), "Strauss's Jewish Heritage," notes that Strauss believed that the persecution of the Jews is an "exemplification of the human problem" - that human beings are divided into groups pitted against each other in mutual hatred and antagonism. Consequently, there is no solution to the Jewish problem. Jews must learn to live with persecution. Ch. 3 (pp. 65-96), "Strauss's German Connection," examines his debt to two German contemporaries - Martin Heidegger and Carl Schmitt. Argues that Strauss did not learn much from the Holocaust, that his political thought is dangerously close to Heidegger's and much more radical than Schmitt's. In Strauss's opinion, the Holocaust was ample proof that faith in the Enlightenment was altogether misplaced. He was opposed to modernity, and supported nationalism and populism. *REK*

0925 Faierstein, Morris M.: Abraham Joshua Heschel and the Holocaust. *Modern Judaism* 19, 3 (Oct 1999) 255-275.

Disputes the impression that the Holocaust was not central to Heschel's theological thought. Heschel confronted the Holocaust directly in his Yiddish writings (intended for survivors), e.g. in his review of poems by Aaron Zeitlin, which appears in translation as appendix 1 (pp. 264-271). His reluctance to theologize directly about the Holocaust may reflect a loss of confidence in Western civilization and an inability to deal with the Holocaust in traditional Jewish terms. His avoidance of speaking of the Holocaust in English reflects the contemporary lack of interest in the Holocaust among American Jewry. Two "eulogies" for East European Jewry were his Yiddish lecture of 1945, which in English became "The

Earth Is the Lord's," and his continuing attempt to memorialize Hasidism, for him the spiritual peak of that Jewry destroyed by the Nazis. Appendix 2 (pp. 272-274) contains the text of an interview conducted by Gershon Jacobson with Heschel, originally published in "The Day-Morning Journal" (13 September 1963). *YC*

0926 Fiero, Petra S.: *Schreiben gegen Schweigen: Grenzerfahrungen in Jean Amérys autobiographischem Werk.* Hildesheim: Georg Olms, 1997. 187 pp. Based on the author's diss. - University of Nebraska, 1994.

Traces Améry's life and discusses his writings on his Holocaust experience (in France, Belgium, and Auschwitz) comparing him with other authors of Holocaust memoirs, particularly Primo Levi. Describes his discovery of his Jewishness at the age of 17; he perceived it as imposed by society and quite early realized that it was a death sentence. He understood the Nuremberg Laws as initiating a process to deprive Jews of their dignity ("Wuerde"). In the camps, the intellect became meaningless; the body was all. Death was humdrum, unaesthetic, impersonal. Discusses Améry's problem in writing about his experience in German, his mother tongue of which the Nazis had dispossessed him, and which, moreover, had been perverted by their misuse of it. He did this (like Celan, Grass, and others) by parody and irony, and by inverting traditional patterns. As a survivor, he felt he had no right to live. He was in despair over the lack of response of the German public to his testimony, the relativization of the Holocaust, and the revival of antisemitism in the shape of left-wing anti-Zionism. In 1978 he committed suicide. *RW*

0927 Friedman, Maurice Stanley: Why Joseph Campbell's Psychologizing of Myth Precludes the Holocaust as Touchstone of Reality. *Journal of the American Academy of Religion* 66, 2 (Sum 1998) 385-401.

Criticizes Campbell's gnosticizing and psychologizing approach to myth. There is a correlation between Campbell's universalist approach to mythology and religion and his negative approach to the historical and unique in general and Judaism in particular. Campbell (1904-1987) rejected history, the event, the unique, and the particular in favor of the internal, the psychological, and the universal. His approach, combined with his antisemitism, reduced the Holocaust to an internal, psychological matter, a sidestream in the great sweep of history. He did not find in it a touchstone of reality. Contrasts Campbell's approach with that of Martin Buber, who saw myth as bound to the existential and historical. See the response by Robert A. Segal, "Joseph Campbell as Antisemite and as Theorist of Myth" [Ibid. 67 (1999) 461-467], in which he contends that Campbell's indifference to the Holocaust must be attributed not to his theory of myth but to his antisemitism. See also Friedman's rejoinder [Ibid. 469-471]. *DR*

0928 Goertz, Karein Kirsten: *Generational Representations of the Holocaust: Trauma, Memory, and the Imagination.* Diss. – University of Texas at Austin, 1997. 198 pp. Unseen.

Historical authenticity and imaginative restraint are commonly upheld as representational guidelines for Holocaust literature. Current research in historiography and psychiatry acknowledges, however, that a stubborn insistence on factual accuracy, logic, and narrative cohesion can be an impediment to recognizing the emotional truth inherent in narratives about trauma. Maintains that imaginative style and description are integral to the authentic and emotionally truthful representation of the Holocaust as both lived and inherited traumatic experience. In the comparative analysis of autobiographical and fictional texts by adult and child survivors, as well as children born after the war, it becomes apparent that while a fictionalizing mechanism is operative in all of the texts, it serves a distinctly different role for each generation. Describes those differences.

0929 Goldschlaeger, Alain; Lemaire, Jacques, eds.: *La Shoah: Témoignage impossible?* Bruxelles: Editions de l'Université de Bruxelles, 1998. 138 pp.

A collection of articles on Holocaust remembrance. Contents: Goldschlaeger, Alain: Paradoxe de la mémoire (9-14); Neray, Ruth Bindefeld: Ma mémoire (15-18); Goldschlaeger, Alain: Problématique de la mémoire: Lire les témoignages des survivants de la "Shoah" (19-39); Lewendel, Isaac: Le témoin et sa mémoire face à l'histoire (41-53); Sephiha, Haïm Vidal: Vie et mort des Judéo-Espagnols [Appeared in "Les Temps Modernes" 394 (1979). On the Sephardi diaspora after 1492; pp. 58-59 discuss the number of European Sephardim killed in the Shoah.] (55-60); Schreiber, Jean-Philippe: Réflexions autour du "Devoir de mémoire" de Primo Levi ["Le devoir de mémoire," comprising an interview with Levi in January 1983 by Anna Bravo and Federico Cereja, was published in France in 1995.] (61-83); Gergely, Thomas: "Tu enseigneras ceci à tes enfants": Shoah et culture de la mémoire (85-88); Teitelbaum-Hirsch, Viviane: Comprendre la mémoire de l'enfant caché (89-94); Romney, Claude: Les témoignages écrits des médecins déportés à Auschwitz (95-112); Hassoun, Jacques: ...D'une souffrance infinie... (113-121); Rosmus, Anna Elisabeth: A l'écran et avec une nomination aux Oscars [Describes the reception of the film "Das schreckliche Maedchen."] (123-133). *HV*

0930 Greenberg, Gershon: Yehudah Leb Gerst's Religious "Ascent" through the Holocaust. *Holocaust and Genocide Studies* 13, 1 (Spr 1999) 62-89.

Gerst (1906-63), a Polish Orthodox Jew, and a thinker close to Agudath Israel, experienced internment in the ghetto of Łódź in 1940-44, then Auschwitz and other camps, and was liberated in Bergen-Belsen. In 1947 he emigrated to Palestine. Already in prewar Poland, Gerst foresaw a catastrophe impending over the Jews. His views were formed under the influence of the religious philosophy underlying the interwar Olim (Ascenders) movement, founded by Nathan Birnbaum. Contrary to other Orthodox thinkers of his time, Gerst saw in the Holocaust neither a prerequisite for redemption, nor God's punishment, nor a tool of religious education. In his view, Nazism was an antithesis to Judaism, and its main goal was the destruction of the "Sinai morality." The responsibility for Nazism and the Holocaust lies in European philosophy (Kant, Schopenhauer) which began to destroy religion. Gerst regarded Nazi racism as secondary to the aim of destroying divine morality. Notes that Gerst's theology was coupled with insight into history. *DR*

0931 Hoffmann, Detlef, ed.: *Das Gedaechtnis der Dinge: KZ-Relikte und KZ-Denkmaeler, 1945-1995.* Frankfurt a.M.: Campus, 1998. 351 pp. (Wissenschaftliche Reihe des Fritz-Bauer-Instituts, 4).

Deals with relics from the concentration camps Auschwitz, Dachau, Buchenwald, Neuengamme, and camps in southern France, where visitors can imagine the events which occurred in these places by viewing the remnants of gates, fences, ramps, and barracks. Contends, however, that only monuments can ensure the events' significance in the future. Includes numerous photographs. Contents: Hoffmann, Detlef: Das Gedaechtnis der Dinge (6-35); Id.: Dachau (36-91); Knigge, Volkhard: Buchenwald (92-173); Wrocklage, Ute: Neuengamme (174-205); Knigge, Volkhard; Hoffmann, Detlef: Die suedfranzoesischen Lager (206-223); Reuber, Werner: Bilder von den suedfranzoesischen Lagern (224-237); Rensinghoff, Ines: Auschwitz-Stammlager - Das Tor "Arbeit macht frei" (240-265); Menzel, Katharina: Auschwitz-Birkenau - Zaunpfaehle (266-277); Wrocklage, Ute: Auschwitz-Birkenau - Die Rampe (278-309); Brink, Cornelia: Die Gedenkstaette Auschwitz in bundesdeutschen Medien zwischen 1989 und 1993 (310-323); Naumann, Klaus: Auschwitz im Gedaechtnisraum der Presse 1995 (324-329); Ruesen, Joern: Ueber den Umgang mit den Orten des Schreckens: Ueberlegungen zur Symbolisierung des Holocaust (330-343). Hoffmann, Detlef: Nachwort (344-350). *ID*

0932 Loew, Andrea; Robusch, Kerstin; Friedlaender, Saul: "Wenn die Erinnerung kommt...": Saul Friedlaender im Gespraech mit Andrea Loew und Kerstin Robusch. *Sachor* 9 (1999) 75-86.

Friedlaender tells how he first heard at the age of 14 from a Jesuit priest about Auschwitz and that his parents probably perished there; how for many years he avoided consulting easily available sources to find out the details of their fate, and devoted himself to research on topics far removed from the Holocaust; and how difficult it was to begin to write his memoirs. In now writing the history of the Holocaust ("Das Dritte Reich und die Juden"), his intention is to integrate the story of the perpetrators and that of the victims; almost all previous historians focused only on one side or the other. *RW*

0933 Marchi, Gian Paolo; Massariello Merzagora, Giovanna, eds.: *Il Lager: Il ritorno della memoria: Atti del Convegno internazionale, 6-7 aprile 1995, Università degli Studi di Verona, Facoltà di Lingue e Letterature Straniere*. Milano: ANED; Trieste: Edizioni Lint, 1997. xii, 171 pp.

Partial contents: Vasari, Bruno: La memoria è ragione: Il valore della testimonianza degli ex-deportati (9-24); Perotti, Berto: Il Lager: Genesi e divenire (25-29); Mortara Ottolenghi, Luisella: Il contributo di un centro di documen-tazione: Il CDEC (31-34); Fargion, Liliana: Il libro della memoria: La ricerca del Centro di Documentazione Ebraica Contemporanea sugli ebrei deportati dall'Italia (35-43); Federzoni, Laura: La geografia dei Lager in Italia: I punti di raccolta (45-62); Tibaldi, Italo: I treni dei "trasporti" dall'Italia ai KZ (63-70); Massariello Merzagora, Giovanna: Il Lager come Babele: Il plurilinguismo nei KZ (127-144); Lamberti, Maria Mimita: Figure e metafore del Lager nell'arte contemporanea (153-167). *MZ*

0934 Maur, Eva auf der: Primo Levi - KZ Nr. 174517. *Freiburger Rundbrief* 5, 4 (1998) 246-251.

Sketches Levi's life and discusses his writings on the Holocaust. Notes that the Holocaust awakened his Jewish consciousness, although he remained non-religious. He suffered from a feeling of guilt for having survived - in general, in his view, only the worst, the selfish, the bullies, and the schemers, survived - even though his survival was not due to any act of his own, but to luck and to the courageous aid of an Italian laborer. He was also depressed by the lack of interest in survivors' testimonies during the first ten postwar years. His writing aims at arousing thought rather than emotion. *RW*

0935 Neiger, Ada, ed.: *Primo Levi: Il mestiere di raccontare, il dovere di ricordare: Atti del Convegno, Trento, 14 maggio 1997*. Fossombrone: Metauro, 1998. 129 pp. At head of title: Comune di Trento, Servizio biblioteca e archivio storico in collaborazione con Università degli studi di Trento.

Partial contents: Sessi, Frediano: La letteratura concentrazionaria tra verità e finzione a partire da Primo Levi (17-30); Tesio, Giovanni: Primo

Levi, scrittore de scrittura (31-38); Carpi, Anna Maria: "La tregua," lingua materna, lingue di babele (39-44); Rizzante, Massimo: Dell'ibrido: Osservazioni su Primo Levi (45-52); Affinati, Eraldo: Primo Levi: La responsabilità della parola (57-62); Neiger, Ada: Il risentimento del sopravvissuto: Una riflessione intorno a Jean Améry e Primo Levi (63-72); Covito, Carmen: L'influenza di Primo Levi su una generazione che non ha conosciuto fame e guerra: Una testimonianza (73-79); De Angelis, Luca: Se questo è scrivere: Una nota il più possibile chiara su Primo Levi (81-104); Pauletto, Patrick: "...E sceglierai la vita..." (109-117); Rutigliano, Enzo: Canetti-Levi: La metamorfosi nel Lager (119-124). *MZ*

0936 Rein, Raanan: *Franco, Israel y los judíos*. Madrid: Consejo Superior de Investigaciones Científicas, 1996. 349 pp. Originally published as "Be-tzel ha-Shoah ve-ha-Inkvizitziyah" (Tel-Aviv: University Publishing Projects, 1995).

0937 Rickman, Gregg J.: Swiss Banks and the Name Lists. *Avotaynu* 14, 2 (Sum 1998) 4-8.

In 1996-97 a team of American researchers conducted a search of accounts dormant in Swiss banks since May 1945. The search, carried out in the U.S. National Archives and other archives around the world, not only demonstrated the extent of the Nazi theft of Jewish assets, but also supplied the researchers with tens of thousands of names of claimants (who had been mostly unsuccessful) and Holocaust victims. It also showed up the total inadequacy of the search conducted by the Swiss Bankers Association in 1962, when only 26 of 500 Swiss banks responded to the Disclosure Decree enacted by the Swiss Federal Council. Discusses the Swiss banks' policies and actions in the postwar period. Describes some of the lists unearthed by the American search team: the accounts of three Swiss banks in New York whose accounts were frozen by the U.S. government in 1941; the list of assets of Polish victims of the Holocaust transferred by Swiss banks to the Polish communist government in 1949; and other lists of assets not connected with Swiss banks. *DR*

0938 Souki, Nádia: *Hannah Arendt e a banalidade do mal*. Belo Horizonte: Editora UFMG, 1998. 152 pp. Based on her M.A. thesis - Universidade Federal de Minas Gerais, 1995.

Explains that Arendt's concept of the "banality of evil," as described in her book "Eichmann in Jerusalem," is supported by her description of totalitarianism as a system in which the exercise of violence does not aim at dominating man but at making him superfluous. The banal acceptance of this order of values by an individual would be based, in Arendt's understanding, on a lack of independent judgment in recognizing the existence of evil. Concludes that Arendt shares with Kant the idea that evil is not intrinsic to man, but a contingency. The difference between them lies in the fact that Kant considered evil to be opposition to moral laws, while Arendt affirms that evil takes place where moral judgment is absent. *MZ*

0939 Teboul, Margaret: Emmanuel Levinas face à Auschwitz. *Tsafon* 32 (Win 1997-1998) 34-53.

Traces the biography and analyzes the writings of Levinas, the French Jewish philosopher, who was born in Lithuania in 1906 and emigrated to France in 1923. The theme of Auschwitz is present in his writings between 1955-60 and after 1967. Levinas underlines the uniqueness of the fate of the Jews during the Second World War, which cannot be compared to other disasters in wartime. Analyzes some of Levinas' philosophical concepts. Many of his views have their sources in his study of the Talmud, which led him to concentrate on the importance of ethics - in relation to the "Other," and in philosophy in general. Levinas criticized Western culture as partly responsible for the disasters of the 20th century. *HV*

0940

וינריב, אברהם: צרור מכתבים של אברהם וינריב. דפים לחקר תקופת השואה 14 (1997) 401-367.

[Weinreb, Avraham: A Collection of Avraham Weinreb's Letters. *Dapim le-Heker Tekufat ha-Shoah* 14 (1997) 367-401.]

Avraham (Abrasha) Weinreb published his memoirs in Hebrew in "Yalkut Moreshet" 27 (1979). Weinreb (Kielce, 1913-Australia, 1993), who functioned as a doctor in the Vilna ghetto until its liquidation, was deported to forced labor camps in Estonia and then transferred to the camps Stutthof and Schomberg; after the war he settled in Australia. Presents a selection of letters written by him during 1987-91 to members of the teaching staff at Beit Lohamei Haghetaot in Israel, where Weinreb had discussed his Holocaust experiences with students during several visits. Weinreb's letters include his comments on the teaching program at the study center, and his reflections on rescue, resistance, and their various manifestations during the Holocaust, as well as his thoughts on his wartime and postwar reactions to his Holocaust experiences. Includes a letter to Weinreb from the Israeli playwright Yehoshua Sobol (pp. 399-401) discussing his play "Ba-martef," which was dedicated to Weinreb. *LFo*

Nazi War Criminals and War Crimes Trials

0941 Cohen-Grillet, Philippe: *Maurice Papon, de la collaboration aux Assises.* Latresne (Bordeaux): Le Bord de l'Eau, 1997. 160 pp.

Describes the role of Papon in the "Aryanization" of the Bordeaux region and in the deportation of 1,690 Jews between 1942-44 when he was secretary-general of the prefecture in the region. Emphasizes the significance of Papon's trial in 1997, not only because he was the only important figure from the Vichy government still alive but also because of his postwar administrative and political career. Discusses the judicial procedures against Papon from the beginning of the accusation in 1981 up to 1997, his defense, and the political background of the trial. Most of the lawyers who defended Papon were close to the Front National. Explains Papon's manipulations, e.g. when he tried to present himself as responsible for saving the lives of 129 persons. The Gestapo documents found by Michel Slitinsky in 1945 in the forest near Bordeaux included papers signed by Papon. Documents relating to the war period and to the trial are appended on pp. 125-148. *HV*

0942 Disselnkoetter, Andreas: Eine Welt von Assoziationen: Orna Birnbach und die NS-Prozesse in Bochum. *Tribuene* 149 (1999) 41-45.

Reports on a visit to Bochum by Birnbach, a Jewish woman from Tarnów now living in Israel, who appeared as a witness at war crimes trials in Bochum between 1964-74 against SS officers connected with the persecution of Jews in her Polish hometown. Among them was Hermann Blache, who shot Jews for sport and taught his 15-year-old son to do the same. The climate in Bochum was hostile to the Jewish witnesses; the court tended to prefer the evidence given by the perpetrators to that of the victims. While Blache and some others were sentenced to life imprisonment, other criminals went free. In her recent visit to Bochum, many institutions were reluctant to invite Birnbach to speak; but where she did speak, audiences were responsive. Often they had not known of the trials and of the crimes which had been committed. *RW*

0943

פרידמן, טוביה: 50 שנה צייד הנאצים: ספר היסטורי-דוקומנטרי. בעריכת משה מייזלס. חיפה: המכון לדוקומנטציה לחקר פעולות הפושעים הנאצים, 1997. 408 עמ'.

[Friedman, Tuviah: *50 Years Nazi Hunter: A Historical-Documentary Book*. Ed.: Mosche Meisels. Haifa: Institute of Documentation in Israel for the Investigation of Nazi War Crimes, 1997. 408 pp.]

Friedman describes his postwar career as a Nazi hunter, based on his autobiography, "15 shanim radafti aharei Eichmann" (Jerusalem: Achiassaf, 1961). Presents the autobiographical material in brief, accompanied by numerous photographs and facsimiles of documents pertaining to his activities as a Nazi hunter, and in particular his hunt for Eichmann. *RE*

0944

פרידמן, טוביה, עורך: 15 הפושעים הנאצים שפעלו בקולומיה בשנות המלחמה 1941-44 והשתתפו ברצח 60 אלף יהודי קולומיה והסביבה, נעצרו בוינה בשנת 1941 ונשלחו לסיביר בשנת 1948 למאסר עולם. חיפה: המכון לדוקומנטציה בישראל לחקר פעולות הפושעים הנאצים, 1999. 41, 20, 16, [43] עמ'.

[Friedman, Tuviah, ed.: *The 15 Nazi War Criminals Who Operated in Kolomyya in 1941-44 and Participated in the Murder of 60,000 Jews from Kolomyya and the Region Were Arrested in Vienna in 1947 and Sent to Siberia in 1948*. Haifa: Institute of Documentation in Israel for the Investigation of Nazi War Crimes, 1999. 41, 20, 16, 43 pp.]

The first two sections of this collection (in Hebrew) present a history of the events of the Holocaust in the town of Kolomyya (Ukraine). The next two sections (in German and Hebrew) contain excerpts from the protocol of the interrogation in Vienna in 1952 of 15 members of the Nazi police ("Schutzpolizei") responsible for atrocities and mass shootings in the Kolomyya ghetto, who were later brought to trial. An earlier version of the last two sections appeared as "Schupo-Kriegsverbrecher in Kolomea vor dem Wiener Volksgericht: Dokumentensammlung" (1986). *SSC*

0945

פרידמן, טוביה: חלומי על אדולף אייכמן: חלקי במבצע אייכמן: העדים החשובים במשפט אדולף אייכמן משנת 1961 בירושלים: אוסף דוקומנטרי. חיפה: המכון לדוקומנטציה בישראל לחקר פעולות הפושעים הנאצים, 1999. מספור עמודים רב.

[Friedman, Tuviah: *My Role in Operation Eichmann: A Documentary Collection*. Haifa: Institute of Documentation in Israel for the Investigation of Nazi War Crimes, 1999. Various pagings.]

Part of the material presented here appeared previously as "Helki be-mivtza Eichmann" (1990). Contains facsimiles of documents – mainly newspaper clippings and letters – describing the author's activities between 1945-60 in searching for the whereabouts of Adolf Eichmann. In addition, presents the testimonies of Holocaust survivors (e.g. Yitzhak Zuckerman, Zivia Lubetkin, Yisrael Gutman) delivered in Hebrew to the Israel Police in 1961, for the Eichmann trial. *SSC*

0946 Jeffrey, Laura S.: *Simon Wiesenthal: Tracking Down Nazi Criminals*. Springfield, NJ: Enslow, 1997. 104 pp. (People to Know).

A biography of Wiesenthal, intended for young people. Describes his youth and his wartime experiences (pp. 13-38); the rest of the book deals with his postwar activities tracking down Nazi war criminals, e.g. Franz Murer, Adolf Eichmann, Franz Stangl. *DR*

0947

מארוס, מיכאל רוברט: שורשיהם של משפטי נירנברג. זמנים 62 (אביב 1998) 28-36.

[Marrus, Michael Robert: The Origins of the Nuremberg Trials. *Zmanim* 62 (Spr 1998) 28-36.]

Discusses the role of the U.S. government in the decision to hold the Nuremberg Trials. Describes a 1942 meeting of representatives of nine European countries, in which they called for punishment of Nazi war criminals. In 1943 the Allies proposed that trials should be held in the countries in which the crimes were committed, but they also discussed other possibilities. The idea of an international tribunal was favored by Stalin and opposed by Churchill, while Roosevelt remained undecided. Roosevelt's Secretary of War, Henry Stimson, advocated an international tribunal and enlisted the help of Jewish lawyer Murray Bernays, who conceived a "conspiracy law" under which those who planned the genocide could be tried although they themselves did not commit murders. The idea became part of official American policy, but it was only implemented after Roosevelt's death by Truman, who convinced the British to cooperate. States that, although the murder of Jews was a major factor in the trials, the Jewish aspect was not relevant in the planning stages. *RE*

0948 Pick, Hella: *Simon Wiesenthal: A Life in Search of Justice*. London: Weidenfeld & Nicolson, 1996. xvii, 349 pp.

A biography of Wiesenthal, based partly on interviews with him and with other persons. Wiesenthal was born in Buczacz in 1908. Ch. 4 (pp. 51-72) deals with the war years. Wiesenthal was interned in prison in Lvov, taken for forced labor, interned in the Lvov ghetto, and then in the Janowska camp. In 1943 he escaped, but he was caught in June 1944 and went through a number of camps, from Płaszów to Mauthausen, where he was liberated. The rest of the book deals with his postwar activities hunting down Nazi criminals. In 1947, Wiesenthal established the Historical Documentation Centre, first in Linz and then in Vienna. Wiesenthal ascribes to himself a key role in hunting down Eichmann, Stangl, and other Nazi criminals, which has led to quarrels with other Nazi hunters. However, it was Wiesenthal who persisted in the search for Eichmann in the late 1940s-50s. He also fought against the statute of limitation in Germany and Austria. During the Waldheim affair, Wiesenthal rejected the accusation that Waldheim was a war criminal, which aroused the ire of the World Jewish Congress. Despite many oddities in Wiesenthal's character, he always has been a man of conscience and a hero of our epoch. *DR*

0949 Sharpe, Robert Barry: *Modesty and Arrogance in Judgment: Hannah Arendt's "Eichmann in Jerusalem."* Westport, CT: Praeger, 1999. xviii, 170 pp. Based on the author's diss. – "Judgment on Trial: Hannah Arendt, Adolf Eichmann and the Judenraete" – University of South Carolina, 1995.

0950 Sprecher, Drexel A.: *Inside the Nuremberg Trial: A Prosecutor's Comprehensive Account. Vol. I-II.* Lanham, MD: University Press of America, 1999. xvii, 1580 pp.

Presents an exhaustive account of the preparations for and conduct of the first Nuremberg war crimes trial, which began in late 1945 and judged 22 Nazi leaders charged with crimes against humanity and peace. The main sections dealing with the Holocaust appear in vol. 1, chs. 33-37 (pp. 323-377): "The Persecution of the Jews before World War II," "The Persecution and Murder of the Jews – the First Years of World War II," "Mass Extermination in the Occupied Soviet Union – the Einsatzgruppen Murders," "The Persecution and Murder of the Jews – from the Wannsee Conference to the German Defeat," and "The Destruction of the Warsaw Ghetto – a Case Study of a Special Killer Group in Action." Other sections deal with specific defendants and their responsibility for the murder of Jews, among them radio propagandist Hans Fritzsche, Alfred Rosenberg, Joachim von Ribbentrop, Hermann Goering, and Julius Streicher. The evidence and documents presented by Sprecher, one of the American prosecutors, are aimed at preventing future denial of the perpetration of these crimes, including the Holocaust. *YC*

0951 Wieviorka, Annette: Le fasi del processo Eichmann: Un processo eccezionale. *Israele – da Mosè agli accordi di Oslo.* Bari: Dedalo, 1999. Pp. 425-445. Appeared in French in "L'Histoire" 108 (1988). This article is followed by an Italian translation of an interview of "L'Histoire" with Idith Zertal dealing with the Eichmann trial (pp. 447-452).

Neo-Nazism and Holocaust Denial

0952 Finkielkraut, Alain: *The Future of a Negation: Reflections on the Question of Genocide.* Trans.: Mary Byrd Kelly. Introd.: Richard J. Golsan. Lincoln: University of Nebraska Press, 1998. xxxiv, 146 pp. Originally published as "L'avenir d'une négation: Réflexion sur la question du génocide" (Paris: Seuil, 1982).

Holocaust denial is not only propagated by neo-Nazi groups; it is also found in extreme-left circles. It has found expression in the activities of the publishing house La Vieille Taupe. The extreme left was the first to publish the revisionist writings of Robert Faurisson. Examines the origins, main ideas, motifs, and psychology of the leftist denial of the Holocaust, contending that it stems from their rigid and simplistic worldview according to which there is only one meaningful social conflict – that between the capitalist and working classes. They contend that the Nazis, as representatives of the former, were no worse than the democracies or the corrupt Soviet regime, that World War II was an ordinary imperialist war, and that the democracies invented the Holocaust myth in order to blacken their former competitor (Germany) and to conceal their own crimes. The Holocaust does not accord with the leftist scheme of world affairs, thus they deny it. Another tendency of the left is to appropriate the language and the imagery of the Holocaust for other victims of worldwide "imperialism," e.g. Palestinians. Pp. xi-xxx contain an introduction by R.J. Golsan, describing, inter alia, the Garaudy-Abbé Pierre affair. *DR*

0953 *Revue d'Histoire de la Shoah* 166 (May-Aug 1999).

This issue is entitled "Négationnisme: Le génocide continué." Partial contents: Igounet, Valérie: Une tradition extrémiste: Le négationnisme [On negationism in France from 1948 to 1998.] (7-43); Tarnero, Jacques: Le négationnisme, ou le symptôme des temps pervers: Une enigme recurrente: Le signe antijuif (44-75); Bensoussan, Georges: Négationnisme et antisionisme: Récurrences et convergences des discours du rejet (76-88); Daeninckx, Didier: De "Treblinka" à Bordeaux... [On Jean-François Steiner, from the publication of his book "Treblinka" (1966) to the evidence he gave at Papon's trial (1997).] (89-99); Lasek, Wilhelm; Mayr, Peter: La propagande "révisionniste" en Autriche: Organisations et publications néo-nazies [Including on the Internet. Appeared in German in "Wahrheit und Auschwitzluege" (1995).] (100-120). *HV*

Anti-Zionism and Anti-Israel

0954 Bodansky, Yossef: *Islamic Anti-Semitism as a Political Instrument.* Shaarei Tikva: Ariel Center for Policy Research; Houston, TX: Freeman Center for Strategic Studies, 1999. 200 pp.

After finding roots of antisemitism in traditional Islamic sources starting with the Qur'an, discusses the use of anti-Jewishness and anti-Judaism as a major political instrument against the State of Israel from the 1950s through the 1990s. Thus, emphasis is placed on political as well as religious dimensions of such Muslim religious concepts as "dhimma" and "jihad." A "global jihad" against the Jews is shown to have ramifications not only among Palestinians and in Iran, but also in more distant Islamic states such as Malaysia and Pakistan. Anti-Jewish stereotypes and the idea of a Jewish conspiracy are widespread. The Hamas movement is discussed for its prominent role in spreading anti-Judaism and antisemitism in a global Islamic context as part of its effort to destroy Israel. *YC*

Christian-Jewish Relations

0955 Elchinger, Leon Arthur: Iglesia y Judaísmo: Un descubrimiento a proseguir (la esperanza de un obispo). *El Olivo* 50 (July-Dec 1999) 99-119.

Based on a lecture delivered at the Community Center of the Synagogue of Peace in Strasbourg in 1993. Asserts that the lack of appreciation by Christians of the permanent values of Judaism, the accusation of deicide, and the teaching of contempt provoked the religious dispute to degenerate into antisemitism and persecution. Relates changes in the Churches after World War II, especially the declaration of Martin Niemoeller and Karl Barth in 1948 recognizing the historical responsibility of the Churches in the genocide of the Jews, and the struggles in the Catholic Church which accompanied the Vatican Council II Declaration of 1965, in which the accusation of deicide was eliminated and the catechism reformed. Mentions efforts of the Churches in France and in the rest of the world to improve Christian-Jewish relations. *MZ*

0956 Hargrove, Katharine T., ed.: *Seeds of Reconciliation: Essays on Jewish-Christian Understanding*. N. Richland Hills, TX: BIBAL Press, 1996. xii, 242 pp.

A collection of essays previously published in "SIDIC." Partial contents: Klenicki, Leon: On Christianity: Towards a Process of Historical and Spiritual Healing - Understanding the Other as a Person of God [Appeared in "SIDIC" 27 (1994).] (3-37); Di Sante, Carmine: The "Old" and the "New" Covenant: How to Relate the Two Testaments (39-52); Baum, Gregory: Chosenness in Christian Tradition [Discusses antisemitic aspects of Christian chosenness from early Christianity until after Vatican II.] (87-97); Fabris, Renzo: Modern Man and the Concept of the Chosen People [Presents some antisemitic views of the "chosen people" idea, then proceeds to discuss its positive elements in Judaism and Christianity.] (99-113); Dubois, Marcel Jacques: Christian Reflection on the Holocaust (157-178); Friedlander, Albert Hoschander: The Shoah and Contemporary Religious Thinking [Appeared in "SIDIC" 22 (1989).] (179-191); Hargrove, Katharine T.: Holocaust Theology (193-201); Klenicki, Leon: Catechism of the Catholic Church: An Interfaith Jewish Reading [Appeared in "PACE" 23 (1994) and "SIDIC" 27 (1994).] (205-221); Fisher, Eugene Joseph: Reflections on the Catechism of the Catholic Church [Appeared in "PACE" 23 (1994) and "SIDIC" 27 (1994).] (223-231). *YC*

0957 Kellenbach, Katharina von: Overcoming the Teaching of Contempt. *A Feminist Companion to Reading the Bible: Approaches, Methods and Strategies*, eds. Athalya Brenner, Carole Fontaine. Sheffield: Sheffield Academic Press, 1997. Pp. 190-202.

Based on her book "Anti-Judaism in Feminist Religious Writings" (1994). Focuses on portrayals of the biblical God as the enemy of feminist values in the writing of feminist theologians. Such writers express ideas in line with traditional Christian anti-Judaism that has portrayed Israel's history as a succession of crimes and failures that climaxed in the rejection and murder of Jesus – making of the Jews paradigmatic sinners and enemies of God as well as the originators of patriarchy. Four dismissive attitudes toward anti-Judaism on the part of feminist scholars involve trivialization, particularization, spiritualization, and universalization. Explains the meaning of these processes, which result in a teaching of contempt toward Jews and Judaism. Calls, instead, for a teaching of respect. *REK*

0958

ליפל, ישראל; רוזן, דוד: הכנסיה הקתולית, העם היהודי ומדינת ישראל (לקט מסמכים). ירושלים: מכון ירושלים למחקר וליחסים עם הדתות, 1996. 69 עמ'.

[Lippel, Israel; Rosen, David: *The Catholic Church, the Jewish People and the State of Israel (a Collection of Documents)*. Jerusalem: Makhon Yerushalayim le-Mehkar u-le-Yahasim im ha-Datot, 1996. 69 pp.]

Contains Hebrew translations of documents issued by the Vatican between 1975-93, relating to Christian-Jewish relations: "Nostra Aetate"; "Guidelines and Suggestions for Implementing the Conciliar Declaration of 'Nostra Aetate' (n. 4)"; "Notes on the Correct Way to Present the Jews and Judaism in Preaching and Catechesis"; a speech delivered by Pope John Paul II at the Rome synagogue on 13 April 1986; a document on the Church and racism by Eugene J. Fisher (1988); "Antisemitism: A Sin against God and Man" (1990); "Teaching on Jews and Judaism in Catholic Education," a lecture delivered by Rémi Hoeckman in 1992; and the agreement between the Vatican and Israel signed in 1993. *RE*

0959 *Sens* 51, 10 (1999).

This issue, entitled "Jacques Maritain et Charles Journet contre l'antisémitisme," also contains articles concerning the Seelisberg Conference of 1947. Partial contents: Charmet, Bruno: Charles Journet, Jacques Maritain: Leur combat contre l'antisémitisme à la lumière du troisième tome de leur correspondance (1940-1949) (403-420); Macina, Menahem R.: Pour une repentance chrétienne: Safran-Journet – Maritain-

Pie XII (421-433); Id.: Le rôle de Paul Démann à Seelisberg (434-439). *HV*

0960 Thoma, Clemens: Das Einrenken des Ausgerenkten: Beurteilung der juedisch-christlichen Dialog-Geschichte seit dem Ende des Zweiten Weltkrieges. *Bulletin der Schweizerischen Gesellschaft fuer Judaistische Forschung* 7 (1998) 2-17.

Discusses the history of Jewish-Christian dialogue since World War II, mentioning failures and achievements of the Catholic Church during and after the war, especially of Popes Pius XI, Pius XII, John XXIII, and John Paul II. Deals with the content of the Vatican document "We Remember," issued on 16 March 1998, on the Catholic Church and the Holocaust. States that the document contains some deceptive passages, because they minimize the responsibility of the Church concerning the annihilation of the Jews, claiming that the roots of Nazi ideology were outside of Christianity. States that the purpose of this document is not apology or repentance, but justification. *ID*

1945-1990: Australia

0961 Berman, Judith: The Lessons and Messages of the Holocaust as Conveyed through Yom haShoah (Holocaust Day) Commemorations in Selected Australian Jewish Communities, 1945-1996. *The European Legacy* 4, 1 (Feb 1999) 54-71.

Demonstrates how, for over 50 years, Australian Jewish commemoration of the Holocaust has focused on lessons pertaining to a Jewish agenda: the importance of the State of Israel, vigilance against antisemitism, Jewish solidarity and aid to Jewish communities (e.g. Soviet Jewry), and reaffirmation of Judaism and Jewish cultural life. These values reflect the demographic make-up of Australian Jewry, 40,000 of whom arrived as prewar refugees or Holocaust survivors. These messages correspond to the goal of Jewish survival in the face of the threats of assimilation and intermarriage. This exclusively Jewish agenda, which has militated against drawing any general, humanitarian lessons from the Holocaust, has been expressed in a policy of a "closed" commemoration for and by Jews and by a consistent choice of speakers who omit reference to more universalistic lessons, such as sympathy for non-Jewish victims of oppression or calls to action on their behalf. *YC*

0962 Berman, Judith: Yom haShoah (Holocaust Day) Commemorations in Melbourne, Perth and Sydney, 1945-1996. *Australian Journal of Jewish Studies* 13 (1999) 94-116.

Attendance at Yom Hashoah commemorations in Melbourne declined after 1963, but increased from the 1980s. This tendency reflects the fact that in the immediate postwar decades the ceremonies were attended mainly by Holocaust survivors, most of whom had settled in Melbourne in the postwar years; in the 1980s-90s, with the increased centrality of the Holocaust in Australian Jewish consciousness, they have been attended by younger people. In Sydney and Perth, the Yom Hashoah ceremonies have attracted smaller audiences than in Melbourne. In order to attract the postwar generation, and to turn it into an event for the whole community, the organizers have changed the format of the ceremony: e.g. its main language is English instead of Yiddish; elements of drama, music, and declamation have been added; members of the Jewish wartime resistance participate in the ceremonies. Other commemoration activities, like exhibitions or Holocaust Awareness Week, are also held in Australia. *DR*

0963 Mendes, Philip: The Jewish Council, Communism and the State of Israel. *Australian Jewish Historical Society Journal* 14, 3 (1998) 459-467.

In the last three decades, studies on the Australian Jewish Council to Combat Fascism and Anti-Semitism have regarded it as a pro-communist and pro-Soviet, and therefore anti-Zionist, organization. Shows that, in fact, during most of the period 1948-70 the Council's stance was pro-Zionist and pro-Israel; this was in contradiction to its denial of Soviet antisemitism and justification of antisemitic trials in the USSR and in Czechoslovakia. The Council's pro-Israel position contrasted with the pro-Arab position adopted by the Communist Party of Australia, and especially with the harshly anti-Zionist position adopted by the radical Left after 1967. Contends that the Council's dependence on the USSR and the CPA has been exaggerated by its adversaries; its reluctance to criticize the USSR was tactical, and does not mean it was a communist puppet. *DR*

1945-1990: Europe

General

0964 Bleich, Erik John Malcolm: *Problem-Solving Politics: Ideas and Race Policies in Britain and France.* Diss. - Harvard University, 1999. 257 pp. Unseen.

Examines the development of British and French race policies since World War II. Describes the differences between the British "race relations" approach and the French "anti-racism" model for countering racism. Despite similar histories, economies, and numbers of ethnic minorities, Britain and France have pursued divergent paths for organizing majority-minority relations and for overcoming racism. British policymakers framed their problems primarily as "color" racism and discrimination, learned much from the North American context, and developed positive beliefs about the effectiveness of "American" policies. In contrast, their French counterparts framed their problems proportionally more in terms of antisemitism and hate speech, drew almost exclusively on their domestic context (without seriously considering other countries' policies), and formulated causal and normative beliefs quite different from those that predominated in Britain. Variation in ideational frames thus emerges as the most important influence shaping the divergent race policies in the two nations.

0965 Ducci, Teo, ed.: *In memoria della deportazione*. Milano: Mazzotta, 1997. 94 pp.

A collection of photographs of monuments and memorials erected by Italian artists in Italy, Austria, Germany, and Poland, to commemorate the Italians (many of them Jews) who were deported by the Nazis. Includes works erected in concentration camps, as well as photographs of the Museo Monumento della Deportazione in Carpi (pp. 40-47). Pp. 7-10 contain an introduction by Bruno Zevi. *MZ*

Austria

0966 Embacher, Helga; Reiter, Margit: *Gratwanderungen: Die Beziehungen zwischen Oesterreich und Israel im Schatten der Vergangenheit*. Wien: Picus, 1998. 387 pp.

Examines the ambivalent relationship between Austria and Israel today, and the impact of the Nazi past on the interaction between the two countries. Surveys various conflicts and crises, e.g. negogations about reparations, relations during the Kreisky era, and the Waldheim Affair. Discusses attitudes of the Austrian media and of the public toward the Six-Day War, wavering between philosemitism and antisemitism. Deals also with Austrian reactions to other Israeli wars, especially its criticism of Israel's invasion of Lebanon in 1982. *ID*

0967 Ramp, Norbert: "Die D.P. bezahlen alle Preise...": Vorurteile und Konflikte zwischen Einheimischen und juedischen DPs in Salzburg und Oberoesterreich. *Flucht nach Eretz Israel: Die Bricha und der juedische Exodus durch Oesterreich nach 1945*, ed. Thomas Albrich. Innsbruck: Studien Verlag, 1998. Pp. 137-160.

Discusses the special position of Jewish DPs in the U.S.-occupied zone in Austria (focusing on the area of Salzburg). They received preferential treatment from the occupying power, while being only reluctantly accepted by the Austrian administration and overwhelmingly rejected by the majority of the local population. Over the years they were the focus of envy, xenophobia, prejudice, and antisemitism. The relationship between the locals and the DPs was determined by their material plight (the Austrians saw the refugees as competitors and a burden when it came to obtaining supplies of vital goods) and by antisemitic stereotypes, already prevalent in Austrian society, which were revitalized by the behavior of the Jewish DPs with their heavy involvement in black-marketeering. With the consolidation of the Austrians' view of themselves as the "first victims of Nazism," the Jews were rejected by the locals who wanted to rid their country of them. *ID*

0968 Ramp, Norbert: Prejudices and Conflicts between Locals and Jewish DPs in Salzburg and Upper Austria. *The Journal of Israeli History* 19, 3 (Fall 1998) 63-82. Appeared in German in "Flucht nach Eretz Israel: Die Bricha und der juedische Exodus durch Oesterreich nach 1945" (Innsbruck: Studien Verlag, 1998).

0969 Wistrich, Robert Solomon: *Austria and the Legacy of the Holocaust*. New York: American Jewish Committee, 1999. viii, 56 pp. (International Perspectives, 44).

Austrian-Jewish relations have been influenced by Austria's wartime record as part of the Nazi Reich and its contribution to the Holocaust. After 1945 Austria adopted the myth of itself as "the first victim of Nazism"; in contrast to postwar Germany, it did not repent for its participation in the Holocaust and denied its historical legacy of antisemitism. Attempts were made to trivialize the Holocaust and to equate Jewish victims with their former persecutors. In the same vein, Austria refused to pay reparations for confiscated Jewish property. Dwells on Bruno Kreisky's personality and his negative stance toward the Jews and Israel. One of Kreisky's goals was to relativize Austria's wartime cooperation with the Nazis, a fact noted by Austrian right-wing radicals. Austrian-Jewish relations were strained even more by the Waldheim Affair. After 1987 Austria attempted to improve its relations with Jews and to admit its responsibility for the Holocaust. However, the same period was marked with the electoral success of Haider's right-wing and xenophobic Freedom Party. *DR*

Belarus

0970 Batvinnik, Marat B.: Pamyatniki genotsida yevreyev na Brestchine [Monuments of the Genocide of Jews in the Brest Region]. *Yevrei Belarusi* 3-4 (1998) 137-154; 5 (2000) 164-181.

Briefly describes the Nazi murder of the Jews in several towns in the Brest region: Brest, Domachevo, Molchad, Baranovichi, et al. In these towns, monuments were erected after the war. Presents a list of the monuments (pp. 150-154), with data on the mass murder commemorated by each monument and the time when it was erected. Pt. 2 of the article deals with events in Kobrin, Luninets, Pinsk, and some other places. *DR*

Czechoslovakia

0971 Nepalová, Šárka: Die juedische Minderheit in Boehmen und Maehren in den Jahren 1945-1948. *Theresienstaedter Studien und Dokumente* (1999) 332-362.

Analyzes the composition of the Jewish population of Czechoslovakia in the first postwar years and the Czech attitude toward them. While the national government issued regulations intended to ensure returning Holocaust survivors their rights, lower-level bureaucrats tended to refuse them these rights - restoration of property and of citizenship - on the grounds that they were "Germanizers": in 1930 they had spoken German or sent their children to German schools. These accusations often were motivated by antisemitism. While some press organs defended the Jews, others portrayed them as greedy capitalists and exploiters who had not learned their lesson from the Holocaust. The suffering undergone by the Holocaust survivors, as well as the high participation of Jews in the Czech army in exile, were trivialized. *RW*

France

0972 Auron, Yair: *Les Juifs d'extrême gauche en mai 68: Une génération révolutionnaire marquée par la Shoah*. Trans.: Katherine Werchowski. Paris: Albin Michel, 1998. 335 pp. Translated from the Hebrew: "Kulanu yehudim germanim: Radikalim yehudim be-Tsarfat bi-shnot ha-shishim ve-ha-shiv`im" (Tel-Aviv: Am Oved; Tel Aviv University; Beer Sheva: Ben-Gurion University of the Negev, 1999).

0973

אורון, יאיר: כולנו יהודים גרמנים: רדיקלים יהודים בצרפת בשנות השישים והשבעים. תל-אביב: עם עובד; אוניברסיטת תל-אביב, המכון לחקר הציונות; באר-שבע: אוניברסיטת בן-גוריון בנגב, המרכז למורשת בן-גוריון, 1999. 288 עמ'.

[Auron, Yair: *We Are All German Jews: Jewish Radicals in France during the Sixties and Seventies*. Tel-Aviv: Am Oved; Tel-Aviv University; Beer-Sheva: Ben-Gurion University of the Negev, 1999. 288 pp.]

Holocaust remembrance was one of the most influential factors which shaped the individual, radical, and Jewish identity of Jewish radicals in France in the 1960s-70s. Many of them were Holocaust survivors or offspring of survivors. These Jewish radicals reflected a specific Jewish-universal memory of the Holocaust. Although they recognized the uniqueness of the Jewish experience of the Holocaust, their perceptions went beyond the Jewish aspect. It was the incentive for their identification with victims of oppression and injustice in general, and victims of racism, fascism, and antisemitism in particular. The Holocaust memory of the Jewish radicals also influenced French collective memory regarding World War II, the Nazi occupation of France, French resistance groups, and the attitude in France toward collaborators. It had, as well, an impact on the Jewish radicals' attitude toward the State of Israel, often extremely leftist. Since Holocaust memory has become an important factor in Jewish identity, it is necessary to shape a collective Holocaust memory which combines the Jewish-universal and Jewish-Zionist aspects. *LFo*

0974 Bellamy, Elizabeth J.: *Affective Genealogies: Psychoanalysis, Postmodernism, and the "Jewish Question" after Auschwitz*. Lincoln, NE: University of Nebraska Press, 1997. 214 pp.

Examines how the Holocaust and Jews have been represented in a wide range of French poststructuralist works. Focuses on the writings of Jacques Derrida (especially on his controversial defense of Heidegger's works and on his "discourse of the Jew" in "De l'esprit" (1987) and other works), Philippe Lacoue-Labarthe, Jean-Luc Nancy, and Jean-François Lyotard. Concludes that recent French thought "encrypts but does not fully confront the trauma of the Holocaust." French postmodernism exposed its inability to mourn the Holocaust, to work through to an understanding of the horror of the Holocaust; this phenomenon is a continuation of the prewar failure to conceptualize Jewish identity and to solve the "Jewish question." Reflects on the question of the "Jewish origins" of psychoanalysis in France and the implications of the Holocaust for psychoanalysis. Despite the impulses of philosemitism that serve as the foundation of post-Holocaust psychoanalytic discourse, the goal for the future of psychoanalysis after Auschwitz "should be the posing of questions that can stand outside the repetition compulsions of both an anti-Semitic and a philo-Semitic (un)conscious." *DR*

0975 Bilange, François: Justin Godart et le Mémorial juif. *Revue d'Histoire de la Shoah* 162 (Jan-Apr 1998) 183-193.

Traces the figure of Godart (1871-1956), a French politician. In the interwar period he was active in the association France Palestine in support of the Zionist cause. From 1933 he was a member of the Comité pour la Défense des Droits des Israélites en Europe Centrale et Orientale.

Focuses on Godart's postwar activities concerning the erection of the Mémorial du Martyr Juif Inconnu in Paris. Godart travelled widely in order to convince various important persons of the necessity for such a monument, and collected money for this purpose as well. The Mémorial was inaugurated on 30 October 1956. *HV*

0976 Leibowitz, Nicole: *L'Affaire Carpentras: De la profanation à la machination*. [Paris]: Plon, 1997. 209 pp.

Chronicles the polemic surrounding the desecration of the Jewish cemetery at Carpentras in 1990, which embroiled France for the following six years. Argues that the social and political climate in the decade preceding the episode was conducive to the growing power of Jean-Marie Le Pen, leader of the Front National. Demonstrates that despite consistent denial of any role by his party in the crime, the confessions by the perpetrators in 1996 proved the contrary. Compares interpretations of the incident in the French press with personal views expressed in interviews with various political personalities and inhabitants of the region. Seeks a response to the question of the ultimate significance of the Carpentras affair for French history and society. *TS*

0977 Meistermann, Tony: Les témoignages de déportés juifs sur le camp d'Auschwitz Birkenau. *Tsafon* 32 (Win 1997-1998) 14-33.

Surveys the history of the memory of Auschwitz in France after World War II. Asserts that in the immediate postwar period only a few testimonies were published, because Holocaust survivors were trying to "forget" the most difficult time of their lives. The Holocaust revisionism and negationism (e.g. the Faurisson affair) which began in France in 1978 caused survivors to realize the necessity to write their memoirs. Stresses the difficulties of relating Holocaust experiences and, at the same time, the need for transmission of the truth, which entails a struggle against ignorance. Surveys common themes found in most of the testimonies about Auschwitz. *HV*

0978 Trotignon, Yves: Quelques réflexions sur les témoignages écrits du système concentrationnaire nazi et la communauté juive. *Revue d'Histoire de la Shoah* 162 (Jan-Apr 1998) 127-140.

Discusses testimonies written by survivors of concentration camps, their literary values, style, and the way the reality was transformed through the vision of the writers. Asserts that in the immediate postwar period the truth about the extermination camps was not made clear to the wider public in France, and Auschwitz was seen mainly as a fascist and anti-communist camp, and not racist. The appearance in the 1970s of negationist publications in France (by Robert Faurisson and Paul

Rassinier) provoked reactions from the Jews, who thought that the Holocaust could never be contested. From the 1980s, the revival of this theme is noticeable in France. *HV*

0979 Wolf, Joan Beth: "Anne Frank Is Dead, Long Live Anne Frank": The Six-Day War and the Holocaust in French Public Discourse. *History & Memory* 11, 1 (Spr-Sum 1999) 104-140.

The Israeli-Arab war in June 1967 evoked a rise of interest in the Holocaust in French society. The press, both Jewish and non-Jewish, Right and Left, used imagery of the Holocaust in interpreting the events. The Holocaust was traumatic for French Jews as it disappointed their belief in full integration in France. It made them feel connected with the State of Israel and responsible for it. Jewish writers identified the June 1967 war as an attempt to perpetrate a new Holocaust. While some non-Jewish writers agreed with the Jewish interpretation of the events, other, pro-Arab ones developed a counter-narrative, in which they dubbed Israelis as the new Nazis and identified the Arabs with the Jewish victims of Nazism. These works show that not only did many non-Jews fail to comprehend the Jewish trauma of the Holocaust, but they sympathized with the Jews only when they were weak; they resented a strong Jewry, embodied in the State of Israel. *DR*

0980 Wolf, Joan Beth: *Negotiating Meaning: Narrative Politics and Memories of the Holocaust in France.* Diss. – University of Chicago, 1997. 429 pp. Unseen.

Explores postwar French public discourse on the Holocaust during a series of controversial events: the Six-Day War (1967), the release of Marcel Ophuls' film "The Sorrow and the Pity," an interview in "L'Express" with former Vichy minister Darquier de Pellepoix, the broadcast of the American TV film "Holocaust," the 1980 attack on the synagogue of rue Copernic, Robert Faurisson's Holocaust revisionist publications, the trials of Klaus Barbie and Paul Touvier, and widespread accusations in the 1990s that François Mitterrand had actively collaborated with the Vichy regime. Argues that Holocaust discourse is constituted by multiple layers of meaning; this discourse both informs and is informed by the constantly shifting relationship between Holocaust narratives, changing ethnic and national identities, and demands for historical responsibility. The Holocaust developed from a primarily Jewish concern in 1967 to a source of diffuse identification in the 1980s-90s. By 1995, the Holocaust had become metonymic for victimization on a broad scale: victimization of the French population by the Vichy regime during World War II, and victimization of the French nation by "the State" in the postwar period.

Germany

0981 Fox, Thomas C.: *Stated Memory: East Germany and the Holocaust.* Rochester, NY: Camden House, 1999. 177 pp.

Examines representative East German responses to the Holocaust from 1945 until Germany's reunification in 1990. The state-organized discourse constructed a Holocaust memory that served political ends, specifically the legitimation of the GDR. This discourse was also designed to celebrate the USSR and communism, as well as to illustrate the vileness of the West German state, where many former Nazis gained secure positions. Discusses the Holocaust master narrative constructed by historians and its popularization in textbooks and other sources. Examines the way in which the official history was concretized in East German concentration camp memorials. Analyzes works by writers and filmmakers who contributed to this discourse. Dwells, also, on the situation of East German Jews, who experienced the contradictions of the state's Holocaust discourse. *DR*

0982 Goldberg, Eckhard: Mahnmale und Denkmale in Moenchengladbach. *Juedisches Leben in Moenchengladbach: Gestern und heute.* Moenchengladbach: Stadtarchiv, 1998. Pp. 42-50.

Describes memorials to the Holocaust erected in Moenchengladbach, Rheydt, Wickrathberg, Odenkirchen, and Neuwerk in the 1950s-90s, including the initiatives for their erection and the dedication ceremonies. *RW*

0983 Hoeppner, Reinhard: Christen, Juden, Israel aus der Perspektive der DDR-Wirklichkeit. *Kirche und Israel* 14, 1 (1999) 3-11.

Traces the history of East German attitudes to Jews, the Holocaust, and the State of Israel, especially the role of the Protestant Church. States that before the separation between the Churches of East and West Germany, East German churchmen played a leading part in sponsoring official confessions of guilt. Subsequently, even in the face of party policy, they cultivated dialogue with Jews and education on Jews and the Holocaust, and protested against the regime's anti-Zionism. *RW*

0984 Hoffmann-Curtius, Kathrin: Denkmaeler fuer das KZ Dachau. *Die Nacht hat zwoelf Stunden, dann kommt schon der Tag: Antifaschismus, Geschichte und Neubewertung,* ed. Claudia Keller. Berlin: Aufbau Taschenbuch Verlag, 1996. Pp. 280-309. Unseen. See the abstract for the revised English version: "Oxford Art Journal" 21 (1998).

0985 Hoffmann-Curtius, Kathrin: Memorials for the Dachau Concentration Camp. *Oxford Art Journal* 21, 2 (1998) 21-44. An earlier German version appeared in "Die Nacht hat zwoelf Stunden, dann kommt schon der Tag" (1996).

Discusses images of Dachau during the Nazi era and the memorials constructed there since the end of World War II. The early postwar memorials tended to employ terms and styles reminiscent of the Nazi era, as well as Christian themes and symbols, with no reference to Jewish victimization. Observes a pattern of open or concealed contempt for Jewish victims in the memorials. After 1960 the prisoners' barracks were torn down, thereby minimizing the camp's relics, to make way for the conversion of the camp into one large memorial. Pp. 39-41 describe the Jewish monument erected by the Jewish community of Bavaria in 1967, which is reminiscent of the death ramp leading to the gas chamber. *REK*

0986 Kattago, Siobhan: *Ambiguous Memory: The Legacy of the Nazi Past in Postwar Germany.* Diss. - New School for Social Research, New York, 1997. 269 pp. Unseen.

Examines official memories of National Socialism in the Federal Republic, the German Democratic Republic, and unified Germany during the 1980s-90s. Investigates the social mechanisms of West German internalization and East German universalization of the Nazi past by examining social frameworks and specific places of memory. Memories of the past provide insight into the society which remembers. The creation of postwar German national identities was not only based on a system of negative mirror-imaging and comparison between the two Germanies, but also on how Nazi Germany was remembered. Unified Germany is framed by a double past which entails a continuation of the West German framework of internationalization accompanied by the "newness" of the restored German nation. Places of memory, such as Bitburg, the "historians' debate," the 50th anniversary commemoration of Kristallnacht, the renarration of Buchenwald, and the proposed Holocaust Memorial in Berlin are examples of the ambiguity of East, West, and unified German memory.

0987 Koselleck, Reinhart: Die Diskontinuitaet der Erinnerung. *Deutsche Zeitschrift fuer Philosophie* 47, 2 (1999) 213-222.

Based on a lecture given at Heidelberg University. Suggests that, in the Nazi period, most Germans had only a fragmentary knowledge of Nazi crimes, acquired in chance experiences; Koselleck, then a soldier in the Wehrmacht, knew nothing of these crimes and felt betrayed when he learned of them after the war. The collective memory is a synthesis of individual memories, but since experience cannot be transmitted, this collective memory is one degree removed – a fact that explains the break between the war generation and the '68ers. The children have no right to blame their parents, since they did not share their experience. Nazi crimes defy all efforts to grasp them scientifically or morally, or atone for them through religion. Argues that memorials should include all the victims of Nazism, not only the Jews. In Germany they should not permit a false identification of Germans with the victims, but point to the perpetrators. See the reply by Gabriel Motzkin, "Moralische Verantwortung und Diskontinuitaet der Erinnerung" [Ibid. 47, 6 (1999) 1023-1031], who remarks that memories cannot be divorced from a moral context, as Koselleck tries to do; a moral dimension also makes empathy possible. The postwar generation of Germans feels guilt not through identification with the perpetrators – which would be unjust – but through empathy with the victims. Holds that a memorial has impact only if it is dedicated to a specific group, in this case the Jews, as the symbolic representatives of all the victims of Nazism. *RW*

0988 Olick, Jeffrey K.; Levy, Daniel: Collective Memory and Cultural Constraint: Holocaust Myth and Rationality in German Politics. *American Sociological Review* 62, 6 (Dec 1997) 921-936.

The Holocaust has long been the standard for evaluating German political activity. Examines the transformation of official representation of the Holocaust in the FRG as a constraint in German political culture. The Holocaust exerts its power in German politics both mythically and rationally; the mechanism of the interplay between Holocaust memory and political behavior is more complicated than simply "holding Germany hostage to Auschwitz." In the Adenauer era, support for Israel and reparations were instrumental in the rehabilitation of Germany as a nation. The post-1967 generation viewed itself as fundamentally different from the older, criminal one and was inclined to abandon support for Israel; both generations, however, were reluctant to admit German collective guilt for the genocide. Dwells on the "historians' debate" of 1985-87. Nolte's arguments evoked a negative reaction because he tried to minimize the power of the Holocaust as a cultural referent. At the same time, the "historians' debate" transformed the Holocaust from a constraint that could only be obeyed or transgressed to one that could be investigated scientifically and debated rationally. *DR*

0989 Reichel, Peter: "Ueber Auschwitz waechst kein Gras": Zur Auseinandersetzung mit dem Holocaust in der westdeutschen Gesellschaft. *Tribuene* 150 (1999) 160-172.

Traces the attitude of Germans to the Holocaust from the end of the war to the present. Argues that in the Nuremberg trials the distinction between genocide and war crimes proper was blurred. The Germans resented being reminded of their guilt, and Adenauer passed the reparations bill against strong opposition. The diary of Anne Frank, the Eichmann and then the Auschwitz trials, forced Germans to face the Holocaust, but also gave rise to a demand to let bygones be bygones. Bundestag debates on extension of the time limit for prosecution of war crimes resulted in severely limiting such cases and in equating genocide with ordinary murder. Finally, the TV series "Holocaust" succeeded, by its mixture of fictional appeal and documentary, in arousing the German public. Since then there has been a lively interest in the Holocaust, but currently it seems to be limited to a debate on its proper commemoration and aesthetic representation. There is also a tendency to identify with the victims, rather than accept the burden of responsibility as offspring of the perpetrators. *RW*

0990 Schulze Wessel, Julia: Die neue Gestalt des Antisemitismus in der deutschen Nachkriegsgesellschaft: Eine Analyse deutscher Polizeiakten der Jahre 1945-1948. *Jahrbuch fuer Antisemitismusforschung* 8 (1999) 177-194.

Mainly on the basis of reports by the German police in Frankfurt and Stuttgart to the American occupation authorities, describes the police's hostility to the DPs, especially the Polish Jews. The police constantly pressed the Americans for greater authority to control the DPs, whom they represented as black marketeers and as criminals who endangered the property and safety of peaceloving, defenseless German civilians. During a police raid on the camp in the Reinsburgstrasse in Stuttgart in March 1946, an Auschwitz survivor was shot dead; public opinion and the press supported the police action. There was no attempt to understand the sensibilities of the DPs. Suggests that this attitude was part of the Germans' endeavor to deny complicity in the Holocaust and to restore the solidarity of German society through exculpation and integration of the perpetrators and the projection of guilt on the victims. *RW*

0991 Vann, Martin Eric: *West German Secondary School Education on the Holocaust.* M.A. thesis - Florida Atlantic University, 1998. 153 pp. Unseen.

After World War II, the Allies attempted to denazify Germany, but were unable to completely reform the educational system. Cold War exigencies dictated that lessons on the Nazi era, particularly the cruelties of the Holocaust, be soft-pedaled. While some German politicians urged greater openness, collective amnesia reigned for over a decade. Early texts showed Germans as mesmerized by Hitler. Gradually, as democracy took root in West Germany, educators responded to the changing political culture by teaching more of the true nature of Nazism. Each decade brought significant improvements in textual coverage as an ethos developed about the need to transmit Germany's recent ugly history.

Teaching methods expanded to include field trips to a growing number of memorial centers and special projects which involved students on a personal and emotional level. Today, Germany's commitment to teaching youth about antisemitism and the Holocaust is to be commended for its thoroughness.

0992 Wolgast, Eike: Die Wahrnehmung der Judenverfolgung und Judenvernichtung in der unmittelbaren Nachkriegszeit. *Trumah* 8 (1999) 97-119.

In 1945-46, Germans preferred to face the future rather than the guilt of the past. When an organization did issue a confession of guilt, it might condemn racism but never named the Jews and the Holocaust. This was true of all the political parties except the Social Democrats, who were led by Kurt Schumacher to recognize the crime against the Jews, although they absolved the working class; whereas the communists recognized only political victims of fascism. The Evangelical Church acted in defense of the Germans against the attribution of collective guilt, and weighed the Holocaust against the wrongs done the Germans. The Catholic Church also avoided mention of the Holocaust, but claimed credit for resistance to Nazism. Neither Church refrained from expressions of traditional anti-Judaism. Speakers at the re-opening of the universities did not mention the fate of Jewish faculty members and students. Explains this silence of the opinion leaders by their preoccupation with the Germans' own misery, their reluctance to raise an unpopular issue, and their shame at their own failure. *RW*

0993

לוין, איתמר: אויבי הוד מלכותו: בריטניה, השואה ורכוש הקורבנות. תל-אביב: משרד הביטחון – ההוצאה לאור, 1999. 293 עמ'.

[Levin, Itamar: *His Majesty's Enemies: Britain, the Holocaust and the Victims' Assets*. Tel-Aviv: Ministry of Defence, 1999. 293 pp.]

A study based on documents of the Custodian of Enemy Property, kept in the Public Record Office in London. Discusses British policy concerning assets which were transferred to British banks by thousands of East European Jews before World War II. Jewish survivors from Germany and other Axis countries were officially considered “enemies” of England, and their assets were used to compensate British subjects for wartime losses. International law supported the expropriation of assets in general; however, there were also rules allowing for their return to the original owners in certain cases, and Britain chose to ignore these rules. Jewish claims after the war were consistently rejected. No British official ever raised the issue that keeping the property of Holocaust victims was immoral. As a result of Levin's research and that of the Holocaust Educational Trust in 1997, the British government agreed to investigate the matter, having also seen the results of the case against the Swiss banks. While the government has admitted wrongdoing, there has been no compensation. *RE*

Great Britain

0994 Miller, Rory: Sir Edward Spears' Jewish Problem: A Leading Anti-Zionist and His Relationship with Anglo-Jewry, 1945-1948. *The Journal of Israeli History* 19, 1 (Spr 1998) 41-60.

Sir Edward Louis Spears (1886-1974) was a leading British anti-Zionist who from 1945 on was an active advocate of the Arab cause. He founded two anti-Zionist organizations in 1945: the Middle East Parliamentary Committee and the Committee for Arab Affairs. Despite his denial of the charge of antisemitism, his vehement anti-Zionism seems to have been motivated by an instinctive dislike of Jews - e.g. while he collaborated with leading non-Jewish anti-Zionists and anti-Zionist bodies throughout the world, he was reluctant to cooperate with the British Jewish anti-Zionist organization Jewish Fellowship. Examination of Spears' life shows that throughout his life he was dogged by accusations that he himself was Jewish. It was Spears' obsession with not appearing Jewish, rather than traditional antisemitism, that made him shun the Jewish organizations and, after 1948, made him more relentlessly anti-Zionist than his colleagues. *DR*

0995 Stone, Dan: The Domestication of Violence: Forging a Collective Memory of the Holocaust in Britain, 1945-6. *Patterns of Prejudice* 33, 2 (1999) 13-29.

Reestablishing the idea of collective memory (rejected by some writers as meaningless), shows the construction of the collective memory of the Holocaust by Anglo-Jewry in 1945-46. Anglo-Jewry was left deeply shocked and with a lasting sense of guilt by the disclosures in 1945 regarding the mass murder of Europe's Jews. Thus, the discourse on the Holocaust was framed within narratives which permitted some form of compensation, or employed terminology which softened the force of this subject and domesticated the horror of what had occurred. These strategies took many forms: use of euphemisms to describe mass murder; condemnation of earlier inactivity in contrast to present activity; characterization of the Nazi genocide as part of a familiar "lachrymose" view of Jewish history; and focus on the Jewish state as compensation for past suffering, resorting to the language of disaster and redemption. Illustrates this forging of collective memory with citations from postwar Jewish writings in Britain. *DR*

Lithuania

0996 Rozina, Judita: K voprosu ob unichtozhenii pamyatnikov istorii i kultury Vilnyusa v poslevoyennyi period [On the Question of the Destruction of Historical and Cultural Monuments of Vilnius in the Postwar Period]. *Yevrei v Rossii: Istoriya i kultura: Sbornik nauchnykh trudov, Vol. 5* [Jews in Russia: History and Culture: Collected Articles, Vol. 5, ed. Dmitri Elyashevitch. Sankt-Peterburg: Peterburgskii Yevreiskii Universitet, 1998. Pp. 246-252.

Traces the efforts of the Soviet authorities in the capital of Lithuania in 1946-63 to erase the vestiges of the Jewish presence in prewar Vilnius. The Great Synagogue was demolished during this period, as well as the old Jewish cemeteries. In 1951, some streets bearing names associated with Jews (Židu, Gaono, Strašunė, etc.) had their names changed. *DR*

0997 Ushpolis, Grigorii: *Trevozhnoye vremya: Razmyshleniya* [Troubled Times: Reflections]. Tel-Aviv: Krugozor, 1997. 414 pp.

Memoirs of a Jewish World War II veteran, who received the award of the Order of a Hero of the Soviet Union. Ch. 7 (pp. 306-389), "Partiinaya ispoved," describes the postwar anti-Jewish campaign in Lithuania, where Ushpolis was born and where he worked from the late 1940s to the 1980s. Dwells on both the Stalinist phase of this campaign (the struggle against cosmopolitans, the "Doctors' Plot" affair) and its Khrushchev-Brezhnev phase (the 1960s-80s), including the "anti-Zionist" campaign. *DR*

Netherlands

0998 Haan, Ido de: *Na de ondergang: De herinnering aan de jodenvervolging in Nederland, 1945-1995* [After the Destruction: The Memory of the Persecution of the Jews in the Netherlands, 1945-1995]. Den Haag: Sdu, 1997. 292 pp.

Deals with various topics relating to the aftermath of the Holocaust in the Netherlands, including Holocaust historiography (both Jewish and non-Jewish); commemoration; the indifference of the Dutch population and Dutch officials to the suffering of the Jews, both during the war and afterwards; postwar alienation of Holocaust survivors from the surrounding society; psychological trauma of the war and its aftermath; the effects of the trials of Eichmann and Demjanjuk in Israel; and recent commemoration of the Holocaust in a national context. *MF*

Poland

0999 Carlberg, Anders: Den sista pogromen i Polen: Den antisemitiska kampanjen 1968-1969 [The Last Pogrom in Poland: The Antisemitic Campaign 1968-1969]. *Rambam* 8 (1999) 48-51. With an English summary.

Summarizes the background to the wave of antisemitism disguised as anti-Zionism which swept Poland in the late 1960s, and describes the actual campaign; it was set off by a speech by Gomułka, the head of the Polish Communist Party, in the wake of the Six-Day War (1967), in which he accused the Polish Jews of constituting a "fifth column." After the student revolt in spring 1968, the campaign intensified and Jews were purged from the party, the army, and the public sector. The majority of the Polish Jews then emigrated to the West or to Israel. *SSC*

1000 Kwiek, Julian: *Żydzi, Łemkowie, Słowacy w województwie krakowskim w latach 1945-1949/50* [Jews, Lemkos, Slovaks in the Kraków Voivodship in the Years 1945-1949/50]. Kraków: Księgarnia Akademicka, 1998. 233 pp.

Pt. I (pp. 13-91), "Żydzi," discusses the situation of the Jewish minority between 1945-50. Anti-Jewish disturbances occurred twice in Kraków in the summer of 1945, on 27 July and 11 August. Pp. 31-46 describe the events of 11 August, which were the most serious. One Jewish woman was killed, a synagogue was burned, tens of Jews were beaten on the streets, and the apartments of some Jews were robbed. Quotes from documents describing the pogrom, produced by various institutions – e.g. the Information and Propaganda Office, the Militia, the Security Service, the Workers' Party, the Jewish Committee. Quotes, as well, from the communist press. Presents various opinions on the causes of the pogrom, among them rumors about "Jews who murdered Polish children" and "a fascist provocation." Supports the view that the communists used these events for their own propaganda aims. Concludes that the anti-Jewish disturbances in Kraków and various forms of hostilities which the Jews experienced in other localities of the Kraków voivodship spurred the emigration of Jews from Poland. *HV*

1001 Michlic-Coren, Joanna: The Troubling Past: The Polish Collective Memory of the Holocaust. *East European Jewish Affairs* 29, 1-2 (Sum-Win 1999) 75-84.

Discussing the Polish collective memory of the Holocaust in the 1980s-90s, ties the ambivalence of Poles toward the Holocaust to their collective identity and their narrative of denial. The myths of Polish innocence and victimization, heroism and grandeur do not make it possible, even today, for Poles to critically examine their failure to offer more help to Jews during the Nazi occupation. The Poles' preoccupation with their own suffering has made them insensitive to that of the Jews. Only a few intellectuals have been willing to question Polish attitudes toward the Jews, who are generally viewed as being implicated in their own deaths, partly for not being as "heroic" as the Poles perceive themselves to be. *YC*

1002

נאדעל, בנימין: אנטיסעמיטיזם און קאמוניזם אין פוילן אין רעטראספעקטיוו (די מערץ געשעענישן פון 1968). אונדזער צייט 670-671 (סעפט'-אקט' 1998) 14-20.

[Nadel, Binyamin: Antisemitism and Communism in Poland in Retrospective (the March-Events of 1968). *Undzer Tsait* 670-671 (Sept-Oct 1998) 14-20.]

Of the ca. 250,000 Jews who returned to live in Poland after World War II, only ca. 30,000 remained in 1960. Several Jews held posts in the communist government of the 1950s, but from 1956 Władysław Gomułka quietly forced them to leave their positions. In June 1967, after the Israeli victory over the communist-supported Arabs in the Six-Day War, Gomułka declared the Jews in Poland an unwanted "fifth column." After this declaration, Jews were quietly forced to leave their positions in the army and in the press. The reaction of the Polish government and press to student unrest in March 1968 was to blame the Jews and to accuse them of dual loyalty. An antisemitic campaign began in the press and other media, disguised as a campaign against Zionism. Discusses the role of Mieczysław Moczar, leader of the "Partisans," in this campaign. Ca. 20,000 Jews, especially those who had been forced out of government and other positions, emigrated at this time. *RE*

1003

רוזנבלום, עמליה: לאצור את הזמן: סיפורה של התצוגה במוזיאון הממשלתי באושוויץ-בירקנאו. פלסטיקה 3 (קיץ 1999) 65-74.

[Rosenblum, Amalia: Displaying History: The Story of the Exhibit at the Auschwitz-Birkenau State Museum. *Plastika* 3 (Sum 1999) 65-74.]

Discusses the history of the Auschwitz-Birkenau State Museum, and how it was affected by local and international politics. When the museum was opened in July 1947, the emphasis was on the suffering of Polish prisoners; the Jews were largely ignored, since the government wanted the museum to be a symbol of the communist defeat of fascism. In 1952 a Jewish exhibit was opened, among exhibits of other countries, but it was inaccessible most of the time; after the Six-Day War (1967) it was closed, and reopened only in 1978 on the 35th anniversary of the Warsaw ghetto uprising. After the fall of the communist regime in 1989, many Jews and

Germans began to visit the museum and brochures pointed out that most of the camp's victims were Jews. Following the renewal of ties between Poland and Israel in 1993, Birkenau became part of the museum and additions were made, including some signs in Hebrew. Due to the large number of Jews visiting the site, further changes are planned. *RE*

1004 Steinlauf, Michael Charles: Polen und die Erinnerung an den Holocaust. *Kirche und Israel* 14, 1 (1999) 20-31.

Analyzes the psychological effects of the Holocaust on its Polish witnesses. Although neither perpetrators nor victims, they felt guilt because of their longstanding animosity toward Jews, their helplessness in the face of the extermination, and the property of the murdered Jews of which they took possession. Their ambivalence was intensified by the communist regime's suppression of all discussion. Immediately after the war, this guilt feeling was a major factor, together with the perception that Jews collaborated with the communists, in acts of aggression both against Jewish survivors and their Polish rescuers. In 1968-70, Moczar and his "Partisans" mobilized the ambivalence toward the Holocaust to arouse Polish nationalism, alleging a German-Jewish (or Nazi-Zionist) conspiracy against Polish historical identity. The new generation of Poles (including those of Jewish descent), who have never known real Jews, have developed a nostalgic interest in them. But problems still exist, as seen in the conflict over the convent at Auschwitz. *RW*

Spain

1005 Rodríguez Jiménez, José Luis: *Antisemitism and the Extreme Right in Spain (1962-1997)*. Jerusalem: Hebrew University, Vidal Sassoon International Center for the Study of Antisemitism, 1999. 27 pp. (Analysis of Current Trends in Antisemitism, 15).

Spain had an antisemitic tradition going back to the time of the persecution of Conversos and the expulsion of the Jews in 1492. The alliance between Franco and Nazi Germany during the Spanish Civil War opened the way for the emergence of racist antisemitism in the Spanish Right. The first Spanish neo-fascist and neo-Nazi groups emerged in the 1960s, during the crisis in the Spanish Right; they gained force in the democratization period (1976-82). The principal extreme right organization was CEDADE (Círculo Español de Amigos de Europa); after its dissolution in 1993, some new rightist organizations adopted its ideas. Antisemitism has not been a central issue for the Spanish extreme right, but it has been far from negligible. Its main elements are racism, the idea of a Jewish world conspiracy, and Holocaust denial. Many rightists combine antisemitism with anti-Zionism. The extreme right tries to influence public opinion through proliferating publications. In recent years laws were enacted in Spain limiting extreme right activities and prohibiting Holocaust denial. *DR*

USSR

(see also Belarus)

1006 Geizer, Matvei M.: *Mikhoels: Zhizn i smert* [Mikhoels: Life and Death]. Moskva: Zhurnalistskoye agentstvo "Glasnost," 1998. 319, 64 pp.

A biography of Solomon Mikhoels (1890-1948). The Holocaust affected the work of the great actor: among his last stage productions were the plays "Freilekhs" and "The Prince Reubeini," inspired by the Nazi genocide. Relates the circumstances of the murder of Mikhoels, and dwells on the Jewish reactions to his death. Mikhoels' murder unleashed the state campaign to liquidate Jewish culture in the USSR. The last 64 pages contain illustrations. *DR*

1007 Heider, Angelika; Rapoport, Tamar: Die "Normalisierung" antisemitischer Erfahrungen: Junge russisch-juedische Migranten erinnern ihre Kindheit in der Sowjetunion. *Jahrbuch fuer Antisemitismusforschung* 8 (1999) 82-106.

43 students from the former USSR who emigrated to Germany, and the same number who emigrated to Israel, most of them in 1990-91, were interviewed in 1995-96. Notes that they grew up in the 1970s-80s, when Jews suffered from social stigma as well as many restrictions (e.g. in admission to universities). They also experienced the beginning of the "perestroika," when radical antisemitic groups became a threat. Nevertheless, in these interviews they recalled only isolated antisemitic incidents, whose importance they minimized. The explanation may lie in the students' intense and proud identification with Russian culture, doubly important to them now in emigration. *RW*

1008 Katerli, Nina S.: *Isk* [The Lawsuit]. Samara: Samarskoye Otdeleniye Rossiiskovo Yevreiskovo Kongressa, 1998. 318 pp.

In 1988, Katerli, in an article for the "Leningradskaya Pravda," accused Aleksandr Romanenko of disseminating Nazi-like antisemitic ideas in his "The Class Essence of Zionism." Romanenko sued her for libel. Describes the trial, which lasted from 1988-90. Dwells, also, on the activities of the "Pamyat" group in Leningrad and Moscow, including the actions in support of Romanenko. Contends that the authorities backed up this aggressively antisemitic group. The district court in Leningrad dismissed Romanenko's suit in 1989; he then appealed this decision and the case passed to the municipal court with a new accusation – defamation of the Party, whose local branch approved Romanenko's book. In 1990, Romanenko dropped the charges, allegedly due to pressure by Party organs in Leningrad. Contends that Russian authorities in the 1990s are passive in the face of antisemitic activities; in 1997 the State Duma voted against a projected law on the prohibition of fascist propaganda in Russia. *DR*

1009 Kolyazin, Vladimir F.: *"Vernite mne svobodu!" Deyateli literatury i iskusstva Rossii i Germanii – zhertvy stalinskovo terrora: Memorialnyi sbornik dokumentov iz arkhivov byvshevo KGB* ["Give Me Back Freedom!" Literary and Artistic Persons in Russia and Germany – Victims of the Stalinist Terror: A Memorial Collection from the Archives of the Former KGB]. Moskva: Medium, 1997. 446 pp.

Pp. 329-360 contain documents of the Jewish Anti-Fascist Committee affair pertaining to Veniamin Zuskin, Mikhoels' successor as manager of the State Jewish Theater in Moscow. Zuskin was arrested in 1948 and executed in 1952. Contains, mainly, protocols of his interrogations; one of the main topics in them was Mikhoels. Pp. 399-403 contain the essay "Veniamin Zuskin i yevo 'delo'" ("Veniamin Zuskin and His 'Case'"), by Maya Turovskaya. *DR*

1010 Kozhinov, Vadim V.: *Rossiya, vek XX, 1939-1964: Opyt bespristrastnovo issledovaniya* [Russia, 20th Century, 1939-1964: An Attempt of Impartial Scrutiny]. Moskva: Algoritm, 1999. 396 pp.

Ch. 7 (pp. 243-308), "Borba s 'antipatriotizmom' i 'yevreiskii vopros'" ("The Struggle against 'Anti-Patriotism' and the 'Jewish Question'"), contends that the "anti-cosmopolitan" campaign, which began in the USSR in 1947, was not directed specifically against Jews. Moreover, many of the artistic intelligentsia of Jewish origin remained unaffected by this campaign. Claims that, contrary to widespread opinion, Stalin was not antisemitic. The anti-Jewish campaign in the USSR in 1949-53 stemmed from Stalin's disappointment with Israel, to which the USSR had given all possible help in 1947-48, and which in 1949 became a satellite of the USA. Holds that the murder of Mikhoels in 1948 and the "Doctors' Plot" affair in 1953 had motives other than antisemitism. *DR*

1011 Kuchenbecker, Antje: Die Juedische Autonome Provinz und der Antisemitismus waehrend der "schwarzen Jahre" (1948-1953): Ein "amerikanisch-zionistisches Spionagezentrum" im Fernen Osten der Sowjetunion. *Jahrbuch fuer Antisemitismusforschung* 8 (1999) 43-59.

One of the victims of the Soviet campaign of 1948-53 against Jewish "nationalism" and "cosmopolitanism" was the Jewish Autonomous

Province of Birobidzhan. It was founded in 1934 to counter international accusations that the USSR discriminated against Jews, as well as internal complaints that the regime favored Jews (resolved by giving them an unfavorable location). Although Birobidzhan was the highest-status Jewish political entity in the USSR, the regime never really meant it to prosper. In successive provincial party conferences, culminating in an extraordinary conference in 1949, the political and cultural leaders of Birobidzhan were accused of links with American Jewish organizations, Zionists, and the Jewish Anti-Fascist Committee. They were sentenced to long terms in prison camps. The Yiddish theater, radio, and literary press were shut down, accused of ignoring Soviet ideology. In the Yiddish newspaper, nothing remained Jewish but the language. Yiddish disappeared from schools, and Jews were afraid to speak it even at home. Birobidzhan then ceased to be Jewish. *RW*

1012 Parrish, Michael: *The Lesser Terror: Soviet State Security, 1939-1953.* Westport, CT: Praeger, 1996. xxii, 424 pp.

Ch. 8 (pp. 197-214), "The Anti-Semitic Campaigns," deals with NKVD/MGB activities against Jews in the postwar period (1945-53). The Jewish victims were relatively more numerous during this period than during the prewar "Great Terror" (1934-40). Although most of the campaigns against various groups of Jews (Jewish intelligentsia, Jews in the army and Soviet apparatus, Jewish wives of prominent party leaders, the victims of the "Doctors' Plot," etc.) were led by V. Abakumov, the Minister of State Security of the USSR, and his henchmen, it was Stalin who initiated them. Neither Abakumov, nor Beria, the Minister of Interior, were innate antisemites. Stalin's antisemitism had complex motivations, and was manifested even before the war. Despite Stalin's antisemitism and the anti-Jewish campaigns of 1946-53, many Jews retained their high positions in various spheres. *DR*

1013 Rutman, Genrikh: Khasidskii romantik (k dvum sobytiyam 1948 g. ili o dvukh shagakh na puti k tragedii) [A Hassidic Romantic: On Two Events of 1948, or about Two Steps on the Way to a Tragedy]. *Yevrei Belarusi* 3-4 (1998) 193-202.

Proposes an explanation for the murder of Solomon Mikhoels, planned and perpetrated by the Soviet political police in 1948. Contends that Mikhoels was killed because he dared to express too much enthusiasm in connection with the establishment of the State of Israel. *DR*

1014 Rybakov, Anatolii N.: "Tyazholyi pesok": Istoriya i sudby geroyev ["Heavy Sand": Its History and the Fate of Its Heroes]. *Russkii Yevrei* 4 (Sum 1997) 22-27.

An excerpt from Rybakov's forthcoming memoirs. He relates the history of his work on this novel dealing with the Nazi Holocaust in the USSR. The novel is based on events which took place in the town of Shchors, in northern Ukraine. Describes his effort to publish the novel in the late 1970s – at the peak of the "anti-Zionist" campaign. The censorship office curtailed the novel and changed its "too Jewish" title "Rakhil" ("Rachel") to "Heavy Sand." Discusses reactions to the novel, including antisemitic ones. *DR*

1015 Vovsi-Mikhoels, Nataliya: *Moi otets – Solomon Mikhoels: Vospominaniya o zhizni i gibeli* [My Father – Solomon Mikhoels: Reminiscences on His Life and Death]. Moskva: Vozvrashcheniye, 1997. 234 pp. An excerpt appeared in "Novyi Mir" 3 (1990).

Relates the circumstances of Mikhoels' murder, as well as the events which followed it in 1948-53: the arrest of Veniamin Zuskin, Mikhoels' "stage partner," the closing of the Jewish theater in Moscow, the trial of members of the Jewish Anti-Fascist Committee, and the "Doctors' Plot" affair. *DR*

1945-1990: Islamic World

Egypt

1016

מירון, יעקב: הפיצוץ ברובע היהודי הקראי בקאהיר ב-19 ביוני 1948. פעמים 70 (חורף 1997) 141-156.

[Meron, Ya`akov: The Explosion in the Karaite Jewish Quarter of Cairo on 19 June 1948. *Pe`amim* 70 (Win 1997) 141-156.]

Presents (in the original Arabic and a Hebrew translation) a report by Ezra ben Yosef Levi, whose two brothers, their wives and their children were among those killed by bombs placed under several houses in the Karaite Quarter on 19 June 1948. The attack was allegedly perpetrated by the Muslim Brethren in reaction to events in Palestine brought on by Israel's War of Independence. The commentary (pp. 154-156) surveys additional incidents related to the anti-Jewish disturbances in Cairo during June-July 1948. *LFo*

Iraq

1017 גת, משה: העליה ההמונית מעיראק: גירוש, הצלה או פתרון בעיה פנימית? יהודות בבל 2 (אביב 1998) 95-109.

[Gat, Moshe: The Mass Aliyah [Immigration to Israel] from Iraq: Expulsion, Rescue or Resolution of a Domestic Problem? *Yahadut Bavel* 2 (Spr 1998) 95-109.]

The mass Jewish emigration from Iraq in 1950-51, during which more than 120,000 Jews left the country, was not a direct result of official Iraqi policy. There were several factors which had contributed to Jewish fears already in the 1930s: the fact that the Iraqi government dealt harshly with other minority groups after 1932, the rise of a pan-Arab movement, the proliferation of Nazi propaganda, the dismissal of many Jewish government employees in 1934 and 1936, and the Arab riots in Palestine in 1936-39. After the pogrom of June 1941 in Baghdad, the situation of the Jews improved temporarily. In the late 1940s, due to an economic decline, the Jews were forced to contribute funds to the government and forbidden to leave the country. In 1949, Zionist activity (which was opposed by the older generation of Jews) was declared a crime, and many activists were arrested and tortured. With the fall of the military regime in December 1949, thousands of Jews began to escape. The Iraqi government then legalized their emigration, thinking that only the activists would leave. Although they had to give up their Iraqi citizenship, the entire Jewish community left the country. *RE*

Turkey

1018 Bali, Rifat N.: The Image of the Jew in the Rhetoric of Political Islam in Turkey. *CEMOTI* 28 (June-Dec 1999) 95-108.

Examines the main antisemitic and anti-Zionist motifs in the rhetoric of Islamist parties in Turkey – the National Order Party founded in 1970 and its successors, the National Salvation Party, Welfare Party, and Virtue Party – mainly as they were pronounced by the leader of these successive parties, Necmettin Erbakan. Among the recurring topics are Israeli expansionism and putative territorial claims to Turkey, Zionist expansionism forcing Turkey to join the Common Market (to counter it), Zionism as the source of anarchy in Turkey, and the Jewish drive for world domination. Remarkably, the "Zionist" and the "international Jew" are abstract, symbolic figures for Erbakan and his followers, and their attitude toward Turkish Jews is different. The thriving Jewish community of Turkey is a demonstration of Islamist tolerance. During his tenure as prime minister (in the 1990s), Erbakan maintained good relations with Israel and with American Jewish organizations. The Turkish Islamists seem to make a distinction between rhetoric and Realpolitik. *DR*

1945-1990: Israel

1019

קפלן, קימי: החברה החרדית בישראל ויחסה לשואה – קריאה חדשה. אלפיים 17 (1999) 176-207.

[Caplan, Kimmy: The Ultra-Orthodox Community in Israel and Its Attitude toward the Holocaust – a New Interpretation. *Alpayim* 17 (1999) 176-207.]

Highlights changes in the Israeli ultra-Orthodox community's postwar reaction to the Holocaust and in its memory of the Holocaust since the 1980s as reflected in the collection edited by Joel Schwartz and Yitzhak Goldstein (published in Hebrew, 1987, and in English, 1990), "Ha-Shoah: Leket dvarim be-nosse' hurban yahadut Europa 5700-5705 mi-tokh aspaklaryah shel Torah" ["Shoah: A Jewish Perspective on Tragedy in the Context of the Holocaust"] and in "Zakhor: Zekhirat ma'ase Amalek she-be-dorenu – Germania ha-Natzit" (Jerusalem, 1993). Instead of putting the blame for the Holocaust on the Zionists and on the alleged effect of their ideology on the inadequate rescue attempts during the Holocaust, these two books probe the theological implications of the Holocaust, including the conduct of the victims, the perpetrators, and the bystanders, using sources and views presented in academic research and in public discourse on the Holocaust. *LFo*

1020 Linn, Ruth: In the Name of the Holocaust: Fears and Hopes among Israeli Soldiers and Palestinians. *Journal of Genocide Research* 1, 3 (Nov 1999) 439-453.

On both sides of the Israeli-Palestinian conflict, Holocaust-related metaphors are used to voice the community's distress and to justify individual actions. In morally problematic campaigns, like the war in Lebanon (1982-85) or the Intifada (1987-94), many Israeli soldiers who refused to serve in the war zone explained their decision in terms of the Holocaust (e.g. they would not follow orders blindly). Contends that by using familiar and powerful Holocaust metaphors to describe their moral dilemmas, the Israelis could facilitate their understanding of the complex situation, organize their experience and interpretation of it, and clarify to themselves and others what stance they ought to take. Above all, they conveyed their moral imagination and moral feelings. Contends that the use of Holocaust imagery by the Palestinians is more pernicious than it is by Israelis. *DR*

1021 Porat, Dina: Some Effects of the Holocaust on Israeli Society and Foreign Policy. *Sino-Judaica: Jews and Chinese in Historical Dialogue: An International Colloquium, Nanjing, 11-19 October 1996,* ed. Aharon Oppenheimer. Tel-Aviv: Tel-Aviv University, 1999. Pp. 159-168.

Discusses the impact of the Holocaust on the demography of Israel and on its ideology and policies. 400,000 Holocaust survivors immigrated to Israel between 1946-1990s; they, their spouses, and their offspring account for ca. one-sixth of the Israeli Jewish population. The Holocaust affected other sectors of society as well. Discusses attitudes in Israeli society towards the survivors. Ben-Gurion was opposed to the Holocaust being a factor in Israeli policy. He was against memorialization of the Holocaust, and somewhat softened his stance only during the Eichmann trial. However, later prime ministers, especially Begin, used Holocaust discourse to justify Israeli foreign policy (e.g. the war in Lebanon which began in 1982). Calls for a more moderate stance: Israeli society should remember the Holocaust, but not be preoccupied with it. *DR*

1022 Shapira, Anita: Autour de la Shoah: Souvenirs privés, mémoire publique. *Les Cahiers du Judaïsme* 4 (Spr 1999) 54-67. Appeared in Hebrew in "Zmanim" 57 (1997); in English in "Jewish Social Studies" 4 (1998).

1023 Sheffi, Na`ama: Cultural Manipulation: Richard Wagner and Richard Strauss in Israel in the 1950s. *Journal of Contemporary History* 34, 4 (Oct 1999) 619-639.

The issue of whether works by Wagner and Strauss should be played in public in Israel was, to a significant degree, a political one. In the 1950s it was related to the initial stages of forming Israel's national identity and pattern of government. The resistance to playing works by these (and other) composers associated with the Nazis came not only from survivors but also from opposition politicians. The latter demanded government intervention, while the government preferred to act behind the scenes. Although the opposition parties used the matter in an attempt to gain "electoral capital," there were also profound questions at stake, especially the nature of emerging relations with West Germany and how the young State of Israel would relate to the Holocaust. The Wagner-Strauss question remained open for many years. *YC*

1024 Sheffi, Na`ama: Who's Afraid of Richard Wagner: The Israeli Press, Politics and Music, 1938-1994. *Patterns of Prejudice* 31, 4 (1997) 35-51.

The virtual ban on performing Wagner's music in Palestine was triggered by the "Kristallnacht" pogrom in Germany in 1938. Many factors combined together to turn Wagner into the epitome of all that Palestinian Jews rejected in German culture: Wagner's antisemitism, the Nazi leaders' adulation of him, Nazi appropriation of the Bayreuth festival, etc. Traces discussions in the Israeli press about lifting the ban on Wagner's music, from 1952 on. Interestingly enough, while the ban on musicians who supported the Nazis (e.g. Richard Strauss, Carl Orff) has been lifted in Israel, the ban on Wagner, who died half a century before the Nazi accession to power, remains. *DR*

1025
שפי, נעמה: טבעת המיתוסים: הישראלים, ואגנר והנאציזם. חיפה: הוצאת הספרים של אוניברסיטת חיפה; תל-אביב: זמורה-ביתן, 1999. 239 עמ'.

[Sheffi, Na`ama: *The Ring of Myths: The Israelis, Wagner and the Nazis*. Haifa: University of Haifa Press; Tel-Aviv: Zmora-Bitan, 1999. 239 pp.]

Discusses the boycott on Wagner's music in Israel, which began in November 1938, after the "Kristallnacht" pogrom, and reviews the arguments over the boycott from the 1950s through the 1990s. Contends that Wagner's criticism of Jewish music and musicians in the 19th century was personal rather than racist. Claims that it is difficult to determine the exact meanings of the antisemitic expressions in his essays and letters, as well as those attributed to him by his wife, Cosima, in her diaries. Concludes that Wagner is wrongly seen by Israelis as a symbol of the Holocaust. Although he was antisemitic and his music was admired by Hitler and used by the Nazis, he did not advocate the destruction of the Jews. *RE*

1945-1990: Latin America

Argentina

1026 *Estudios Migratorios Latinoamericanos* 43 (Dec 1999).

This issue is entitled "Inmigrantes, refugiados y criminales de guerra en la Argentina de la segunda posguerra." Partial contents: Meding, Holger M.: La emigración a la República Argentina de los nacional-socialistas buscados: Una aproximación cuantitativa (241-259); Buchrucker, Cristián; Brown, Fabián; Jozami, Gladys: Los judíos en el ejército: Ausencias y presencias (303-322); Calceglia, Inés M.: La inserción de agentes de origen judío en el Poder Judicial de la Nación (323-348);

Klich, Ignacio: El impacto de nazis y otros criminales de guerra en la incorporación de judíos a la diplomacia, 1946-1955 (349-375). *MZ*

1027 *The Jewish Quarterly* 173 (Spr 1999).

This issue of the journal contains a section on the establishment of the Commission of Enquiry into the Activities of Nazism in Argentina (CEANA) in Buenos Aires in May 1997, and the influence of Nazism and of Nazi war criminals in Argentine society and culture. Contents: Klich, Ignacio: Argentina's Past under Scrutiny [P. 43 contains a summary of the interim findings of the CEANA research.] (42-44); Sosnowski, Saúl: Counting Nazis in Argentina: A Cultural Perspective (45-48); Buchrucker, Cristián: Whitewashing the 'New Order': Nazi and Fascist Ideological Influence in Argentina after the Second World War (49-52). *SSC*

1945-1990: North America

USA

1028 Chmiel, Mark J.: The Political Varieties of Sacred Remembrance: Elie Wiesel and U.S. Foreign Policy. *Journal of Church and State* 40, 4 (Fall 1998) 827-846.

In 1978 Elie Wiesel was appointed chairman of the President's Commission on the Holocaust (later the U.S. Holocaust Memorial Council). Contends that, despite his proclaimed distancing from politics and determination to preserve the purity of Holocaust remembrance, Wiesel utilized Holocaust memory in relation to U.S. foreign policy in the Reagan era. In his "Sanctuary" speech, held in 1985, Wiesel used Holocaust symbols in order to express solidarity with the refugees from Central America. He exploited Holocaust remembrance to legitimate U.S. political actions, both in the past and in the present. In 1984-85 Wiesel clashed with Reagan on the Bitburg Affair, protesting against the equation of SS-men who died in the war with the victims of the Holocaust. However, after receiving a gold medal from Reagan's aides during the President's visit to Germany, Wiesel responded with a conciliatory speech. The examined cases show that Holocaust remembrance ceased to be the ultimate end for Wiesel. *DR*

1029 Flanzbaum, Hilene: The Americanization of the Holocaust. *Journal of Genocide Research* 1, 1 (Mar 1999) 91-104.

Criticizes the representation of the Holocaust in American culture, which interpreted the Holocaust story as a survival story (which can be illustrated by the literary and scenic interpretation of Anne Frank's diary), thus providing the tragedy with a happy end. It also trivialized the Holocaust, using it as a paradigm for other cases of persecution of minorities. Moreover, it commercialized it. Reflects on why in the 1990s there has been an explosion of representations of the Holocaust in American culture. This may be connected, inter alia, with ethnic Americans' growing interest in their roots. Concludes that if the Holocaust is to be remembered at all, American historians, writers, movie directors, teachers, etc. must "re-see" and "re-say" it. *DR*

1030 Novick, Peter: *The Holocaust in American Life*. Boston: Houghton Mifflin, 1999. 373 pp.

The Holocaust, marginalized in the USA during World War II and in the first postwar years, has recently taken a central place in American life. It was consciously made a part of the American collective memory. The initiative to put the Holocaust on the American agenda came from the Jews, but non-Jews proved to be receptive to the memorialization of the Holocaust. During the war, both Jewish and non-Jewish America could do nothing on behalf of European Jews; the genocide of the Jews was not perceived as singular among all the Nazi crimes. In the postwar years, the memory of the Holocaust fell victim to the Cold War. The revival of interest in the Holocaust took place in the context of the rise of an ethnic particularist ethos in the USA and of some other shifts in the public consciousness. At the same time, the USA needed to improve her self-image, damaged in the wake of the Vietnam War. The Middle East crisis and the racial conflict also offered an opportunity to deploy the Holocaust as a political weapon. Raises questions on the consequences for the Jews of founding their identity on the Holocaust, and whether the Holocaust is, in fact, a useful American memory. *DR*

1031 Raden, David: Is Anti-Semitism Currently Part of an Authoritarian Attitude Syndrome? *Political Psychology* 20, 2 (June 1999) 323-343.

Discusses the results of the 1990 General Social Survey, with the participation, among others, of 1,119 White non-Jewish respondents. Contrary to conclusions reached by the Adorno group in 1950, and by two national surveys conducted in 1964 and 1981, this survey revealed only weak correlations between antisemitism and a broader attitudinal syndrome including conservative, racist, authoritarian, and isolationist attitudes. There were also no clear-cut differences between respondents with higher or lower education levels. The results suggest a diminished role for antisemitism in authoritarian attitude syndromes. *DR*

1032 Shandler, Jeffrey Alan: *While America Watches: Televising the Holocaust.* New York: Oxford University Press, 1999. 316 pp. Based on the author's diss. - Columbia University, New York, 1995.

1033 Sherman, Franklin: Luther and the Jews: An American Perspective. *Internationales Josephus-Kolloquium, Muenster 1997: Vortraege aus dem Institutum Judaicum Delitzchianum,* eds. Folker Siegert, Juergen U. Kalms. Muenster: Institutum Judaicum Delitzschianum, 1998. Pp. 210-217.

Discusses efforts to contextualize the American publication of Luther's antisemitic treatise of 1543, "Von den Juden und ihren Luegen," in order to neutralize potentially negative social effects. Parts of Luther's works had been used by hate groups such as the KKK. As editor of Vol. 47 of "Luther's Works" (1971), Sherman persuaded the publisher to repudiate the "infamous" treatise and express concern about its "possible misuse." In the 1990s the general public became aware of Luther's role as "a font of antisemitism in the West." In response, on 18 April 1994, the Evangelical Lutheran Church in America passed a resolution directed to the Jewish community, which expressed sorrow over the tragic consequences of Luther's ideas, especially in the Holocaust, and over the failure of Lutheran Churches to resist Nazi antisemitism, as well as resolve to oppose such bigotry today within the Church and without. The text of the Declaration is included. Notes positive Jewish reactions to this historic repudiation. *YC*

1034

שיף, עופר: ארבעה ניסיונות תיווך בין טכסטים יהודיים ואמריקניים: סופרים יהודים-אמריקנים לנוכח איום האנטישמיות בתקופת מלחמת העולם ה-II. הציונות 21 (1998) 355-369.

[Shiff, Ofer: Four Attempts at Mediation between Jewish and American Texts: Jewish-American Writers Facing the Threat of Antisemitism during World War II. *Ha-Tzionut* 21 (1998) 355-369.]

Examines the responses of American-Jewish intellectuals to encounters with antisemitism in the USA and to the dissonance between their Jewish and American identities, as reflected in "A Symposium on Racial and Religious Prejudice," published in February and May 1945 in the "Journal of Social Issues," edited by Gene Weltfish; in the novel by Ruth Seid (alias Jo Sinclair), "Wasteland" (1946); in Saul Bellow's novel "The Victim" (1947); and in Arthur Miller's novel "Focus" (1945). The first three works depict the Jews' situation as that of the "ghetto," whose walls must be torn down in order for the Jew to be accepted in American society. Miller, in contrast, presaged that the symbol of the Jew as victim, rather than disappear, would take on a universal and all-American significance, making the Jewish condition, via the Shoah, a universal rather than a particularistic issue. *LFo*

1035 Webb, Clive: Closing Ranks: Montgomery Jews and Civil Rights, 1954-1960. *Journal of American Studies* 32, 3 (Dec 1998) 463-481.

Jews in Montgomery, Alabama were hostile to the bus boycott initiated in December 1955, led by Martin Luther King. Their motivation was self-preservation. The support of Northern Jews for the boycott was resented by Montgomery's Jews because they feared that such support would stimulate antisemitism among local anti-integrationist Whites. In fact, the Alabama segregationist movement split over antisemitism in 1956. Even segregationists who were not antisemitic threatened to expose those Jews who were insufficiently zealous in upholding White supremacy. Any expression by Jews in favor of integration would have aroused an antisemitic reaction. Jews who paid dues to the White Citizens' Council did so out of self-protection, not conviction. Until 1960 Jews were able to avoid confrontation with either African-Americans or Whites. Events in Montgomery provide a microcosmic portrait of the situation of all Southern Jews during this period. *REK*

1036 Weiler, N. Sue: Keeping Their Memory Alive: 50+ Years of Effort, Chicago, Illinois, 1944-Present. *Western States Jewish History* 32, 1 (Fall 1999) 62-81.

Sketches the history of Chicago's commemoration of the Warsaw ghetto uprising and the Holocaust. Shows how handing on the legacy of Jewish memory to younger generations has been combined with concern to prevent racism and to encourage justice and peace in the USA and the world. The Chicago tradition of uninterrupted commemoration precedes that of New York by 20 years. Discord among Jewish groups in the community (including about communism) sometimes prevented a unified ceremony from being held. After 50 years individual synagogues began holding their own commemorations. Provides information about organizers, speakers, and contents of programs (dramatic works, songs, personal accounts of survivors) and exhortations to honor the memory of the six million Jews and the millions of non-Jewish victims of World War II by involvement in fighting antisemitism and racism and by improving society. *YC*

1991-1999: General

1037 Newman, Amy: The Idea of Judaism in Feminism and Afrocentrism. *Insider/Outsider: American Jews and Multiculturalism,* eds. David Biale, Michael Galchinsky, Susannah Heschel. Berkeley: University of California Press, 1998. Pp. 150-181.

Writers of the 1980s-90s who criticize the European culture, worldview, and behavior from a postmodernist stance, ascribe the faults of European culture to the Judaic roots of Western civilization. Afrocentrist writers (e.g. Marimba Ani, Cheikh Anta Diop) point to Judaism as the source of anti-Black racism and as a teaching which, using a paradigm such as the Hebrew conquest of Canaan, justifies conquest and genocide. Feminist writers (e.g. Nel Noddings, Gerda Lerner, Julia Kristeva) contend that Judaic monotheism, in its war against paganism, ousted the worship of goddesses which was widespread in the ancient Near East and established a patriarchal religion. Both groups of scholars dismiss Judaism as a religion hostile to nature, which sanctions slavery and worships an especially wrathful and vengeful God. These ideas stem from an oversimplified image of Judaism as a uniform religion, from poor knowledge of ancient Near Eastern cultures, and from the adoption of Christian anti-Judaic stereotypes. There is systematic use of a negative concept of Judaism to legitimate criticism of whatever ideas and practices are viewed as most corrupt and oppressive. *DR*

1038 Porat, Dina; Stauber, Roni; Vago, Raphael, eds.: *Anti-Semitism Worldwide – 1996/7.* Tel Aviv: Tel Aviv University, Stephen Roth Institute for the Study of Contemporary Anti-Semitism and Racism, 1997. xii, 372 pp. On title page also: ADL – Anti-Defamation League; World Jewish Congress.

An annual report giving details of antisemitic activities and trends in various countries and regions worldwide. Contains sections on Western Europe, Eastern Europe and the Former Soviet Union, the Middle East, North America, Latin America, and Australia, Japan and South Africa. For each country or region gives information on the Jewish community, extremist movements and hate groups, attitudes towards the Holocaust and the Nazi era, Holocaust denial, and actions against racism and antisemitism. This issue includes a section entitled "Research Topics," with the following articles: Sitman, David: Anti-Semitism on the Internet (293-308); Beker, Avi: Jewish Property in Europe: A Re-Evaluation of National Myths (309-328). *SSC*

Responses to the Holocaust

1039 *Antisemitism Research* 3, 1 (Fall 1999).

This issue, entitled "Jewish Assets," deals with restitution and compensation for Jewish property confiscated during the Holocaust and how the matter was dealt with in various countries after World War II, and discussion of this issue in recent years. Contents: Sanders, Joop: Legal Rehabilitation in the Netherlands (5-8); Mitten, Richard: The Price of Historical Responsibility: "Wiedergutmachung" and the Austrian Historians' Commission [On a commission appointed in 1998 to investigate expropriations in Austria during the Holocaust and restitution or compensation.] (9-15); Reisel, Berit; Bruland, Bjarte: Rediscovering History: The Restitution Process in Norway [Includes information on the fate of the Norwegian Jewish community in the Holocaust.] (16-21); Teitelbaum-Hirsch, Viviane: Restitutions in Belgium (22-26); Hajdenberg, Henri: Summary of the Mattéoli Commission's Work on the Spoliation of the French Jews [Includes information on the confiscation of Jewish property in France.] (27-32); Lillie, Catherine A.: Holocaust Claims Processing Office in the U.S. (33-36); Webber, Anne: The International Network of the European Commission on Looted Art (37-40). Pp. 41-42, "The Washington Principles: Principles with Respect to Nazi-Confiscated Art," contain guidelines approved in December 1998 at the Washington Conference on Holocaust-Era Assets. *RE*

1040

ברזל, נעימה: הערות לחקר חסידי אומות העולם. דפים לחקר תקופת השואה 14 (1997) 187-199.

[Barzel, Neima: Comments on the Study of the "Righteous among the Nations." *Dapim le-Heker Tekufat ha-Shoah* 14 (1997) 187-199.]

Discusses the complexities involved in conducting research on Righteous Gentiles. The characterization of these rescuers according to the legal guidelines laid down by Yad Vashem for recognition of "Righteous among the Nations" is not adequate for the multiplicity of cases and circumstances of rescue. In fact, it is impossible to document and research all rescue activities since not all the rescued, or rescuers, survived, or reported their acts. A statistical approach does not reflect aspects that cannot be expressed by quantitative means, such as psychological, sociological, and ethical factors. Reflects on whether research should focus on the image of the rescuers or their acts. *LFo*

1041 Bernstein, Michael André: Hommage à l'extrême: La Shoah et l'hyperbole de la catastrophe. *Le Débat* 101 (Sept-Oct 1998) 183-192. Appeared in English in the "Times Literary Supplement" (6 March 1998).

A critique of the current direction of the historiographical search for meaningful language proportionate to the enormity of the Shoah. Whether drawn from theological or psychoanalytical sources, this terminology reduces language to clichés. Focuses on the motif of 19th-century predictions of an apocalypse, culminating in Auschwitz. Attributes the delay of 50 years in confronting the Shoah to a desire to view it as unique rather than a need to reflect on its effects. Warns against the temptation to see this event as the only model of barbarity in Jewish and world history. Refers to certain contradictions, such as the emphasis on its indescribability and the exploitation of the Shoah by opposing political forces to further justify their ideals. Concludes with an appeal for meditation on the Shoah and its significance for Jewry and humanity. *TS*

1042 Bindenagel, J.D., ed.: *Washington Conference on Holocaust-Era Assets, November 30-December 3, 1998: Proceedings.* [Washington, DC]: U.S. Government Printing Office, 1999. xix, 1146 pp.

Proceedings of the international conference, hosted by the United States Department of State and the United States Holocaust Memorial Museum, held in Washington, DC. Presents, first, the opening remarks (pp. 39-81) and concluding statements (pp. 83-143). Pp. 145-438 contain delegation statements from 29 participating countries and organizations. Pp. 441-908 present the discussions centered around specific topics: Nazi-Confiscated Art Issues; Holocaust-Era Insurance Claims; Communal Property; Archives, Books and the Role of Historical Commissions; Holocaust Education, Remembrance and Research. Pp. 909-1146 contain appendices. Appendix O (pp. 1067-1078), Pierre Blet's "The Myth in the Light of the Archives: The Recurring Accusations against Pope Pius XII," is a translation of his article published in Italian in "La Civiltà Cattolica" (March 1998), defending Pius XII against accusations that he helped Nazi criminals to evade punishment after the war. Appendix P (pp. 1079-1146) contains a sample of Nazi documents from Russian archives. *DR*

1043 Birn, Ruth Bettina; Riess, Volker: Das Goldhagen-Phaenomen oder: Fuenfzig Jahre danach. *Geschichte in Wissenschaft und Unterricht* 49, 2 (Feb 1998) 80-95.

Discusses Goldhagen's book and its critics, as well as the role it plays in Jewish-American and in German self-perception. The book was celebrated in the mass media and almost unanimously disqualified by scholars. Remarks on the difference between survivors' complex, differentiated descriptions of the Holocaust and the one-sided view of young American Jews, who count themselves among the "victims," and for whom the Holocaust is often the only source of their Jewish identity; Goldhagen's book gratified this group. It also gratified ordinary Germans, because Goldhagen, in a footnote, exonerates present-day Germany; and also because he presents the perpetrators of the Holocaust – without any consideration of the circumstances, the gradual habituation, the potential for evil in all human beings – as acting out of a single-minded "eliminationist antisemitism." The reader who is not an eliminationist antisemite can thus feel secure that he himself would never have perpetrated such crimes. *RW*

1044 Braham, Randolph L.: Remembering and Forgetting: The Vatican, the German Catholic Hierarchy, and the Holocaust. *Holocaust and Genocide Studies* 13, 2 (Fall 1999) 222-251.

The Vatican document "We Remember: A Reflection on the Shoah" is the culmination of Pope John Paul II's activities on behalf of the normalization of Catholic-Jewish relations. Examines the positive and negative components of this document, focusing on the latter. "We Remember" tends to absolve the Church of responsibility for the Holocaust. It distinguishes between Christian anti-Judaism and Nazi antisemitism, and interprets positively the role played by Pope Pius XII and the German Catholic Church during the Nazi era. A close examination of the latter issues disproves the claimed positive role. Pius XII was pro-German and anti-liberal in his worldview. Both he and the German Catholic hierarchy tended toward reaching a compromise with Hitler, and they shared some of his anti-Jewish views. Concludes that despite its weaknesses, "We Remember" reflects the Church's desire for reconciliation with the Jews. *DR*

1045 Brecher, Robert: Understanding the Holocaust: The Uniqueness Debate. *Radical Philosophy* 96 (July-Aug 1999) 17-28.

Takes issue with several of the contributors to Alan Rosenbaum's edited collection "Is the Holocaust Unique?" (1996), especially David Stannard, arguing that the Holocaust should be viewed not as "unique" but as "unprecedented." In terms of the scope of genocides, notes the more comprehensive one of North American Indians and that of Blacks due to slavery. Opposes the view of the Holocaust's incomprehensibility, as it leads to a removal of the event from history and, therefore, to the impossibility of judging the perpetrators. What was unprecedented was the Nazis' "internal genocide," their intention to annihilate creators of part of their own European culture. Expresses concern that preoccupation with "uniqueness" will prevent efforts to avoid future holocausts. *YC*

1046 Brocke, Edna; Henrix, Hans Hermann: Zur Heiligsprechung Edith Steins: Ein Briefwechsel. *Kirche und Israel* 14, 1 (1999) 54-67.

An exchange of letters by the two authors in November 1998, following the beatification of Edith Stein in October. Brocke questions the Pope's characterization of Stein as "eminent daughter of Israel and faithful daughter of the Church," and as a martyr both for her people and her faith. She was no longer a Jew according to Jewish understanding; and she died not as a martyr for a cause, but, like all the other "non-Aryans," because of Nazi racial policy. Contends that although the Pope sincerely strives for reconciliation with the Jews, his emphasis on the Jewish background of this Catholic saint smacks of triumphalism. Henrix admits that the Pope's wording was unfortunate, but points out that in Stein, Christians are meant to remember all the victims of the Holocaust. The triumph is not of the Church over the Synagogue but of the victims over the perpetrators. *RW*

1047 *Cahier International sur le Témoignage Audiovisuel* 3 (June 1999).

Papers presented at the Third International Meeting on the Audiovisual Testimony of Survivors from Nazi Concentration and Extermination Camps, Brussels, June 1998. Partial contents: Martin-Chauffier, Manette: Septante témoignages: Bilans et réflexions (37-42); Blikstein, Izidoro: Analyse sémiotique et linguistique des témoignages de survivants juifs non-allemands, résidents au Brésil: Etude comparative entre les survivants allemands et non-allemands (53-57); Gelbin, Cathy: Concluding Remarks on Potsdam's "Archive of Memory" [On a project of the Moses Mendelssohn-Zentrum, Universitaet Potsdam, carried out in 1995-97, recording 79 testimonies of Holocaust survivors.] (59-63); Discussion (75-87); Walter, Jacques: Pour une périodisation des témoignages de survivants à la télévision [On French television programs relating to the Shoah.] (91-102); Simon, Roger I.: "What Happens When We Press Play?" Future Research on the Substance and Use of Holocaust Audiovisual Testimony (103-111); Rudof, Joanne Weiner: Present Research and Future Challenges [Report on the use of the Fortunoff Video Archive for Holocaust Testimonies at Yale University.] (113-117); Fargion, Liliana: Expériences et réflexions sur le témoignage audiovisuel en Italie [On the work of the Fondazione Centro di Documentazione Ebraica Contemporanea, Milano. Includes an English summary of her film "Memoria" on the deportation of Italian Jews to Auschwitz, directed by Ruggero Gabbai.] (119-125); Berlatzky, Iris: Characteristic Features of Child-Survivor Testimonies as They Appear in Their Narration [Focuses on coping mechanisms of children in the Holocaust.] (139-146); Tarsi, Anita: Integration of Oral Testimony in a Planned Curricula: Two Examples (157-159); Rudof, Joanne Weiner: Beyond Research: Education and Popular Culture [On the current popularization of and fascination with the Holocaust in the U.S.] (161-163); Hartman, Geoffrey H.: Survivor Videotestimony: Challenges and Limits (169-172); Vernay, Denise: "Mémoires de la Déportation": Un Cédérom sur la déportation partie de France (173-176); Zarka, Josette: Pollution humaine: Promiscuité et proximité [On psychological aspects of life in the concentration camps.] (177-184); Blikstein, Izidoro: La crédibilité des témoignages des survivants et le négationnisme au Brésil: Le cas des publications de la "Revisão Editoria" [On S.E. Castan's adaptation of the Leuchter Report denying the Shoah.] (185-189); Discussion (199-217). *TS*

1048 Chrostowski, Waldemar: Intellectual and Religious Life after Auschwitz and Gulag. *Collectanea Theologica* [Fasciculus specialis] (1998) 115-124.

The Holocaust (as part of Nazi totalitarianism) and communism (epitomized in the Gulag camp system) challenge Catholic theology, demanding that it take a stand toward totalitarian regimes. Argues that Christianity does not bear responsibility for the Nazi genocide; Christian dogmatic anti-Judaism has nothing in common with Nazi racism. But Catholic theology did not relate to Auschwitz or to the Gulag, despite two encyclicals, against communism and against Nazism, issued by Pope Pius XI in 1937. Contends that the only correct stance the Church can take toward totalitarianism is to side with the persecuted. *DR*

1049 Dethlefsen, Knut; Hebler, Thomas B., eds.: *Bilder im Kopf: Auschwitz – einen Ort sehen = Obrazy w głowie: Oświęcim – ujęcia pewnego miejsca.* Berlin: Hentrich, 1997. 176 pp. In German and Polish.

A collection of essays published in conjunction with a traveling exhibition organized by the Fritz Bauer Institut, and based on photographs by participants in workshops at the International Youth Center in Auschwitz. The photographs and comments of the young photographers accompany the essays. Partial contents: Loewy, Hanno: Das Tor. Der Zaun. Der Ofen. Das Tor... Die Bilder = Brama. Ogrodzenia. Piec. Brama... Obrazy (12-27); Koessler, Gottfried: Fotografien als Medium des Erinnerns und der Verstaendigung = Fotografie jako medium pamięci i porozumienia (32-43); Hoffmann, Detlef: Auschwitz im Kopf = Auschwitz w głowie (48-60); Lueer, Joerg: Auschwitz heute oder die Menschheit erinnert sich = Auschwitz dzisiaj albo ludzkość przypomina sobie (66-77); Dethlefsen, Knut: Vom schwierigen Gedenken = O trudnej pamięci (84-99); Frysztacki, Marek: Oświęcim, eine lebendige Stadt = Oświęcim, żywe miasto [Includes discussion of the attitude of Oświęcim residents to the camp.] (104-118); Dethlefsen, Knut: Spurensuche in O. = Ślady w O. (124-135); Id.: Fuenfzig Jahre danach = 50 lat później [On the commemorations.] (152-156). *RW*

1050 Gaudard, Pierre-Yves: *Shoah: L'impasse des explications monocausales.* Dijon: Centre Regional de Documentation Pedagogique de Bourgogne, 1998. 32 pp. (Savoir et transmettre, 3).

Addresses the issue of efforts by historians to understand the underlying motivation for the extermination of the Jews in the Holocaust. Contends that various hypotheses formulated to date have been counterproductive;

they have exacerbated the feeling of incapacity to comprehend the meaning of the Holocaust. Rejects such theories as the intentionalist idea, which emphasizes the psychopathology of Hitler, and the functionalist theory, which underscores the conflicts within the Nazi Party. Refutes Goldhagen's hypothesis for lack of concrete evidence, and argues that the search for a singular cause for the Holocaust clouds our reasoning and could lead to misleading interpretations. *TS*

1051 Gephart, Werner; Saurwein, Karl-Heinz, eds.: *Gebrochene Identitaeten: Zur Kontroverse um kollektive Identitaeten in Deutschland, Israel, Suedafrika, Europa und im Identitaetskampf der Kulturen*. Opladen: Leske + Budrich, 1999. 169 pp.

Partial contents: Gephart, Werner: Das Gedaechtnis und das Heilige: Zur identitaetsstiftenden Bedeutung der Erinnerung an den Holocaust fuer die Gesellschaft der Bundesrepublik (29-46); Zuckermann, Moshe: Die Parzellierung der Shoah-Erinnerung im heutigen Israel: Vom historischen Ereignis zum Gegenstand ideologischer Projektion (47-60); Saurwein, Karl-Heinz: Antisemitismus als nationales Identitaetsprojekt? Der soziologische Gehalt der Goldhagen-These (61-96). *ID*

1052 Hartmann, Rudi: Dealing with Dachau in Geographic Education. *Land and Community: Geography in Jewish Studies,* ed. Harold Brodsky. Bethesda, MD: University Press of Maryland, 1997. Pp. 357-369.

A paper presented at a conference held in 1995, at the University of Maryland and at the Library of Congress. Maintains that geography instructors can deal with the topic of the Holocaust and contribute to a program of Holocaust education. Proposes five perspectives from which a teacher can discuss the concentration camp Dachau in the classroom. Dachau can be approached as an object of political and economic geography, population geography (how the camp population changed between 1933-45), social geography (showing the diversity of people detained in Dachau), as a historic and tourist site, or from a humanistic viewpoint – the meaning of the site for camp survivors, for the current residents of the city of Dachau, and worldwide. *DR*

1053 Holocaust Studies: The Universal Aspect. *Jewish Studies* 39 (1999) 57-85.

A "round table" discussion held at the 12th World Congress of Jewish Studies, 1997, chaired by Yehuda Bauer, with the participation of Gershon Greenberg, Yisrael Gutman, Deborah Lipstadt, and Moshe Zimmermann. The questions posed were: 1) How can we explain the profound interest of world culture in the subject of the Holocaust? 2) Is there a link, or should academics make one, between the Holocaust and other cases of genocide? 3) Since it is claimed that there is a correspondence between the Holocaust and other instances of mass murder, is the universal message of these catastrophes more important than the Jewish element? Greenberg states that researchers should study Jewish life, especially Judaism, before attempting to compare genocides. Lipstadt believes that people are racist when they are indifferent to non-European genocides. For Zimmermann, interest in the history of the Holocaust is related to the fact that both Germans and Jews are "champions of memory." He claims that there is no history without comparative research. Gutman attributes the focus on the Holocaust among both scholars and the public to the perception that it marked the breach of civilization. Bauer stresses the "unprecedented" nature of the Holocaust, and that the uniqueness of the Holocaust makes it incumbent on Jews to try to oppose all genocides. *YC*

1054 Le Maître, Philippe: *Unicité de la Shoah*. Dijon: Centre Régional de Documentation Pédagogique de Bourgogne, 1998. 28 pp. (Savoir et transmettre, 1).

Interprets the concept of uniqueness of the Shoah. Argues that the debate on this issue was connected to internal debates within the American Jewish community and was extended to the European controversy in the "Historikerstreit." Considers the dilemma of the historian, who is expected to be objective, and the inevitable subjectivity which arises when studying the Shoah. Dwells on the obstacles faced by educators who teach in multiethnic institutions, where overemphasis on uniqueness might be misinterpreted by students as denigration of the suffering of other persecuted peoples. Delineates four criteria for defining the uniqueness of the Shoah: its bureaucratic and industrial nature; its dehumanization of Jews; its goal to murder the Jews and eliminate their culture; and its plan to erase the Jewish impact on other cultures. Greater emphasis on these criteria by teachers could enrich the discussion by including both the specific and universal message of the Shoah. *TS*

1055 Loew, Andrea: "Allgemeine Geschwaetzigkeit": Anmerkungen zur Goldhagen-Debatte. *Sachor* 7 (1997) 134-143.

Evaluates Goldhagen's "Hitler's Willing Executioners" and its influence in Germany. Criticizes the "free-for-all" attacks against the book and its author, which sometimes descended to levels worthy of the yellow press. However, there was also much serious criticism. Goldhagen's achievement is that he refocused the historical discourse on the central question about the Holocaust: what sort of society, what sort of people, were capable of such a crime? He also presented valuable new material, e.g. on the labor camps and the death marches. On the other hand, he oversimplified and drew inferences from insufficient evidence; and his claim to a wholly new understanding of the Holocaust is unfounded. *RW*

1056 Macho, Thomas: Erinnertes Vergessen: Denkmaeler als Medien kultureller Gedaechtnisarbeit. *Harmonie und Gewalt: Oesterreich, Europa und die Zukunft der Vergangenheit,* eds. Wilfried Graf, Roswitha Riepl. Muenster: agenda, 1997. Pp. 157-169.

Contends that "memorials" are a sign that what they stand for is being forgotten; otherwise there would be no need for them. In Germany and Austria, Holocaust memorials are a substitute for living memory; they replace the failed dialogue between the generations; vying with monuments to other tragedies, they contribute to the relativization of the Holocaust; and they tend to express an official "political correctness" or to induce a cheap emotional reaction. Mentions "negative" memorials aimed to challenge the viewer to deeper reflection. Argues that the debates over a planned memorial contribute far more to thinking about the Holocaust than the memorial itself. In this spirit, discusses the plans for the Berlin Holocaust memorial and that on the Judenplatz in Vienna. *RW*

1057 Mény, Didier: *Shoah: Limites de l'histoire.* Dijon: Centre Régional de Documentation Pédagogique de Bourgogne, 1998. 30 pp. (Savoir et transmettre, 2).

Evaluates French sources for Holocaust historiography. Contends that the Shoah, symbol of evil, is relevant to all of humanity rather than only those directly affected by it. French historians confronted the moral implications of the Shoah only decades after the war. Despite the difficulty for postwar generations to imagine the survivors' experiences, it is vital for educators to delve into the reasons that such sophistication and efficiency as were utilized by the Nazis served the cause of racism. *TS*

1058 Noi ricordiamo: Una riflessione sulla "Shoah." *Amicizia Ebraico-Cristiana* 33, 1-2 (Jan-June 1998).

Pp. 3-12 present the text, in Italian, of the Vatican document "We Remember" (1998) and the letter addressed by the Pope to Cardinal Cassidy. Pp. 13-18 contain reflections on the document by Gian Carlo Zoli, André Chouraqui, Daniel Vogelmann, and Manuela Sadun. *MZ*

1059

פלדיאל, מרדכי: 'חסידי אומות העולם' – הראויים והזכאים. דפים לחקר תקופת השואה 14 (1997) 201-246.

[Paldiel, Mordecai: "The Righteous among the Nations" – Those Who Are Worthy and Those Who Are Entitled. *Dapim le-Heker Tekufat ha-Shoah* 14 (1997) 201-246.]

Describes and explains the activities of the public committee at Yad Vashem which grants the distinction of "Righteous among the Nations" to non-Jews who endangered their lives to save Jews during the Holocaust. Discusses the criteria guiding the committee and explains the attitude toward controversial cases, such as those of collaborators, antisemites, Germans who worked for the Nazi establishment, and Jews who converted. Characterizes, in general, the rescue acts of the Righteous Gentiles. Pp. 224-246 contain excerpts from testimonies regarding Righteous Gentiles taken from the files of the public committee. *LFo*

1060 Pecherskaya, Natalia, ed.: *Bogosloviye posle Osventsima i GULaga i otnosheniye k yevreyam i iudaizmu v Pravoslavnoi Tserkvi bolshevistskoi Rossii: Materialy mezhdunarodnoi nauchnoi konferentsii 26-29 yanvarya 1997 goda Sankt-Peterburg, Rossiya* [Theology after Auschwitz and the GULag and the Attitude toward the Jews and Judaism in the Orthodox Church of Bolshevik Russia: Proceedings of the International Scholar Conference 26-29 January 1997 St. Petersburg, Russia]. Sankt-Peterburg: Vysshaya Religiozno-Filosofskaya Shkola, 1997. 186 pp. Appeared simultaneously in English (except for the article by Yershov).

Partial contents: Levin, Isidor: Aushvits/Osventsim: Kak eto moglo, a potomu mozhet opyat sluchitsya [Auschwitz: How It Could Have Happened - and Why It Could Happen Again] (29-57); Stoehr, Martin: Bogosloviye posle Osventsima i GULaga [Theology after Auschwitz and GULag] (58-72); Pawlikowski, John T.: Kholokost: Vliyaniye na khristianskuyu mysl i etiku [The Holocaust: Its Impact on Christian Thought and Ethics] (76-93); Benevitch, Grigori: "Teologiya posle Osventsima" - pravoslavnaya tochka zreniya ["Theology after Auschwitz" - the Russian Orthodox View] (94-113); Levine, Hillel: Bogosloviye posle Osventsima i GULaga: Priglasheniye k novomu iudeo-pravoslavnomu dialogu [Theology after Auschwitz and after the GULag: An Invitation to a New Jewish-Russian Orthodox Dialogue] (160-167); Yershov, Gleb: Iskusstvo posle Osventsima: O vystavke rabot Andreasa Felgera (Germaniya) i Iyegudy Bakona (Izrail) [Art after Auschwitz: On the Exhibition of Art Works by Andreas Felger (Germany) and Yehuda Bacon (Israel).] (179-186). *DR*

1061 Ries, Henry: *Auschwitz: Pruefstein des deutschen Gewissens*. Berlin: Aufbau-Verlag, 1997. 96 pp.

A compilation of interviews with visitors and employees at Auschwitz (Jews and non-Jews), conducted in 1994-96, interspersed with photographs of the camp and excerpts from the memoirs of Rudolf Hoess and other SS camp officials. *RW*

1062 Rudin, Arnold James: Reflections on the Vatican's "Reflection on the Shoah." *Cross Currents* 48, 4 (Win 1998-1999) 518-529.

Expresses reservations about the long-awaited document issued by the Vatican in March 1998. The accompanying letter to Cardinal Cassidy expresses Pope John Paul II's profound sorrow about Jewish suffering during the Shoah and can serve as a response to Holocaust denial. However, the document "raised more questions than it answered and created more problems than it solved." It fails to provide adequate answers, e.g. concerning the responsibility of the Church for prejudices that led to the Holocaust and the largely indifferent behavior of Christians to the genocide of the Jews. The Vatican document is compared, unfavorably, with more outspoken statements of European and American bishops. In view of the limitations noted, concludes that there is a need for further study of Catholicism and the Shoah. *YC*

1063 *Sens* 51, 3 (1999).

This issue, entitled "Le document du Vatican sur la Shoah," contains commentaries on the 1998 Vatican declaration "We Remember" on the Holocaust and on relations between Jews and Christians. Partial contents: Cassidy, Edward Idris: "Nous nous souvenons, une réflexion sur la Shoah": Suites (99-110); Réponse au document du Vatican "Nous nous souvenons: Une réflexion sur la Shoah" [par le] Comité juif international pour les consultations interreligieuses (111-119); Willey, Diane: "Pas d'avenir sans mémoire" (120-122); Grosman, Robert-G.: La déclaration du Vatican ne deviendra-t-elle qu'un souvenir évanescent? (123-126). *HV*

1064 Shafir, Shlomo: Die Stimmung wird besser: Amerikas Juden und Deutschland. *Tribuene* 151 (1999) 122-132.

Traces the ups and downs of American Jewry's attitude towards Germany since 1945. Among the negative influences mentions the Holocaust, continuing German antisemitism and its recent reemergence, the drawn-out negotiations over compensation to forced laborers, and the delay in the erection of the Berlin Holocaust memorial. *RW*

1065 Staffa, Christian; Klinger, Katherine, eds.: *Die Gegenwart der Geschichte des Holocaust: Intergenerationelle Tradierung und Kommunikation der Nachkommen*. Berlin: Institut fuer Vergleichende Geschichtswissenschaften, 1998. 255 pp.

Proceedings of an international conference, held in Berlin in January 1997, on the effects of the Holocaust on the descendants of victims and perpetrators, and on Holocaust remembrance. Partial contents: Rosenthal, Gabriele; Voelter, Bettina: Trennende und bindende Vergangenheiten:

Zur familienbiographischen Arbeit und Dynamik in Ehen zwischen Nachkommen von Ueberlebenden der Shoah und von Nazi-Taetern (17-46); Bar-On, Dan: Politische und pragmatische Konsequenzen des Holocaust im intergenerationell en Uebergang (47-57); Einfuehrung in die Arbeit der Nachkommen von Opfern und Taetern: Podiumsdiskussion (59-70); Staffa, Christian: Leben in einer Welt nach der Shoah (71-76); Strauss, Daniel: Die Gegenwaertigkeit der Vergangenheit (77-87); Heenen-Wolff, Susann: Die versteckten Kinder des Holocaust (103-109); Begegnung als Bearbeitungsform der nachtraeglichen Wirksamkeit des Holocaust: Podiumsdiskussion (111-142); Stein, Ernst M.: Tradierung und Kommunikation der Nachkommen aus der Perspektive der juedischen Gemeinde (143-151); Gruenberg, Kurt: Schweigen und Ver-Schweigen: Zur Differenz der Bearbeitungsformen in Opfer- und Taeterzusammenhaengen (153-176); Strobl, Ingrid: Antisemitismus, Antizionismus und die deutsche Linke (177-182); Ehmann, Annegret: Paedagogische Arbeit in Gedenkstaetten: Das Haus der Wannsee-Konferenz: Gedenk- und Bildungsstaette (183-190); Mumme, Petra; George, Uta: Entwicklung und Entstehung von schulischen und ausserschulischen Bildungsmaterialien zur Geschichte und Wirkung des Holocaust (Fritz-Bauer-Institut) und "TaeterInnen vor Gericht" - eine Methode der Gedenkstaette Hadamar (191-204); Bearbeitungsformen der Nachkommen und ihre laenderspezifischen Kontexte: Beitraege aus Deutschland, Gross-britannien, Israel, den Niederlanden, Oesterreich, Tschechien und den USA (205-249). *ID*

1066 Todorov, Tzvetan: Dix ans sans Primo Levi. *Esprit* 240 (Feb 1998) 125-139.

Discusses Levi's last work, "The Drowned and the Saved," as a major contribution to understanding the role of remembrance decades after the Shoah. Compares two essential components in Levi's view of humanity: the capacity for evil coupled with the quest for justice. Interprets the concepts of "memory" and "crime" through Levi's perspective on Auschwitz. Subscribes to the author's creed of humanity's fallibility regarding memory, but distinguishes between the right of the victim to forget and the obligation of the criminal to remember. Reflects on Levi's view that man plays one of four roles: the benefactor, the receiver, the criminal, or the victim. Contrasts justice and vengeance, and contends that forgiveness for crimes such as the Holocaust requires society's participation in the stages of moral education. *TS*

1067 Vatykan i Katastrofa [The Vatican and the Holocaust]. *Yehupets* 5 (1999) 192-215.

A Ukrainian translation of four documents relating to the "We Remember" declaration: Pope John Paul II's letter to Cardinal Edward Idris Cassidy, the text of "We Remember: A Reflection on the Shoah," the presentation of the document by Cassidy before the American Jewish Congress, and the response to Cassidy by Martin S. Kaplan. *DR*

1068 Weiss, Gershon: *The Holocaust and Jewish Destiny: Jewish Heroism, Human Faith, and Divine Providence*. Southfield, MI: Targum Press; Nanuet, NY: Feldheim, 1999. xvi, 275 pp. On title page also: In collaboration with Yehuda Sorscher.

Presents an ultra-Orthodox approach to understanding the Holocaust. Ch. 1 (pp. 5-38) relates the experiences of the author and his family in the Holocaust. Weiss was a child when his family escaped from Czechoslovakia to Hungary in 1940. They survived the war in Budapest, in hiding and in the ghetto, but half of the extended family perished in the Holocaust. Stresses the spiritual resistance of religious Jews, giving examples of the heroic behavior of rabbis and ordinary people. The rest of the book discusses persecution of the Jews in biblical times as a guide to understanding the Holocaust and Divine providence. Draws extensive parallels between Pharaoh and Hitler. Emphasizes the role of God as the Master of History, pointing out a number of miracles which occurred during World War II, including the Nazis' failure to conquer Eretz-Israel. Attributes antisemitism and Jewish suffering to the sins of Reform and assimilated Jews. Germany is equated with the traditional mythical enemies of the Jews, Esau and Amalek. Exile and the Holocaust are seen as "iron crucibles," in which the Jews were purified and from which they emerged a greater nation. *YC*

1069 Wiesel, Elie: *Ethics and Memory = Ethik und Erinnerung*. Berlin: Walter de Gruyter, 1997. 53 pp. In English and German.

A lecture held at the Wissenschaftskolleg in Berlin, June 1996. Discusses the importance of memory to human life. Unlike history, it has an ethical aspect; and it must be inclusive – of good and evil, the painful and the encouraging, Nazi crimes and those few who heroically resisted them. Points to the ghetto and concentration camp inmates who wrote journals, and to Dubnow who continued to write his history, so that we might remember. We should remember the Jews in the photographs – the old Jew whose beard is being cut off by laughing German soldiers, the starving children begging for bread. But as the generation of the Holocaust passes away, who will remember or want to remember? While Germany is now a praiseworthy democracy, there are troubling manifestations of insensitivity toward the Jewish past. Pp. 1-10 contain intoductory remarks by Wolf Lepenies on the occasion of the lecture. *RW*

1070 Wistrich, Robert Solomon: The Pope, the Church, and the Jews. *Commentary* 107, 4 (Apr 1999) 22-28.

Criticizes the document issued by the Vatican in 1998, "We Remember: A Reflection on the Shoah." This document, over a decade in the making, failed to point out the responsibility of the Church for preparing the ground for 19th-20th-century antisemitism. Elevating some individual churchmen who deplored Nazi antisemitism, it fails to mention that the Catholic Church acquiesced in Nazi racism and cooperated with the Nazis in many ways. Moreover, it presents Pope Pius XII as an active opponent of Nazi antisemitism. Pius XII, who indeed attempted to intervene at times to halt deportations of Jews, failed to take a firm stand against the Nazi genocide, and thus did not meet the moral obligations of the papacy. The inaction of the Church hierarchy contrasts with the heroism of those ordinary Catholics who helped Jews, but it is the Church as an institution which "We Remember" tries to declare blameless. The imperfections of "We Remember" are a matter of great concern. *DR*

1071 Zuckermann, Moshe: *Zweierlei Holocaust: Der Holocaust in den politischen Kulturen Israels und Deutschlands*. Goettingen: Wallstein, 1998. 182 pp.

Discusses ideological forms of Holocaust-reception within the political cultures of Israel and Germany. Contends that each type of commemoration has imprinted the cultures of both countries, and points to correlations between them. Emphasizes the historical constellation, but also the need for critical analysis of both ideologies, and especially of the relationship between collective memory and ideology. Israel, the "state of the victims," constantly demands historical responsibility from Germany, the "state of the perpetrators." Israel claims a monopoly on the meaning of Holocaust commemoration, and Germany, feeling under obligation, avoids questioning it. Concludes that Israel's commemoration emanates from its Zionist ideology, while Germany opts for a more relativistic, dread-normalizing reception. *ID*

Nazi War Criminals and War Crimes Trials

1072 Violet, Bernard: *Le dossier Papon*. Paris: Flammarion, 1997. 290 pp.

Surveys the life and career of Maurice Papon (b. 1910). Ch. 2 discusses the war period, when he was secretary-general of the Gironde region. The rest of the book deals with his postwar career. Seeks to account for the 16-year lapse from the publication of the first incriminating evidence until his trial in 1997. Chs. 13-15 (pp. 195-237) cover the period 1981-97. Assesses the responses of the defense to the charge of Papon's guilt of crimes against humanity. Pp. 241-269 contain documents from the war period and the list of participants in the civil suit against Papon. *TS*

1073 Wood, Nancy: Memory on Trial in Contemporary France: The Case of Maurice Papon. *History & Memory* 11, 1 (Spr-Sum 1999) 41-76.

Papon, a former functionary in the Vichy regime, responsible for the deportation of ca. 1,600 Jews from Bordeaux, and the last living symbol of collaborationism in France, was convicted of complicity in crimes against humanity at his trial held in Bordeaux in 1997-98. Examines problems with which the French judicial system had to deal during this trial. French legislation does not enable the conviction of a person who acted due to career interests rather than ideological motives. A crucial question was whether and to what extent Papon knew of the fate awaiting those whom he deported. The historians called upon as witnesses were unable to assess, unambiguously, the measure of knowledge possessed by a Vichy functionary, nor could they assess to what extent Papon represented the typical Vichy functionary. Current historiography of the Vichy period tends to depict the Vichy official as simply an obedient executor. The Papon case shows the discrepancy between historical and judicial knowledge, and between those and collective memory. *DR*

Neo-Nazism and Holocaust Denial

1074 Ayass, Wolfgang; Krause-Vilmar, Dietfrid: *Mit Argumenten gegen die Holocaust-Leugnung: Die Leugnung der nationalsozialistischen Massenmorde als Herausforderung fuer Wissenschaft und politische Bildung*. Wiesbaden: Hessische Landeszentrale fuer Politische Bildung, 1996. 30 pp.

Traces the history of Holocaust denial and analyzes its specific arguments and its tactics. Surveys the documentary and research evidence which refutes the revisionists' allegations, warning, however, against exaggerated or unfounded statements, which play into their hands. Asserts that Holocaust denial, with its pretensions to scientific proof, influences credulous young people, and should not be ignored as in the past, but refuted on the basis of the facts. *RW*

1075 Lipstadt, Deborah Esther: Holocaust Denial: Will It Cast a Shadow on Holocaust Memory in the New Millennium? *Justice* 22 (Win 1999) 3-6.

A paper delivered at the second "Remember Berlin" conference in Berlin, June 1999. Discusses the gap between Jewish perceptions of the extent of antisemitism and the reality (indicated by surveys) that Holocaust deniers have apparently had a limited impact. Poses the question of whether one should react to or ignore Holocaust denial, especially since overreaction publicizes these views. Counsels fighting Holocaust denial, since in the future it will be easier to raise doubts concerning the Holocaust, when there are no longer living survivors. Advocates targeting those susceptible to denial arguments (but not deniers themselves, whose prejudice is impervious to argument) by exposing the absurdity of those arguments through facts, survivor testimony, and especially documents by perpetrators of the Holocaust, which are the hardest to deny. Concludes that Holocaust denial should be opposed, but should not be overestimated. *YC*

1076 Maegerle, Anton: Antisemitische Esoterik: Die Kreise um "Jan van Helsing." *Tribuene* 152 (1999) 124-132.

Notes that esoteric teachings are enjoying great popularity in German-speaking countries, and that many esoteric publications and ca. one in four of esoteric societies belong to the radical right. The most-read radical-right esoteric writer is Helsing (pseudonym of Jan Udo Holey, b. 1967), author of the two-volume work "Geheimgesellschaften" (Ewert Verlag, 1993-95). This book, though banned in Germany, is widely sold under the counter and is also available through the "Thule-Netz" on the internet. A videotape interview with Helsing is also widely distributed. Helsing propagates the theory of the Jewish world conspiracy and Holocaust denial. Traces his connections with the network of writers, organizations, publishers and other agencies that disseminate esoteric literature laced with neo-Nazism and Holocaust denial. *RW*

1077 Maegerle, Anton: Holocaustleugner. *Tribuene* 151 (1999) 102-106.

Traces the worldwide connections between Germar Rudolf – publisher of the "Vierteljahrshefte fuer freie Geschichtsforschung," the central organ of German Holocaust deniers, and author of a "scientific" tract demonstrating the impossibility of gas chambers in Auschwitz – and other authors and publishers propagating Holocaust denial. *RW*

Christian-Jewish Relations

1078 *Collectanea Theologica* 69, 2 (1999).

Partial contents: Chrostowski, Waldemar: Ku nowej chrześcijańskiej teologii judaizmu: X Jubileuszowe Sympozjum Teologiczne "Kościół a Żydzi i judaizm," Warszawa, 14 V 1998 [Towards a New Christian Theology of Judaism: The Xth Theological Symposium "The Church, Jews and Judaism," Warsaw, May 14, 1998] (5-10); Id.: Czy jest potrzebna nowa chrześcijańska teologia judaizmu? [Is a New Christian Theology of Judaism Needed?] (11-36); Kamykowski, Łukasz: Podstawowe elementy nowej chrześcijańskiej teologii judaizmu [The Basic Elements of a New Christian Theology of Judaism] (37-59). *HV*

1079 Erklaerung der Evangelisch-Lutherischen Kirche in Bayern zum Thema "Christen und Juden." *Kirche und Israel* 14, 1 (1999) 74-80. Appeared also in the "Amtsblatt fuer die Evangelisch-Lutherische Kirche in Bayern" 1 (1999) and in the "Freiburger Rundbrief" 6, 3 (1999).

This declaration states that Christian anti-Judaism, and particularly that of Luther, was one of the causes of the Holocaust, and therefore the Church is called upon to make a new beginning. Theology should deal more with the Jewish roots of Christianity and what the two religions have in common, rather than use Judaism as a negative foil to Christianity. Congregations should examine their behavior toward Jews during the Nazi period; they should teach about Judaism and establish ties with the Jewish community. Rolf Rendtorff (pp. 71-73) summarizes the important insights of the declaration, and points out several whose consequences might have been stated more clearly and fully. *RW*

1080 Kraus, Wolfgang, ed.: *Christen und Juden: Perspektiven einer Annaeherung*. Guetersloh: Chr. Kaiser, Guetersloher Verlagshaus, 1997. vi, 202 pp.

A collection based on lectures delivered at a series of symposia organized by theologians and clergy of the Evangelical-Lutheran Church of Bavaria, 1992-95. Partial contents: Kraus, Wolfgang: Christologie ohne Antijudaismus? Ein Ueberblick ueber die Diskussion (21-48); Hirschberg, Peter: Probleme und Chancen bei der Rezeption des Themas Judentum in den Gemeinden (95-106); Roloff, Juergen: Die Predigt ueber Texte der Passionsgeschichte und der Antijudaismus (107-118); Rothgangel, Martin: Empirische Ueberlegungen zur gegenwaertigen Behandlung der Passionsgeschichte im evangelischen Religionsunterricht (119-141). *RW*

1081 Pecherskaya, Natalia, ed.: *Bogosloviye posle Osventsima i yevo svyaz s bogosloviyem posle Gulaga: Sledstviya i vyvody: Materialy vtoroi mezhdunarodnoi nauchnoi konferentsii, Sankt-Peterburg, 26-28 yanvarya 1998 g*. [Theology after Auschwitz and Its Relation to the Theology after GULag: Deductions and Conclusions: Proceedings of the Second International Scholar Conference, Sankt-Peterburg, 26-28 January 1998]. Sankt-Peterburg: Vysshaya Religiozno-Filosofskaya Shkola, 1999. 175 pp. Originally published as "Proceedings of the Second International Conference – Theology after Auschwitz and Its Correlation with Theology after the GULag: Consequences and Conclusions" (St. Petersburg: St. Petersburg School of Religion and Philosophy, 1998).

Partial contents: Tabak, Yuri: Kholokaust kak istoricheski zakonomernoye voploshcheniye ekskliuzivistskoi modeli religioznovo soznaniya [The Holocaust as a Historically Determined Realization of the Exclusion Model of Religious Consciousness] (11-20); Matzikh, Leonid: Iudeiskiye bogoslovy o katastrofe [Jewish Theologians on the Holocaust] (21-39); Hackel, Sergii: Kak sootnositsya zapadnoye bogosloviye posle Osventsima s myshleniyem i bogosluzhebnoi zhiznyu Russkoi Pravoslavnoi Tserkvi? [The Relevance of Western Post-Holocaust Theology to the Thought and Practice of the Russian Orthodox Church] (79-96); Pollefeyt, Didier: Iudeo-khristianskiye otnosheniya posle Osventsima s katolicheskoi tochki zreniya [Jewish-Christian Relations after Auschwitz from a Catholic Viewpoint] (114-134); Manemann, Juergen: Dvizheniye k anamnesticheskoi kulture: Politiko-bogoslovskii aspekt bogosloviya [The Movement towards an Anamnestic Culture: The Political and Theological Aspect of Theology] (135-152); Spektor, Roman: Zakat grazhdanskovo miroporyadka [The Decline of Social Order] (153-159). *SSC*

1082 Zanicky, Robert Martin: *Anti-Judaism in Contemporary Christian Preaching.* DMin. - Princeton Theological Seminary, Princeton, NJ, 1997. 219 pp. Unseen.

Contends that there is a continued presence of anti-Judaism in contemporary Christian preaching. Presents an overview of historic anti-Judaism and briefly notes its relationship with antisemitism. Anti-Judaism in preaching is determined by using six categories: 1. The Pharisees and other Jewish leaders are referenced without noting the polemical context; 2. The scriptural text is used without qualifying anti-Judaic terms; 3. Judaism is denigrated as being inferior or passé; 4. Christianity is presented as being superior to Judaism; 5. The Jews are depicted as not knowing how to correctly interpret their scriptures; 6. The Jews and/or their religious leaders are charged with the death of Jesus. Four preaching journals were examined for the period 1993-96: "The Christian Ministry," "Journal for Preachers," "Preaching," and "Pulpit Digest." The research findings support the hypothesis that anti-Judaism continues to be preached from the Christian pulpit, and that people can be made aware of its presence and potential harm.

1991-1999: Africa

South Africa

1083 Saks, David; Kopelowitz, Seymour: Antisemitism in South Africa 1998: Report for the Institute of Contemporary Antisemitism and Racism, Tel Aviv University. *Jewish Affairs* 54, 1 (Fall 1999) 61-65.

As in previous years, in 1998 traditional right-wing antisemitism, almost entirely confined to the White community, became an increasingly marginal factor, whereas antisemitism in the Muslim community increased. The latter phenomenon may be attributed partly to the faltering peace process in the Middle East under the Netanyahu government and to the U.S. bombing of Iraq. Dwells briefly on antisemitic propaganda and other activities conducted by various right-wing or Muslim organizations, acts of violence, anti-Israel demonstrations with heavy anti-Jewish overtones, and antisemitic harassment. *DR*

1991-1999: Australia

1084 Berman, Judith: Australian Representation of the Holocaust: Jewish Holocaust Museums in Melbourne, Perth, and Sydney, 1984-1996. *Holocaust and Genocide Studies* 13, 2 (Fall 1999) 200-221.

Examines the exhibitions and accompanying narrative at the three Holocaust museums in Australia – the Jewish Holocaust Museum in Melbourne (opened in 1984), the Holocaust Institute of Western Australia (1990), and the Sydney Jewish Museum (1992). These museums were established by Holocaust survivors and financed by private Jewish funds, thus they do not serve a goal of "Australianization" of the Holocaust, in contrast to the Holocaust museums in the USA and Israel. The museums in Australia present neither a humanist narrative, as does the United States Holocaust Memorial Museum, nor a Zionist narrative, as does Yad Vashem. They were established primarily to educate the broader Australian public about the Holocaust, to show "how it really was." This task is all the more vital given the growing Holocaust denial in Australia. Criticizes, also, some shortcomings in the museums' narratives. *DR*

1991-1999: Europe

General

1085 Colijn, G. Jan; Pollefeyt, Didier: Geen Duitsers, geen Sjoa? De receptie van Goldhagen in Nederland en België [No Germans, No Shoah? The Reception of Goldhagen in the Netherlands and Belgium]. *Ter Herkenning* 25, 4 (1997) 238-253.

Dutch critics claim that Goldhagen's "Hitler's Willing Executioners" (a Dutch translation appeared in 1996) has no value as it shows no new historical insights. Notes that Goldhagen repeatedly asks his critics to show him where he is wrong. The Dutch newspaper "Vrij Nederland" states that Goldhagen blames all the Germans for having been willing executioners, a statement that is almost racist. However, the much-hoped-for wish on the part of the Germans that the reunification of Germany and the recent commemoration services for the end of World War II would cover up the Nazi past was ruined. The Goldhagen affair shows once again that "there is no business like Shoah business." *MF*

1086 Wiedmer, Caroline Alice: *The Claims of Memory: Representations of the Holocaust in Contemporary Germany and France*. Ithaca, NY: Cornell University Press, 1999. x, 244 pp.

Ch. 1 (pp. 11-31), "Maus," discusses the work by Art Spiegelman, concluding that his approach to the Holocaust manifests a dynamic relationship between personal focus and historical accounting. The remainder of the book deals with the relationship between individual fate and national history, particularly in terms of how two countries, France and Germany, decide to memorialize, or not to memorialize, the Holocaust. Ch. 2 (pp. 32-78), "Reoccupations," reveals how French monuments commemorate the Holocaust but do not relate to the Vichy regime and French collaboration with the Nazis. Changes in France's "remembrance" of the Holocaust are reflected in discussions on the kind of memorials appropriate for Drancy and for the site of the Vélodrome d'Hiver. Ch. 3 (pp. 79-199), "The German Question," discusses the tension between German remembering and "forgetting" as revealed in several cases: discussions on Michael Verhoeven's film "Das schreckliche Maedchen," problems connected with memorializing the Sachsenhausen concentration camp, and controversies over Holocaust memorial projects in Berlin. Memorial projects that were rejected are noted in terms of what messages Germany did not want transmitted. *YC*

Austria

1087 Zeit zur Umkehr: Die Evangelischen Kirchen in Oesterreich und die Juden. *Freiburger Rundbrief* 6, 2 (1999) 127-130.

The text of a declaration of the General Synod of the Evangelical Church in Austria, published on 29 October 1998 in Vienna. Based on the Synod's previous declarations of 1965 and 1996, it states that 60 years after the "Kristallnacht" pogrom, Evangelical Christians feel obliged once more to cope with the history of the annihilation of European Jewry. It is impossible to deny the Church's and Christians' share of guilt in the Holocaust. The Church commemorates all the persecuted und murdered victims, and undertakes to always remember the Holocaust, to clear all sermons, liturgy, and teachings from antisemitism, to fight against manifestations of social or personal antisemitism, and to establish a new attitude toward Jews and Judaism. *ID*

France

1088 Bar-Zohar, Michael: *Perfume amargo: O caso da L'Oréal, dos Nazistas e do boicote árabe*. Trans.: A.B. Pinheiro de Lemos. Rio de Janeiro: Imago, 1997. 289 pp. Originally published as "Bitter Scent: The Case of L'Oréal, Nazis, and the Arab Boycott" (Harmondsworth, Mdsx: Dutton, 1996).

1089 Freedman, Robert O.: Jean Marie LePen: The Threat to the Jews of France. *Midstream* 44, 5 (July-Aug 1998) 9-12.

Antisemitism is an important component of the beliefs of Le Pen's National Front party, which won 15% of the vote in both the 1995 presidential elections and the 1997 legislative elections. The popularity of the National Front can be explained by the rise in unemployment, the influx of Muslim immigrants from North Africa, and growing unease about the subordination of France to the European Union. Antisemitism is not the central focus of the party but it is one of the major issues and is often expressed in National Front newspapers and by Le Pen himself. Describes efforts made by the Conseil Représentif des Institutions Juives de France to combat Le Pen. States that the National Front is the ideological successor of the anti-Dreyfus forces and of Pétain, but it lacks popular support; the Church and right-wing intellectuals, though traditionally anti-Jewish, do not support the party. *REK*

1090 Grand, Philippe: Le fichier juif: Un malaise (réponse au rapport Rémond remis au premier ministre le 3 juillet 1996). *Revue d'Histoire de la Shoah* 167 (Sept-Dec 1999) 53-101.

Traces the "fichier juif affair," the debate over the list of the census taken of the Jewish community of Paris which was ordered by the French police from 1940 until 1944. The affair developed into a controversy following the list's discovery by Serge Klarsfeld in 1991, and the appointment of the Rémond Commission in 1992 to protest its transfer to the Memorial of the Unknown Jewish Martyr rather than its preservation in the National Archives. Grand, one of the archivists, details the events of 1991-93 and examines the arguments contesting the authenticity of the list, which was thought to have been destroyed during the war. Criticizes the final report of the Rémond Commission, published in 1996. Rejects the argument that the original list was destroyed and opposes the distinction in the report between an "individual" and "family" list. See the response by Jean-Pierre Azéma [Ibid. 169 (2000) 249-256], one of the editors of the final report of the Rémond Commission, refuting Grand's hypothesis that the extant "individual" list is part of the original one, because there are fragments dated after 1940, the publication date of the lost list. Outlines his proposed categories of lists: 1) the original census of 1940, destroyed in 1948; 2) the list of Jews who were arrested within and outside Paris; 3) the list of police raids where at least one Jewish family member was arrested; 4) the list of deportations, especially from Drancy. *TS*

Germany

1091 Baier, Lothar: Un romancier reconverti en oracle national: Essai d'explication de l'"Affaire Walser" en Allemagne. *Les Temps Modernes* 603 (Mar-Apr 1999) 1-19.

Describes discussions in Germany regarding the public speech made by German writer Martin Walser (b. 1927), winner of the Peace Prize for 1998 at the Frankfurt Book Fair. Walser opposed the continuation of memorialization of Nazi crimes in Germany. He advocated use of the word "shame" instead of "crime"; this changes the accent put up to now on objective facts (crime) to a subjective judgment of these facts (shame). Walser does not wish to live in a condemned country; he rejects the idea of a Holocaust memorial in Berlin. His speech was applauded by the German elite, and the only person who objected was the leader of the German Jewish community, Ignatz Bubis. Analyzes reactions in the press and the television confrontation between Walser and Bubis. Reflects on possible consequences of acceptance by Germans of his point of view, which could lead to a rejection of remembrance of the Nazi period and to a growth in nationalist ideology and antisemitism in unified Germany. *HV*

1092 Bunk, Bernd: *Das Holocaust-Mahnmal: Eine Ueberforderung fuer deutsche Kulturtraeger: Betrachtung und alternative Loesungsansaetze.* Sandhatten: Loog Verlag, 1999. 24 pp.

Addressed to the German Minister for Culture, Michael Naumann, criticizes the plan to establish a Holocaust monument in Berlin, including a museum and a research institute. Views the concept of such a museum as an intellectual solution to an artistic task not yet fulfilled. Argues that there is no correlation between the large expense involved and the site's ability to engender remembering. Calls for a re-orientation; suggests a Holocaust memorial as a place of Jewish spirituality, or a grove of remembrance, where Jewish culture and living Jewish spirit and art would be presented, showing also the Jewish influence in Occidental European culture. Pp. 9-22 contain pictures of various monuments throughout the world, with explanations of their background. *ID*

1093 Cullen, Michael S., ed.: *Das Holocaust-Mahnmal: Dokumentation einer Debatte.* Zuerich: Pendo, 1999. 295 pp.

A collection of articles which appeared in various German newspapers in 1995-98, dealing with the Holocaust memorial debate. Partial contents:

Korn, Salomon: In der Schieflage: Nachtrag zum Denkmal fuer die ermordeten Juden Europas (31-35); Id.: Monstroese Platte: Zur Debatte um das Holocaust-Denkmal (36-41); Baecher, Max: Der ganze Wettbewerb war ein Missgriff: Phrasenhafte Ausschreibung, falscher Standort: So muss das Holocaust-Mahnmal scheitern (42-48); Korn, Salomon: Die Tafeln sind zerbrochen: Ueber die Darstellung des Unvorstellbaren, das Vergessen und den Streit um das Holocaust-Denkmal in Berlin (49-60); Kittsteiner, Heinz Dieter: Der Angriff der Gegenwart auf die Vergangenheit: Ueber das vermeintliche Recht, ein Denkmal fuer die ermordeten Juden zu bauen (61-70); Korn, Salomon: Der Tragoedie letzter Teil – das Spiel mit der Zeit: Anmerkungen zum Holocaust-Denkmal in Berlin, bevor ueber das Mahnmal fuer die ermordeten Juden Europas entschieden wird (71-84); Koselleck, Reinhart: Vier Minuten fuer die Ewigkeit: Das Totenreich vermessen – Fuenf Fragen an das Holocaust-Denkmal (85-92); Id.: Erschlichener Rollentausch: Das Holocaust-Denkmal im Taeterland (97-102); Meier, Christian: Zweierlei Opfer (103-109); Kittsteiner, Heinz Dieter: Der Denker denkt – die Lenkungsgruppe lenkt: "Holocaust-Denkmal": Dritte Expertenrunde (110-114); Mittig, Hans-Ernst: Von der Schuld zur Verantwortung (115-123); Korn, Salomon: Durch den Reichstag geht ein Riss: Wenn Kunst die Barbarei gestalten soll: Das geplante Berliner Holocaust-Mahnmal und die nationale Identitaet (124-129); Buddensieg, Tilmann: Berliner Symmetrie-Wahn: Ueberlegungen zum Holocaust-Mahnmal (130-134); Meier, Christian: Das Problem eines Berliner Denkmals (135-157); Beaucamp, Eduard: Kunst in der Falle: Das Holocaust-Denkmal und das Scheitern der Kuenstler (158-164); Broder, Henryk M.: Wer ein Menschenleben rettet, rettet die Welt (165-170); Korn, Salomon: Mit falschem Etikett: Verhindert das Denkmal fuer die ermordeten Juden ein zentrales Mahnmal gegen Tat und Taeter? (171-177); Kudielka, Robert: Das falsche Gewicht der Betroffenheit (178-185); Broder, Henryk M.: "Auf der Hoehe der Zeit": Vier neue Entwuerfe stehen fuer das Berliner Holocaust-Mahnmal bereit, die Enstscheidung soll schon im Januar fallen – in jedem Fall wird es eine Fehlentscheidung (186-190); Konrád, György: Abschied von der Chimaere: Wider das Holocaust-Denkmal (191-197); Jens, Walter: In letzter Minute: Mein Widerruf zum Holocaust-Mahnmal (206-210); Grasskamp, Walter: Das Mahnmal als Kanzlerfalle: Das Denkmal fuer die ermordeten Juden Europas – Auswege aus einem verfahrenen Verfahren (214-220); Koselleck, Reinhart: Die falsche Ungeduld: Wer darf vergessen werden? Das Holocaust-Mahnmal hierarchisiert die Opfer (221-229); Schoeps, Julius Hans: Niemand kann den Ermordeten ihre Wuerde zurueckgeben: Im Streit ueber das Holocaust-Mahnmal geht so viel durcheinander, dass ein Innehalten keine Schande waere (233-237). *ID*

1094 Gehle, Holger: Gedaechtniswechsel: Martin Walsers Essay "Unser Auschwitz" im Werkkontext. *Peter Weiss Jahrbuch* 8 (1999) 114-140.

Traces the development behind Walser's notorious remark, in his speech accepting the Peace Prize of the German Book Trade in 1998, that it was time to stop asking Germans constantly to confront the Holocaust. Already in 1965, after attending several sessions of the Auschwitz trial, Walser rebelled against the sensational representation of the Holocaust as the work of inhuman brutes rather than the systematic national program it was. In his essay "Unser Auschwitz" he said that this left ordinary Germans like himself feeling uninvolved. Thereafter, he avoided the Holocaust in his literary work, although he returned to it in many essays, speeches, and interviews. In his eyes, the only "morally beautiful" representations of the Holocaust were accurate and critical accounts of personal experience, by both victims and perpetrators or bystanders. German national history includes all of the 20th century and all Germans, Jews and non-Jews, who in this wider context are victims too. *RW*

1095 *Goldhagen und die deutsche Linke oder, Die Gegenwart des Holocaust.* Berlin: Elefanten Press, 1997. 192 pp. At head of title page: Ulrike Becker, Frank Behn, Clara Fall, Matthias Kuentzel, Wladimir Schneider, Juergen Starck, Klaus Thoerner, Rolf Woltersdorf.

A book written by eight authors, some of them affiliated with various left-wing organizations before 1989, which emerged from discussions held by them following the publication of Goldhagen's book. States that the left-wing press in Germany reacted to Goldhagen's book and its supposed allegation of collective guilt with similar arguments as the moderate and right-wing press, though it defended the working class rather than all the Germans: Nazism was a capitalist plot against labor, and labor solidly opposed it. Shows that in fact the workers and their parties were also antisemitic and pro-Nazi. Goldhagen's chief merit was to show that the individual was free to decide how to act despite social pressure. However, he did not recognize the vital role of biological antisemitism as the basis of German identity and unity; and though he correctly linked forced labor with the Nazis' image of Jews as unproductive parasites, he did not see that they were wiping out the Jews as the representatives of "abstract" capital in favor of a utopia of "blood, soil, and labor": Nazism was an anti-capitalist, not a capitalist, revolution. Antisemitism is still rife on both the Right and the Left. German liberals' appreciation for Goldhagen and their willingness to deal with the memory of the Holocaust is due to a feeling of discontinuity between then and now: the blacker Germany's past crimes, the brighter the "new Germany." *RW*

1096 Hasian, Marouf, Jr.; Frank, Robert E.: Rhetoric, History, and Collective Memory: Decoding the Goldhagen Debates. *Western Journal of Communication* 63, 1 (Win 1999) 95-114.

The controversy surrounding Goldhagen's "Hitler's Willing Executioners" is used as a case study to examine the ideological structures at work when rhetoric, history, and collective memory intersect. Academic histories and collective memories are not antithetical concepts; together, they serve three rhetorical functions: they help create society's identities, maintain its traditions, and allow it to forget. Goldhagen's book became a bestseller in Germany because it invited readers to change their views of their own national identity, traditions, and collective memories. Goldhagen showed the limits of German "normalization," and also showed Germans how to change their national identity. His arguments tapped into the reservoir of feelings and beliefs that existed in the collective memories of his readers. Many scholars criticized Goldhagen for his non-academic approach and style, as well as for his dangerous denigration of the Germans. Right as they are, Goldhagen's book possesses great rhetorical value. *DR*

1097 Heer, Hannes: The Difficulty of Ending a War: Reactions to the Exhibition "War of Extermination: Crimes of the Wehrmacht 1941 to 1944." *History Workshop Journal* 46 (Fall 1998) 187-203. Appeared in German in "Zeitschrift fuer Geschichtswissenschaft" 45, 12 (1997).

1098 Jasper, Willi; Vogt, Bernhard: Integration und Selbstbehauptung: Russische Juden in Deutschland. *Tribuene* 149 (1999) 157-172.

Describes problems of integration of Jewish immigrants from the former Soviet Union in German society and in the Jewish communities. Although the number of these immigrants is negligible in relation to the German population, Germans feel they are being overrun. Center parties attempt (vainly) to attract voters by adopting the anti-immigrant stance of the right-wing parties. The media present purportedly sympathetic but actually stereotypical and folkloristic images of Jews, or seize on conflicts and scandals in the Jewish community. Immigrant absorption and employment agencies discriminate between ethnic German and Jewish immigrants. *RW*

1099 Kautz, Fred: *Gold-Hagen und die "Huernen Sewfriedte": Die Holocaust-Forschung im Sperrfeuer der Flakhelfer*. Berlin: Argument, 1998. 133 pp.

Contends that the condemnation of Goldhagen by leading German Holocaust historians as emotional and unscholarly stems from their own use of scholarship as a defense mechanism. Socialized in the Nazi period, they cannot accept German responsibility for the Holocaust, which Goldhagen bases on the voluntary participation of ordinary Germans in the killing of Jews. Perpetrators and victims disappear from their view; they concentrate on impersonal processes. They flee to scholarship in order not to face the concrete reality of the Holocaust, which can be adequately described only as Goldhagen describes it: with the help of literary methods, involving emotion and imagination. *RW*

1100 Petersdorff, Ulrich von: Das Berliner Holocaustdenkmal: Zur Namensnennung der NS-Opfer auf Denkmaelern. *Geschichte in Wissenschaft und Unterricht* 49, 2 (1998) 115-118.

Deals with the debate surrounding the construction of a Holocaust memorial monument in Berlin on which the names of all the victims are to be engraved. Discusses concepts of data protection rights, based on a report of a case in which a person was not willing to see the names of his murdered Jewish relatives on the monument, and the question of whether is is right to take a decision concerning millions of victims whose relatives cannot complain because they are no longer alive. Contends that there is a common interest that the victims not be forgotten, especially in the face of revisionist statements, and that the personal dignity of relatives living today is insignificant when compared with the general purpose. *ID*

1101 Rembiszewski, Sarah: *Die Endloesung des Holocaust: Die Leugnung des Holocaust in Deutschland: Fallstudie eines Leugners der Nachkriegsgeneration.* Trans.: Ela Moszkowicz. [Tel Aviv]: Tel Aviv University, Stephen Roth Institut fuer das Studium des zeitgenoessichen Rassismus und Antisemitismus, 1999. vi, 138 pp. Originally published as "The Final Lie: Holocaust Denial in Germany: A Second-Generation Denier as a Test Case" (Tel Aviv: Tel Aviv University, Project for the Study of Anti-Semitism, 1996).

1102 Rensmann, Lars: Holocaust Memory and Mass Media in Contemporary Germany: Reflections on the Goldhagen Debate. *Patterns of Prejudice* 33, 1 (1999) 59-76.

Goldhagen's "Hitler's Willing Executioners" provoked an unprecedented controversy in German society over the antisemitic origins of the Holocaust. Contends that, despite its shortcomings, this book has brought important new perspectives to Holocaust studies – e.g. in that it focused on the motives of the perpetrators, and on the average Nazi rather than the elites. The first media reaction, and even academic reactions, to the book in Germany was completely negative. In their portrayal of Goldhagen's alleged anti-German hostility, many participants in the debate exhibited covert anti-Jewish prejudices. Suggests a psychoanalytical explanation of such "secondary antisemitism": unmastered guilt feelings are projected against "imaginary accusers" who remind the public of the Nazi past. Goldhagen has destroyed all efforts of the Germans to recreate a positive national identity, to regain collective narcissism. *DR*

1103 Rubinstein, Alvin Z.; Rubinstein, Frankie: Coping with the Holocaust: Germans and Jews. *Midstream* 45, 3 (Apr 1999) 21-25.

Contends that memorialization of the Holocaust in present-day Germany has taken a wrong direction. It perpetuates a sense of national guilt and shame among Germans and dooms Germany to a continued pariah status. This type of memorialization may cause a backlash, a kind of "Holocaust fatigue," and resentment among Germans. Suggests that commemoration of the Holocaust in Germany must shift its emphasis from victimization of the Jews to celebration of Jewish contributions to German life and culture. Present-day Germany has become a thriving democratic country and has coped with the Nazi past. *DR*

1104 Schneider, Richard Chaim: *Fetisch Holocaust: Die Judenvernichtung – verdraengt und vermarktet*. Muenchen: Kindler, 1997. 287 pp.

Examines the emotional factors involved in the debate taking place in Germany on the annihilation of the Jews. Claims that although information on the mass murders was available to German society at the time, wide sections of the population were not able to cope with it emotionally and grasp the extent of the crime. Deals with the Goldhagen debate; despite the massive criticism of his book (condemning it as unscientific, superficial, inexact), it is precisely this emotional approach to the Holocaust which enables it to reach the German postwar generation. Goldhagen succeeded in clearing away the haze of the last 50 years, enabling people to feel and enter into the event. *ID*

1105 Zacher, Hans-Juergen: Die Reichspogromnacht: Erfahrungen mit einem Unterrichtsprojekt. *Freiburger Rundbrief* 6, 2 (1999) 120-124.

In connection with commemoration of the "Kristallnacht" pogrom, a project was initiated in a German school in November 1997, in which the pupils were asked to speak to their grandparents about their recollections of the events on 9 November 1938. Relates some of the pupils' experiences, discussing especially cases in which the grandparents would not answer the children's questions. States that young people may react with fears, and with feelings of profound insecurity or even crisis. Therefore, pupils should be prepared in advance for possible silence on the topic. They should be given some basic knowledge of Jewish history, antisemitism, and racism. The project is also intended to warn the younger generation about the dangers of right-wing radical tendencies. *ID*

Great Britain

1106 Bray, Martin: Antisémitisme en Angleterre: Formes anciennes et nouvelles. *Tsafon* 35-36 (Fall-Win 1998-1999) 119-125.

Describes some manifestations of antisemitism in present-day England, as noted in the Jewish press and in books, based on personal experiences. States that old forms of antisemitic prejudice still exist; Jewish students hear pejorative remarks concerning their origin. Concludes that there is a more dangerous new form of antisemitism on university campuses. There, Muslim (most of them Pakistani) students manifest hostility toward Jewish students and toward the State of Israel. *HV*

1107 Cooke, Steven: Beth Shalom: Re-thinking History and Memory. *Journal of Holocaust Education* 8, 1 (Sum 1999) 21-41.

Discusses Britain's first Holocaust memorial center, opened in 1995, focusing on two features. Firstly, the Beth Shalom Holocaust education center was established by a non-Jew, Stephen Smith, and basically funded by non-Jews. Against the background of a lack of interest in the Holocaust by Britons, including British Jews, Beth Shalom conveys the message that all should be interested and that Christians need to atone for Christianity's role in the Holocaust. The second issue considered is the out-of-the-way location of the center in Laxton, Nottinghamshire, and its symbolism. For England, the site (representing the country's unfenced-in, natural past) is marginal, just as the Holocaust is to today's English/Anglo-Jewish consciousness. Beth Shalom may serve as a site of pilgrimage, where those who visit are transformed. *YC*

1108 Dale, Ruth: An Evaluation of CD-ROM Resources as a Tool for Teaching the Holocaust. *Journal of Holocaust Education* 8, 1 (Sum 1999) 58-75.

Discusses how CD-ROM material may be used to teach the Holocaust to 15-16-year-olds in Britain. Stresses the need for a balance between the extreme of too many facts and statistics (that leads to desensitization) and too graphic an approach (that leads to traumatization). Recommends that teachers make use of CD programs by assigning focused tasks that require personal reflection and involvement; provides sample questions as models. Evaluates three CD-ROMs as Holocaust resources. Recommends the encyclopedia "Encarta '97" as a research tool, but notes its lack of emotional impact. "World War II: Sources and Analysis" is not recommended; it is misleading in a number of areas and adds nothing to previous sources. "Lest We Forget," which is wholly devoted to the Holocaust, is highly recommended. It combines statistics with faces of victims, and discusses topics ignored elsewhere, e.g. responses by Jews and non-Jews to the persecution of Jews, and Holocaust denial. *YC*

1109 Robinson, Alan: War Crimes, Old Soldiers and Fading Memories: The Serafinowicz Case. *Journal of Holocaust Education* 8, 1 (Sum 1999) 42-57.

Discusses the first trial (1997) in Britain under the 1991 War Crimes Act, which ended with the defendant Szymon Serafinowicz (b. 1910) being found senile and, hence, unfit for trial. Serafinowicz was accused, inter alia, of involvement, as a Belorussian militia officer, in German mass murders of Jews in Belarus (Mir, Turets, and Dolmatowszczyna) in 1941-42. Argues that although there is little doubt of the defendant's guilt, he would probably not have been convicted due to the difficulty of meeting standards of proof of the commission of war crimes. Comparison is made with similar outcomes in two Canadian war crimes trials, those of Imre Finta (from Hungary) in 1989-92 and of Pawlowski (accused of crimes in Belarus) in 1993. *YC*

Hungary

1110 Braham, Randolph L.: Offensive contre l'histoire: Les nationalistes hongrois et la Shoah. *Les Temps Modernes* 606 (Nov-Dec 1999) 123-141.

After the collapse of communist rule in Hungary in 1989, the Holocaust – now a legitimate historical topic in the country – became an embarrassing issue for the new governments. While the number of outright Holocaust deniers and their influence are negligible in Hungary, the number of "history cleansers" who try to distort and denigrate the catastrophe of Hungarian Jewry is fairly large. Their methods include transferring responsibility for the Holocaust to the Germans or, at least, to the Arrow Cross organization and regime; blaming the Jews themselves for their tragedy; relativizing the Holocaust, e.g. equating it with Soviet and communist atrocities; and deflecting public attention from the Hungarian perpetrators of the Holocaust to the tiny group of rescuers of Jews (in whose number some historians even include Horthy). The leaders of Hungary have so far failed to acknowledge publicly their country's responsibility for the Holocaust. *DR*

1111 Enyedi, Zsolt: Erklaerungsmodelle des Antisemitismus: Zwei ungarische Meinungsumfragen. *Jahrbuch fuer Antisemitismusforschung* 8 (1999) 228-247.

Analyzes the results of two polls: one conducted in 1994 of a representative sample of 1,000 out of the entire Hungarian adult population, and one in 1998 of 400 fathers (occasionally mothers) of teenagers in two cities. Examining the correlation of antisemitism with the chief factors usually tested by researchers, finds that, of socioeconomic factors, the most influential are social status and schooling; of the psychological factors, authoritarianism and ethnocentrism; of political factors, mainly an orientation toward the right. On the whole, the socioeconomic factors are relatively weak; the strongest factors are the psychological ones, especially ethnocentrism. *RW*

1112 Kovács, András: *Antisemitic Prejudices in Contemporary Hungary.* Jerusalem: Hebrew University, Vidal Sassoon International Center for the Study of Antisemitism, 1999. 49 pp. (Analysis of Current Trends in Antisemitism, 16).

Analyzes the results of a survey conducted in Hungary in March 1995, based on questionnaires distributed to ca. 1,500 subjects aged 18+ years. 29% of the population are not antisemitic; 25% are antisemitic (8% of them extreme antisemites); 32% accept traditional antisemitic stereotypes, but are not antisemitic. Antisemitism in Hungary is currently a Budapest phenomenon; prejudice is scant in other localities. No significant correlation was found with age, education, and socioeconomic status. Feelings of frustration and anomie, in combination with religious-conservative ideology, lead to xenophobia. Compares this survey with one conducted in 1991 (by the American Jewish Committee), in 1995 and 1996 (by Vienna University), and others by Hungarian scholars. *DR*

1113 Kovács, András: Antisemitismus im heutigen Ungarn: Ein Forschungsbericht. *Jahrbuch fuer Antisemitismusforschung* 8 (1999) 195-227. Translated from the English: "Antisemitic Prejudices in Contemporary Hungary" (Jerusalem: Hebrew University, Vidal Sassoon International Center for the Study of Antisemitism, 1999).

1114 Postma, Koos: *Changing Prejudice in Hungary: A Study on the Collapse of State Socialism and Its Impact on Prejudice against Gypsies and Jews.* Amsterdam: Thesis Publishers, 1996. 200 pp. Based on the author's diss. - University of Groningen.

Based on sociological surveys conducted in Hungary in 1987, 1992, and 1993, tests a theory which attributes the growth of ethnic prejudices against Gypsies and Jews in the post-communist period to the impact of economic transformations in the 1990s which brought about a deterioration in the quality of life. Hypothesizes that an individual faced with uncompensated deterioration of social production functions will blame a minority group (scapegoat) as well as find an alternative source of social approval, which leads him to overestimate his ingroup (Hungarians) and underestimate outgroups (Gypsies, Jews). However, the survey data do not support this hypothesis. The level of the respondents' social losses does not correlate significantly with the extent of their prejudice against Jews. Concludes that national identity is not used in

Hungary as a vehicle to enhance social identity or produce alternative social approval. Also, groups other than Jews and Gypsies are victims of scapegoating. Therefore, the hypothesis must be rejected and other studies be taken up to account for the phenomenon of prejudice. *DR*

Netherlands

1115 Konig, Ruben Peter: *Christelijke religie en antisemitisme in Nederland 1990: Een sociaal-wetenschappelijk onderzoek* [Christianity and Antisemitism in the Netherlands 1990: A Social-Scientific Study]. Kampen: Kok, 1997. 169 pp. Based on the author's diss. - Katholieke Universiteit Nijmegen, 1997. With an English summary (pp. 132-136).

A study based on data from the national survey "Social and Cultural Developments in the Netherlands 1990," a project conducted by the universities of Nijmegen and Tilburg. During the winter of 1990-91, 1,185 respondents were asked questions about, among other topics, anti-Jewish prejudice and Christian religion. The final statistical analysis, however, involved 783 respondents. The results show that Christian religion is still a source of antisemitism in the Netherlands at the beginning of the 1990s. Christian belief also leads, indirectly, to secular antisemitism. Explains the relationship between Christian religion and antisemitism in terms of "breadth of perspective." People with a narrow perspective on social reality often have a narrow perspective on Jews as well, whether they are religious or secular. In the secularized Dutch society of the nineties, religious antisemitism appears to be less persistent than secular antisemitism. Christian religion does correlate with antisemitism, but this should be attributed mainly to non-religious factors. To sum up, people subscribe to traditional Christian tenets and to anti-Jewish prejudice because of a narrow perspective on social reality. *MF*

Poland

1116 Ambrosewicz-Jacobs, Jolanta; Orla-Bukowska, Annamaria: After the Fall: Attitudes towards Jews in Post-1989 Poland. *Nationalities Papers* 26, 2 (June 1998) 265-282.

Jews disappeared as a visible group in Poland after World War II. The fact that the country was hermetically closed under the communist regime resulted in the conservation of Jewish stereotypes. Jews have become a symbol, a political code in contemporary Poland. Having no contacts with real Jews, many Poles use the word "Jew" (or the Star of David) as a code signifying every possible evil. Incidents involving the use of anti-Jewish code language by prominent persons are rare and are met with a negative reaction. A survey conducted among 14-year-old students in six public schools in Kraków shows that young Poles have stereotyped ideas concerning Jews and are often indifferent to anti-Jewish behavior or graffiti. However, Poland is not seen as more antisemitic or xenophobic than other Western countries. The negative stereotypes can be broken; one of the ways to do this is through personal contact with Jews. *DR*

1117 Muszyński, Henryk: *Die polnisch-juedischen Beziehungen 30 Jahre nach "Nostra aetate."* Ed.: Volker Drehsen. Tuebingen: Mohr Siebeck, 1998. 73 pp.

A lecture delivered at the University of Tuebingen on the occasion of the author receiving the Lucas Prize, 1997. Affirms that until the Holocaust, Poland was a real home for the majority of the world's Jews, the cradle of a rich Jewish culture; there was antisemitism, but it cannot be compared to the racist antisemitism in Nazi Germany. Protests against allegations that the concentration camps were possible in Poland because of Polish antisemitism and collaboration; this accusation, along with perceptions of Jewish participation in the hated communist regime, made postwar reconciliation between Jews and Poles almost impossible. Part of Germany's guilt is the taint left on Poland because of the Holocaust; therefore, Germany must take part in the effort for Polish-Jewish reconciliation. Sees hopeful beginnings of this process. Voelker Drehsen's speech at the ceremony (pp. 53-73) touches on the problematic aspects of remembrance of the Holocaust in both Germany and Poland, and lauds Archbishop Muszyński for promoting Christian-Jewish dialogue and the study of Judaism in Poland. *RW*

1118 Tych, Feliks: Obraz zagłady Żydów w potocznej świadomości historycznej w Polsce [The Perception of the Holocaust of the Jews in Common Historical Consciousness in Poland]. *Biuletyn Żydowskiego Instytutu Historycznego* 189 (Mar 1999) 3-13.

Analyzes the reasons for the Poles' moral indifference and lack of reflection on the Holocaust. One of them is that postwar official propaganda praised the monoethnic Polish state without any reflection on why it became monoethnic after the war. History textbooks gave information about the extermination of Polish citizens, without mentioning the Jews as the main victims of the Holocaust. There has been an improvement in knowledge about the Jews in the post-communist period, but it is still not sufficient. Because a negative stereotype of the Jew is still present in Poland, especially among non-educated people, it is impossible to make a real moral evaluation of the Holocaust, as well as understand the real role of the Jews in the history of Poland. Mentions that school textbooks still do not show the difference between the fate of Poles and the fate of Jews under the Nazi occupation, but concludes that changing the consciousness through books and other media will surely bring results in the next decade. *HV*

1119 Tych, Feliks: Problem Holokaustu w polskich podręcznikach szkolnych do nauczania historii [The Problem of the Holocaust in Polish Textbooks for Teaching History]. *Biuletyn Żydowskiego Instytutu Historycznego* 188 (Dec 1998) 34-43.

Criticizes the way the Holocaust is presented in Polish textbooks for schoolchildren published in the last two years. States that the experts' report published by the Jewish Historical Institute in Warsaw was not taken into consideration and many of the same textbooks were approved last year by the Ministry of Education. Analyzes textbooks, stating that most of them fail to explain differences in the situation of the Poles and the Jews under Nazi occupation. Usually, the Holocaust is represented as part of the Nazi policy in Poland and the writers fail to show its universal aspects. In some textbooks the situation of Poles during World War II is described without mentioning the fate of the Jews. Another distortion is in the attitude of Poles toward Jews; according to the textbooks, sympathy and help predominated, which is not true. Concludes that despite many faults and inaccuracies, the newer textbooks are closer to the truth than the previous ones. *HV*

Russia

1120 Butkevich, Nickolai, ed.: *Antisemitism, Xenophobia and Religious Persecution in Russia's Regions, 1998-1999*. Washington, DC: Union of Councils for Soviet Jews, 1999. xi, 204 pp.

The first in a series of annual reports on the subject. Antisemitism was on the rise in Russia in 1998-99. Presents results of a year-long monitoring of antisemitism in 62 of Russia's 89 administrative regions. Reporting on each region, relates antisemitic incidents and the activities of the Nazi-style Russkoye Natsionalnoye Yedinstvo (Russian National Unity) group. Pp. 13-31 contain reports on the RNU, on antisemitism in the Communist Party of the Russian Federation, on the role of the Russian Orthodox Church in promoting antisemitism, and on other extremist groups. Pp. 189-198 contain an appendix: "Chronicle of Antisemitism in Russia: May 1998-August 1999." *DR*

1121

האם קיימת ברוסיה "השאלה היהודית"? יהודי ברית-המועצות במעבר 3 [18] (1997) 71-90.

[Is There a "Jewish Question" in Russia? *Yehudei Brit ha-Mo`atzot be-Ma`avar* 3 [18] (1997) 71-90.] First published in Russian in "Rossiiskaya Federatsiya" 12 (1995).

Presents remarks made at a discussion group which meets regularly in Moscow. The ten participants are professors, writers, philosophers, etc.: V. I. Tolstykh, D.E. Furman, A.S. Salutskii, B.F. Slavin, G. Pomerants, A. Gerber, A.V. Buzgalin, G.D. Gachev, F.B. Beleliubskii, and A. Guseinov. Some of the participants condemn antisemitism and claim that it is decreasing in Russia, but some point out that there is a reason for its existence. The latter argue that the Jews have fluctuated between support for communism and capitalism, since they have no connection to the land. Many Jews were active in the 1917 revolution and were then blamed for the problems of communism. In the 1990s many Jews were influential in the democratic government and are now being blamed for the problems of capitalism. *RE*

1122 Klier, John Doyle: The Dog That Didn't Bark: Anti-Semitism in Post-Communist Russia. *Russian Nationalism, Past and Present,* eds. Geoffrey Hosking, Robert Service. Basingstoke, Hants: Macmillan, 1998. Pp. 129-147.

Specialists on the USSR and post-communist Russia have been awaiting an outburst of antisemitism in Russia, but it is not happening. Analyzing the evolution of Russian-Soviet official policy toward the Jews, as well as that of Russian antisemitism, offers an explanation for this lack of antisemitism in contemporary Russia. Contends that there was never a discrete entity which may be characterized as "Russian antisemitism." Popular and pre-revolutionary official Judeophobia was built on the perception that the Jews were distinctively alien, not on racism. More sophisticated anti-Jewish ideologies did not have wide support. At present, neither antisemitic doctrines of the extreme right, nor those of conservative communists, of neo-Eurasianism, or of the Russian Church hold sway over Russian society as a whole. Meanwhile, the Jews have progressively lost distinctive traits which designated them as "aliens" and in many ways have become part of the Russian people. *DR*

1123 Krichevskii, Lev Yu.: *Russian Jewish Elites and Anti-Semitism.* New York: American Jewish Committee, 1999. vi, 51 pp. (International Perspectives, 43).

Focuses on antisemitism from 1989 to 1999 in Russia in terms of reactions to the success of a number of Jews in business and government. This selective antisemitism has been exploited politically not only by marginal groups but also by the renascent Communist Party, whose Duma representative, Albert Makashov, the Russian parliament refused to censure for his antisemitic remarks. Makashov suggested that Jewish participation in government be limited. The prominence of Jews in Yeltsin's government and of successful Jewish tycoons, like Boris Berezovsky, has not only stirred up antisemitism but also created ambivalence among Russian Jews. Some of the latter retain stereotypes (e.g. economic fields are appropriate for Jews), but believe that aspiring to top political or military posts could lead to an antisemitic backlash. Among those in Russian society that are prone to antisemitism are members of the Soviet-era elite who have lost their influence and people who envy Jews who have succeeded in a more open economy. Concludes that antisemitism is a serious test case for Russian democracy and for a society with a long tradition of antisemitism. *YC*

1124 Lyosov, Sergei V.: "Yevreiskii vopros" v russkoi intellektualnoi zhizni, 1985-1995 [The "Jewish Question" in Russian Intellectual Life, 1985-1995]. *Znamya* 9 (1996) 182-187.

Before the collapse of communism in the USSR in 1989-91 the "Jewish question" was a focus of official Soviet ideology; antisemitism was popular, including among oppositional intellectuals. Unexpectedly, after aborted attempts to revive old antisemitic motifs (Jews as the instigators of the Russian revolution, etc.) in the transitionary period of 1988-91, the intensity of antisemitism fell drastically, and the Jews became an ethnic minority like all others, with no distinctions. The Jew was no longer the archenemy of the Russian radical right. However, many anti-Jewish stereotypes persisted, as did ideological antisemitism. The latter exists as a way to express nostalgia for the "good old days" (as does communist propaganda) and due to the abolition of the taboo on hate speech and on expressions of ethnic phobias. *DR*

1125 Mathyl, Markus: Staatlicher Antisemitismus in Russland – Renaissance oder Aufloesung? *Jahrbuch fuer Antisemitismusforschung* 8 (1999) 60-81.

Describes the growth in the Russian Federation of an anti-Yeltsin coalition of radically nationalist and antisemitic political parties and organizations, uniting neo-communists with neo-fascists and racists. Citing the "Protocols of the Elders of Zion," this coalition propagates theories of a "Zionist world conspiracy" to depose the Russian elite (under pretense of equality and democracy), and thus lead to the ruin of the Russian people. They claim that Jews are behind the October Revolution, "perestroika," and Yeltsin's attack on the Parliament in 1993. The coalition includes leading members of the Communist Party in the Duma and many regional governors. Russian National Unity, a neo-Nazi organization, one of many, which openly calls for death to the Jews, is allowed to run paramilitary training camps and to patrol the streets in many localities. They are protected by nationalists in the Duma. Points to the danger that with the population's waning faith in the state, this coalition might come to power. *RW*

1126 Mondry, Henrietta: A Note on the Invocation of Russian Classics in the Present-Day Nationalist and Philosemitic Russian Press. *East European Jewish Affairs* 29, 1-2 (Sum-Win 1999) 129-139.

Recalls that Russian nationalism, which has reemerged in post-Soviet Russia, has always made use of antisemitism. Since the Russian literary press will probably continue to feature nationalistic debate in mass circulation publications, considerable use will probably be made of antisemitic statements by classical Russian authors (e.g. charges of ritual murder made in 1913 by Vasilii Rozanov) or of myths, e.g. that the poet Sergei Yesenin was murdered by Jews. Russian-Jewish journals, like "Russkii Yevrei," present articles about Jews and Russian literature that attempt to refute antisemitic views, sometimes overreacting in an overly philosemitic vein. This apologetic approach even exonerated of antisemitism both Dostoevsky and the philosopher Aleksei Losev (1893-1988). Nationalist publications accuse Jewish literary critics, both past and present, of lacking true understanding of the Russian soul and of being parasites of Russian culture who exploit it for their own benefit. *YC*

Ukraine

1127 Dymerskaya-Tsigelman, Liudmila; Finberg, Leonid: *Antisemitism of the Ukrainian Radical Nationalists: Ideology and Policy*. Trans.: Yisrael Cohen. Jerusalem: Vidal Sassoon International Center for the Study of Antisemitism, Hebrew University, 1999. 32 pp. Translated from the Russian manuscript.

In the first years of Ukrainian independence (from 1991), antisemitism, as part of the Soviet legacy, was rejected by the democratic- and Western-oriented leaders of the country. The economic crisis and growing disappointment with democratic rule and its principles led to a revival of radical nationalist parties in Ukraine, e.g. the Ukrainian National Assembly (UNA), led by Korchinskii, and State Independence of Ukraine, led by Roman Koval. Their antisemitic ideas are influenced by the writings of Dontsov, a Ukrainian extreme-right nationalist émigré and by Nazi writings, but most of all by contemporary Russian antisemitic publications. Nationalist theoreticians succeeded in transplanting the Russian-Aryan myth (with the image of the Jew as antagonistic to the Aryan) and neopaganism onto Ukrainian soil. However, antisemitism in Ukraine is marginal, and is perceived as an imported phenomenon. "Asemitism," i.e. indifference toward the Jews and their problems, is more prevalent. *DR*

1128 Paniotto, Volodymyr: The Level of Anti-Semitism in Ukraine. *International Journal of Sociology* 29, 3 (Fall 1999) 66-75.

Examines the data from a KIIS-Omnibus sociological poll conducted in Ukraine in 1994-98 by the Kiev International Institute of Sociology with regard to determining the attitude of Ukraine's population toward given ethnic groups. The survey comprised ca. 9,000 respondents throughout Ukraine. The results show that the index of social distance from Jews is intermediate, similar to the distance from Americans or Canadians. Concludes that there has been a general rise in the level of xenophobia in Ukraine between 1994-98, with relative stability in attitudes toward Jews. Antisemitism drops as the educational level rises, and ethnic Ukrainians evince, on average, much more antisemitism than ethnic Russians. Cautions that these data should be used as relative data; more conclusive results require a deeper study than the one carried out here. *DR*

1991-1999: Far East

Malaysia

1129 Kessler, Clive S.: A Malay Diaspora? Another Side of Dr Mahathir's Jewish Problem. *Patterns of Prejudice* 33, 1 (1999) 23-42.

Over the last decade, there has been much academic and journalistic probing of the Malaysian prime minister's expressed negative attitudes toward Jews. He stereotyped Jews negatively as early as 1970 in his book "The Malay Dilemma." Examination of Mahathir's views may show, however, that his antisemitism is motivated mainly by anti-Zionism and solidarity with Muslim Palestinians (as well as with other oppressed Muslim minorities). He holds many ethnic prejudices, which, however, do not influence his policies, thus he cannot be regarded as a classical political antisemite. Moreover, a growing interest in Jewish issues among the Malaysian political elite may be attributed to the desire to discover, or invent a "Malay diaspora" and thus to a kind of "diaspora-envy." *DR*

1991-1999: Islamic World

Palestinian National Authority

1130 Israeli, Raphael: Education, Identity, State Building and the Peace Process: Educating Palestinian Children in the Post-Oslo Era. *Ariel Center for Policy Research: Policy Papers* 58 (1999) 7-31. Appeared also in "Terrorism and Political Violence" 12, 1 (Spr 2000) 79-94.

Despite the fact that in the Oslo I and II Accords (1993 and 1995) the Palestinian Authority (PA) agreed to remove hate texts from Palestinian textbooks, defamation of Jews, Israel, and Zionism is prevalent in contemporary (1995-98) textbooks for which the PA is responsible. There is widespread denigration of Judaism, dehumanization of Jews, delegitimization of Israel, and denial of any connection of Jews with the Land of Israel. Jews/Israelis are blamed for all Arab miseries and for universal evils. These attitudes, plus calls in textbooks for "jihad" and martyrdom, undermine hope for a peaceful settlement between Israelis and Palestinians. Part of the antisemitic rhetoric reflects an Islamic fundamentalist approach to Jews and the Land/State of Israel. *YC*

1131 Sharan, Shlomo: Israel and the Jews in the Schoolbooks of the Palestinian Authority. *Ariel Center for Policy Research: Policy Papers* 58 (1999) 33-61.

Based on a study of 140 textbooks authorized by the Palestinian Authority, concludes that they contain a great deal of antisemitic propaganda that presents Jews/Israelis as enemies of Palestinians, Arabs, Islam, and all humanity. Points out the danger of the political socialization of children by inculcating such attitudes in a totalitarian society with values taken from an Islamic religious civilization. Palestinian television features similar antisemitic stereotypes of Jews, Judaism, Israel, and Zionism. Public ceremonies, as well as textbooks, reflect a historical revisionism that denies a Jewish connection with the Land of Israel. All of these media encourage a "children's crusade" against Jews and Israelis by stressing the obligations of "jihad" and martyrdom. Such antisemitic attitudes will require profound change on all levels of Palestinian education if there is to be a basis for Israeli-Palestinian coexistence. *YC*

Turkey

1132

שמואלביץ, אריה: עמדת העיתונות האסלאמית בתורכיה כלפי ישראל. המזרח החדש 39 (1997-1998) 114-124.

[Shmuelevitz, Aryeh: The Attitude of the Islamic Press in Turkey toward Israel. *Hamizrah Hehadash* 39 (1997-1998) 114-124.]

From 1961, when a law upholding freedom of speech was passed by the Turkish government, there have been radical movements in Turkey advocating a return to Islamic fundamentalism and there have been expressions in their periodicals of antisemitism and anti-Zionism. The "Jewish world conspiracy" is mentioned often and the Zionists are blamed for every international occurrence, even for planning the Holocaust in order to increase immigration to Israel. In 1993 the fundamentalist press reacted to Israel's giving refuge to Bosnian Muslim children by insisting that the Jews wanted to use their body parts and their blood. Many articles claim that the Jews have harbored a hatred for Islam from its very beginning, and that Israel is conspiring to rule the entire Middle East. *RE*

1991-1999: Israel

1133

גולני, מוטי: לחנך על זה או להסתייג מזה? על הקשר שבין הקמת המדינה לבין השואה. בשביל הזיכרון 21 (מרץ-אפר' 1998) 19-25.

[Golani, Motti: To Teach about It or to Object to It? On the Connection between the Revival of Israel and the Holocaust. *Bishvil ha-Zikaron* 27 (Mar-Apr 1998) 19-25.]

Rejects the simplistic and historically untrue link between the Holocaust and the revival of Israel which is prevalent in the teaching of the Holocaust in Israel. Contends that the impact of the Holocaust on the international political discourse which led to the foundation of the State of Israel was marginal. Israel was founded despite the Holocaust, not because of it. The integration of the Holocaust into the Israeli collective identity should not focus on identification with victimizing experiences. Suggests to discontinue at once the "March of the Living" (at Auschwitz) and the trips of pupils to Poland and to reconsider the ways of integrating the Holocaust into the Israeli national identity. *LFo*

1134 Marcou, Léa: Cinquante ans après, des enfants cachés se racontent. *Bulletin du Centre de Recherche Français de Jérusalem* 2 (Spr 1998) 77-81. In French and English.

Describes the aims of "Aloumim," the Israeli Association of Children Hidden in France during the Shoah. The Association was founded in 1993 and has ca. 500 members. It organizes meetings for the members to talk about their traumatic childhoods, and it collects their testimonies. Other goals include obtaining the award of "Righteous of the Nations" for non-Jews who helped Jews to survive, and fighting Holocaust denial through publications about the survivors' wartime experiences. The Association publishes a French-Hebrew bulletin, "Mémoire vive." *HV*

1135 Marcou, Léa: Fifty Years After, Hidden Children Speak. *Bulletin du Centre de Recherche Français de Jérusalem* 2 (Spr 1998) 149-153. In French and English. See the abstract for the French version.

1136

מישקינסקי, משה: בשולי מעגל ההיסטוריוגרפיה היהודית והישראלית-יהודית. הציונות 21 (1998) 9-20.

[Mishkinsky, Moshe: In the Margins of the Cycle of Jewish and Israeli-Jewish Historiography. *Ha-Tzionut* 21 (1998) 9-20.]

Based on a lecture prepared for the 32nd Internationale Tagung der Historiker der Arbeiterbewegung, convened in Linz in July-August 1996. Raises various issues in connection with recent Jewish and Israeli historiography of the Holocaust, e.g. responses to the Holocaust in Israel and the role of Yad Vashem in Holocaust memorialization. Also discusses some works on the Holocaust, especially Goldhagen's "Hitler's Willing Executioners." *LFo*

1991-1999: North America

USA

1137 Abrams, Elliott: *Faith or Fear: How Jews Can Survive in a Christian America.* New York: Free Press, 1997. xii, 237 pp.

Discusses the crisis engendered in the American Jewish community by a high intermarriage rate and low fertility rate amongst Jews. Calls for a decision to place Judaism (i.e. religion) back in the center of Jewish life. Chs. 3-4 (pp. 37-98) deal with the attitudes of Christians toward Jews in the post-World War II period, including antisemitism and the development of Christian-Jewish dialogue. Notes that the Catholic and mainline Protestant Churches abandoned the "theology of contempt" toward Jews. Regrettably, within the American Jewish community there is considerable suspicion, even hostility toward Christianity. Posits that mutual respect and approbation for each other's religion would give Jews more of a reason and desire to preserve Judaism. Pp. 152-161 reflect on the Holocaust and antisemitism as factors cementing Jewish identity, but these will not hold the Jewish community together – only Judaism can do that, meaning religious faith. *DR*

1138 Azoulay, Katya Gibel: *Black, Jewish, and Interracial: It's Not the Color of Your Skin, but the Race of Your Kin, & Other Myths of Identity.*

Durham, NC: Duke University Press, 1997. xii, 219 pp. Based on the author's diss. – Duke University, 1995.

In her exploration of the merging of Black and Jewish identities, the author (whose mother was Jewish and father Jamaican) challenges common assumptions about identity and racism. She uses historical, theoretical, and personal perspectives (her own, plus those of other Black-Jewish individuals she interviewed) to discuss different ways interracial identities are experienced and understood in America today. Includes implications regarding prejudices against Jews, Blacks, and others in contemporary America. Concludes that being biracial undermines "the authority of categorization" in a race-conscious society. *YC*

1139 Bischoping, Katherine: Responses to Holocaust Denial: A Case Study at the University of Michigan. *Contemporary Jewry* 18 (1997) 44-59.

A study based on 40 in-depth interviews conducted in 1992 with undergraduate students at the University of Michigan College of Literature, Sciences, and Arts. The survey aimed to assess the level of success of the Holocaust denial message among students; to determine where the students locate Holocaust denial in the debates on civil liberties, academic freedom, and protection of minorities against hate speech; and to identify resources that enable students to reject the arguments of Holocaust deniers. The results show that none of the respondents accepted the message that the Holocaust never happened; however, the basis for rejecting Holocaust denial was not historical knowledge, but stemmed from the profound mythologization of the Holocaust in the collective consciousness. The majority believed that Holocaust deniers should be allowed to speak freely on campus. *DR*

1140 Bischoping, Katherine; Kalmin, Andrea: Public Opinion about Comparisons to the Holocaust. *Public Opinion Quarterly* 63, 4 (1999) 485-507.

In 1992 a survey was conducted in the USA, in which 491 respondents answered at least one question concerning the Holocaust. The survey indicates that only a minority of the population believes the Holocaust to be unique. The greater part of the comparisons concur with the United Nations' definition of genocide; a minority compares the Holocaust to war or to prejudice and its resultant conflicts. Examines three hypotheses on the motivation for Holocaust comparisons: influences of education and gender on ways of knowing, effects of race and political orientation on ethnoracial comparisons, and generational differences in collective memory. Comparisons to the Holocaust are best predicted by education, gender, race, and current events. *DR*

1141 Brabham, Edna Greene: Holocaust Education: Legislation, Practices, and Literature for Middle-School Students. *The Social Studies* 88, 3 (May-June 1997) 139-142.

Half a century after the end of World War II, seven states in the USA have passed laws supporting Holocaust education and ten states have developed curricula to be used for Holocaust studies. Examines Holocaust education legislation, points out shortcomings in Holocaust education in various states, and surveys newly-published books, both fiction and non-fiction, which can be used in the learning process. *DR*

1142 Dershowitz, Alan M.: *The Vanishing American Jew: In Search of Jewish Identity for the Next Century.* Boston: Little, Brown, 1997. x, 395 pp.

With the marginalization of antisemitism in the USA, the most powerful factor which unified the Jews and buttressed Judaism seems to have disappeared. Pt. 2 (pp. 69-165), "Why So Many Jews Are Drifting Away," shows that institutional antisemitism (of governmental institutions, churches, corporations, etc.) has come to an end. There are, however, some forms of antisemitism that may persist in the 21st century, directed at the mainstream position of the Jews or at the establishment Jews. These forms include antisemitism of the radical right, Holocaust denial, Black antisemitism, and Islamic fundamentalism. Suggests some tactics to counteract Holocaust denial and Black antisemitism. Believes that a special danger is posed by a covert antisemitic source, the Christian right, because it tries to attract the state to its cause. Contends that Judaism does not need "an antisemitic tsar" in order to survive, and proposes solutions for withstanding assimilation. *DR*

1143 Flanzbaum, Hilene, ed.: *The Americanization of the Holocaust.* Baltimore: Johns Hopkins University Press, 1999. 261 pp.

This book grew out of a panel of the same name delivered at the 1996 Chicago Modern Language Association Convention. Partial contents: Flanzbaum, Hilene: Introduction: The Americanization of the Holocaust [Appeared in "Journal of Genocide Research" 1 (1999).] (1-17); Flanzbaum, Hilene: The Imaginary Jew and the American Poet [Appeared in "ELH - English Literary History" 65 (1998).] (18-32); Shandler, Jeffrey Alan: Aliens in the Wasteland: American Encounters with the Holocaust on 1960s Science Fiction Television (33-44); Greenspan, Henry (Hank): Imagining Survivors: Testimony and the Rise of Holocaust Consciousness (45-67); Young, James Edward: America's Holocaust: Memory and the Politics of Identity [Appeared in "Survey of Jewish Affairs" (1991) and in "The Texture of Memory" (1993).] (68-82); Furman, Andrew: Inheriting the Holocaust: Jewish American Fiction and the Double Bind of the Second-Generation Survivor [Focuses on Thane

Rosenbaum's collection of stories "Elijah Visible" (1996).] (83-101); Hungerford, Amy: Surviving Rego Park: Holocaust Theory from Art Spiegelman to Berel Lang (102-124); Antler, Joyce: "Three Thousand Miles Away": The Holocaust in Recent Works for the American Theater [Appeared in "American Theatre" 12 (1995). With particular focus on Arthur Miller's "Broken Glass" and Cynthia Ozick's "Blue Light."] (125-141); Horowitz, Sara Reva: The Cinematic Triangulation of Jewish American Identity: Israel, America, and the Holocaust (142-166); Steinweis, Alan E.: Reflections on the Holocaust from Nebraska [On Holocaust consciousness and commemoration in Nebraska.] (167-180); Michaels, Walter Benn: "You Who Never Was There": Slavery and the New Historicism – Deconstruction and the Holocaust [Appeared in "Narrative" 4 (1996).] (181-197); Thomas, Laurence Mordekhai: Suffering as a Moral Beacon: Blacks and Jews [Criticizing the rivalry between Blacks and Jews over who has suffered more, contends that both groups became pawns of American society, played against each other, but both can become moral beacons if they stop arguing.] (198-210). *YC*

1144 Patraka, Vivian M.: Spectacles of Suffering: Performing Presence, Absence, and Historical Memory at U.S. Holocaust Museums. *Performance and Cultural Politics,* ed. Elin Diamond. London: Routledge, 1996. Pp. 89-107. Unseen.

1145 Pipes, Daniel: America's Muslims against America's Jews. *Commentary* 107, 5 (May 1999) 31-36.

In recent years, fundamentalist Islam has become the most virulently antisemitic movement in the U.S. The Islamists control the American Muslim community of six million. Their antisemitic credo combines elements of Christian anti-Judaism, Nazi-style racism, ideas from "The Protocols of the Elders of Zion," and the theories of the Nation of Islam. Their political successes, as well as the number of victims of their violence, surpass by far those of the traditional extreme right. If the Nation of Islam has become the country's leading font of anti-Jewish rhetoric, mainstream Islamist organizations have a near monopoly on anti-Jewish violence. The fundamentalist Islamic movements present a much greater danger for Jews than the Christian far right. *DR*

1146 Rockaway, Robert Allen: The Nation of Islam and the Jews. *Jewish Journal of Sociology* 40, 1-2 [1998] (1999) 48-78.

Notes that, contrary to the image of benevolent cooperation, Black-Jewish relations in the U.S. have been strained for several generations. Traces the history of the Nation of Islam from its founding, ca. 1931, to its present version under the leadership of the charismatic Louis Farrakhan. The movement's antisemitism is attributed to Christian roots, general prejudice against Jews in the U.S., Jews being viewed as Whites, Black experience of Jewish landlords, etc. Jews are divided about the significance of Farrakhan and his movement: some do not consider them a major influence, others view them as real dangers. American Jewish organizations refuse to dialogue with Farrakhan until he apologizes for antisemitic statements, which he is not likely to do. Concludes that Farrakhan can be expected to find a receptive audience until the "terrible plight" of African Americans is considerably improved. *YC*

1147 Schoenfeld, Gabriel: Auschwitz and the Professors. *Commentary* 105, 6 (June 1998) 42-46.

Questions whether Holocaust studies is a legitimate academic discipline and whether it is good that Holocaust studies have proliferated in American universities in recent decades. The process of "academizing" the Holocaust has resulted, inevitably, in distancing, objectifying, and naturalizing its horrors. Some of the worst excesses and distortions of the Holocaust have emerged from "gender studies" which exploit the Holocaust to promote a feminist agenda which aims to sever Jewish women from their families and from the larger Jewish community. Feminist voices pursuing such goals are becoming increasingly prominent in Holocaust museums, resource centers, and university enclaves. *REK*

1148 Sharlet, Jeff: How the Holocaust Came to America: Peter Novick's Challenge to Received Wisdom. *The Jewish Quarterly* 174 (Sum 1999) 65-68.

Discusses Novick's view on the misuse of Holocaust memory, as expressed in his book "The Holocaust in American Life" (1999) and in an interview Sharlet conducted with him. Novick objects to the fixation on Holocaust memory in present-day American Jewish and non-Jewish discourse, and to the misuse of it. He takes issue with many commonly-held assumptions about the Holocaust and its consequences, some of which are presented here. Notes, also, reactions to Novick's views by other American Jewish scholars, some approving and some critical. *DR*

1149 Stern, Kenneth S.: *Hate and the Internet*. New York: American Jewish Committee, 1999. 53 pp.

An examination of American racist and antisemitic web sites. In the 1990s hate groups found a good outlet for their propaganda on the Internet. Hate web sites are especially dangerous because they are available to young people, they can be accessed by millions and their call for violence is not subject to punishment. Describes some antisemitic web sites (including Holocaust denial). Discusses methods for combating hate speech on the Internet and projects (e.g. the Nizkor project) set up for that purpose. *DR*

ANTISEMITISM IN LITERATURE AND IN THE ARTS

1150 Adamczyk-Garbowska, Monika: The Return of the Troublesome Bird: Jerzy Kosiński and Polish-Jewish Relations. *Polin* 12 (1999) 284-294.

Examines the reception of Jerzy Kosinski's novel "The Painted Bird" (1965) by Polish critics; it was received negatively both in communist Poland and by Polish émigrés. The second wave of criticism came after 1988, when the novel was translated into Polish, and especially after the publication of Joanna Siedlecka's "Czarny ptasior" (1994), in which the critic fiercely attacks both Kosinski and his novel. Notes that the attacks in the 1960s and in 1988-90s have much in common. Two recurring motifs are the "ingratitude of the Jew" to those who rescued him during the war and a conspiracy of Americans, Germans, and Jews against Poland and its people. Generally, the novel's reception in the 1990s has been more favorable; some critics admit that the novel may reflect reality, that its author had a traumatic Holocaust experience, and that the work is, after all, fiction. The polemics surrounding Kosinski's novel reflect the present state of Polish-Jewish relations and Polish political thought. *DR*

1151 Afoumado, Diane: Les dessins des concentrationnaires français: Témoins de la résistance spirituelle dans les camps nazis. *Revue d'Histoire de la Shoah* 162 (Jan-Apr 1998) 96-126.

Analyzes drawings made in the concentration camps by French prisoners, both Jews and non-Jews. Reflects on the inner need of some people to create art, even under the very difficult conditions of the camps. Describes how the prisoners got the materials (paper, pencils, etc.), and deals with themes in the drawings, which mainly depicted human beings, work, the presence of death. Concludes that doing any kind of art work in the camps was a way of surviving. Pp. 107-126 present drawings by artists from Buchenwald, Ravensbrueck, Dora, and Bergen-Belsen. *HV*

1152

ברטוב, עומר: קיטש וסדיזם בפלנטה של ק. צטניק: נוער ישראלי מדמיין את השואה. אלפיים 17 (1999) 148-175.

[Bartov, Omer: Kitsch and Sadism in Ka-Tzetnik's Planet: Israeli Youth Imagine the Holocaust. *Alpayim* 17 (1999) 148-175.] Appeared in English in "Jewish Social Studies" 3 (1997).

1153 Bayerdoerfer, Hans-Peter; Schoenert, Joerg, eds.: *Theater gegen das Vergessen: Buehnenarbeit und Drama bei George Tabori*. Tuebingen: Max Niemeyer, 1997. ix, 385 pp.

Partial contents: Bayerdoerfer, Hans-Peter: Stimmen in der Wueste: Juedischer Intellekt und dramatische Gestalt in Taboris Stuecken (3-44); Dahlke, Karin: 'Ueberrumpelte Katastrophen': Taboris 'Witz' im Schatten der Shoah (123-154); Pott, Sandra; Sander, Marcus: Abdankung des Dokumentarischen? Taboris "Die Kannibalen" als Gegenstueck zu "Die Ermittlung" von Peter Weiss (155-182); Sander, Marcus: Friedhofs-Monologe: George Taboris "Jubilaeum" (183-217); Id.: Der Tod der Juedischen Frau: Bertolt Brechts Einakter des Exils, 'nach Auschwitz' bearbeitet von George Tabori [On Tabori's use of Brecht's one-act play "Die juedische Frau" (1944) in his "Jubilaeum."] (218-241); Pott, Sandra: "Ecce Schlomo": "Mein Kampf" - Farce oder theologischer Schwank? (248-269); Stammen, Silvia: Geschichte der Zerstoerung - Zerstoerung der Geschichte: "Nathans Tod" im Textvergleich mit "Nathan der Weise" (283-318); Pott, Sandra; Schoenert, Joerg: Tabori unter den Deutschen: Stationen einer 'authentischen Existenz'? (346-377). *RW*

1154 Birati, Rachel: Looking at the Holocaust through Hebrew and other Jewish Literature. *Australian Journal of Jewish Studies* 11 (1997) 71-94.

Discusses works on the Holocaust, written by Jews, in Hebrew and other languages. Suggests that fiction may be more effective than historical documentation in describing the devastating truth. Discusses poetry by Sharon Kessler, Abba Kovner, and Dan Pagis, and novels or stories by Aharon Appelfeld, Elie Wiesel, and Cynthia Ozick. States that survivor-writers can transcend their Holocaust experiences, and those who did not experience the Holocaust can identify with it through their works. Discusses, also, works by Aharon Megged, Gershon Shaked, and Savyon Liebrecht, which relate to the Israeli response to the Holocaust. *REK*

1155 Boeschenstein, Bernhard; Weigel, Sigrid, eds.: *Ingeborg Bachmann und Paul Celan: Poetische Korrespondenzen*. Frankfurt a.M.: Suhrkamp, 1997. 269 pp.

Partial contents: Briegleb, Klaus: Ingeborg Bachmann, Paul Celan: Ihr (Nicht-)Ort in der Gruppe 47 (1952-1964/65): Eine Skizze [While Bachmann distanced herself from the group, Celan was ostracized because of the group's taboo (under the leadership of Hans Werner Richter) on dealing with the Nazi past, and its prejudice against Jews who would not let it rest.] (29-81); Erdle, Birgit R.: Bachmann und Celan treffen Nelly Sachs: Spuren des Ereignisses in den Texten (85-115); Gehle, Holger: Poetologien nach Auschwitz: Bachmanns und Celans Sprechen ueber Dichtung zwischen 1958 und 1961 (116-130);

Boeschenstein, Renate: Der Traum als Medium der Erkenntnis des Faschismus (131-148). *RW*

1156 Briegleb, Klaus: *Bei den Wassern Babels: Heinrich Heine – juedischer Schriftsteller in der Moderne*. Muenchen: Deutscher Taschenbuch Verlag, 1997. 439 pp.

Contends that Judaism is not merely one topic among others in Heine's works, but the main source of his writing. Discusses Heine's works, focusing on an as yet unappreciated text, "Shylock" (1838), an essay parodying Shakespeare's characterization, in which Heine expressed his ambivalent attitude toward Judaism, Christianity, conversion, and antisemitic stereotypes. Argues that in "Der Rabbi von Bacherach," and additional texts in which he wrote about pogroms, persecution of Jews, Crusades, Conversos, etc., Heine coped with his own identity problem. *ID*

1157 Burmeister, Heike A.: *Der 'Judenknabe': Studien und Texte zu einem mittelalterlichen Marienmirakel in deutscher Ueberlieferung*. Goeppingen: Kuemmerle Verlag, 1998. 378, [60] pp. Based on the author's diss. – Philipps-Universitaet Marburg, 1997.

Examines the German versions of the legend, already found in Greek and Latin patristic literature, of the little Jewish boy who goes to church with Christian friends and there takes the sacrament. For this, the Jews (in most versions, his own father) throw him into a fiery oven, but the Virgin Mary saves him through a miracle and he is not burned. Places the story in the context of medieval adoration of the Virgin and anti-Judaism. Discusses the texts in chronological order, from the 13th-century "Juedel" to 19th-century Catholic devotional literature, including variations in the portrayal of the Jews (mostly hostile). Includes the texts, with translations into modern German. The last 60 pages contain facsimiles. *RW*

1158 Chazan, Robert: The Hebrew First-Crusade Narratives and Their Intertextual Messages. *Ki Baruch Hu: Ancient Near Eastern, Biblical, and Judaic Studies in Honor of Baruch A. Levine,* eds. Robert Chazan, William W. Hallo, Lawrence H. Schiffman. Winona Lake, IN: Eisenbrauns, 1999. Pp. 467-481.

A literary analysis of three Hebrew narratives on the First Crusade (1096), showing how they used biblical allusions in two important ways. The Crusaders were basically identified with ancient enemies of the Jews who, despite their successes, were ultimately overcome. The suffering of the Jewish victims was explained either traditionally, as punishment for sin, or, innovatively, since the Jews of Worms and Mainz were viewed as undeserving of suffering, they were seen as atoning for the sins of their ancestors. Concludes that, in either case, Jewish theology is affirmed as countering Christian claims of the divine rejection of the Jews. *YC*

1159 Delany, Sheila: Chaucer's Prioress, the Jews, and the Muslims. *Medieval Encounters* 5, 2 (1999) 198-213.

States that in the "Prioress's Tale," Chaucer is representing real, not symbolic, Jews. Despite the expulsion of Jews from England, Chaucer had the opportunity to meet and know about Jews abroad. He often viewed Jews as allied to or even interchangeable with Muslims, who threatened Europe and controlled Asia, the setting for the tale. The Prioress told about Jews who were killed for allegedly murdering a Christian child. Since Islamic society treated Jews far better than Christians did, socially and religiously, the Prioress projected European attitudes onto "Asia." The Christian clergy (whom the Prioress represents) resented the Judeo-Islamic symbiosis. Some modern Chaucer scholars are shown to share antisemitic Christian attitudes. Two views are possible in regard to Chaucer himself. If one identifies him with the Prioress, he shares her antisemitism. However, if one detects irony in her representation, she is an ignorant bigot whose views should be rejected. Suggests that Chaucer may have had different attitudes toward Jews, Muslims, et al. at different times. *YC*

1160 Edelman, Charles: Which Is the Jew That Shakespeare Knew? Shylock on the Elizabethan Stage. *Shakespeare Survey* 52 (1999) 99-106.

Questions assumptions about the image of Shylock. Labels as myth the idea that there existed an Elizabethan stereotype of the villainous stage Jew or that everyone in Elizabethan England was antisemitic. Notes that none of the traditional antisemitic charges against Jews (Christ-killers, ritual murderers, host-desecrators, etc.) were referred to in "The Merchant of Venice." Further, the view that Shylock was condemned for being a usurer is open to doubt. Argues that the Elizabethan horror of usury is another myth. Shakespeare himself lent money for interest, which is mainly called "usance" in the play. Finally, rather than answering the question in his title, Edelman raises a number of points that undermine the widely accepted view that both Shakespeare and his contemporaries were without doubt antisemites. *YC*

1161 Flanzbaum, Hilene: The Imaginary Jew and the American Poet. *ELH* 65 (1998) 259-275.

Discusses the identity of American poet Karl Shapiro, an assimilated Jew, and his immediate postwar identification as an American soldier, in contrast to the metaphorical identification of non-Jewish poets like John Berryman (author of the story "The Imaginary Jew," 1945) with the Jew as outsider and victim. Randall Jarrell and Robert Lowell also identified with Jews in distancing themselves from middle-brow America. In his 1958 "Poems of a Jew," Shapiro criticizes himself for his past indifference to the Holocaust and accepts his Jewish identity. Shapiro's development symbolizes a process seen in the 1960s, when American Jews began to move beyond assimilation and identify with the suffering of their fellow Jews in the Holocaust. *YC*

1162 Fuchs, Anne; Krobb, Florian, eds.: *Ghetto Writing: Traditional and Eastern Jewry in German-Jewish Literature from Heine to Hilsenrath.* Columbia, SC: Camden House, 1999. 231 pp.

A collection of articles. The following deal with antisemitism: Bourke, Eoin: The Frankfurt Judengasse in Eyewitness Accounts from the Seventeenth to the Nineteenth Century (11-24); Krobb, Florian: Reclaiming the Location: Leopold Kompert's Ghetto Fiction in Post-Colonial Perspective [Views German cultural imperialism vis-à vis the Jews as proto-colonialism.] (41-53); Beug, Joachim: Pogroms in Literary Representation (83-96); Thunecke, Joerg: Philo-Semitic Tendencies in Wilhelm Jensen's Historical Novel "Die Juden von Coelln" [Written in 1869, set in the Middle Ages, at the time of the Black Death, Jensen presents the Jewish side of the case in as positive a light as possible.] (97-110); Horrocks, David: The Construction of Eastern Jewry in Joseph Roth's "Juden auf Wanderschaft" [Discusses Roth's creation of a positive image of the Ostjude to counter negative images.] (126-139); Kane, Michael: From Ghetto to Nation: Hofmannsthal's Poetics of Assimilation [Discusses the poet's assimilation as a reaction to antisemitic German culture.] (140-155); Pedersen, Ena: Persecution, Exile, and the Mental Ghetto in Henry William Katz's Novel "Die Fischmanns" (156-170); Fuchs, Anne: Edgar Hilsenrath's Poetics of Insignificance and the Tradition of Humour in German-Jewish Ghetto Writing [Relates to the black humor in Hilsenrath's novel "Jossel Wassermanns Heimkehr" (1993) on the Holocaust and the shtetls which were destroyed.] (180-194). *YC*

1163 Gazey, Katja: *"Which Is the Merchant Here? and Which the Jew?" Aufzeichnungen zum "Kaufmann von Venedig" in der Inszenierung der Bremer Shakespeare Company.* Bremen: Uni-Druck; Raphael-Lemkin-Institut fuer Xenophobie- und Genozidforschung, 1997. 30 pp.

Discusses the question of fresh interpretations of Shakespeare's texts, in particular in the production of "The Merchant of Venice" in 1995 in Bremen. Scholars have pointed to medieval passion plays as the inspiration for Shakespeare's confrontation between good Christian and evil Jew. In the trial scene in the Bremen production, Antonio enters the courtroom as a grotesque impersonation of Christ; in the middle of the scene he and Shylock exchange roles and costumes, and Shylock appears much more credible as victim and martyr than the fat Antonio. The scene mocks the Oberammergau passion play, which still portrays Judas as the traitor and the Jews as deicides; the whole production mocks the stereotype of the Jew still prevalent in the minds of many in the German audience. *RW*

1164 Gidion, Heidi: Haupt- und Nebensache in Arthur Schnitzlers Roman "Der Weg ins Freie." *Text + Kritik* 138-139 (1998) 47-60.

Discusses Schnitzler's novel "Der Weg ins Freie" (1908), claiming that, contrary to appearance, its main topics are the "Jewish question" and antisemitism, expressed in the dialogues between various characters in the novel. Traces the different types of Jews in fin-de-siècle Vienna and in the novel, e.g. the intellectual writer, conscious of the antisemitism around him, including his own self-hatred; the Zionist, who despises any compromise; the assimilated Jew, who makes efforts to be accepted by the aristocracy. Examines the attitudes of Jews toward other Jews and non-Jews, and those of non-Jews toward Jews. Notes the collective ideology of "difference" based on racial origin as depicted in this novel about the late 19th century as a dangerous potential of social catastrophe, which was then not yet foreseeable. *ID*

1165 Görög, Veronika: Stéréotypes ethniques et littérature orale: L'image du Juif à travers deux contes merveilleux hongrois. *Cahiers de Littérature Orale* 44 (1998) 173-190. An earlier version appeared in Hungarian in "Vérvádak üzenete" (1996).

Examines the image of the Jew in oral folklore through the interpretation of two Hungarian folktales (AT 561 and AT 592) found in several printed collections of the 19th-20th centuries. While both portray the Jew in the principal role, one describes the Jew as victim, the other as the villain. Argues that distrust of the Jew is manifest in the various versions of these tales. The antisemitic images presented in these tales are a reflection of the hostility of the Church toward Jews throughout history. *TS*

1166 Grunfeld, Uriel Jeremiah: *Representing the Holocaust on Film: "Schindler's List" and the Pedagogy of Popular Memory.* Diss. – Pennsylvania State University, 1997. 182 pp. Unseen.

Aims to identify conditions under which narrative constructions of popular memory either enable or disable the exercise of critical awareness and agency. This perspective raises the pedagogical challenge of learning how to distinguish between commemorative practices which function as mechanisms of denial and those which are conducive to the working through of collective trauma. Examines the narrative representation of the Holocaust in Steven Spielberg's film "Schindler's List."

1167 Gubser, Martin: *Literarischer Antisemitismus: Untersuchungen zu Gustav Freytag und anderen buergerlichen Schriftstellern des 19. Jahrhunderts.* Goettingen: Wallstein, 1998. 327 pp. Based on the author's diss. - Universitaet Fribourg (Schweiz), 1995.

Notes that opinions differ on whether Freytag was an antisemite. Proposes criteria for antisemitism in literature, giving examples from other German authors: portrayal of a fictional character solely by negative stereotypes; ridicule of Yiddish elements in his speech; stigmatization by linguistic devices; Manichaean plots – i.e. the villainous Jew is a foil for the ideal German hero; and authors' comments pointing out the ludicrousness or villainy of a Jew. Antisemitism is expressed on various grounds: religious (Judaism is immoral and unenlightened), economic (Jews live only to make money), political (they are cosmopolitan, not patriotic), and biological ("voelkisch" and racist). Jewish types include the East European "Schacherjude"; the partially assimilated, ridiculous parvenu; the coldblooded intellectual; and the Beautiful Jewess. In early articles in his journal "Grenzboten," Freytag used religious and political antisemitic arguments. The play "Die Journalisten" (1852) and his best-known novel, "Soll und Haben" (1855) both have Manichaean plots, ridiculous and scheming Jews and, in the novel, an arch-villain, Veitel Itzig. After 1869, Freytag published articles against antisemitism. *RW*

1168 Hollender, Elisabeth; Berzbach, Ulrich: Einige Anmerkungen zu biblischer Sprache und Motiven in Piyyutim aus der Kreuzzugszeit. *Frankfurter Judaistische Beitraege* 25 (Dec 1998) 63-74.

Based on a lecture delivered at the 12th World Congress for Jewish Studies, Jerusalem, July 1997. Shows that the authors of the piyyutim (liturgical poems) used phrases from the Bible to describe the Crusader pogroms: e.g. from Jeremiah and Lamentations on the destruction of the Temple as analogous to the destruction of the Rhineland communities. But many situations had no equivalents in the Bible. The many references to the "akedah" stress that Isaac was ready to be sacrificed by his father (a sacrifice that in the end did not take place); but the medieval martyrs actually did sacrifice themselves. They also sacrificed their children, like burnt offerings in the Temple. Many allusions are used with an opposite meaning: thus a reference to Esther 9:22, describing the month in which grief turned to joy, becomes a month in which joy turned into grief. *RW*

1169 Jackson, Deirdre Elizabeth: *Marian Antisemitism in Medieval Life and Legend: A Study Based on Alfonso X's "Cantigas de Santa Maria."* M.A. thesis - University of Victoria (Canada), 1997. 138 pp. Unseen.

Examines three miracle tales with anti-Jewish leitmotivs from Alfonso X's "Cantigas de Santa Maria" (Escorial, Biblioteca Real, MS T.I.1). In each one, a Jewish male is confronted by an image of the Virgin and Child, which provokes a strong reaction from him corresponding to his acceptance or rejection of the doctrine of the Incarnation. The Jewish characters who interact with the image of the Virgin in a positive way are integrated into the Christian community; Jews who respond negatively are punished. These narratives (and another involving a Moor, with which they are compared) not only demonstrate the power of the Virgin over non-Christians, but the power of the Marian image over Muslims and Jews, whose rejection of religious imagery was a source of concern for medieval Christians. Draws attention also to the role played by sacred images in medieval life.

1170 Judaken, Jonathan: The Mirror Image and the Politics of Writing: Reflections on 'the Jew' in Sartre's Early Thought. *Historical Reflections/Réflexions Historiques* 25, 1 (Spr 1999) 33-59.

An earlier and shorter version of this article appeared in the "Proceedings of the Annual Meeting of the Western Society for French History" 23 (1997). Discusses two works of fiction relating to the Jews in France written by Sartre in the 1930s and shows, contrary to the accepted view, that his work was already political before World War II. In both "La Nausée" ("Nausea") and "L'Enfance d'un chef" ("The Childhood of a Leader"), the instability of identity of the protagonists Roquentin and Fleurier are stabilized through their respective images of "the Jew," as is the path of their "engagement." Roquentin associates "the Jew" with the redemption, attainable via writing; Fleurier views "the Jew" as other or outsider and becomes an antisemite, first in worldview, then in action. These characters reflect "competing versions of how intellectuals and the [French] nation could respond to the crisis of the 1930s... for both Sartre and France, this response depended on differing representations of Jews and Judaism." *YC*

1171 Koester, Rudolf: Jakob Wassermann, Anti-Semitism and German Politics. *Orbis Litterarum* 53, 3 (1998) 179-190.

Antisemitism pursued and preoccupied Wassermann (1873-1934) from his childhood to the end of his life. Characterizations of antisemitism and the Jewish experience can be traced throughout his novels. However, it is in his essays – e.g. "Das Los der Juden" (1904), "Mein Weg als Deutscher und Jude" (1921) – that his concern with this problem receives its most poignant expression. In most of his essays, he derives his conclusions from an interfusion of personal experience, historical events, and contemporary politics. His outlook on the problem of antisemitism was pessimistic. Public ostracism in the Nazi period precipitated his death. *DR*

1172 Kolitz, Zvi: *Yosl Rakover Talks to God.* Trans.: Carol Brown Janeway. New York: Vintage International, 1999. 99 pp. On title page also: From the edition established by Paul Badde.

The story "Yossel Rakover Speaks to God" was first published in Yiddish in Buenos Aires in "Di Yiddishe Tsaytung" (September 1946). It appeared in German as "Jossel Rakovers Wendung zu Gott" (Moehlin: Rauhreif, 1994), edited and translated by Paul Badde. This English edition contains the text of the story (pp. 3-25) and essays about it. Contents: Badde, Paul: Zvi Kolitz (27-77); Levinas, Emmanuel: Loving the Torah More than God [First published in English in F.J. van Beeck's "Loving the Torah More than God?" (Chicago: Loyola University Press, 1989); in French: "Yossel Rakover s'adresse à Dieu" (Paris: Calmann-Lévy, 1963).] (79-87); Wieseltier, Leon: A Privation of Providence [On Levinas' commentary.] (89-99). *SSC*

1173 Kornblatt, Judith Deutsch: "Ladies' Tailor" and the End of Soviet Jewry. *Jewish Social Studies* 5, 3 (Spr-Sum 1999) 180-195.

Relates the plot of the Russian film "Ladies' Tailor" (1990), directed by Leonid Gorovets. It shows one day in a Jewish home in Kiev, the day before the massacre at Babii Yar. The film places equal blame for the horrors of this century on Stalin as on Hitler, and is as much on the fate of the Jews in the 20th century as on the fate of Soviet society. The Jews are presented as a humanizing factor, "a light unto the nations," vital for the moral resurrection of Soviet society which was deformed and dehumanized by Stalinism. With the end of Soviet Jewry, whether at Babii Yar or through assimilation or emigration, nobody would be left who, like the protagonist of the film, could repair the torn fabric of Soviet society. *DR*

1174 Le Roy, Eric: "Les Corrupteurs" ou le cinéma français à l'heure nazie. *Revue d'Histoire de la Shoah* 163 (May-Aug 1998) 203-226.

Discusses the film "Les Corrupteurs," directed by Pierre Ramelot, produced in France in 1941 as an accompanying propaganda tool to the exposition "Le Juif et la France." The film was produced on the initiative of the propaganda service of the Commissariat Général aux Questions Juives. It was comprised of three parts – "Crime," "Déchéance," and "Scandale" – and its goal was to convince the spectators that the Jews were responsible for the high degree of juvenile delinquency before the war, for prostitution, and for scandalous financial affairs. Describes the plot of the film and how themes in the plot (e.g. crime) were manipulated for antisemitic propaganda. Notes that well-known directors and producers of films from the interwar period refused to take part in the production of such films. *HV*

1175 Leeson, Daniel N.: Antisemitism in Richard Wagner's Music Dramas. *Midstream* 44, 7 (Nov-Dec 1998) 9-12.

There exists a view that Wagner's antisemitism did not affect his operas, which, after all, have no Jewish characters. Contends that Wagner's operas, notably "The Master Singers of Nuremberg" and "The Ring of the Niebelung" cycle, contain covertly "Jewish," i.e. antisemitic images. In "Master Singers" it is the villain Sixtus Beckmesser, and in "The Ring" it is the gnomes. Objects to recent attempts in some countries, including Germany and Israel, to whitewash Wagner's antisemitism. *DR*

1176 Leveson, Marcia: The Mineral Revolution, Fiction and the Jewish Image in South Africa. *Patterns of Migration, 1850-1914,* eds. Aubrey Newman, Stephen W. Massil. London: Jewish Historical Society of England, in association with the Institute of Jewish Studies, University of London, 1996. Pp. 187-196.

A paper delivered at a conference in London, 1993. A common image in South African fiction since the late 19th century has been the antisemitic stereotype linking Jews with criminality, ugliness, and animal-like behavior. This image of the Jew, which has remained operative to the present, was modeled in part on "conspicuous eccentrics" (e.g. Barney Barnato, who arrived in South Africa from England in 1873) and stereotypes from the British stage. *REK*

1177 *Medieval Encounters* 5, 3 (1999).

The section "Anti-Judaism, Fantasy, and Medieval Literature" contains four essays which derive from an NEH Seminar entitled "Absence and Presence: The Jew in Early English Literature," which took place during a seven-week period on the campus of SUNY, Stony Brook, summer 1996. Contents: Rosenthal, Judith: Introduction [Pp. 359-363 contain a combined bibliography.] (358-363); Stone, Carole: Anti-Semitism in the Miracle Tales of the Virgin [Examines three tales: "The Child Slain by Jews," "The Jewish Boy," and "The Merchant's Surety."] (364-374); Gaynor, Stephanie: He Says, She Says: Subjectivity and the Discourse of the Other in the "Prioress's Portrait" and "Tale" (375-390); Weinhouse, Linda: Faith and Fantasy: The Texts of the Jews [Examines "kinot" (lamentations) written by Jews following the First and Second Crusades, and juxtaposes the self-image of the Jews with the image depicted in Chaucer's "Prioress's Tale."] (391-408); Rosenthal, Judith: Margery Kempe and Medieval Anti-Judaic Ideology [Focuses on the depiction of Jews by Kempe (1373-1440), an English nun, as Christkillers in her autobiographical "Book."] (409-420). *SSC*

1178 Monicelli, Francesco; Saletti, Carlo, eds.: *Il racconto della catastrofe: Il cinema di fronte ad Auschwitz*. Verona: Società Letteraria di Verona, 1998. 126 pp.

Based on papers delivered at a conference in Verona, 1998, with additional material. Contents: Mesnard, Philippe: Memorie cinematografiche della Shoah (15-25); Pezzetti, Marcello: Rappresentare la Shoah, trasmetterne la memoria [Appeared in "Educare dopo Auschwitz" (1995).] (27-33); Sivan, Eyal: Memoria, archivi: Gli archivi della memoria (35-39); Ertel, Rachel: Il nero miracolo [Appeared in French in "Au sujet de Shoah, le film de Claude Lanzmann" (1990).] (43-49); Sessi, Frediano: Attorno a un capolavoro: Annotazione su "Shoah" di Claude Lanzmann (51-57); Chevrier, Marc; Le Roux, Hervé: Il luogo e la parola [An interview with Claude Lanzmann. Appeared in French in "Cahiers du Cinéma" (1985) and in "Au sujet de Shoah, le film de Claude Lanzmann" (1990).] (59-72); Belpoliti, Marco: Le occasioni di Primo Levi (77-83); Saletti, Carlo: Superstes Primo: Aporie della testimonianza (85-89); Toaff, Daniel; Ascarelli, Emanuele: Ritorno ad Auschwitz [An interview with Primo Levi on Italian television in 1982.] (91-101). *MZ*

1179 Mosès, Stéphane: Paul Celans Inskription der Vernichtung. *Der Exodus aus Nazideutschland und die Folgen: Juedische Wissenschaftler im Exil*, eds. Marianne Hassler, Juergen Wertheimer. Tuebingen: Attempto, 1997. Pp. 264-273.

Discusses the only-partially scarred-over wound of the memory of the Holocaust, a wound which keeps erupting anew, as expressed in the poems of Paul Celan. Asserts that this wound is one of all humanity, and not, as some critics allege, merely of Celan personally or of the Jewish people. *RW*

1180 Ochse, Katharina: *Joseph Roths Auseinandersetzung mit dem Antisemitismus*. Wuerzburg: Koenigshausen & Neumann, 1999. 254 pp. Based on the author's diss. - Freie Universitaet, Berlin, 1995.

Ch. 2 (pp. 14-39) discusses the role of Jews in German culture between 1918-39, and how they coped with antisemitism. The rest of the book examines Roth's writings. Focuses on his first two novels: "Das Spinnennetz" (1923) and "Tarabas: Ein Gast auf dieser Erde" (1934). The first, discussed on pp. 40-104, is a visionary outline of the crisis of the early Weimar Republic and a psychologically persuasive representation of the promotion of an antisemitic careerist who, from the position of a simple teacher, becomes the Imperial Head of Security. The second, discussed on pp. 148-212, also focuses on the figure of an antisemite. Discusses, also, the journalistic work "Juden auf Wanderschaft" (1925-27), which deals with victims of antisemitism, and Roth's novel "The

Antichrist" (1934) which discusses Jew-hatred. Roth's interpretation of antisemitism as anti-Christianity must be understood in the context of his conversion to Catholicism, while his explication of Nazi antisemitism does not take into account the modern components of Jew-hatred. *ID*

1181 Pakendorf, Gunther: Lies That Set One Free: On Jurek Becker's Novel "Jacob the Liar." *Jewish Affairs* 53, 1 (Fall 1998) 18-24.

Discusses Jurek Becker's life and his novel "Jacob the Liar," which depicts life in a ghetto during World War II. Becker (b. 1937) is a Jew who survived the Łódź ghetto and the Ravensbrueck and Sachsenhausen concentration camps. Discusses the character of Jacob, the novel's hero, who feeds the ghetto's inhabitants "news" which he has fabricated about a Russian advance, thereby giving them hope and allaying their despondency. Jacob feels that he must lie in order to help the Jews; when he finally tells the truth, the Jews lose all hope and some commit suicide. The novel challenges reality, questioning which is more correct and morally acceptable - the truth or the lie. Jacob's fantasy can be understood as a means of preserving an area of personal autonomy in a world dominated by oppression. *REK*

1182 Radevagen, Thomas Til: Juden im Film: Eine Retrospektive. *Tribuene* 150 (1999) 173-181.

Surveys the Holocaust films produced in many countries from the end of the war to the present. Few of these were made by Germans, and those few were largely ignored by the West German public through the 1960s; so were films made in Germany by Jews, and foreign and East German films on the theme (few of which were imported in any case). The American TV series "Holocaust" was a turning point that elicited a great emotional response. Since then the German public has shown interest in the growing number of excellent Holocaust films, both fictional and documentary, produced elsewhere. But in Germany itself few such films are produced, and even fewer that do not draw illegitimate parallels between Jewish suffering and that of the Germans. *RW*

1183 Rosen, Roee: The Visibility and Invisibility of Trauma: Traces of the Holocaust in the Work of Moshe Gershuni and in Israeli Art. *The Jerusalem Review* 2 (1997-1998) 98-118.

Although the Holocaust has a central place in Israeli consciousness, until the 1990s it was almost invisible in Israeli art. Until then there was no major exhibition on the theme of the Holocaust. Most Israeli artists opposed depiction of the Holocaust in artworks, mainly because the dimensions of the tragedy made its depiction seem unsuitable. Works by young Israeli artists in the 1990s suggest that explicit articulation of the

Holocaust has become possible. Discusses the work of Moshe Gershuni, who began focusing on the Holocaust already in the 1980s. Gershuni employs covert devices in his works on the Holocaust, which also contain juxtapositions of Jewish and Nazi identity. Many of Gershuni's works could be perceived as "improper" within the context of Holocaust reflection. However, the coexistence of the proper and the improper is an essential part of his art. *REK*

1184 Ruszniewski-Dahan, Myriam: La Shoah, entre mémoire et roman. *Revue d'Histoire de la Shoah* 162 (Jan-Apr 1998) 141-163.

Analyzes French literature on the Holocaust – its specificity, problems, and values. Deals with writings by André Schwarz-Bart, Anna Langfus, Elie Wiesel, Romain Gary, Georges Perec, Patrick Modiano, and others. In the 1950s it was mainly memoirs and testimonies which appeared; only in the 1960s were the first novels and stories published. Discusses the difficulties of writing fiction on the Shoah, and differences between various authors in approaching the memory of the atrocities. *HV*

1185 Scherr, Lilly: Cinéma et antisémitisme dans les années 30: Propagande et films de fiction. *Yod* 4-5 (1998) 175-186.

Discusses the image of the Jew in propaganda and fictional films in the 1930s in Germany, France, and the USA. The first antisemitic films of the Nazi regime were designed to warn the spectator of the danger of the assimilated Jew in German society. Contrasts this idea with French cinema, which was hostile to all foreign influences. The production of three French films, financed by the Catholic Church, was representative of the significance of theological anti-Judaism. Regards two major films of Jean Renoir, both philosemitic, as exceptions that proved the rule. During the Occupation, the image of the Jew disappeared from French cinema. Deplores the attitude of Hollywood producers, who, precisely due to their Jewish origins, refused to create anti-fascist films for fear of inciting a backlash to America's neutral stance. Concludes that these three countries, despite their ideological and political differences, contributed to the forging of images of the Jew in film which remain with us today. *TS*

1186 Schurr, Maureen M. O'Bannon: *Luba Krugman Gurdus: An Artist's Holocaust Remembrance.* M.A. thesis – University of Louisville, 1996. 133 pp. Unseen.

Neither language nor visual images can make comprehensible Adolf Hitler's carefully orchestrated plan for the destruction of world Jewry. The question has often been asked: 'How do you turn the execution and dehumanization of 6,000,000 human beings into art?' Artists from many countries and backgrounds have reacted to the Holocaust through drawings, paintings, and sculpture. Whatever their styles, techniques or media, artists have portrayed every aspect of Hitler's Final Solution. Examining the life, art, and poetry of a Holocaust survivor, artist, and art historian, Luba Krugman Gurdus, situates her work in the context of the major aspects of the Holocaust to which she has borne eloquent witness.

1187 Seillan, Jean-Marie: Huysmans, un antisémite fin-de-siècle. *Romantisme* 95 (1997) 113-126.

Discusses the antisemitism of the French art critic and novelist Joris-Karl Huysmans (1848-1907). Analyzing various works, contends that his antisemitism has three aspects: socioeconomic, theological, and racial. Reveals Huysmans' ties with Zola until the publication of "J'accuse," after which he accused Zola of hatred of the Church and love of money. Demonstrates that Huysmans' criticism of modern Christian art echoes his belief in a Jewish plot to harm Christianity. Argues that Huysmans' anti-Jewish sentiments portray not only the conventional prejudices of late 19th-century France, but also his own creed of the Jew as heir to and accomplice of the Devil. Concludes that Huysmans' writing is inseparable from his antisemitism, which is an integral part of his literary ideology. *TS*

1188 Sheffi, Na`ama: The Jewish Expulsion from Spain and the Rise of National Socialism on the Hebrew Stage. *Jewish Social Studies* 5, 3 (Spr-Sum 1999) 82-103. An expanded version of an article which appeared in Hebrew in "Ha-Tzionut" 21 (1998).

1189

שפי, נעמה: תיאטרון, זיכרון היסטורי וזהות לאומית: הצגת 'האנוסים', התקבלותה בארץ-ישראל והייחס לרייך השלישי. הציונות 21 (1998) 157-173.

[Sheffi, Na`ama: Theater, Historical Memory and National Identity: The Performance of "Marranos," Its Reception in Eretz-Israel and the Attitude toward the Third Reich. *Ha-Tzionut* 21 (1998) 157-173.]

In December 1938, Max Zweig's play "Die Marranen," dealing with the expulsion of the Jews from Spain, was performed (in Hebrew) by the Habimah Theater in Tel-Aviv. The production was well received by critics and by the public. In this play, Jews in the Yishuv saw interpretations and explanations of contemporary events, primarily in Germany. The theme of the Spanish expulsion had attracted the interest of German-speaking Jewish intellectuals from the Jewish Enlightenment on – works of fiction and historical works were devoted to it. After 1933, the Inquisition and the expulsion were used as metaphors for the persecution of Jews in the diaspora, especially in Nazi Germany. The parallels were obvious, and both the events were regarded as crucial points in diaspora history. After September 1939 any historical comparison with current events became out of place. *DR*

1190 Solomon-Godeau, Abigail: Mourning or Melancholia: Christian Boltanski's "Missing House." *Oxford Art Journal* 21, 2 (1998) 1-20.

"The Missing House," by the French artist Christian Boltanski, produced for and in the city of Berlin in the autumn of 1990, is located in a building in former East Berlin which was damaged in an Allied bombing raid on 3 February 1945. It consists of black and white plaques indicating the family name, profession, and period of residency of each family that lived in the building when it was bombed. Unbeknownst to the artist when he began the project, prior to 1942 many of the building's residents were Jews. By the time of the bombing these tenants had been evicted, displaced, deported, and presumably killed. Though "The Missing House" and other of Boltanski's works referring to death have been interpreted (especially by Jewish viewers) as referring to Jews, in fact they do not. Consequently, they convey a generalized sense of melancholy, but can have no role in mourning the Jewish victims of the Holocaust. *REK*

1191 Spence, Rachel: Made in Italy: Local Responses to Roberto Benigni's "Life Is Beautiful." *The Jewish Quarterly* 173 (Spr 1999) 57-60.

Discusses reactions of Italian Jews to Benigni's award-winning film "La vita è bella" (1997), about the life of an assimilated Italian Jew before World War II, and his later deportation, along with his young son (and his wife who voluntarily joined the transport), to a concentration camp. Briefly describes the situation in fascist Italy before and during the war, and the subject of rescue vs. collaboration activities on the part of the non-Jewish population. The film was approved by Italy's Jews, with some reservations regarding the light mood of the film and the unrealistic depiction of the concentration camp. States that Benigni has delivered a film which touches on the dark side of the country's war record, yet he has done it in a way that does not leave the Jewish audience feeling angry and alienated from their homeland. *SSC*

1192 Stancic, Mirjana: Der Churban oder die unfassbare Gewissheit: Manès Sperbers juedische Themen. *Sachor* 9 (1999) 60-74.

Sperber grew up in a traditional Jewish milieu, but in his youth he loosened his ties with Judaism in favor of a universalist socialism. The Holocaust (which, after a flight through Czechoslovakia and France, he survived in Switzerland) caused him to return to his Jewish identity and his solidarity with the fate of the Jewish people. In "Der Churban oder die unfassbare Gewissheit," a collection of essays published in 1979, he deals with Jewish themes, including a psychological interpretation of antisemitism, and with the Holocaust. In the third section, "Wolyna" (published separately, first in French, 1952), of the third volume of his fictional trilogy "Wie eine Traene im Ozean" he tells the story of a town very similar to his own hometown Zablotów, and its destruction in the Holocaust. *RW*

1193 Stargardt, Ute: "Rassenpolitik" in National Socialist Cinema. *Shofar* 16, 3 (Spr 1998) 1-27.

During the Third Reich, the German film industry enthusiastically put itself at the service of the Nazi murderous "Rassenpolitik." The Nazis regarded the cinema as the most effective propaganda device in the state's efforts to win popular approval of its destruction of "life undeserving of life." From 1933 on, they conducted a policy of "dejudaization" of the German film industry. In 1936-41 the films were commissioned to promote public endorsement of the euthanasia program. The "Kristallnacht" pogrom of 1938 marked the beginning of a new genre, the antisemitic film (e.g."Jud Suess," "Die Rothschilds," "Der Ewige Jude"). The main message of these productions was that Germany must destroy the Jews before they destroy Germany, i.e. the necessity for genocide. The film industry justified its activities as a timely response to popular mandates when in fact no such mandates had been granted. The directors, actors, and actresses most prominently engaged in this enterprise got away with it unscathed. *DR*

1194

סתיו, אריה: השלום – קריקטורה ערבית: עיונים בדיוקן האנטישמי. תל-אביב: זמורה-ביתן, 1996. 248 עמ'.

[Stav, Arie: *The Peace – Arab Caricature: Studies in the Antisemitic Image*. Tel-Aviv: Zmora-Bitan, 1996. 248 pp.]

At a time when Israel is engaged in conducting a peace process with the Arab world, emphasizes the prevalence of antisemitism in the Arab press and media. Chs. 1-4 (pp. 11-64) survey expressions of antisemitism in artworks and caricatures from the Middle Ages through the Nazi period. Chs. 5-9 (pp. 65-234) discuss antisemitism in Arab or Islamic countries. Reproduces caricatures from many journals. Compares Arab and Nazi antisemitic caricatures, stating that both have no element of humor, use anti-Jewish stereotypes, and advocate the physical destruction of the Jews. *RE*

1195 Stav, Arie: *Peace – the Arabian Caricature: A Study of Anti-Semitic Imagery*. Trans.: Martin Kett. Jerusalem: Gefen, 1999. 287 pp. Originally published as "Ha-shalom – karikaturah aravit: Iyyunim bi-dyukan ha-antishemi" (Tel-Aviv: Zmora-Bitan, 1996).

1196
שיינטוך, יחיאל: ק' צטניק — מבוא לחקר יצירתו בשבעים השנים האחרונות. דפים לחקר תקופת השואה 14 (1997) 109-148.

[Szeintuch, Yechiel: Ka-Tzetnik - an Introduction to the Study of His Work in the Last Seventy Years. *Dapim le-Heker Tekufat ha-Shoah* 14 (1997) 109-148.]

Contends that the Israeli writer Ka-Tzetnik's fictional works on the Holocaust were intended to document in literary form issues that documents do not relate to, such as various reactions of individuals to the Holocaust reality as actually experienced by Ka-Tzetnik. Proposes to research Ka-Tzetnik's works in their historical context, which may be reconstructed from testimonies and historical writings. *LFo*

1197 *Text + Kritik* 144 (Oct 1999).

This issue is entitled "Literatur und Holocaust." Partial contents: Struempel, Jan: Im Sog der Erinnerungskultur: Holocaust und Literatur - "Normalitaet" und ihre Grenzen (9-17); Beyer, Marcel: Kommentar: Holocaust - Sprechen (18-24); Korte, Hermann: "Es ist in aller Trauer der tiefste Hang zur Sprachlosigkeit": Der Holocaust in der Lyrik nach 1945 (25-47); Feinberg, Anat: Die Splitter auf dem Boden: Deutschsprachige juedische Autoren und der Holocaust (48-58); Brumlik, Micha: Messianischer Blick oder Wille zum Glueck: Die Kryptotheologie der Walser-Bubis-Debatte (59-66); Luehe, Irmela von der: Wie bekommt man "Lager"? Das Unbehagen an wissenschaftlicher Zurichtung von "Holocaust-Literatur" - mit Blick auf Carl Friedmans Erzaehlung "Vater" [Friedman is a Dutch Jewish female writer.] (67-78); Braese, Stephan; Gehle, Holger: Literaturwissenschaft und Literaturgeschichte nach dem Holocaust [Mainly on Klaus Briegleb and Saul Friedlaender.] (79-95). *RW*

1198 Trepte, Hans-Christian: Juedische Themen in der polnischen Exilliteratur nach 1945 in den USA. *Mnemosyne* 21 (Dec 1996) 18-25.

Polish writers living abroad (focuses mainly on Jewish writers in the U.S.) could write more freely than those remaining in Poland about themes such as the Holocaust, including collaboration of Poles, and about postwar Polish antisemitism. However, writers in Poland also wrote on these themes, even during the Holocaust itself, as well as in the postwar years, and especially since the 1980s. Touches on fictional and non-fictional works by many writers, both in Poland and in the U.S. Also mentions two Polish-Jewish authors who write in English: Jerzy Kosinski and Eva Hoffman. *RW*

1199 Vaughan, Larry: Heinrich Boell's "Aufenthalt in X": Safeguarding the Tracks. *Germanisch-Romanische Monatsschrift* 48, 3 (1998) 375-379.

On the surface, Boell's short story, situated in World War II, relates the encounter of a young German soldier with a girl at a train station in a small Hungarian town, their connection, and their parting. Below the surface, however, is the unspoken but all-pervasive fact of the deportation of provincial Hungarian Jewry to Auschwitz. *DR*

1200 Zaenker, Karl A.: The Bedeviled Beckmesser: Another Look at Anti-Semitic Stereotypes in "Die Meistersinger von Nuernberg." *German Studies Review* 22, 1 (Feb 1999) 1-20.

Shows how the songs of the unsuccessful singer Beckmesser in Wagner's comic opera mark him as a foreigner who misunderstands German and hint at the later persecution of Jews in Europe. Discusses present-day approaches to staging this opera, and concludes that antisemitic aspects of the stereotyped character should not be exorcised or covered up but rather confronted honestly. Evidence is cited linking Wagner's image of Beckmesser, who is associated with evil and the devil, with folktales and other writings of the Brothers Grimm. *YC*

AUTHOR INDEX

SUBJECT INDEX

Bei Fragen zur Produktsicherheit wenden Sie sich bitte an:
If you have any questions regarding product safety,
please contact:

Walter de Gruyter GmbH
Genthiner Straße 13
10785 Berlin
productsafety@degruyterbrill.com